The Whitestone Building Maintenance And Repair Cost Reference 2004-2005

Ninth Annual Edition

Peter S. Lufkin
Chad Turner
Jon Miller

Notes

Building maintenance and repair (M&R) is the collection of activities necessary for keeping a building in good working order. Other tasks associated with building operation such as custodial services, landscaping, waste disposal, and the provision of central utilities are not included in our definition of M&R.

For the purposes of this book, M&R activities are divided into three types:

1. **Preventive maintenance (PM) and minor repair** consists of scheduled tasks that sustain a component's level of service during a prescribed lifetime.

2. **Unscheduled maintenance** consists of service calls, emergency response, and other tasks that cannot be individually anticipated.

3. **Renewal and replacement** consists of component overhaul or major replacement tasks. These tasks extend a component lifetime, and reset the schedule of PM and minor repair tasks.

Building size is expressed in gross square feet (Gsft.), and includes the entire area within the building perimeter.

All costs throughout the book are expressed in mid-year 2004 dollars, unless noted otherwise. All costs for Canadian cities are reported in Canadian dollars.

Preface

The *Whitestone Building Maintenance and Repair Cost Reference* is a comprehensive source of building maintenance and repair cost statistics. It provides a collection of unique tools for answering simple but important questions:

How much does it cost to maintain a building over its service lifetime?

How many trade labor hours are required?

What is the historic inflation rate of M&R construction costs?

How do M&R costs vary across different metropolitan areas?

What is the lifetime of a specific building component?

The *Reference* is intended for two audiences. The first group has a common need to know the long-term M&R costs of specific buildings. This group consists of analysts, developers, architects, bankers, investors, and others who must account for M&R costs that, over a 50-year building lifetime, can easily exceed the costs of construction. Forecasts of these costs are critical for justifying the expenditure of hundreds of millions of dollars—even a rudimentary business plan or benefit/cost study must recognize M&R—yet such forecasts are usually based on *ad hoc* approximation or ill-defined rules of thumb. To meet this need, we provide detailed estimates of long-term M&R requirements for selected building models.

The second audience consists of facility managers and all those responsible for estimating and justifying facility maintenance budgets. Acutely aware of immediate needs, this group often does not put funding requests in a larger context: How much does funding need to rise just to keep up with inflation? What is the appropriate funding level for a facility with an inventory of particular building types? Such information provides a benchmark, a point of reference, for demonstrating M&R requirements. For this purpose, we provide national and local M&R indexes, and forecasts of in-house and contract maintenance costs for major U.S. metropolitan areas.

Perhaps the most unique information presented in this book is the set of 50-year cost profiles for individual buildings. These two-page reports are a product of the Whitestone MARS system, an M&R cost forecast tool capable of providing building-specific cost profiles for any region of the country. The profiles are representative of 51 common building types, in terms of use and construction, and provide detailed estimates of M&R costs per square foot, and as a percentage of replacement value. Whitestone forecasts can be directly incorporated into life cycle cost analyses of similar buildings, or modified to include alternative components. A complete listing of the components and life cycle data used in estimating the profiles is included. This listing is organized according to the UNIFORMAT II classification system published by the American Society of Testing and Materials (ASTM).

Also unique to the *Reference* are local and national cost indexes calculated specifically for the maintenance construction industry. A forecast of M&R costs for the same building model for 210 metropolitan areas provides convenient measures for interregional comparisons. Detailed contract labor rates for both in-house staff and contract trades are presented for the same 210 areas. Much of this information is taken from original surveys and forecasts not available elsewhere.

Whitestone M&R Cost Indexes are the only statistics we know of for analyzing M&R cost and funding trends. These are weighted composite indexes reflecting the prices of goods and services specific to commercial and public, residential, and military facilities M&R construction. Component and comparable indexes of selected goods and services related to M&R construction are also included.

Acknowledgments

Each year, completing this book depends on information, advice, and support from a wide variety of sources. Much of the employment, wage, and price data were provided by federal agencies, and these are credited specifically throughout the book. The Maintenance Task Database developed by the Army Corps of Engineers continues to be an invaluable resource for defining repair tasks, as are the various Engineering Performance Standards (EPS) Handbooks provided jointly by the military services. Many of the building models were defined with the generous assistance of a panel of engineers from Jacobs Facilities, Inc. Survey results from the Building Owners and Managers Association (BOMA) International's *Experience Exchange Report* were used in discussing the sensitivity of our forecasts. Our estimates of labor rates for both contract labor and facility staff were made with information provided by R.S. Means, Inc. and the Construction Labor Research Council (CLRC). To understand new technologies and discover new data sources, we continued to rely on the patience and expertise of a wide variety of construction trade organizations and professional associations; to validate material, labor, and equipment costs we had innumerable conversations with manufacturers, wholesalers, and tradesmen. To all these individuals and organizations, we express our gratitude.

Our knowledge of facilities finance issues has grown greatly with the project support provided by the Office of the Deputy Under Secretary of Defense for Installations and Environment. Facility service life estimates presented in the Appendix are one product of this work. This Office has also supported basic research addressing estimating facility recapitalization costs—obsolescence, change in use, building safety and code, etc.—not included in this book, but a related topic of importance for capital asset management.[1]

We continue to be surprised at the ways in which the data from this book can be used. A recent study for Johnson Controls used the building cost profiles as a starting point in estimating the market for facility M&R in North America. Combined with county-level building inventories, they provided a detailed map of the demand for services and materials valued at over $210 billion.

Much of the *Reference* is produced using the *MARS Facility Maintenance Cost Forecast System*, a software tool soon to be in its sixth version. MARS was designed by Tony Pepitone and continues to evolve under the direction of Jeff Seawards.

Gage Design provided the cover design.

Printing and production assistance was provided by Haagen Printing.

Disclaimer

The information presented in this book is intended to provide reasonable and objective estimates of building M&R costs and labor requirements. Such estimates should be considered only approximations—points of departure from which local knowledge of M&R practices and costs must be used to produce refined estimates. We encourage the reader to review the definitions and methodologies (Chapter 6) that underlie the information presented in this book. We make no warranty and assume no liability for the accuracy or completeness of the data presented.

---

[1] Lufkin, Desai, and Janke. *Estimating the Restoration and Modernization Costs of Infrastructure and Facilities.* Whitestone Research: June 2004.

About Whitestone Research

Whitestone Research is an employee-owned corporation with principals in economics, computer science, and operations research. Whitestone specializes in applied economic research and software development for both the public sector and corporate clients. Whitestone also publishes periodic technical reports and develops proprietary data products.

Whitestone Consulting & Training

- Life-Cycle Cost Studies
- Facility Operations Cost Modeling
- Index Definition and Benchmarking
- Economic Impact Studies
- Manpower Productivity and Efficiency Measurement
- National Defense Economics

MARS **MARS 5.0 Facility Maintenance Cost Forecast System**

MARS 5.0 is the most recent version of the Whitestone system for forecasting facility maintenance and repair costs and projecting facility funding requirements. Based on component specific tasks, MARS provides an unmatched level of detail and credibility. MARS is a fully configured turnkey system and comes pre-loaded with the building models and extensive component library used in *The Whitestone Building Maintenance and Repair Cost Reference*. MARS capabilities include:

- Life-Cycle Cost Analysis
- Benchmarking
- Deferred Maintenance Estimates
- Level of Service Analysis
- Budget and Staffing Forecasts

MARS is a capital asset management and budget tool used by many government agencies, corporations, and large individual facilities. MARS is used by government agencies and other large facility portfolio managers to report deferred maintenance based on component life-cycle cost principles. Actual experience has shown MARS estimates fall within 10 percent of traditional assessment inspections, yet are less than half the costs of inspections. MARS meets both Federal Accounting Standards Advisory Board (FASAB) Standard Number 6 and Department of Energy (DOE) Memorandum 430.1B deferred maintenance reporting requirements.

Whitestone on the Web

Visit us at **www.whitestoneresearch.com** for free monthly updates of many of the unique statistics shown in Chapter 4. This site also offers Whitestone News and Reports, a set of case studies and technical notes on M&R cost forecasting and related issues. Product description and ordering information are also provided.

Contents

Chapter 5 Component Data

Chapter 6 Definitions and Methods

Additional Reading 279

Appendix

Tables

Boxes

Figures

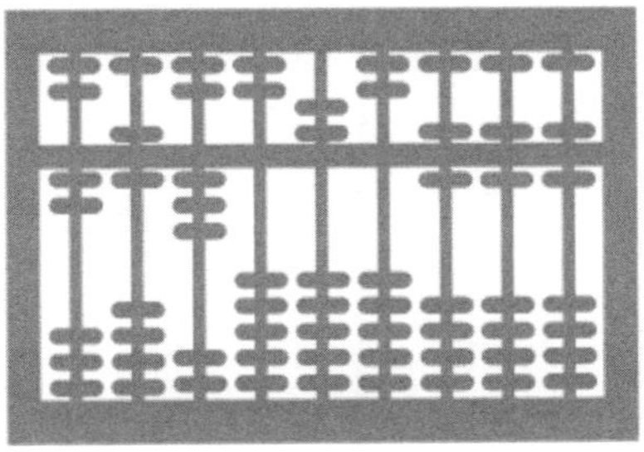

Chapter 1
Getting Started

How to Use This Book

Frequently Asked Questions

1. Getting Started

How to Use This Book:

Start at the beginning and work to the back.
This book is organized around cost profiles for the building models shown in Chapter 2. The subsequent chapters present the detailed data—labor rates, material costs, and component life cycle tasks—that go into these estimates. We suggest you work your way through the book according to your required level of detail.

Everything costs more in New York.
The model cost forecasts shown in Chapter 2 are made for the Washington DC area. Before these forecasts are used for other areas, they should be modified to reflect local labor costs and life cycles. The local maintenance cost index, shown in Chapter 3.1, is based on a comparison of costs across 210 metro areas for the same building model—the Two-Story Office Building—using Washington DC as a reference point.

A facility is just a collection of buildings.
The building model costs can be used to approximate the M&R budget for an entire facility. Categorize your facility inventory according to the 51 Whitestone building models in Chapter 2, and then multiply the actual total square footage in each category by the appropriate model's estimated cost (expressed in costs per square feet). Be sure to adjust the costs for your area using the local maintenance cost index.

Don't hesitate to mix and match.
The model cost forecasts can be used as a point of departure for making more specific estimates. All of the information necessary is provided: adjust for a different location using the local cost index in Chapter 3.1; substitute different components with the data provided in Chapter 5; or put the entire forecast in a different base year using the appropriate index from Chapter 4.

Use the right inflation rate.
The costs of roof repairs and HVAC maintenance have not grown at the same rate as milk and eggs, yet many contractors, facility managers, and even cost analysts still use a consumer price index when estimating M&R cost inflation. National-level cost indexes for three types of M&R construction—commercial and public, residential, and military—are computed in Chapter 4. These indexes are updated monthly and available for no charge at www.whitestoneresearch.com.

Remember, it's warmer in Phoenix.
M&R requirements for HVAC equipment vary substantially across cities. More specifically, an air conditioner in Alaska is likely to experience much less demand, and have lower M&R costs, than one in Arizona. This explains the extensive listing of repair frequencies by climate zones shown in Chapter 5.2.

Frequently Asked Questions:

What is the lifetime M&R funding requirement for an individual building?
A model two-story office building in the Washington DC metro area will cost an annual average of $2.13 per gsft. or 2.26 percent of replacement cost over a 50-year service life. Average costs for other building types varied from $10.74 to $0.52 per gsft. Detailed cost profiles for 51 model buildings are shown in Chapter 2.

How can M&R costs be estimated for an actual building?
The model buildings shown in Chapter 2 can be used as a point of departure for making more specific estimates. All of the information necessary is provided: adjust for a different location using the local cost index in Chapter 3.1; substitute different components with the data provided in Chapter 5; or put the entire forecast in a different base year using the appropriate index from Chapter 4.1.

How do M&R costs vary across metropolitan areas?
An index of local maintenance costs shows New York, NY has the highest average M&R cost in our 210-area sample, at $3.39 per gsft. for a typical office building; Beaufort, SC has the lowest rate at $1.17 for the same building. For a cost comparison among metro areas, see Chapter 3.1.

How will regional M&R costs change in 2004?
Labor costs account for over 40 percent of M&R costs. Recent wage agreements indicate that the highest wage increases in 2004, averaging 4.8 percent, will be in New England. Construction wage changes are summarized by region in Chapter 6.

What will 2004 rates for construction trades be in specific metropolitan areas?
The New York metro area has the highest contract labor rates of the 210 metro areas studied. A contract electrician in New York will cost an estimated $86.23 per hour, while a contract plumber will cost an estimated $81.58. A listing of contract labor costs by trade for each metro area is shown in Chapter 3.

What is the inflation rate for commercial M&R costs?
In 2004, M&R costs for commercial and public buildings will increase at an annual rate of 2.2 percent, military facilities costs will increase by 2.5 percent, and residential buildings costs will increase by 3.0 percent. By comparison, new construction costs will increase by 0.8 percent and consumer prices by 1.8 percent during the same period. Whitestone M&R construction indexes are shown in Chapter 4.

How are M&R tasks distributed between contract workers and in-house staff?
For estimation purposes, we assume in-house staff members at shop rates shown in Chapter 3 perform preventive maintenance, minor repairs and unscheduled maintenance. We assume repair and replacement tasks are done by contract workers at rates also shown in Chapter 3.

What are the labor requirements for building M&R?
Labor requirements, like costs, vary by building type, size, and location. A model two-story office building in the Washington DC metro area requires an annual average of 5,469 hours over a 50-year service life. Labor requirements for other building types are shown in Table A-2 in the Appendix.

How sensitive are Whitestone estimates?
There is no straightforward answer, as estimates of M&R costs for entire buildings necessarily vary by location, components, and use. Also important are unscheduled maintenance multipliers, in-house markup rates and types of utilization. Different values for these factors can change M&R costs substantially, as shown in the discussion of estimate sensitivity in Chapter 6.

Where does Whitestone data come from?
Data sources, key assumptions, and a complete explanation of the MARS forecast model are presented in Chapter 6.

Chapter 2
Building M&R Cost Profiles

2. Building M&R Cost Profiles

This chapter presents estimates of 50-year maintenance cost profiles for 51 building models. Each two-page profile includes a description of the model building, a list of major components, and forecasts of maintenance and repair (M&R) costs at various levels of aggregation. The profile estimates were made with the Whitestone MARS forecast system, calibrated for the Washington DC metropolitan area. The profiles can be adjusted for other metro areas using the Local Maintenance Cost Index shown in Chapter 3, and modified to include different components shown in Chapter 5.

Table 2-1
Summary of M&R Cost Profiles

Building Type	Gross Sqft.	Annual M&R Cost per Gsft.*	Annual M&R Cost as % of Repl. Value
Car Wash	800	$10.74	4.99%
Garage, Service Station	1,400	7.93	6.54
Apartments, 1-3 story	22,500	6.93	6.68
Apartments, 4-7 story	60,000	6.64	6.30
Motel, 18 Units	8,000	6.48	6.66
Fire Station	6,000	5.92	5.66
Fast Food Restaurant	4,000	5.85	4.93
Bank	4,100	5.69	3.54
Telephone Exchange	5,000	5.39	4.91
Laundromat	3,000	5.39	4.43
Motel, 40 Unit	18,000	5.30	4.85
Restaurant	10,000	5.29	4.19
Country Club	6,000	5.13	3.74
Religious Education	10,000	4.99	4.50
Self-storage Warehouse	24,000	4.56	6.98
Store, Convenience	4,000	4.51	6.02
Medical Clinic	13,000	4.36	2.48
Community Center	10,000	4.33	4.36
Research Hospital	540,200	4.26	1.53
General Hospital	125,000	4.25	3.33
Movie Theater	10,000	4.23	3.85
Dormitory, 50 Room	25,000	4.19	4.85
Store, Retail	8,000	4.08	5.23
Bus Terminal	12,000	4.04	4.22
Funeral Home	10,000	3.99	4.46
Church	17,000	3.92	3.31
Town Hall, 1 Story	11,000	3.88	4.28
Post Office	13,000	3.67	4.25
Court House, 1 Story	30,000	3.67	2.71
Public Library, 3 Story	60,000	3.61	3.28
Auditorium	24,000	3.61	3.29
College Student Union	25,000	3.57	3.35
Social Club	22,000	3.28	3.36
Apartments, 24 Story	220,000	3.26	4.01
Gymnasium	40,000	3.21	3.37
Rink, Hockey	30,000	3.11	2.78
College Classroom	90,000	3.05	2.84
Elementary School	47,000	2.88	3.95
Childcare Center	12,000	2.87	2.44
Bowling Center	20,000	2.71	4.09
Garage, Auto Sales	21,000	2.68	3.75
County Jail	318,455	2.56	0.64
Light Manufacturing Plant	45,000	2.49	4.18
Office Park	65,000	2.41	4.96
Department Store	94,000	2.28	3.30
Supermarket	96,000	2.24	3.13
Office Building, 2 Story	83,000	2.13	2.26
Aircraft Hangar	32,000	2.01	2.50
Office Building, 15 Story	250,000	1.95	1.60
Warehouse	80,000	1.90	4.02
Garage, Parking	110,808	0.52	1.57

*Average costs over 50-year lifetime, Washington DC metro area

From a cost analysis perspective, the most useful information in these profiles is the year-by-year total shown under the "Cost per Sqft. by System" section. A projection of M&R costs is required in the financial evaluation of virtually all large construction or renovation projects. Often this trend is estimated with a simple approximation (typically, 2 to 4 percent of replacement value) that obscures the actual oscillations in M&R requirements, and misstates costs when expressed in terms of present value. In comparison, Whitestone estimates are based on component life cycles that provide a more realistic and defensible projection of M&R costs.

For the purposes of the facility manager, average values for M&R costs may be more useful than detailed year-to-year estimates. Conversations about facility funding and budgeting usually dwell on average costs per square foot, or average costs as a percentage of replacement value. Among our building models, the highest average cost per gsft. is for the car wash ($10.74), while the parking garage had the lowest average cost ($0.52).

The reader may note the rankings in order of cost are different when expressed in terms of replacement value. The highest average M&R cost from this perspective was for the self-storage warehouse—6.98 percent of replacement value—a result due primarily to a low estimated replacement cost of $65.38 per square foot. A complete list of replacement costs is shown in the Appendix. In general, we are wary of costs expressed in terms of replacement values because of the great variation in new construction costs and the difficulty of determining replacement costs for older buildings.

Profile estimates are sensitive to a variety of factors such as unscheduled maintenance rates, in-house shop rates, and types of utilization. These sensitivities are discussed in Chapter 6, Definitions and Methods.

Aircraft Hangar

Gross Sqft:	32,000
Height in Ft:	30
Exterior:	Steel
Floor coverings:	Vinyl Tile/Carpet/Finished Concrete
HVAC:	Heat Pump, Gas Boiler, Unit Heaters
Occupancy:	150
Replacement Cost:	$2,563,618

50-Year M&R Cost Summary

	50 Year Total	Annual Cost per Sqft	Annual cost as % replacement
PM & Minor Repair	$607,434	$0.38	0.47%
Unscheduled Maintenance	$565,440	$0.35	0.44%
Renewal & Replacement	$2,035,457	$1.27	1.59%
Total M&R Costs	$3,208,331	$2.01	2.50%

Components

Component	Quantity	Unit
B20 Exterior Enclosure		
Steel, Painted, Exterior Finish, 1st Floor	3050	Sq Ft
Steel, Painted, Exterior Finish, 2nd Floor	3050	Sq Ft
Steel, Painted, Exterior Finish, 3rd+ Floor	3050	Sq Ft
Steel Frame, Painted, Operable Window, 12 sf, 1st Floor	122	Each
Steel, Painted, Exterior Door	4	Each
Steel 20'x20', Painted, Roll-up Door	5	Each
B30 Roofing		
Metal Roof	32960	Sq Ft
C10 Interior Construction		
Wood, Solid Core, Painted, Interior Door	8	Each
Steel, Painted, Interior Door	10	Each
C30 Interior Finishes		
Sheetrock, Unstippled, Interior Wall Finish	4760	Sq Ft
Vinyl Interior Wall Finish	480	Sq Ft
Steel, Painted, Interior Wall Finish	7880	Sq Ft
Concrete, Finished Flooring	28160	Sq Ft
Vinyl Sheet Flooring	1280	Sq Ft
Carpet, Nylon 20 oz., Low Traffic	2560	Sq Ft
Sheetrock, Unstippled, Ceiling	4800	Sq Ft
Metal Ceiling, Painted	32000	Sq Ft
D20 Plumbing		
Tankless Water Closet	8	Each
Urinal, Vitreous China	3	Each
Lavatory, Vitreous China	14	Each
Service Sink, Iron Enamel	1	Each
Shower, Fiberglass	2	Each
Drinking Fountain, Refrigerated	3	Each
Pipe & Fittings, 3/4" Copper, Cold Water	1.1	K Ln Ft
Pipe & Fittings, 3/4" Copper, Hot Water	1.3	K Ln Ft
Pipe & Fittings, 2" Copper, Cold Water	1.5	K Ln Ft
Pipe Insulation, Cold Water	1.4	K Ln Ft
Pipe Insulation, Hot Water	1.2	K Ln Ft
Circulator Pump, 1/12 HP, Hot Water	1	Each
Water Heater, Gas/Oil 175 Gph	1	Each
Water Storage Tank, 250 Gal.	1	Each
Pipe & Fittings, 6" Cast Iron	0.7	K Ln Ft
Pipe & Fittings, 10" Cast Iron	0.4	K Ln Ft
Roof Drain, 4-6"	9	Each
Pipe & Fittings, 4" DWV PVC	0.6	K Ln Ft
D30 HVAC		
Pipe & Fittings, 4" Steel, Gas	0.8	K Ln Ft
Chemical Feed System	1	Each
Circulation Pump 5 HP, Hot Water	2	Each
Expansion Tank, 60 Gal.	1	Each
Gas Boiler, 750 Mbh	2	Each
Heat Pump, 10 Ton	1	Each
Suspended Heater, Hot Water, 4000 Cfm, 180 Mbh	8	Each
D40 Fire Protection		
Fire Sprinkler Head	372	Each
D50 Electrical		
Distribution Switch, Fused, <600 V	5	Each
Main Switchgear, <1200 Amp.	1	Each
Motor Starter, 5-20 HP, <600 V	2	Each
Incandescent Lighting Fixture, Basic, 100w	20	Each
Fluorescent Lighting Fixture, 160w	75	Each
Metal Halide Lighting Fixture, High Bay, 1000w	41	Each
Wiring Device, Switch	40	Each
Receptacle, 120V, 15 Amp	160	Each
Annunciation Panel	1	Each
Fire Alarm Bell, 6"	6	Each
Fire Alarm Control Panel	1	Each
Manual Pull Station	6	Each
Smoke Detector	64	Each

Distribution of Repair Costs

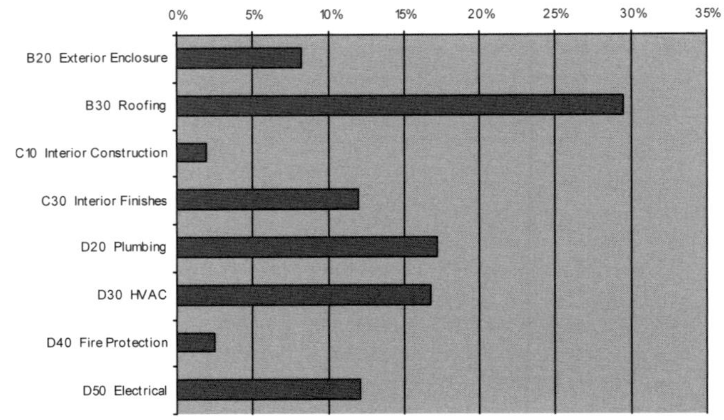

Thirty Most Costly Repair Tasks

Major Repair Task	Task Cost*	Pct.**
Remove & Replace Metal Roof	14.95	18.1%
Minor Repair, Metal Roof	5.66	6.9%
Maintain Metal Roof	3.98	4.8%
Replace Pipe & Fittings, 2" Copper, Cold Water	3.17	3.8%
Inspect & Test Gas Boiler, 750 Mbh	3.07	3.7%
Replace Pipe & Fittings, 3/4" Copper, Hot Water	2.41	2.9%
Replace Pipe & Fittings, 3/4" Copper, Cold Water	2.05	2.5%
Maintain Suspended Heater, Hot Water, 4000 Cfm, 180 Mbh	1.81	2.2%
Refinish Metal Ceiling, Painted	1.77	2.1%
Replace Metal Halide Lighting Fixture, High Bay, 1000w	1.75	2.1%
Replace Suspended Heater, Hot Water, 4000 Cfm, 180 Mbh	1.57	1.9%
Replace Lamp, Metal Halide Lighting Fixture, High Bay, 1000w	1.46	1.8%
Refinish Steel, Painted, Interior Wall Finish	1.46	1.8%
Check Operation, Smoke Detector	1.33	1.6%
Replace Vinyl Sheet Flooring	1.28	1.5%
Minor Repair, Metal Ceiling, Painted	1.21	1.5%
Maintain Steel 20'x20', Painted, Roll-up Door	1.21	1.5%
Refinish Concrete, Finished Flooring	1.20	1.5%
Replace Carpet, Nylon 20 oz., Low Traffic	1.16	1.4%
Repair Fire Sprinkler Head	1.12	1.4%
Maintain Heat Pump, 10 Ton	1.09	1.3%
Replace Ballast, Metal Halide Lighting Fixture, High Bay, 1000w	1.07	1.3%
Replace Heat Pump, 10 Ton	1.06	1.3%
Replace Gas Boiler, 750 Mbh	1.05	1.3%
Replace Steel 20'x20', Painted, Roll-up Door	1.04	1.3%
Refinish Steel, Painted, Exterior Finish, 3rd+ Floor	.99	1.2%
Minor Repair, Concrete, Finished Flooring	.97	1.2%
Replace Water Heater, Gas/Oil 175 Gph	.91	1.1%
Refinish Sheetrock, Unstippled, Ceiling	.91	1.1%
Refinish Sheetrock, Unstippled, Interior Wall Finish	.90	1.1%

**Task cost ($2004) per gross square foot over 50 years.*

***Percent of total M&R costs.*

Cost per Sqft. by System

Building Age	Exterior Closure	Roofing	Interior Constrution	Stairways	Interior Finishes	Conveying Systems	Plumbing	HVAC Systems	Fire Protection	Electrical	Equipment	Total per Sqft.
1	.07	.17	.03				.07	.23	.01	.08		**.65**
2	.07	.17	.03				.18	.24	.01	.08		**.77**
3	.07	.17	.03				.07	.23	.01	.08		**.65**
4	.07	.17	.03		.27		.18	.24	.01	.08		**1.05**
5	.15	.80	.03		.18		.11	.23	.01	.34		**1.85**
6	.07	.17	.03				.18	.24	.01	.08		**.77**
7	.07	.17	.03				.09	.34	.01	.08		**.77**
8	.07	.17	.03		.57		.31	.24	.01	.08		**1.48**
9	.07	.17	.03		.01		.07	.23	.01	.08		**.66**
10	.74	.80	.13		.20		.35	.43	.39	1.01		**4.04**
11	.07	.17	.03				.07	.23	.01	.08		**.65**
12	.07	.17	.03		.27		.18	.25	.01	.08		**1.07**
13	.07	.17	.03				.13	.23	.01	.08		**.71**
14	.11	.17	.03				.19	.35	.01	.08		**.93**
15	.15	.80	.03		.89		.10	.84	.01	.63		**3.45**
16	.07	.17	.03		.57		.31	.24	.01	.08		**1.48**
17	.07	.17	.03				.07	.23	.01	.08		**.65**
18	.07	.17	.03		.64		.18	.24	.01	.10		**1.43**
19	.07	.17	.03				.07	.23	.01	.08		**.65**
20	.77	1.14	.14		.66		.39	.74	.32	1.49		**5.65**
21	.07	.17	.03				.10	.34	.01	.08		**.79**
22	.07	.17	.03				.18	.24	.01	.08		**.77**
23	.07	.17	.03				.07	.23	.01	.10		**.67**
24	.07	.17	.03		.57		.31	.25	.01	.08		**1.50**
25	.15	.80	.03		.78		4.37	.30	.01	.42		**6.86**
26	.07	.17	.03				.20	.24	.01	.08		**.79**
27	07	.17	.03		.01		.08	.23	.01	.08		**.67**
28	.11	.17	.03		.27		.18	.35	.01	.10		**1.22**
29	.07	.17	.03				.10	.23	.01	.08		**.68**
30	.74	14.95	.19		.91		.34	2.01	.39	1.54		**21.06**
31	.07	.17	.03				.08	.23	.01	.08		**.66**
32	.07	.17	.03		.57		.31	.24	.01	.08		**1.47**
33	.07	.17	.03				.08	.23	.01	.10		**.68**
34	.07	.17	.03				.17	.24	.01	.08		**.76**
35	1.13	.80	.03		.18		.46	.23	.01	.32		**3.16**
36	.07	.17	.03		.91		.11	.25	.01	.10		**1.66**
37	.07	.17	.03				.15	.34	.01	.08		**.84**
38	.07	.17	.03				.13	.24	.01	.08		**.72**
39	.07	.17	.03				.16	.23	.01	.08		**.74**
40	.66	.80	.13		.95		.39	.81	.32	1.59		**5.66**
41	.07	.17	.03				.14	.23	.01	.10		**.74**
42	.11	.17	.03				.13	.24	.01	.08		**.76**
43	.07	.17	.03				.14	.23	.01	.08		**.72**
44	.07	.17	.03		.27		.11	.35	.01	.08		**1.09**
45	.26	.80	.03		.90		.29	.84	.01	.61		**3.74**
46	.07	.17	.03				.11	.24	.01	.10		**.73**
47	.07	.17	.03				.14	.23	.01	.08		**.72**
48	.07	.17	.03		.57		.25	.25	.01	.08		**1.43**
49	.07	.17	.03				.15	.23	.01	.08		**.73**
50	.62	1.14	.13		.80		4.51	.38	.39	.89		**8.87**
Total	8.18	29.53	1.96		11.97		17.25	16.80	2.48	12.09		100.26

50-Year Total M&R Cost Profile per Sqft.

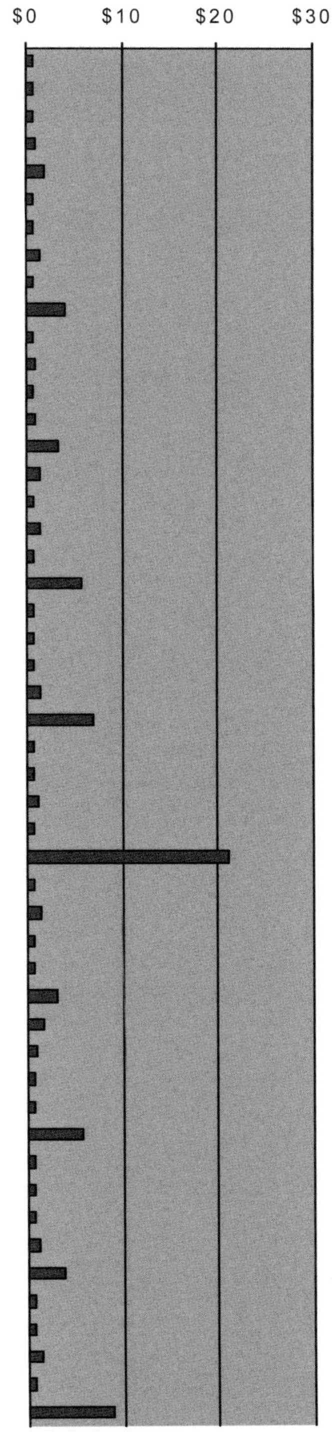

A value of "0.00" means a cost of more than $.000 but less than $.005 per gross square foot.

Apartments, 1-3 Story

Gross Sqft:	22,500
Height in Ft:	30
Exterior:	Brick Veneer
Floor coverings:	Carpet/Ceramic Tile/Vinyl Tile
HVAC:	Chilled Water, Oil Boiler, Baseboard Rad.
Occupancy:	78
Replacement Cost:	$2,333,481

50-Year M&R Cost Summary

	50 Year Total	Annual Cost per Sqft	Annual cost as % replacement
PM & Minor Repair	$1,824,327	$1.62	1.56%
Unscheduled Maintenance	$1,825,438	$1.62	1.56%
Renewal & Replacement	$4,142,398	$3.68	3.55%
Total M&R Costs	$7,792,163	$6.93	6.68%

Components

Component	Quantity	Unit
B20 Exterior Enclosure		
Clay Brick, Exterior, 1st Floor	3520	Sq Ft
Clay Brick, Exterior, 2nd Floor	3520	Sq Ft
Clay Brick, Exterior, 3rd+ Floor	3520	Sq Ft
Aluminum Operable Window, 24 sf, 1st Floor	32	Each
Aluminum Operable Window, 24 sf, 2nd Floor	32	Each
Aluminum Operable Window, 24 sf, 3rd+ Floor	32	Each
Steel w/ Safety Glass, Painted, Exterior Door	4	Each
B30 Roofing		
Built-up Roof	7500	Sq Ft
C10 Interior Construction		
Wood, Hollow Core, Painted, Interior Door	281	Each
C20 Stairs		
Metal Interior Steps, Painted	600	Sq Ft
C30 Interior Finishes		
Sheetrock, Stippled, Interior Wall Finish	31500	Sq Ft
Concrete Block, Painted, Interior Wall Finish	9600	Sq Ft
Wallpaper Interior Wall Finish	11250	Sq Ft
Ceramic Tile Flooring	2250	Sq Ft
Vinyl Tile Flooring	6750	Sq Ft
Carpet, Nylon 20 oz., High Traffic	13500	Sq Ft
Sheetrock, Unstippled, Ceiling	22500	Sq Ft
D10 Conveying		
Elevator, Hydraulic, 2500 lbs, 3-5 Floor, 200 fpm	1	Each
D20 Plumbing		
Flush Tank Water Closet	60	Each
Lavatory, Vitreous China	60	Each
Bathtub & Shower Enclosure, Fiberglass	42	Each
Pipe & Fittings, 3/4" Copper, Hot Water	3.06	K Ln Ft
Pipe & Fittings, 3/4" Copper, Cold Water	6.14	K Ln Ft
Pipe & Fittings, 2" Copper, Cold Water	5.33	K Ln Ft
Pipe & Fittings, 4" Steel	0.92	K Ln Ft
Pipe Insulation, Cold Water	6.6	K Ln Ft
Pipe Insulation, Hot Water	3.32	K Ln Ft
Water Heater, Electric 52 Gal.	24	Each
Pipe & Fittings, 6" Cast Iron	1.96	K Ln Ft
Pipe & Fittings, 10" Cast Iron	0.2	K Ln Ft
Roof Drain, 4-6"	5	Each
Pipe & Fittings, 4" DWV PVC	0.15	K Ln Ft
D30 HVAC		
Chemical Feed System	1	Each
Circulation Pump 5 HP, Hot Water	2	Each
Expansion Tank, 60 Gal.	2	Each
Gas Boiler, 1000 Mbh	1	Each
Condenser, Air-Cooled, 60 Ton	1	Each
Circulation Pump, 5 HP, Chiller & Condenser Water	2	Each
Chiller, Reciprocal Air-Cooled Hermetic, 60 Ton	1	Each
Exhaust Fan, Ceiling, 200-500 Cfm	65	Each
Fan Coil, Two-Pipe, 400 Cfm	100	Each
D40 Fire Protection		
Fire Sprinkler Head	175	Each
D50 Electrical		
Safety Switch, Fused, 400 Amp., 3 Ph.	6	Each
Distribution Panel Board	44	Each
Main Switchgear, <1200 Amp.	1	Each
Exit Lighting Fixture, w/ Battery	6	Each
Emergency Horn & Strobe	10	Each
Incandescent Lighting Fixture, Basic, 200w	135	Each
Wiring Device, Switch	168	Each
Grounded Fault Circuit Interruptor, 15 Amps	100	Each
Receptacle, 120V, 15 Amp	225	Each
Annunciation Panel	1	Each
Manual Pull Station	10	Each
Heat Detector	2	Each
Smoke Detector	60	Each
Security System Panel	1	Each

Distribution of Repair Costs

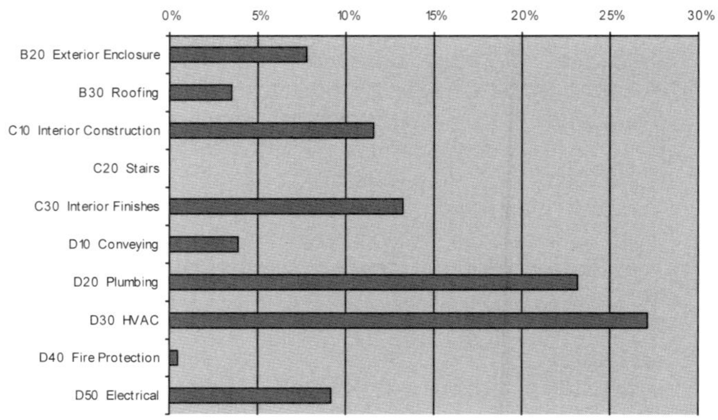

Thirty Most Costly Repair Tasks

Major Repair Task	Task Cost*	Pct.**
Maintain Fan Coil, Two-Pipe, 400 Cfm	20.40	7.7%
Replace Fan Coil, Two-Pipe, 400 Cfm	17.77	6.7%
Replace Carpet, Nylon 20 oz., High Traffic	17.41	6.6%
Replace Pipe & Fittings, 3/4" Copper, Cold Water	16.26	6.1%
Replace Pipe & Fittings, 2" Copper, Cold Water	16.04	6.0%
Maintain Elevator, Hydraulic, 2500 lbs, 3-5 Floor, 200 fpm	11.53	4.3%
Maintain Wood, Hollow Core, Painted, Door Locks	10.42	3.9%
Annual PM, Distribution Panel Board	9.22	3.5%
Refinish Sheetrock, Stippled, Interior Wall Finish	8.51	3.2%
Replace Pipe & Fittings, 3/4" Copper, Hot Water	8.05	3.0%
Replace Wood, Hollow Core, Painted, Door Locks	6.30	2.4%
Refinish Sheetrock, Unstippled, Ceiling	6.07	2.3%
Replace Chiller, Reciprocal Air Cooled Hermetic, 60 Ton	4.91	1.9%
Maintain Built-up Roof	4.87	1.8%
Replace Bathtub & Shower Enclosure, Fiberglass	4.77	1.8%
Maintain Chiller, Reciprocal Air Cooled Hermetic, 60 Ton	4.45	1.7%
Replace Wallpaper Interior Wall Finish	3.87	1.5%
Replace Wood, Hollow Core, Painted, Interior Door	3.74	1.4%
Clean & Reseal Clay Brick, Exterior, 3rd+ Floor	3.49	1.3%
Repair Fan Coil, Two-Pipe, 400 Cfm	3.45	1.3%
Refinish Concrete Block, Painted, Interior Wall Finish	3.27	1.2%
Maintain Fan & Motor, Exhaust Fan, Ceiling 200-500 Cfm	3.27	1.2%
Repair Chiller, Reciprocal Air Cooled Hermetic, 60 Ton	3.16	1.2%
Replace Faucet Washer & Clean Trap, Lavatory, Vitreous China	2.94	1.1%
Inspect & Test Gas Boiler, 1000 Mbh	2.91	1.1%
Clean & Reseal Clay Brick, Exterior, 2nd Floor	2.91	1.1%
Replace Fan & Motor, Exhaust Fan, Ceiling 200-500 Cfm	2.67	1.0%
Replace Condenser, Air-Cooled, 60 Ton	2.42	0.9%
Clean & Reseal Clay Brick, Exterior, 1st Floor	2.32	0.9%
Repair Diverter Valve, Bathtub & Shower Enclosure, Fiberglass	2.24	0.8%

**Task cost ($2004) per gross square foot over 50 years.*

***Percent of total M&R costs.*

Cost per Sqft. by System

Building Age	Exterior Closure	Roofing	Interior Construtions	Stairways	Interior Finishes	Conveying Systems	Plumbing	HVAC Systems	Fire Protection	Electrical	Equipment	Total per Sqft.
1	.01	.20	.57			.23	.03	1.19	.01	.48		**2.71**
2	.01	.20	.57		.02	.23	.48	1.19	.01	.48		**3.18**
3	.01	.21	.57			.23	.04	1.19	.01	.48		**2.74**
4	.01	.20	.73		1.50	.23	.94	1.19	.01	.49		**5.29**
5	.01	.20	.57		.11	.23	.47	1.20	.01	.62		**3.43**
6	.01	.21	.57		.02	.23	.49	1.19	.01	.48		**3.21**
7	.01	.20	.57			.23	.08	1.28	.01	.48		**2.85**
8	.01	.20	.73		4.39	.23	1.18	1.19	.01	.49		**8.43**
9	.01	.21	.57	.02	.01	.23	.04	1.19	.01	.48		**2.76**
10	4.21	.20	2.15		.79	.23	1.80	3.51	.26	.94		**14.09**
11	.01	.20	.57			.23	.03	1.19	.01	.48		**2.71**
12	.01	.21	.73		1.50	.23	.95	1.19	.01	.48		**5.31**
13	.01	.20	.57			.23	.17	1.19	.01	.48		**2.85**
14	.07	.78	.57		.02	.23	.53	1.29	.01	.49		**3.98**
15	.01	.21	.57	.01	.12	.23	1.36	7.56	.01	1.14		**11.22**
16	.01	.20	.73		4.39	.23	.94	1.19	.01	.48		**8.18**
17	.01	.20	.57			.23	.03	1.19	.01	.48		**2.71**
18	.01	.21	.57	.02	.65	.23	.49	1.19	.01	.49		**3.87**
19	.01	.20	.57			.23	.03	1.19	.01	.48		**2.71**
20	4.31	.20	2.30		3.65	.23	3.39	4.72	.22	1.84		**20.85**
21	.01	.21	.57			.23	.09	1.28	.01	.48		**2.88**
22	.01	.20	.57		.02	.23	.48	1.19	.01	.48		**3.18**
23	.01	.20	.57			.23	.27	1.19	.01	.48		**2.95**
24	.01	.21	.73		4.59	.23	.95	1.19	.01	.49		**8.42**
25	2.64	.20	.57		.11	1.27	22.65	2.45	.01	1.61		**31.51**
26	.01	.20	.57		.02	.23	.55	1.19	.01	.48		**3.25**
27	.01	.21	.57	.02	.01	.23	.07	1.19	.01	.48		**2.79**
28	.07	1.15	.73		1.50	.23	.96	1.29	.01	.49		**6.42**
29	.01	.20	.57			.23	.06	1.19	.01	.48		**2.74**
30	4.21	.20	3.87	.01	.79	.23	2.65	9.42	.26	1.17		**22.80**
31	.01	.21	.57			.23	.06	1.19	.01	.48		**2.76**
32	.01	.20	.57		4.39	.23	.91	1.19	.01	.48		**7.99**
33	.01	.20	.57			.23	.07	1.19	.01	.48		**2.76**
34	.01	.21	.73		.02	.23	.45	1.19	.01	.49		**3.33**
35	.01	.20	.57		.11	.23	3.32	1.20	.01	.62		**6.28**
36	.01	.20	.57	.02	2.13	.23	.59	1.19	.01	.48		**5.43**
37	.01	.21	.57			.23	.38	1.28	.01	.48		**3.17**
38	.01	.20	.73		.02	.23	.41	1.19	.01	.49		**3.28**
39	.01	.20	.57			.23	.49	1.19	.01	.48		**3.17**
40	4.31	.21	2.15		6.54	.23	2.61	5.97	.22	2.13		**24.37**
41	.01	.20	.57			.23	.38	1.19	.01	.48		**3.07**
42	.07	.78	.73		.02	.23	.19	1.19	.01	.48		**3.69**
43	.01	.21	.57			.23	.38	1.19	.01	.48		**3.08**
44	.01	.20	.57		1.50	.23	.58	1.29	.01	.49		**4.88**
45	.01	.20	.57	.03	.13	.23	1.95	7.56	.01	1.14		**11.83**
46	.01	.21	.73		.02	.23	.12	1.19	.01	.48		**3.00**
47	.01	.20	.57			.23	.38	1.19	.01	.48		**3.06**
48	.01	.20	.57		4.59	.23	.59	1.19	.01	.49		**7.88**
49	.01	.21	.57			.23	.44	1.19	.01	.48		**3.14**
50	6.84	.20	2.30		2.06	1.27	23.54	2.32	.26	1.33		**40.12**
Total	27.04	12.14	40.15	.12	45.75	13.60	80.04	94.07	1.66	31.73		346.32

A value of "0.00" means a cost of more than $.000 but less than $.005 per gross square foot.

50-Year Total M&R Cost Profile per Sqft.

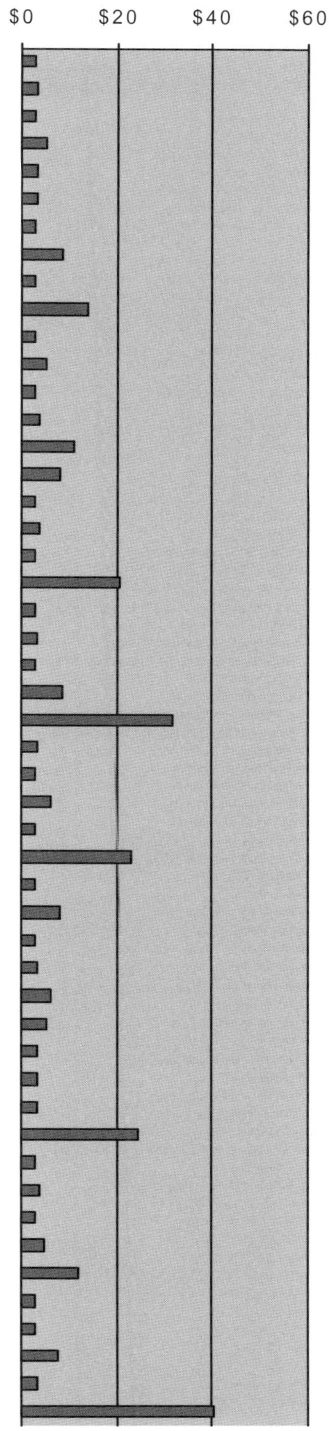

Apartments, 4-7 Story

Gross Sqft:	60,000
Height in Ft:	62
Exterior:	Brick Veneer
Floor coverings:	Carpet/Ceramic Tile/Vinyl Tile
HVAC:	Chilled Water, Oil Boiler, Baseboard Rad.
Occupancy:	150
Replacement Cost:	$6,326,889

Components

Component	Quantity	Unit
B20 Exterior Enclosure		
Clay Brick, Exterior, 1st Floor	8884	Sq Ft
Clay Brick, Exterior, 2nd Floor	8884	Sq Ft
Clay Brick, Exterior, 3rd+ Floor	8884	Sq Ft
Aluminum Operable Window, 24 sf, 1st Floor	97	Each
Aluminum Operable Window, 24 sf, 2nd Floor	97	Each
Aluminum Operable Window, 24 sf, 3rd+ Floor	97	Each
Steel w/ Safety Glass, Painted, Exterior Door	2	Each
Aluminum Frame, Fully Glazed, Exterior Door	2	Each
B30 Roofing		
Built-up Roof	10000	Sq Ft
C10 Interior Construction		
Wood, Hollow Core, Painted, Interior Door	750	Each
C20 Stairs		
Metal Interior Steps, Painted	1600	Sq Ft
C30 Interior Finishes		
Sheetrock, Stippled, Interior Wall Finish	108465	Sq Ft
Wallpaper Interior Wall Finish	38800	Sq Ft
Ceramic Tile Wall Finish	7748	Sq Ft
Vinyl Tile Flooring	18000	Sq Ft
Carpet, Nylon 20 oz., Low Traffic	36000	Sq Ft
Sheetrock, Unstippled, Ceiling	60000	Sq Ft
D10 Conveying		
Elevator, Geared, 3500 lbs, 6-15 Floor, 350 fpm	2	Each
D20 Plumbing		
Flush Tank Water Closet	122	Each
Lavatory, Vitreous China	122	Each
Sink, Stainless Steel	81	Each
Service Sink, Iron Enamel	5	Each
Bathtub & Shower Enclosure, Fiberglass	81	Each
Pipe & Fittings, 3/4" Copper, Cold Water	16	K Ln Ft
Pipe & Fittings, 4" Steel	1.048	K Ln Ft
Pipe & Fittings, 2" Copper, Cold Water	8.831	K Ln Ft
Pipe & Fittings, 3/4" Copper, Hot Water	9.1	K Ln Ft
Pipe Insulation, Cold Water	16.5	K Ln Ft
Pipe Insulation, Hot Water	9.4	K Ln Ft
Water Heater, Gas/Oil 75 Gph	32	Each
Water Heater, Electric 52 Gal.	83	Each
Pipe & Fittings, 6" Cast Iron	3.884	K Ln Ft
Pipe & Fittings, 10" Cast Iron	0.2	K Ln Ft
Pipe & Fittings, 4" DWV PVC	0.358	K Ln Ft
Roof Drain, 4-6"	5	Each
D30 HVAC		
Gas Boiler, 1000 Mbh	2	Each
Chemical Feed System	1	Each
Circulation Pump 5 HP, Hot Water	2	Each
Expansion Tank, 60 Gal.	2	Each
Circulation Pump, 5 HP, Chiller & Condenser Water	2	Each
Chiller, Reciprocal Water-Cooled Hermetic, 100 Ton	2	Each
Condenser, Air-Cooled, 60 Ton	4	Each
Exhaust Fan, Ceiling, 200-500 Cfm	200	Each
Fan Coil, Two-Pipe, 400 Cfm	200	Each
D40 Fire Protection		
Fire Standpipe System	1	Each
Fire Sprinkler System	1	Each
Fire Sprinkler Head	420	Each
Fire Suppression System Water Pump	1	Each
D50 Electrical		
Safety Switch, Fused, 400 Amp., 3 Ph.	6	Each
Motor Starter, 5-20 HP, <600 V	6	Each
Distribution Panel Board	81	Each
Grounded Fault Circuit Interruptor, 15 Amps	320	Each
Emergency Horn & Strobe	20	Each
Incandescent Lighting Fixture, Basic, 100w	490	Each
Exit Lighting Fixture, w/ Battery	18	Each
Wiring Device, Switch	810	Each
Receptacle, 120V, 15 Amp	810	Each
TV Cable Outlet	42	Each
Annunciation Panel	1	Each
Fire Alarm Control Panel	1	Each
Manual Pull Station	15	Each
Heat Detector	7	Each
Smoke Detector	175	Each

50-Year M&R Cost Summary

	50 Year Total	Annual Cost per Sqft	Annual cost as % replacement
PM & Minor Repair	$4,386,881	$1.46	1.39%
Unscheduled Maintenance	$4,713,047	$1.57	1.49%
Renewal & Replacement	$10,816,936	$3.61	3.42%
Total M&R Costs	$19,916,864	$6.64	6.30%

Distribution of Repair Costs

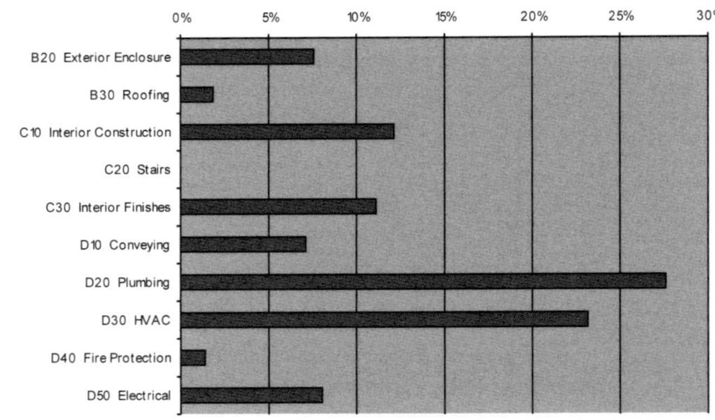

Thirty Most Costly Repair Tasks

Major Repair Task	Task Cost*	Pct.**
Maintain Elevator, Geared, 3500 lbs, 6-15 Floor, 350 fpm	18.62	7.3%
Replace Pipe & Fittings, 3/4" Copper, Cold Water	15.89	6.3%
Maintain Fan Coil, Two-Pipe, 400 Cfm	15.30	6.0%
Replace Fan Coil, Two-Pipe, 400 Cfm	13.33	5.3%
Refinish Sheetrock, Stippled, Interior Wall Finish	10.98	4.3%
Maintain Wood, Hollow Core, Painted, Door Locks	10.43	4.1%
Replace Pipe & Fittings, 2" Copper, Cold Water	9.96	3.9%
Replace Pipe & Fittings, 3/4" Copper, Hot Water	8.98	3.5%
Replace Carpet, Nylon 20 oz., Low Traffic	8.71	3.4%
Annual PM, Distribution Panel Board	6.36	2.5%
Replace Water Heater, Gas/Oil 75 Gph	6.33	2.5%
Replace Wood, Hollow Core, Painted, Door Locks	6.31	2.5%
Refinish Sheetrock, Unstippled, Ceiling	6.07	2.4%
Clean Water Heater, Gas/Oil 75 Gph	5.13	2.0%
Replace Wallpaper Interior Wall Finish	5.00	2.0%
Renovate Elevator, Geared, 3500 lbs, 6-15 Floor, 350 fpm	5.00	2.0%
Replace Chiller, Reciprocal Water-Cooled Hermetic, 100 Ton	4.90	1.9%
Maintain Fan & Motor, Exhaust Fan, Ceiling 200-500 Cfm	3.78	1.5%
Replace Wood, Hollow Core, Painted, Interior Door	3.74	1.5%
Replace Condenser, Air-Cooled, 60 Ton	3.62	1.4%
Replace Bathtub & Shower Enclosure, Fiberglass	3.45	1.4%
Maintain Chiller, Reciprocal Water-Cooled Hermetic, 100 Ton	3.31	1.3%
Clean & Reseal Clay Brick, Exterior, 3rd+ Floor	3.30	1.3%
Replace Fan & Motor, Exhaust Fan, Ceiling 200-500 Cfm	3.09	1.2%
Repair Chiller, Reciprocal Water-Cooled Hermetic, 100 Ton	2.79	1.1%
Clean & Reseal Clay Brick, Exterior, 2nd Floor	2.75	1.1%
Replace Water Heater, Electric 52 Gal.	2.66	1.0%
Repair Fan Coil, Two-Pipe, 400 Cfm	2.59	1.0%
Maintain Built-up Roof	2.44	1.0%
Replace Faucet Washer & Clean Trap, Lavatory, Vitreous China	2.24	0.9%

**Task cost ($2004) per gross square foot over 50 years.*

***Percent of total M&R costs.*

Cost per Sqft. by System

Building Age	Exterior Closure	Roofing	Interior Constrution	Stairways	Interior Finishes	Conveying Systems	Plumbing	HVAC Systems	Fire Protection	Electrical	Equipment	Total per Sqft.
1	.00	.10	.57			.37	.33	.93	.07	.37		**2.74**
2	.00	.10	.57			.37	.76	.94	.07	.37		**3.18**
3	.00	.11	.57			.37	.34	.93	.07	.37		**2.76**
4	.00	.10	.73		1.42	.37	1.09	.94	.07	.37		**5.08**
5	.00	.10	.57		.16	.37	.92	.94	.09	.50		**3.66**
6	.00	.11	.57			.37	.78	.94	.07	.37		**3.20**
7	.00	.10	.57			.37	.36	1.00	.07	.37		**2.84**
8	.00	.10	.73		1.42	.37	2.24	.94	.07	.37		**6.23**
9	.00	.11	.57	.02	.01	.37	.34	.93	.07	.37		**2.79**
10	3.98	.10	2.15		1.17	.37	1.89	2.88	.31	.98		**13.83**
11	.00	.10	.57			.37	.33	.93	.07	.37		**2.74**
12	.01	.11	.73		1.42	.37	1.11	.94	.07	.37		**5.11**
13	.00	.10	.57			.37	.69	.93	.07	.37		**3.10**
14	.01	.39	.57			.37	.80	1.00	.07	.37		**3.59**
15	.00	.11	.57	.01	3.05	.37	1.70	6.16	.09	1.25		**13.32**
16	.00	.10	.73		1.42	.37	1.92	.94	.07	.37		**5.92**
17	.00	.10	.57			.37	.33	.93	.07	.37		**2.74**
18	.00	.11	.57	.02	.63	.37	.78	.94	.07	.40		**3.88**
19	.00	.10	.57			.37	.33	.93	.07	.37		**2.74**
20	4.09	.10	2.31		4.14	.37	3.05	4.67	.27	1.96		**20.96**
21	.00	.11	.57			.37	.63	1.00	.07	.37		**3.12**
22	.00	.10	.57			.37	.76	.94	.07	.37		**3.18**
23	.00	.10	.57			.37	.64	.93	.07	.40		**3.08**
24	.01	.11	.73		1.42	.37	1.94	.94	.07	.37		**5.95**
25	2.49	.10	.57		.16	2.87	19.76	1.95	.09	1.19		**29.19**
26	.00	.10	.57			.37	.79	.94	.07	.37		**3.21**
27	.00	.11	.57	.02	.01	.37	.37	.93	.07	.37		**2.82**
28	.01	.58	.73		1.42	.37	1.10	1.00	.07	.40		**5.68**
29	.00	.10	.57			.37	.61	.93	.07	.37		**3.02**
30	3.98	.10	3.87	.01	4.05	.37	2.89	7.69	.31	1.24		**24.52**
31	.00	.11	.57			.37	.36	.93	.07	.37		**2.78**
32	.00	.10	.57		1.42	.37	1.89	.94	.07	.37		**5.73**
33	.00	.10	.57			.37	.37	.93	.07	.40		**2.82**
34	.00	.11	.73			.37	.73	.94	.07	.37		**3.31**
35	.00	.10	.57		.16	.37	2.88	.94	.09	.47		**5.59**
36	.01	.10	.57	.02	2.05	.37	.82	.94	.07	.40		**5.35**
37	.00	.11	.57			.37	.86	1.00	.07	.37		**3.35**
38	.00	.10	.73			.37	.84	.94	.07	.37		**3.41**
39	.00	.10	.57			.37	.69	.93	.07	.37		**3.10**
40	4.09	.11	2.15		4.14	.37	3.72	5.68	.27	2.42		**22.95**
41	.00	.10	.57			.37	.61	.93	.07	.40		**3.05**
42	.01	.39	.73			.37	.53	.94	.07	.37		**3.40**
43	.00	.11	.57			.37	.62	.93	.07	.37		**3.03**
44	.00	.10	.57		1.42	.37	.81	1.00	.07	.37		**4.71**
45	.00	.10	.57	.03	3.06	.37	2.43	6.16	.09	1.22		**14.04**
46	.00	.11	.73			.37	.48	.94	.07	.40		**3.09**
47	.00	.10	.57		-	.37	.61	.93	.07	.37		**3.02**
48	.01	.10	.57		1.42	.37	1.65	.94	.07	.37		**5.49**
49	.00	.11	.57			.37	.65	.93	.07	.37		**3.07**
50	6.47	.10	2.31		1.17	2.87	20.63	1.89	.31	.75		**36.51**
Total	25.29	6.07	40.19	.12	36.75	23.62	91.74	76.70	4.60	26.86		331.95

A value of "0.00" means a cost of more than \$.000 but less than \$.005 per gross square foot.

50-Year Total M&R Cost Profile per Sqft.

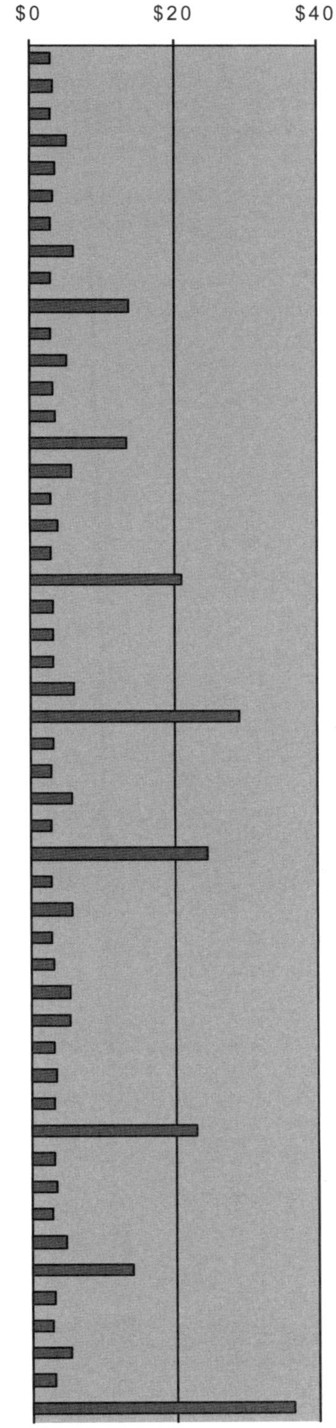

Apartments, 24 Story

Gross Sqft:	220,000
Height in Ft:	240
Exterior:	Stucco
Floor coverings:	Finished Concrete/Vinyl Sheet/Carpet
HVAC:	Chilled Water, Gas Boiler, Multizone Air
Occupancy:	450
Replacement Cost:	$17,883,074

Components

Component	Quantity	Unit
B20 Exterior Enclosure		
Stucco, Exterior, 1st Floor	2860	Sq Ft
Stucco, Exterior, 2nd Floor	2860	Sq Ft
Stucco, Exterior, 3rd+ Floor	62941	Sq Ft
Aluminum Fixed Window, 24 sf, 1st Floor	4	Each
Aluminum Fixed Window, 24 sf, 2nd Floor	482	Each
Aluminum Operable Window, 24 sf, 3rd+ Floor	482	Each
Steel, Painted, Exterior Door	18	Each
Aluminum Frame, Fully Glazed, Exterior Door	302	Each
B30 Roofing		
Single-Ply Thermosetting Roof	9167	Sq Ft
C10 Interior Construction		
Wood, Solid Core, Painted, Interior Door	164	Each
Wood, Hollow Core, Painted, Interior Door	872	Each
Wood, Hollow Core, Painted, Sliding Interior Door	436	Each
Steel, Painted, Interior Door	96	Each
C20 Stairs		
Metal Interior Steps, Painted	7200	Sq Ft
Metal Interior Railing, Painted	1920	Ln Ft
C30 Interior Finishes		
Sheetrock, Stippled, Interior Wall Finish	563230	Sq Ft
Concrete Block, Painted, Interior Wall Finish	27500	Sq Ft
Ceramic Tile Wall Finish	18750	Sq Ft
Concrete, Finished Flooring	11000	Sq Ft
Vinyl Sheet Flooring	33750	Sq Ft
Carpet, Nylon 20 oz., Low Traffic	59000	Sq Ft
Carpet, Wool 45 oz., Premium	116250	Sq Ft
Sheetrock, Stippled, Ceiling	100000	Sq Ft
Textured, Spray, Ceiling	120000	Sq Ft
D10 Conveying		
Elevator, Geared, 3500 lbs, 16-25 Floor, 800 fpm	2	Each
D20 Plumbing		
Flush Tank Water Closet	306	Each
Lavatory, Vitreous China	304	Each
Sink, Stainless Steel	152	Each
Shower, Fiberglass	154	Each
Bath Tub, Enameled Steel	150	Each
Booster Pump, 10 HP	2	Each
Pipe & Fittings, 3/4" Copper, Cold Water	38.9	K Ln Ft
Pipe & Fittings, 2" Copper, Cold Water	35.3	K Ln Ft
Pipe & Fittings, 3/4" Copper, Hot Water	23.6	K Ln Ft
Pipe Insulation, Cold Water	39.5	K Ln Ft
Pipe Insulation, Hot Water	24.5	K Ln Ft
Water Heater, Electric 52 Gal.	152	Each
Pipe & Fittings, 6" Cast Iron	9.8	K Ln Ft
Pipe & Fittings, 10" Cast Iron	0.2	K Ln Ft
Sewage Ejector Pump, 1 HP	1	Each
Pipe & Fittings, 4" DWV PVC	1.3	K Ln Ft
Roof Drain, 4-6"	3	Each
Sump Pump, 3 HP	1	Each
D30 HVAC		
Pipe & Fittings, 4" Steel, Gas	1.8	K Ln Ft
Circulation Pump 25 HP, Hot Water	1	Each
Gas Boiler, 2000 Mbh	1	Each
Expansion Tank, 250 Gal.	1	Each
Circulation Pump, 75 HP, Chiller & Condenser Water	3	Each
Chiller, Centrifugal Hermetic, Bundled w/ Condenser, 200 Ton	1	Each
Chiller, Centrifugal Hermetic, Bundled w/ Condenser, 300 Ton	1	Each
Cooling Tower, 500 Ton	1	Each
Air Handler, Multizone, 15,000 Cfm	13	Each
D40 Fire Protection		
Fire Sprinkler Head	1100	Each
D50 Electrical		
Primary Transformer, Dry, >500 KVA	1	Each
Secondary Transformer, Dry, 150 KVA	15	Each
Main Switchgear, >1200 Amp.	1	Each
Distribution Switch, Fused, <600 V	32	Each
Motor Starter, 21-50 HP, <600 V	8	Each
Incandescent Lighting Fixture, Basic, 100w	1650	Each
Fluorescent Lighting Fixture, 160w	50	Each
Wiring Device, Switch	1600	Each
Receptacle, 120V, 15 Amp	4360	Each
Annunciation Panel	1	Each
Fire Alarm Control Panel	1	Each
Manual Pull Station	48	Each
Smoke Detector	164	Each

50-Year M&R Cost Summary

	50 Year Total	Annual Cost per Sqft	Annual cost as % replacement
PM & Minor Repair	$4,162,273	$0.38	0.47%
Unscheduled Maintenance	$5,024,232	$0.46	0.56%
Renewal & Replacement	$26,622,943	$2.42	2.98%
Total M&R Costs	$35,809,448	$3.26	4.00%

Distribution of Repair Costs

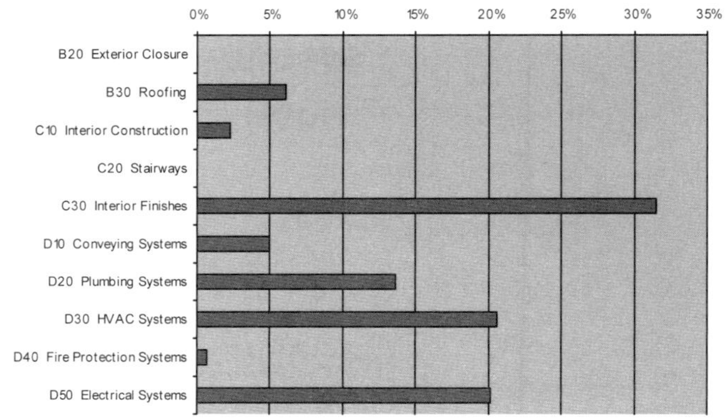

Thirty Most Costly Repair Tasks

Major Repair Task	Task Cost*	Pct.**
Refinish Sheetrock, Stippled, Interior Wall Finish	15.55	11.1%
Replace Carpet, Wool 45 oz., Premium	12.28	8.8%
Replace Pipe & Fittings, 2" Copper, Cold Water	10.86	7.8%
Replace Pipe & Fittings, 3/4" Copper, Cold Water	10.54	7.5%
Maintain Elevator, Geared, 3500 lbs, 16-25 Floor, 800 fpm	6.63	4.7%
Replace Pipe & Fittings, 3/4" Copper, Hot Water	6.35	4.5%
Replace Vinyl Sheet Flooring	4.90	3.5%
Replace Carpet, Nylon 20 oz., Low Traffic	3.89	2.8%
Maintain Wood, Hollow Core, Painted, Door Locks	3.31	2.4%
Replace Air Handler, Multizone, 15,000 Cfm	2.89	2.1%
Refinish Stucco, Exterior, 3rd+ Floor	2.85	2.0%
Refinish Sheetrock, Stippled, Ceiling	2.77	2.0%
Maintain Air Handler, Multizone, 15,000 Cfm	2.06	1.5%
Renovate Elevator, Geared, 3500 lbs, 16-25 Floor, 800 fpm	2.03	1.5%
Replace Wood, Hollow Core, Painted, Door Locks	2.00	1.4%
Replace Chiller, Centrifugal Hermetic, Bnd'd w/ Condenser, 300 Ton	1.71	1.2%
Maintain Wood, Hollow Core, Painted, Sliding Door Locks	1.65	1.2%
Minor Repair, Sheetrock, Stippled, Interior Wall Finish	1.53	1.1%
Replace Faucet Washer & Clean Trap, Lavatory, Vitreous China	1.52	1.1%
Repair Chiller, Centrifugal Hermetic, Bnd'd w/ Condenser, 300 Ton	1.45	1.0%
Replace Water Heater, Electric 52 Gal.	1.33	0.9%
Replace Diverter Valve, Bath Tub, Enameled Steel	1.32	0.9%
Replace Incandescent Lighting Fixture, Basic, 100w	1.32	0.9%
Replace Receptacle	1.25	0.9%
Replace Pipe Insulation, Cold Water	1.19	0.9%
Replace Wood, Hollow Core, Painted, Interior Door	1.19	0.8%
Maintain Aluminum Frame, Fully Glazed, Exterior Door Locks	1.17	0.8%
Replace Chiller, Centrifugal Hermetic, Bnd'd w/ Condenser, 200 Ton	1.16	0.8%
Repair Aluminum Frame, Fully Glazed, Exterior Door	1.11	0.8%
Replace Wood, Hollow Core, Painted, Sliding Door Locks	1.02	0.7%

**Task cost ($2002) per gross square foot over 50 years.*

***Percent of total M&R costs.*

Cost per Sqft. by System

Building Age	Exterior Closure	Roofing	Interior Construtrion	Stairways	Interior Finishes	Conveying Systems	Plumbing	HVAC Systems	Fire Protection	Electrical	Equipment	Total per Sqft.
1	.06	.00	.33			.13	.01	.15	.01	.03		**.72**
2	.06	.00	.33			.13	.31	.15	.01	.03		**1.02**
3	.06	.01	.33			.13	.02	.15	.01	.03		**.73**
4	.06	.00	.42		1.61	.13	.52	.15	.01	.03		**2.93**
5	.06	.00	.33		.14	.13	.31	.15	.01	.13		**1.27**
6	.06	.01	.33			.13	.32	.15	.01	.03		**1.04**
7	.06	.00	.33	.01		.13	.04	.16	.01	.03		**.77**
8	.06	.00	.42		1.61	.13	.72	.15	.01	.03		**3.13**
9	.06	.01	.33	.02	.03	.13	.02	.15	.01	.03		**.78**
10	.68	.03	1.26		.24	.13	1.27	.89	.17	.39		**5.07**
11	.06	.00	.33			.13	.01	.15	.01	.03		**.72**
12	.34	.01	.42		1.61	.13	.53	.15	.01	.03		**3.22**
13	.06	.00	.33			.13	.08	.15	.01	.03		**.78**
14	.09	.00	.33	.01		.13	.34	.16	.01	.03		**1.10**
15	.06	.01	.33	.02	1.45	.13	.86	1.47	.01	.46		**4.79**
16	.06	.00	.42		1.61	.13	.56	.15	.01	.03		**2.97**
17	.06	.00	.33			.13	.01	.15	.01	.03		**.72**
18	.06	.01	.33	.02	2.45	.13	.32	.15	.01	.08		**3.55**
19	.06	.00	.33			.13	.01	.15	.01	.03		**.72**
20	.79	.12	1.36		3.55	.13	1.78	1.54	.14	1.37		**10.78**
21	.06	.00	.33	.01		.13	.04	.16	.01	.03		**.77**
22	.06	.00	.33			.13	.31	.15	.01	.03		**1.02**
23	.06	.01	.33			.13	.17	.15	.01	.04		**.89**
24	.34	.00	.42		1.67	.13	.57	.15	.01	.03		**3.32**
25	.06	.00	.33		6.28	1.15	15.27	.23	.01	.35		**23.68**
26	.06	.01	.33			.13	.32	.15	.01	.03		**1.04**
27	.06	.00	.33	.02	.03	.13	.04	.15	.01	.03		**.80**
28	.09	.00	.42	.01	1.61	.13	.52	.16	.01	.04		**2.99**
29	.06	.01	.33			.13	.03	.15	.01	.03		**.75**
30	.68	.03	1.99	.02	1.55	.13	1.79	2.25	.17	1.33		**9.96**
31	.06	.00	.33			.13	.03	.15	.01	.03		**.74**
32	.06	.01	.36		1.61	.13	.54	.15	.01	.03		**2.89**
33	.06	.00	.33			.13	.04	.15	.01	.04		**.76**
34	.06	.00	.39			.13	.29	.15	.01	.03		**1.06**
35	.06	.01	.33	.01	.14	.13	2.06	.15	.01	.12		**3.02**
36	.34	.00	.36	.02	4.06	.13	.10	.15	.01	.08		**5.25**
37	.06	.00	.33			.13	.24	.16	.01	.03		**.96**
38	.06	.01	.39			.13	.28	.15	.01	.03		**1.05**
39	.06	.00	.33			.13	.48	.15	.01	.03		**1.19**
40	.79	.12	1.53		3.55	.13	1.16	1.62	.14	1.59		**10.63**
41	.06	.00	.33			.13	.24	.15	.01	.04		**.96**
42	.09	.00	.39	.01		.13	.12	.15	.01	.03		**.92**
43	.06	.01	.33			.13	.44	.15	.01	.03		**1.15**
44	.06	.00	.36		1.61	.13	.09	.16	.01	.03		**2.46**
45	.06	.00	.33	.04	1.48	.13	1.45	1.47	.01	.45		**5.42**
46	.06	.01	.39			.13	.09	.15	.01	.04		**.87**
47	.06	.00	.33			.13	.44	.15	.01	.03		**1.15**
48	.34	.00	.36		1.67	.13	.14	.15	.01	.03		**2.83**
49	.06	.01	.33	.01		.13	.26	.15	.01	.03		**.98**
50	.68	.03	1.32		6.38	1.15	15.75	.81	.17	.16		**26.45**
Total	7.62	.62	23.14	.19	45.90	8.67	51.35	16.53	1.07	7.69		162.77

50-Year Total M&R Cost Profile per Sqft.

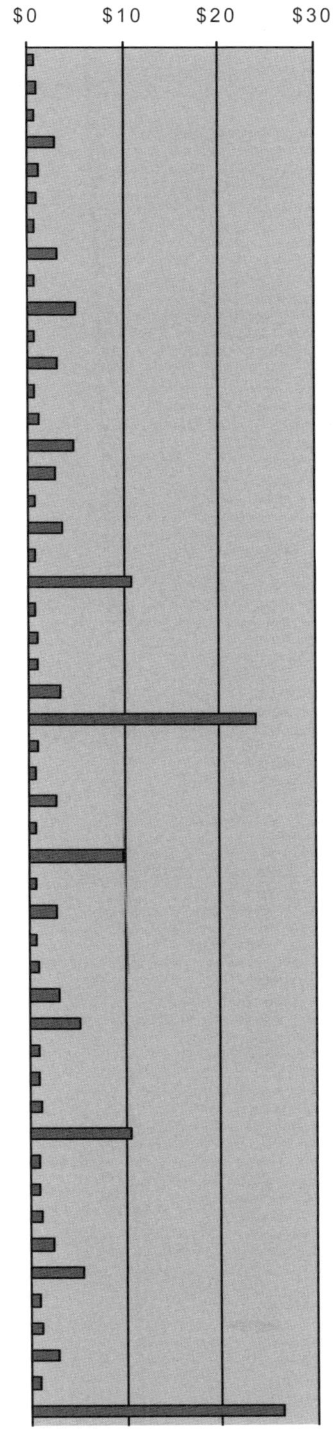

A value of "0.00" means a cost of more than $.000 but less than $.005 per gross square foot.

Auditorium

Gross Sqft:	24,000
Height in Ft:	24
Exterior:	Precast Concrete Panel
Floor coverings:	Carpet/Hardwood
HVAC:	Electric Cooling, Gas Heating
Occupancy:	3600
Replacement Cost:	$2,636,255

Components

Component	Quantity	Unit
B20 Exterior Enclosure		
Concrete, Painted, Exterior, 1st Floor	14900	Sq Ft
Aluminum Fixed Window, 12 sf, 1st Floor	31	Each
Steel, Painted, Exterior Door	6	Each
Aluminum Frame, Fully Glazed, Exterior Door	6	Each
B30 Roofing		
Built-up Roof	24000	Sq Ft
Concrete Decking	1000	Sq Ft
Concrete Steps	600	Sq Ft
C10 Interior Construction		
Steel, Painted, Interior Door	48	Each
C20 Stairs		
Concrete Interior Steps	360	Sq Ft
Masonry Interior Steps, Painted	12288	Sq Ft
C30 Interior Finishes		
Wood Parquetry Flooring	16800	Sq Ft
Carpet, Nylon 20 oz., High Traffic	7200	Sq Ft
Acoustical Tile Ceiling	24000	Sq Ft
D20 Plumbing		
Tankless Water Closet	17	Each
Urinal, Vitreous China	6	Each
Lavatory, Iron, Enamel	23	Each
Lavatory, Vitreous China	23	Each
Sink, Stainless Steel	2	Each
Shower, Fiberglass	4	Each
Drinking Fountain, Refrigerated	6	Each
Pipe & Fittings, 3/4" Copper, Hot Water	0.88	K Ln Ft
Pipe & Fittings, 2" Copper, Cold Water	2.25	K Ln Ft
Pipe & Fittings, 4" Steel	1.439	K Ln Ft
Pipe Insulation, Cold Water	2.61	K Ln Ft
Pipe Insulation, Hot Water	1.29	K Ln Ft
Water Heater, Gas/Oil 850 Gph	2	Each
Pipe & Fittings, 6" Cast Iron	1.065	K Ln Ft
Pipe & Fittings, 10" Cast Iron	0.31	K Ln Ft
Roof Drain, 4-6"	14	Each
Pipe & Fittings, 4" DWV PVC	0.694	K Ln Ft
D30 HVAC		
Air Conditioner, Rooftop, 20 Ton	1	Each
Air Conditioner, Rooftop, 50 Ton	1	Each
D40 Fire Protection		
Fire Standpipe System	1	Each
Fire Sprinkler System	1	Each
Fire Sprinkler Head	185	Each
Fire Suppression System Water Pump	1	Each
D50 Electrical		
Distribution Panel Board	2	Each
Primary Transformer, Dry, >500 KVA	1	Each
Main Switchgear, <1200 Amp.	1	Each
Motor Starter, 5-20 HP, <600 V	4	Each
Exit Lighting Fixture, w/ Battery	12	Each
Emergency Horn & Strobe	10	Each
Light Dimming Panel	2	Each
Incandescent Lighting Fixture, Basic, 100w	50	Each
Metal Halide Lighting Fixture, High Bay, 400w	240	Each
Wiring Device, Switch	48	Each
Receptacle, 120V, 15 Amp	192	Each
Annunciation Panel	1	Each
Fire Alarm Bell, 6"	8	Each
Fire Alarm Control Panel	1	Each
Manual Pull Station	8	Each
Heat Detector	12	Each
Smoke Detector	12	Each
Sound System	1	Each

50-Year M&R Cost Summary

	50 Year Total	Annual Cost per Sqft	Annual cost as % replacement
PM & Minor Repair	$982,382	$0.82	0.75%
Unscheduled Maintenance	$938,397	$0.78	0.71%
Renewal & Replacement	$2,410,909	$2.01	1.83%
Total M&R Costs	$4,331,688	$3.61	3.29%

Distribution of Repair Costs

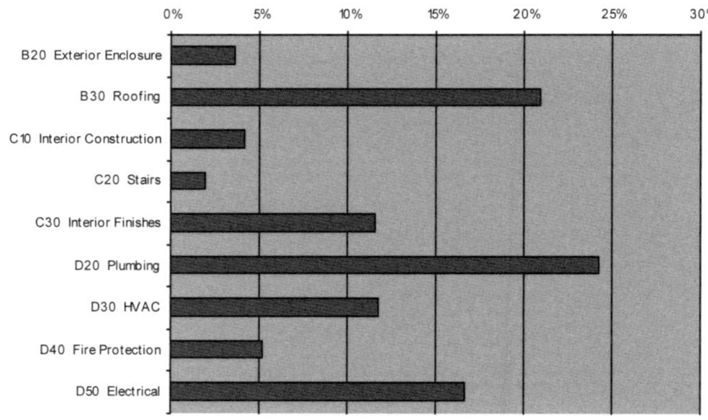

Thirty Most Costly Repair Tasks

Major Repair Task	Task Cost*	Pct.**
Maintain Built-up Roof	14.62	10.3%
Replace Carpet, Nylon 20 oz., High Traffic	8.71	6.2%
Replace Metal Halide Lighting Fixture, High Bay, 400w	8.16	5.8%
Replace Wood Parquetry Floor	8.16	5.8%
Replace Air Conditioner, Rooftop, 50 Ton	6.43	4.5%
Replace Pipe & Fittings, 2" Copper, Cold Water	6.35	4.5%
Replace Ballast, Metal Halide Lighting Fixture, High Bay, 400w	5.31	3.8%
Replace Water Heater, Gas/Oil 850 Gph	5.20	3.7%
Refinish Concrete, Painted, Exterior, 1st Floor	4.90	3.5%
Replace Lamp, Metal Halide Lighting Fixture, High Bay, 400w	4.22	3.0%
Repair Air Conditioner, Rooftop, 50 Ton	3.93	2.8%
Clean & Service Water Heater, Gas/Oil 850 Gph	3.90	2.8%
Replace Air Conditioner, Rooftop, 20 Ton	3.80	2.7%
Fire Sprinkler System, Annual PM	3.60	2.5%
Place New Membrane Over Existing, Built-up Roof	3.49	2.5%
Remove & Replace Membrane, Built-up Roof	3.45	2.4%
Refinish Masonry Interior Steps, Painted	2.57	1.8%
Replace Steel, Painted, Door Locks	2.56	1.8%
Replace Pipe & Fittings, 3/4" Copper, Hot Water	2.17	1.5%
Sand & Refinish Wood Parquetry Flooring	2.11	1.5%
Maintain Air Conditioner, Rooftop, 50 Ton	1.83	1.3%
Minor Repair, Acoustic Tile Ceiling	1.78	1.3%
Replace Primary Transformer, Dry, >500 KVA	1.75	1.2%
Maintain Steel, Painted, Door Locks	1.70	1.2%
Repair Air Conditioner, Rooftop, 20 Ton	1.52	1.1%
Maintain Air Conditioner, Rooftop, 20 Ton	1.42	1.0%
Annual PM, Fire Suppression System Water Pump Motor	1.17	0.8%
Maintain Roof Drain, 4-6"	1.07	0.8%
Replace Faucet Washer & Clean Trap, Lavatory, Vitreous China	1.06	0.7%
Replace Faucet Washer & Clean Trap, Lavatory, Iron, Enamel	1.05	0.7%

**Task cost ($2004) per gross square foot over 50 years.*

***Percent of total M&R costs.*

Cost per Sqft. by System

Building Age	Exterior Closure	Roofing	Interior Construtiion	Stairways	Interior Finishes	Conveying Systems	Plumbing	HVAC Systems	Fire Protection	Electrical	Equipment	Total per Sqft.
1	.02	.59	.09				.09	.12	.16	.11		**1.17**
2	.02	.59	.09		.01		.98	.12	.16	.11		**2.07**
3	.02	.64	.09				.09	.12	.16	.11		**1.22**
4	.02	.59	.12		.01		.98	.12	.16	.11		**2.10**
5	.02	.59	.09				.17	.12	.21	.87		**2.06**
6	.02	.64	.09		.01		.98	.12	.16	.11		**2.12**
7	.02	.66	.09	.37			.13	.12	.16	.11		**1.66**
8	.02	.59	.12		1.45		1.30	.12	.16	.11		**3.87**
9	.02	.64	.09		.36		.09	.12	.16	.11		**1.58**
10	1.00	.59	.60		.54		1.48	1.94	.45	2.74		**9.34**
11	.02	.59	.09				.09	.12	.16	.11		**1.17**
12	.08	.64	.12		.01		.98	.12	.16	.11		**2.20**
13	.02	.59	.09				.22	.12	.16	.11		**1.30**
14	.10	2.41	.09	.37	.01		1.02	.12	.16	.11		**4.39**
15	.02	.64	.09	.03			.13	3.41	.21	1.24		**5.77**
16	.02	.59	.12		1.45		1.30	.12	.16	.11		**3.87**
17	.02	.59	.09				.09	.12	.16	.11		**1.17**
18	.02	.64	.09		.36		.98	.12	.16	.17		**2.53**
19	.02	.59	.09				.09	.12	.16	.11		**1.17**
20	1.02	.96	.63		.54		1.54	.12	.41	4.99		**10.20**
21	.02	.71	.09	.37			.17	.12	.16	.11		**1.75**
22	.02	.59	.09		.01		.98	.12	.16	.11		**2.07**
23	.02	.59	.09				.09	.12	.16	.16		**1.22**
24	.08	.64	.12		1.45		1.30	.12	.16	.11		**3.97**
25	.02	.59	.09				5.47	1.94	.21	.92		**9.23**
26	.02	.59	.09		.01		1.05	.12	.16	.11		**2.15**
27	.02	.64	.09	.80	.36		.10	.12	.16	.11		**2.39**
28	.11	3.53	.12	.37	.01		1.01	.12	.16	.16		**5.58**
29	.02	.59	.09				.13	.12	.16	.11		**1.22**
30	1.00	.59	.60	.03	.54		1.47	3.41	.45	4.71		**12.79**
31	.02	.64	.09				.10	.12	.16	.11		**1.23**
32	.02	.59	.12		1.45		1.29	.12	.16	.11		**3.85**
33	.02	.59	.09				.10	.12	.16	.16		**1.23**
34	.02	.64	.09		.01		.97	.12	.16	.11		**2.11**
35	.02	.66	.09	.37			1.03	.12	.21	.82		**3.31**
36	.08	.59	.12		.36		.85	.12	.16	.17		**2.43**
37	.02	.64	.09				.25	.12	.16	.11		**1.39**
38	.02	.59	.09		.01		.86	.12	.16	.11		**1.95**
39	.02	.59	.09				.30	.12	.16	.11		**1.39**
40	1.02	1.02	.63		9.61		1.86	1.94	.41	5.09		**21.56**
41	.02	.59	.09				.21	.12	.16	.16		**1.35**
42	.10	2.41	.09	.37	.01		.88	.12	.16	.11		**4.24**
43	.02	.64	.09				.21	.12	.16	.11		**1.35**
44	.02	.59	.12		.01		.85	.12	.16	.11		**1.97**
45	.02	.59	.09	.03	.36		.53	3.41	.21	1.19		**6.42**
46	.02	.64	.09		.01		.85	.12	.16	.16		**2.04**
47	.02	.59	.09				.23	.12	.16	.11		**1.32**
48	.08	.59	.12		1.45		1.17	.12	.16	.11		**3.79**
49	.02	.71	.09	.37			.24	.12	.16	.11		**1.82**
50	1.00	.59	.60		.54		6.49	.12	.45	2.59		**12.38**
Total	6.48	37.73	7.46	3.44	20.92		43.72	21.27	9.41	30.05		180.49

A value of "0.00" means a cost of more than $.000 but less than $.005 per gross square foot.

50-Year Total M&R Cost Profile per Sqft.

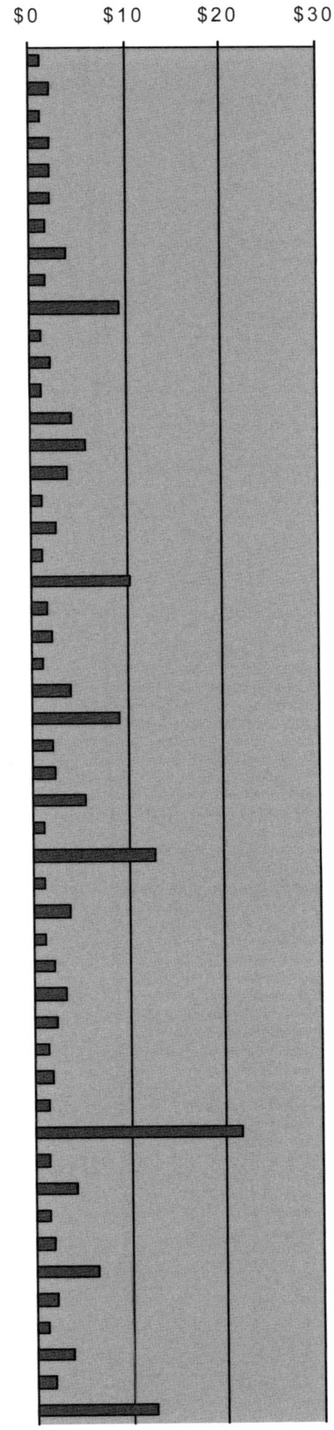

Bank

Gross Sqft:	4,100
Height in Ft:	14
Exterior:	Brick Veneer
Floor coverings:	Carpet Tile/Vinyl Tile
HVAC:	Electric Cooling, Gas Heating
Occupancy:	120
Replacement Cost:	$658,879

Components

Component	Quantity	Unit
B20 Exterior Enclosure		
Stucco, Exterior, 1st Floor	2867	Sq Ft
Steel, Painted, Exterior Door	3	Each
Aluminum & Glass Revolving Door	1	Each
B30 Roofing		
Built-up Roof	4100	Sq Ft
Concrete Decking	200	Sq Ft
C10 Interior Construction		
Aluminum, w/ Safety Glass, Plain or Anodized, Interior Door	21	Each
C30 Interior Finishes		
Quarry Tile Flooring	2050	Sq Ft
Carpet, Nylon 20 oz., High Traffic	2050	Sq Ft
Acoustical Tile Ceiling	4100	Sq Ft
D10 Conveying		
Pneumatic Tube Station	4	Each
D20 Plumbing		
Tankless Water Closet	5	Each
Urinal, Vitreous China	2	Each
Lavatory, Iron, Enamel	6	Each
Sink, Stainless Steel	2	Each
Drinking Fountain, Refrigerated	2	Each
Pipe & Fittings, 3/4" Copper, Cold Water	0.62	K Ln Ft
Pipe & Fittings, 3/4" Copper, Hot Water	0.25	K Ln Ft
Pipe & Fittings, 2" Copper, Cold Water	0.673	K Ln Ft
Pipe & Fittings, 4" Steel	0.612	K Ln Ft
Pipe Insulation, Cold Water	0.926	K Ln Ft
Pipe Insulation, Hot Water	0.428	K Ln Ft
Water Heater, Electric 52 Gal.	1	Each
Pipe & Fittings, 6" Cast Iron	0.364	K Ln Ft
Pipe & Fittings, 10" Cast Iron	0.128	K Ln Ft
Roof Drain, 4-6"	3	Each
Pipe & Fittings, 4" DWV PVC	0.36	K Ln Ft
D30 HVAC		
Air Conditioner, Rooftop, 20 Ton	1	Each
Pneumatic Thermostat	2	Each
D40 Fire Protection		
Fire Sprinkler Head	32	Each
D50 Electrical		
Distribution Panel Board	3	Each
Main Switchgear, <1200 Amp.	1	Each
Exit Lighting Fixture, w/ Battery	4	Each
Emergency Horn & Strobe	2	Each
Fluorescent Lighting Fixture, 160w	41	Each
Wiring Device, Switch	9	Each
Receptacle, 120V, 15 Amp	30	Each
TV Cable Outlet	1	Each
Annunciation Panel	1	Each
Fire Alarm Bell, 6"	1	Each
Manual Pull Station	2	Each
Smoke Detector	5	Each
Security Camera System	1	Camera

50-Year M&R Cost Summary

	50 Year Total	Annual Cost per Sqft	Annual cost as % replacement
PM & Minor Repair	$304,371	$1.48	0.92%
Unscheduled Maintenance	$230,045	$1.12	0.70%
Renewal & Replacement	$631,740	$3.08	1.92%
Total M&R Costs	$1,166,156	$5.69	3.54%

Distribution of Repair Costs

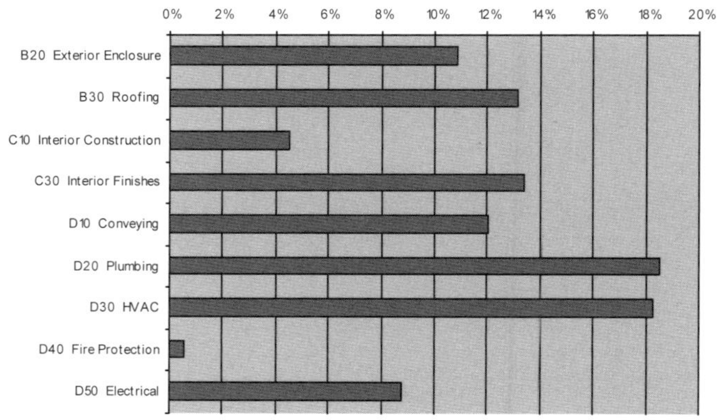

Thirty Most Costly Repair Tasks

Major Repair Task	Task Cost*	Pct.**
Maintain Pneumatic Tube Station	26.91	11.8%
Replace Air Conditioner, Rooftop, 20 Ton	22.25	9.7%
Replace Quarry Tile Flooring	21.18	9.3%
Replace Aluminum & Glass Revolving Door	20.55	9.0%
Maintain Built-up Roof	14.62	6.4%
Replace Carpet, Nylon 20 oz., High Traffic	14.51	6.4%
Replace Pipe & Fittings, 2" Copper, Cold Water	11.11	4.9%
Replace Pipe & Fittings, 3/4" Copper, Cold Water	9.01	3.9%
Repair Air Conditioner, Rooftop, 20 Ton	8.92	3.9%
Maintain Air Conditioner, Rooftop, 20 Ton	8.30	3.6%
Refinish Stucco, Exterior, 1st Floor	5.63	2.5%
Maintain Aluminum, w/ Safety Glass, Plain or Anodized, Door Locks	4.37	1.9%
Replace Pipe & Fittings, 3/4" Copper, Hot Water	3.61	1.6%
Place New Membrane Over Existing, Built-up Roof	3.49	1.5%
Remove & Replace Membrane, Built-up Roof	3.45	1.5%
Annual PM, Distribution Panel Board	3.45	1.5%
Replace Pneumatic Thermostat	2.78	1.2%
Replace Ballast & Lamps, Fluorescent Lighting Fixture, 160w	2.76	1.2%
Annual PM, Pneumatic Thermostat	2.07	0.9%
Replace Fluorescent Lighting Fixture, 160w	1.96	0.9%
Repair Main Switchgear, <1200 Amp.	1.92	0.8%
Minor Repair, Acoustic Tile Ceiling	1.78	0.8%
Replace Drinking Fountain, Refrigerated	1.75	0.8%
Replace Faucet Washer & Clean Trap, Lavatory, Iron, Enamel	1.61	0.7%
Maintain Pipe & Fittings, 4" DWV PVC	1.60	0.7%
Replace Pipe Insulation, Cold Water	1.50	0.7%
Replace Main Switchgear, <1200 Amp.	1.35	0.6%
Maintain Roof Drain, 4-6"	1.34	0.6%
Install New Gasket & Bolts, Pipe & Fittings, 4" Steel	1.25	0.5%
Replace Coolant & Adjust Drinking Fountain, Refrigerated	1.23	0.5%

**Task cost ($2004) per gross square foot over 50 years.*

***Percent of total M&R costs.*

Cost per Sqft. by System

Building Age	Exterior Closure	Roofing	Interior Construction	Stairways	Interior Finishes	Conveying Systems	Plumbing	HVAC Systems	Fire Protection	Electrical	Equipment	Total per Sqft.
1	.08	.59	.24			.68	.16	.39	.01	.25		**2.39**
2	.08	.59	.24		.01	.68	.55	.39	.01	.25		**2.80**
3	.08	.64	.24			.68	.16	.39	.01	.25		**2.44**
4	.09	.59	.24		.01	.68	.55	.39	.01	.25		**2.80**
5	.08	.59	.24			.68	.21	.39	.01	.64		**2.82**
6	.08	.64	.24		.01	.68	.55	.39	.01	.25		**2.85**
7	.08	.64	.24			.68	.21	.39	.01	.25		**2.49**
8	.09	.59	.24		2.42	.68	.64	.39	.01	.25		**5.29**
9	.08	.64	.24		.36	.68	.16	.39	.01	.25		**2.80**
10	1.21	.59	.44		.01	.68	1.17	3.83	.26	1.68		**9.86**
11	.08	.59	.24			.68	.16	.39	.01	.25		**2.39**
12	.09	.64	.24		.01	.68	.55	.39	.01	.25		**2.86**
13	.08	.59	.24			.68	.38	.39	.01	.25		**2.61**
14	.32	2.39	.24		.01	.68	.60	.39	.01	.25		**4.88**
15	6.88	.64	.24		.10	.68	.36	7.50	.01	1.27		**17.68**
16	.09	.59	.24		2.42	.68	.58	.39	.01	.25		**5.24**
17	.08	.59	.24			.68	.16	.39	.01	.25		**2.39**
18	.08	.64	.24		.37	.68	.55	.39	.01	.25		**3.20**
19	.08	.59	.24			.68	.16	.39	.01	.25		**2.39**
20	1.22	.86	.44		.01	.68	1.27	.86	.22	2.44		**8.00**
21	.08	.69	.24			.68	.21	.39	.01	.25		**2.55**
22	.08	.59	.24		.01	.68	.55	.39	.01	.25		**2.80**
23	.08	.59	.24			.68	.22	.39	.01	.25		**2.45**
24	.09	.64	.24		2.42	.68	.58	.39	.01	.25		**5.29**
25	.08	.59	.24			.68	14.30	3.36	.01	.75		**20.00**
26	.08	.59	.24		.01	.68	.73	.39	.01	.25		**2.97**
27	.08	.64	.24		.36	.68	.18	.39	.01	.25		**2.82**
28	.33	3.51	.24		.01	.68	.58	.39	.01	.25		**5.99**
29	.08	.59	.24			.68	.18	.39	.01	.25		**2.41**
30	8.01	.59	.44		.11	.68	1.30	7.97	.26	2.20		**21.56**
31	.08	.64	.24			.68	.18	.39	.01	.25		**2.46**
32	.09	.59	.24		2.42	.68	.56	.39	.01	.25		**5.22**
33	.08	.59	.24			.68	.18	.39	.01	.25		**2.41**
34	.08	.64	.24		.01	.68	.53	.39	.01	.25		**2.83**
35	.08	.64	.24			.68	1.16	.39	.01	.64		**3.84**
36	.09	.59	.24		.37	.68	.53	.39	.01	.25		**3.14**
37	.08	.64	.24			.68	.18	.39	.01	.25		**2.46**
38	.08	.59	.24		.01	.68	.61	.39	.01	.25		**2.85**
39	.08	.59	.24			.68	.38	.39	.01	.25		**2.61**
40	1.22	.91	.44		2.42	.68	1.64	3.83	.22	2.56		**13.92**
41	.08	.59	.24			.68	.18	.39	.01	.25		**2.41**
42	.32	2.39	.24		.01	.68	.55	.39	.01	.25		**4.84**
43	.08	.64	.24			.68	.18	.39	.01	.25		**2.46**
44	.09	.59	.24		.01	.68	.53	.39	.01	.25		**2.78**
45	6.88	.59	.24		.46	.68	.67	7.50	.01	1.27		**18.29**
46	.08	.64	.24		.01	.68	.53	.39	.01	.25		**2.83**
47	.08	.59	.24			.68	.21	.39	.01	.25		**2.44**
48	.09	.59	.24		2.42	.68	.56	.39	.01	.25		**5.22**
49	.08	.69	.24			.68	.20	.39	.01	.25		**2.54**
50	1.21	.59	.44		21.19	.68	15.05	.86	.26	1.56		**41.84**
Total	30.86	37.39	12.79		38.03	34.18	52.61	51.91	1.66	24.98		284.43

A value of "0.00" means a cost of more than $.000 but less than $.005 per gross square foot.

50-Year Total M&R Cost Profile per Sqft.

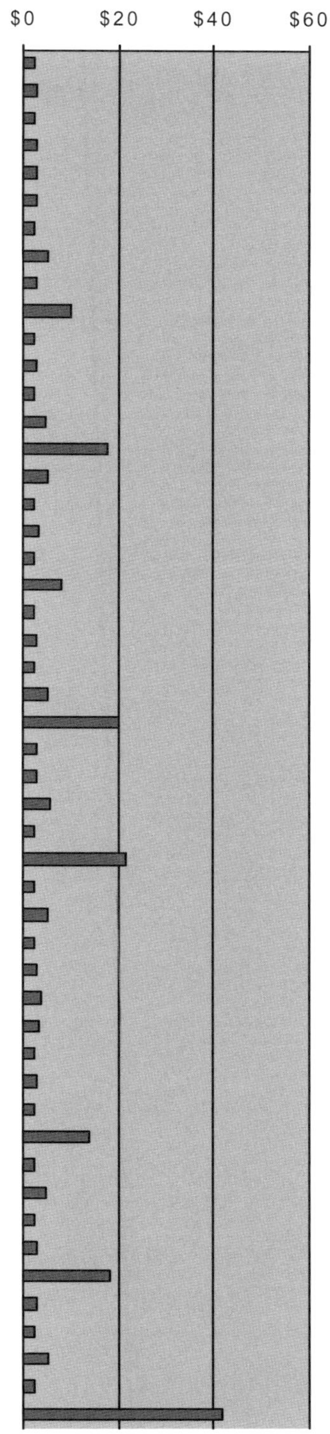

Bus Terminal

Gross Sqft:	12,000
Height in Ft:	14
Exterior:	Brick Veneer
Floor coverings:	Finished Concrete/Vinyl Tile/Quarry Til
HVAC:	Electric Cooling, Gas Heating
Occupancy:	1200
Replacement Cost:	$1,147,611

Components

Component	Quantity	Unit
B20 Exterior Enclosure		
Clay Brick, Exterior, 1st Floor	5096	Sq Ft
Steel w/ Safety Glass, Painted, Exterior Door	6	Each
Steel, Painted, Exterior Door	2	Each
Aluminum Frame, Fully Glazed, Sliding Exterior Door	2	Each
B30 Roofing		
Built-up Roof	12000	Sq Ft
C10 Interior Construction		
Steel, Painted, Interior Door	80	Each
C20 Stairs		
Masonry Interior Steps, Painted	5096	Sq Ft
C30 Interior Finishes		
Ceramic Tile Wall Finish	22400	Sq Ft
Concrete, Finished Flooring	6000	Sq Ft
Quarry Tile Flooring	3000	Sq Ft
Vinyl Tile Flooring	3000	Sq Ft
Acoustical Tile Ceiling	12000	Sq Ft
D20 Plumbing		
Tankless Water Closet	8	Each
Urinal, Vitreous China	2	Each
Lavatory, Iron, Enamel	9	Each
Sink, Stainless Steel	2	Each
Drinking Fountain, Refrigerated	2	Each
Pipe & Fittings, 3/4" Copper, Cold Water	0.8	K Ln Ft
Pipe & Fittings, 3/4" Copper, Hot Water	0.34	K Ln Ft
Pipe & Fittings, 2" Copper, Cold Water	0.914	K Ln Ft
Pipe & Fittings, 4" Steel	0.976	K Ln Ft
Pipe Insulation, Cold Water	1.288	K Ln Ft
Pipe Insulation, Hot Water	0.609	K Ln Ft
Water Heater, Electric 120 Gal.	2	Each
Pipe & Fittings, 6" Cast Iron	0.5	K Ln Ft
Pipe & Fittings, 10" Cast Iron	0.219	K Ln Ft
Roof Drain, 4-6"	2	Each
Pipe & Fittings, 4" DWV PVC	0.173	K Ln Ft
D30 HVAC		
Exhaust Fan, Ceiling, 200-500 Cfm	6	Each
Air Conditioner, Rooftop, 20 Ton	1	Each
Air Conditioner, Rooftop, 50 Ton	1	Each
Electric Cabinet Heater	6	Each
Pneumatic Thermostat	1	Each
D40 Fire Protection		
Fire Sprinkler Head	90	Each
D50 Electrical		
Distribution Panel Board	2	Each
Primary Transformer, Dry, >500 KVA	1	Each
Main Switchgear, <1200 Amp.	1	Each
Exit Lighting Fixture, w/ Battery	4	Each
Fluorescent Lighting Fixture, 160w	120	Each
Grounded Fault Circuit Interruptor, 15 Amps	6	Each
Receptacle, 120V, 15 Amp	60	Each
Annunciation Panel	1	Each
Smoke Detector	5	Each
Time Control Clock	1	Each

50-Year M&R Cost Summary

	50 Year Total	Annual Cost per Sqft	Annual cost as % replacement
PM & Minor Repair	$504,728	$0.84	0.88%
Unscheduled Maintenance	$549,180	$0.92	0.96%
Renewal & Replacement	$1,369,333	$2.28	2.39%
Total M&R Costs	$2,423,241	$4.04	4.22%

Distribution of Repair Costs

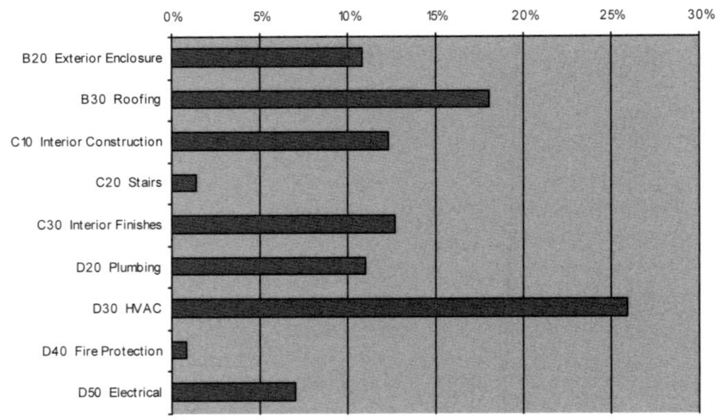

Thirty Most Costly Repair Tasks

Major Repair Task	Task Cost*	Pct.**
Maintain Built-up Roof	14.62	9.4%
Replace Air Conditioner, Rooftop, 50 Ton	12.85	8.2%
Minor Repair, Ceramic Tile Wall Finish	10.74	6.9%
Replace Quarry Tile Flooring	10.59	6.8%
Replace Steel, Painted, Door Locks	8.52	5.5%
Repair Air Conditioner, Rooftop, 50 Ton	7.87	5.0%
Replace Air Conditioner, Rooftop, 20 Ton	7.60	4.9%
Clean & Reseal Clay Brick, Exterior, 1st Floor	6.31	4.0%
Maintain Steel, Painted, Door Locks	5.68	3.6%
Replace Pipe & Fittings, 2" Copper, Cold Water	5.16	3.3%
Replace Pipe & Fittings, 3/4" Copper, Cold Water	3.97	2.5%
Repoint (50% surface) Clay Brick, Exterior, 1st Floor	3.77	2.4%
Maintain Air Conditioner, Rooftop, 50 Ton	3.66	2.3%
Replace Primary Transformer, Dry, >500 KVA	3.50	2.2%
Place New Membrane Over Existing, Built-up Roof	3.49	2.2%
Remove & Replace Membrane, Built-up Roof	3.45	2.2%
Replace Electric Cabinet Heater	3.23	2.1%
Repair Air Conditioner, Rooftop, 20 Ton	3.05	2.0%
Maintain Air Conditioner, Rooftop, 20 Ton	2.84	1.8%
Replace Ballast & Lamps, Fluorescent Lighting Fixture, 160w	2.76	1.8%
Inspect & Maintain Electric Cabinet Heater	2.21	1.4%
Refinish Masonry Interior Steps, Painted	2.13	1.4%
Replace Fluorescent Lighting Fixture, 160w	1.96	1.3%
Minor Repair, Acoustic Tile Ceiling	1.78	1.1%
Replace Pipe & Fittings, 3/4" Copper, Hot Water	1.68	1.1%
Replace Vinyl Tile Flooring	1.05	0.7%
Refinish Steel, Painted, Interior Door	1.03	0.7%
Replace Faucet Washer & Clean Trap, Lavatory, Iron, Enamel	.82	0.5%
Non-Destructive Moisture Inspection	.81	0.5%
Repair Clay Brick, Exterior, 1st Floor	.81	0.5%

**Task cost ($2004) per gross square foot over 50 years.*

***Percent of total M&R costs.*

Cost per Sqft. by System

Building Age	Exterior Closure	Roofing	Interior Constrution	Stairways	Interior Finishes	Conveying Systems	Plumbing	HVAC Systems	Fire Protection	Electrical	Equipment	Total per Sqft.
1	.03	.59	.31				.03	.36	.01	.07		**1.39**
2	.03	.59	.31				.20	.36	.01	.07		**1.56**
3	.03	.64	.31				.03	.36	.01	.07		**1.44**
4	.04	.59	.39				.20	.36	.01	.07		**1.65**
5	.03	.59	.31				.05	.36	.01	.21		**1.55**
6	.03	.64	.31				.20	.36	.01	.07		**1.61**
7	.03	.59	.31	.30			.05	.36	.01	.07		**1.72**
8	.04	.59	.39				.25	.36	.01	.07		**1.70**
9	.03	.64	.31		.37		.03	.36	.01	.07		**1.81**
10	3.03	.59	2.01		2.15		.48	4.08	.25	1.16		**13.74**
11	.03	.59	.31				.03	.36	.01	.07		**1.39**
12	.05	.64	.39				.20	.36	.01	.07		**1.72**
13	.03	.59	.31				.14	.36	.01	.07		**1.50**
14	.25	2.33	.31	.30			.22	.36	.01	.07		**3.85**
15	.03	.64	.31		.23		.25	7.93	.01	.28		**9.68**
16	.04	.59	.39				.21	.36	.01	.07		**1.66**
17	.03	.59	.31				.03	.36	.01	.07		**1.39**
18	.03	.64	.31		.88		.20	.36	.01	.07		**2.49**
19	.03	.59	.31				.03	.36	.01	.07		**1.39**
20	3.04	.59	2.10		2.15		.52	.65	.21	1.54		**10.79**
21	.03	.64	.31	.30			.05	.36	.01	.07		**1.77**
22	.03	.59	.31				.20	.36	.01	.07		**1.56**
23	.03	.59	.31				.07	.36	.01	.07		**1.43**
24	.05	.64	.39				.21	.36	.01	.07		**1.73**
25	2.32	.59	.31		.34		6.60	3.99	.01	.23		**14.38**
26	.03	.59	.31				.29	.36	.01	.07		**1.65**
27	.03	.64	.31	.66	.37		.04	.36	.01	.07		**2.48**
28	.25	3.45	.39	.30			.21	.36	.01	.07		**5.05**
29	.03	.59	.31				.04	.36	.01	.07		**1.40**
30	3.03	.59	2.01		2.38		.66	8.01	.25	4.69		**21.63**
31	.03	.64	.31				.04	.36	.01	.07		**1.45**
32	.04	.59	.39				.20	.36	.01	.07		**1.65**
33	.03	.59	.31				.04	.36	.01	.07		**1.40**
34	.03	.64	.31				.19	.36	.01	.07		**1.60**
35	.03	.59	.31	.30			.53	.36	.01	.21		**2.32**
36	.05	.59	.39		.88		.19	.36	.01	.07		**2.54**
37	.03	.64	.31				.04	.36	.01	.07		**1.45**
38	.03	.59	.31				.23	.36	.01	.07		**1.60**
39	.03	.59	.31				.15	.36	.01	.07		**1.51**
40	3.04	.64	2.10		2.15		.64	4.29	.21	1.56		**14.62**
41	.03	.59	.31				.04	.36	.01	.07		**1.40**
42	.25	2.33	.31	.30			.20	.36	.01	.07		**3.82**
43	.03	.64	.31				.04	.36	.01	.07		**1.45**
44	.04	.59	.39				.19	.36	.01	.07		**1.64**
45	.03	.59	.31		.60		.41	7.93	.01	.28		**10.15**
46	.03	.64	.31				.19	.36	.01	.07		**1.60**
47	.03	.59	.31				.06	.36	.01	.07		**1.42**
48	.05	.59	.39				.20	.36	.01	.07		**1.67**
49	.03	.64	.31	.30			.05	.36	.01	.07		**1.76**
50	5.32	.59	2.01		13.08		6.92	.44	.25	1.14		**29.74**
Total	21.85	36.43	24.88	2.79	25.57		22.27	52.28	1.60	14.28		201.94

A value of "0.00" means a cost of more than $.000 but less than $.005 per gross square foot.

50-Year Total M&R Cost Profile per Sqft.

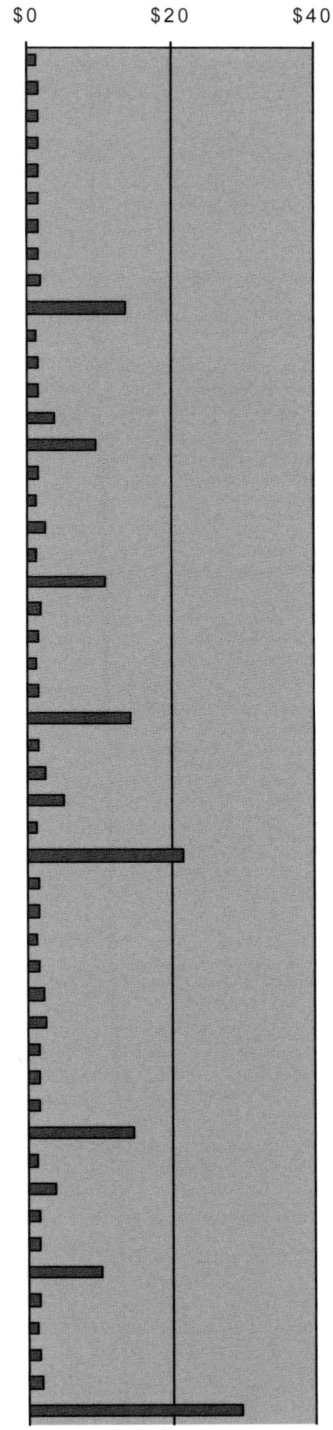

Car Wash

Gross Sqft:	800
Height in Ft:	12
Exterior:	Brick Veneer
Floor coverings:	Finished Concrete
HVAC:	Unit Heaters
Occupancy:	
Replacement Cost:	$172,111

Components

Component	Quantity	Unit
B20 Exterior Enclosure		
Clay Brick, Exterior, 1st Floor	958	Sq Ft
Aluminum Operable Window, 12 sf, 1st Floor	8	Each
Steel, Painted, Exterior Door	2	Each
Steel 20'x20', Painted, Roll-up Door	2	Each
B30 Roofing		
Built-up Roof	800	Sq Ft
C10 Interior Construction		
Steel, Painted, Interior Door	3	Each
C20 Stairs		
Concrete Interior Steps	12	Sq Ft
C30 Interior Finishes		
Concrete Block, Painted, Interior Wall Finish	960	Sq Ft
Concrete, Unfinished Flooring	800	Sq Ft
D20 Plumbing		
Tankless Water Closet	4	Each
Lavatory, Iron, Enamel	4	Each
Drinking Fountain, Refrigerated	1	Each
Pipe & Fittings, 3/4" Copper, Cold Water	0.4	K Ln Ft
Pipe & Fittings, 3/4" Copper, Hot Water	0.12	K Ln Ft
Pipe & Fittings, 2" Copper, Cold Water	0.442	K Ln Ft
Pipe & Fittings, 4" Steel	0.326	K Ln Ft
Pipe Insulation, Cold Water	0.564	K Ln Ft
Pipe Insulation, Hot Water	0.227	K Ln Ft
Water Heater, Gas/Oil 65 Gph	1	Each
Pipe & Fittings, 6" Cast Iron	0.198	K Ln Ft
Pipe & Fittings, 10" Cast Iron	0.57	K Ln Ft
Roof Drain, 4-6"	2	Each
Pipe & Fittings, 4" DWV PVC	0.012	K Ln Ft
D30 HVAC		
House Furnace, Gas/Oil, 100 Mbh	1	Each
Pipe & Fittings, 3/4" Copper, High Temp.	0.2	K Ln Ft
Pneumatic Thermostat	1	Each
D50 Electrical		
Main Switchgear, <1200 Amp.	1	Each
Fluorescent Lighting Fixture, 160w	12	Each
Grounded Fault Circuit Interruptor, 15 Amps	6	Each
Receptacle, 120V, 15 Amp	13	Each
Smoke Detector	4	Each

50-Year M&R Cost Summary

	50 Year Total	Annual Cost per Sqft	Annual cost as % replacement
PM & Minor Repair	$104,465	$2.61	1.21%
Unscheduled Maintenance	$123,119	$3.08	1.43%
Renewal & Replacement	$202,226	$5.06	2.35%
Total M&R Costs	$429,810	$10.75	4.99%

Distribution of Repair Costs

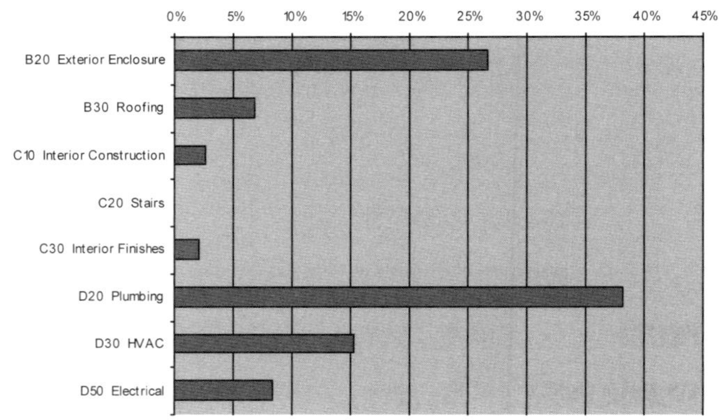

Thirty Most Costly Repair Tasks

Major Repair Task	Task Cost*	Pct.**
Replace Pipe & Fittings, 2" Copper, Cold Water	37.40	9.8%
Replace Pipe & Fittings, 3/4" Copper, Cold Water	29.79	7.8%
Maintain House Furnace, Gas/Oil, 100 Mbh	22.74	5.9%
Maintain Steel 20'x20', Painted, Roll-up Door	19.29	5.0%
Clean & Reseal Clay Brick, Exterior, 1st Floor	17.79	4.6%
Replace Steel 20'x20', Painted, Roll-up Door	16.69	4.4%
Replace Water Heater, Gas/Oil 65 Gph	14.84	3.9%
Replace Pipe & Fittings, 3/4" Copper, High Temp.	14.81	3.9%
Maintain Built-up Roof	14.64	3.8%
Clean Water Heater, Gas/Oil 65 Gph	12.05	3.1%
Repoint (50% surface) Clay Brick, Exterior, 1st Floor	10.62	2.8%
Repair Main Switchgear, <1200 Amp.	9.85	2.6%
Refinish Concrete Block, Painted, Interior Wall Finish	9.21	2.4%
Replace Pipe & Fittings, 3/4" Copper, Hot Water	8.88	2.3%
Install New Gasket & Bolts, Pipe & Fittings, 10" Cast Iron	8.14	2.1%
Repair Steel 20'x20', Painted, Roll-up Door	7.28	1.9%
Replace Pneumatic Thermostat	7.11	1.9%
Replace House Furnace, Gas/Oil, 100 Mbh	7.04	1.8%
Replace Main Switchgear, <1200 Amp.	6.90	1.8%
Refinish Steel 20'x20', Painted, Roll-up Door	6.21	1.6%
Replace Faucet Washer & Clean Trap, Lavatory, Iron, Enamel	5.49	1.4%
Annual PM, Pneumatic Thermostat	5.29	1.4%
Replace Steel, Painted, Door Locks	4.79	1.3%
Replace Pipe Insulation, Cold Water	4.69	1.2%
Maintain Roof Drain, 4-6"	4.59	1.2%
Replace Drinking Fountain, Refrigerated	4.49	1.2%
Replace Ballast & Lamps, Fluorescent Lighting Fixture, 160w	4.13	1.1%
Place New Membrane Over Existing, Built-up Roof	3.49	0.9%
Remove & Replace Membrane, Built-up Roof	3.45	0.9%
Maintain Main Switchgear, <1200 Amp.	3.42	0.9%

**Task cost ($2004) per gross square foot over 50 years.*

***Percent of total M&R costs.*

Cost per Sqft. by System

Building Age	Exterior Closure	Roofing	Interior Construtión	Stairways	Interior Finishes	Conveying Systems	Plumbing	HVAC Systems	Fire Protection	Electrical	Equipment	Total per Sqft.
1	1.10	.59	.17				.99	1.07		.27		**4.19**
2	1.10	.59	.17				2.07	1.07		.27		**5.27**
3	1.10	.64	.17				.99	1.07		.27		**4.24**
4	1.13	.59	.22		.77		2.07	1.07		.27		**6.12**
5	1.79	.59	.17				1.72	1.07		2.05		**7.39**
6	1.10	.64	.17				2.07	1.07		.27		**5.32**
7	1.10	.59	.17				1.09	1.07		.27		**4.29**
8	1.13	.59	.22		.77		3.82	1.07		.27		**7.86**
9	1.10	.64	.17				.99	1.07		.27		**4.24**
10	12.09	.59	1.13				3.95	3.27		3.57		**24.59**
11	1.10	.59	.17				.99	1.07		.27		**4.19**
12	1.13	.64	.22		.77		2.07	1.07		.27		**6.17**
13	1.10	.59	.17				2.80	1.15		.27		**6.07**
14	1.92	2.33	.17				2.17	1.07		.27		**7.93**
15	1.79	.64	.17	.03	.37		1.13	2.56		2.50		**9.18**
16	1.13	.59	.22		.77		3.82	1.07		.27		**7.86**
17	1.10	.59	.17				.99	1.07		.27		**4.19**
18	1.10	.64	.17				2.07	1.07		.27		**5.32**
19	1.10	.59	.17				.99	1.07		.27		**4.19**
20	12.24	.59	1.18		.77		4.28	2.28		5.96		**27.30**
21	1.10	.64	.17				1.68	1.07		.27		**4.93**
22	1.10	.59	.17				2.07	1.07		.27		**5.27**
23	1.10	.59	.17				.99	1.07		.27		**4.19**
24	1.13	.64	.22		1.34		3.82	1.07		.27		**8.48**
25	8.24	.59	.17				49.14	9.46		2.19		**69.79**
26	1.10	.59	.17				3.18	1.07		.27		**6.38**
27	1.10	.64	.17				1.05	1.07		.27		**4.30**
28	1.95	3.45	.22		.77		2.11	1.07		.27		**9.84**
29	1.10	.59	.17				1.64	1.07		.27		**4.84**
30	12.09	.59	1.13	.03	.37		3.88	3.77		3.87		**25.72**
31	1.10	.64	.17				1.05	1.07		.27		**4.30**
32	1.13	.59	.22		.77		3.76	1.07		.27		**7.81**
33	1.10	.59	.17				1.05	1.07		.27		**4.25**
34	1.10	.64	.17				2.02	1.07		.27		**5.27**
35	17.49	.59	.17				4.04	1.07		2.05		**25.41**
36	1.13	.59	.22		.77		2.02	1.07		.27		**6.06**
37	1.10	.64	.17				1.64	1.07		.27		**4.89**
38	1.10	.59	.17				2.06	1.15		.27		**5.34**
39	1.10	.59	.17				2.22	1.07		.27		**5.41**
40	10.42	.64	1.18		.77		6.51	3.27		6.10		**28.88**
41	1.10	.59	.17				1.05	1.07		.27		**4.25**
42	1.92	2.33	.17				2.02	1.07		.27		**7.78**
43	1.10	.64	.17				1.05	1.07		.27		**4.30**
44	1.13	.59	.22		.77		2.02	1.07		.27		**6.06**
45	3.61	.59	.17	.03	.37		2.86	2.56		2.50		**12.67**
46	1.10	.64	.17				2.02	1.07		.27		**5.27**
47	1.10	.59	.17				1.15	1.07		.27		**4.34**
48	1.13	.59	.22		1.34		3.76	1.07		.27		**8.38**
49	1.10	.64	.17				1.05	1.07		.27		**4.30**
50	16.72	.59	1.13				51.15	9.69		3.42		**82.69**
Total	143.30	36.43	14.00	.08	11.45		205.12	81.99		44.90		537.26

A value of "0.00" means a cost of more than $.000 but less than $.005 per gross square foot.

50-Year Total M&R Cost Profile per Sqft.

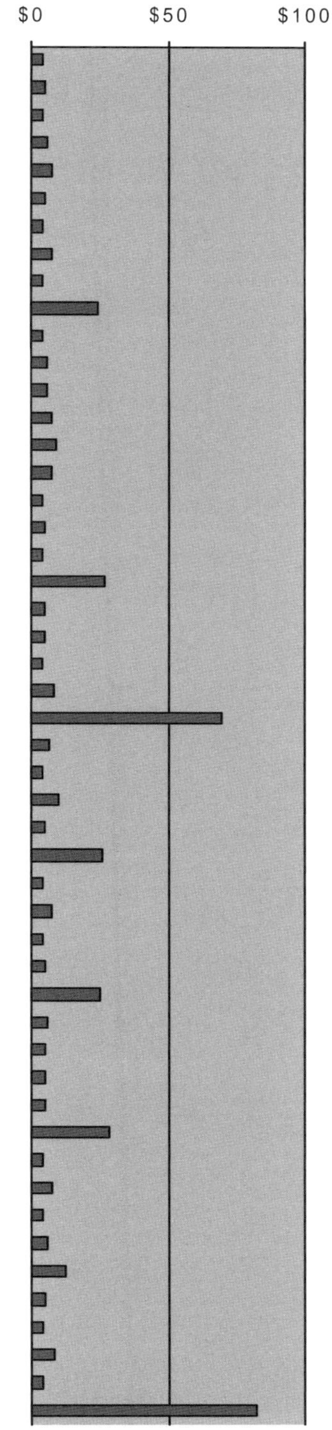

Bowling Center

Gross Sqft:	20,000
Height in Ft:	16
Exterior:	Concrete Block
Floor coverings:	Finished Concrete/Rubber Tile/Carpet
HVAC:	Single Zone Rooftop Units
Occupancy:	520
Replacement Cost:	$1,324,651

Components

Component	Quantity	Unit
B20 Exterior Enclosure		
Concrete Block, Painted, Exterior, 1st Floor	8560	Sq Ft
Aluminum Fixed Window, 24 sf, 1st Floor	21	Each
Steel, Painted, Exterior Door	3	Each
Aluminum Frame, Fully Glazed, Exterior Door	5	Each
Steel Single 12'x12', Painted, Roll-up Door	1	Each
B30 Roofing		
Built-up Roof	20000	Sq Ft
C10 Interior Construction		
Wood, Solid Core, Painted, Interior Door	12	Each
C30 Interior Finishes		
Sheetrock, Unstippled, Interior Wall Finish	11200	Sq Ft
Concrete Block, Painted, Interior Wall Finish	8156	Sq Ft
Acoustical Interior Wall Tile	2000	Sq Ft
Concrete, Finished Flooring	2500	Sq Ft
Rubber Tile Flooring	5000	Sq Ft
Carpet, Nylon 20 oz., High Traffic	2500	Sq Ft
Metal Ceiling, Painted	2500	Sq Ft
Acoustical Tile, Dropped, Ceiling	17500	Sq Ft
D20 Plumbing		
Tankless Water Closet	7	Each
Urinal, Vitreous China	2	Each
Lavatory, Vitreous China	4	Each
Sink, Stainless Steel	2	Each
Service Sink, Iron Enamel	1	Each
Drinking Fountain, Refrigerated	1	Each
Pipe & Fittings, 3/4" Copper, Cold Water	0.6	K Ln Ft
Pipe & Fittings, 2" Copper, Cold Water	0.7	K Ln Ft
Pipe & Fittings, 3/4" Copper, Hot Water	0.3	K Ln Ft
Pipe Insulation, Cold Water	0.6	K Ln Ft
Pipe Insulation, Hot Water	0.3	K Ln Ft
Pipe & Fittings, 6" Cast Iron	0.5	K Ln Ft
Pipe & Fittings, 10" Cast Iron	0.3	K Ln Ft
Pipe & Fittings, 4" DWV PVC	0.2	K Ln Ft
Roof Drain, 4-6"	6	Each
D30 HVAC		
Pipe & Fittings, 4" Steel, Gas	0.3	K Ln Ft
Air Conditioner, Rooftop, 50 Ton	2	Each
D40 Fire Protection		
Fire Sprinkler Head	100	Each
D50 Electrical		
Main Switchgear, <1200 Amp.	1	Each
Distribution Switch, Fused, <600 V	12	Each
Motor Starter, 5-20 HP, <600 V	24	Each
Fluorescent Lighting Fixture, 160w	250	Each
Wiring Device, Switch	20	Each
Receptacle, 120V, 15 Amp	80	Each
Fire Alarm Control Panel	1	Each
Manual Pull Station	5	Each
Heat Detector	10	Each
Smoke Detector	30	Each

50-Year M&R Cost Summary

	50 Year Total	Annual Cost per Sqft	Annual cost as % replacement
PM & Minor Repair	$617,036	$0.62	0.93%
Unscheduled Maintenance	$574,933	$0.57	0.87%
Renewal & Replacement	$1,516,616	$1.52	2.29%
Total M&R Costs	$2,708,585	$2.71	4.09%

Distribution of Repair Costs

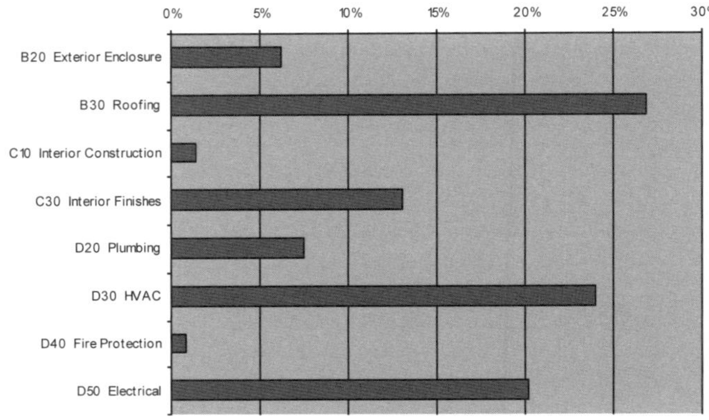

Thirty Most Costly Repair Tasks

Major Repair Task	Task Cost*	Pct.**
Replace Air Conditioner, Rooftop, 50 Ton	15.42	14.5%
Maintain Built-up Roof	14.62	13.7%
Repair Air Conditioner, Rooftop, 50 Ton	9.44	8.9%
Inspect & Clean Motor Starter, 5-20 HP, <600 V	5.43	5.1%
Maintain Air Conditioner, Rooftop, 50 Ton	4.39	4.1%
Replace Rubber Tile Flooring	4.17	3.9%
Refinish Concrete Block, Painted, Exterior, 1st Floor	3.75	3.5%
Replace Carpet, Nylon 20 oz., High Traffic	3.63	3.4%
Place New Membrane Over Existing, Built-up Roof	3.49	3.3%
Remove & Replace Membrane, Built-up Roof	3.45	3.2%
Replace Ballast & Lamps, Fluorescent Lighting Fixture, 160w	3.44	3.2%
Refinish Sheetrock, Unstippled, Interior Wall Finish	3.40	3.2%
Refinish Concrete Block, Painted, Interior Wall Finish	3.13	2.9%
Replace Fluorescent Lighting Fixture, 160w	2.45	2.3%
Replace Pipe & Fittings, 2" Copper, Cold Water	2.37	2.2%
Repoint (50% surface) Concrete Block, Painted, Exterior, 1st Floor	2.29	2.1%
Replace Pipe & Fittings, 3/4" Copper, Cold Water	1.79	1.7%
Replace Coil, Motor Starter, 5-20 HP, <600 V	1.73	1.6%
Replace Distribution Switch, Fused, <600 V	1.38	1.3%
Replace Motor Starter, 5-20 HP, <600 V	1.23	1.1%
Check Operation, Smoke Detector	1.00	0.9%
Refinish Acoustical Interior Wall Tile	.93	0.9%
Replace Pipe & Fittings, 3/4" Copper, Hot Water	.89	0.8%
Non-Destructive Moisture Inspection	.81	0.8%
Maintain Roof Drain, 4-6"	.55	0.5%
Minor Repair, Acoustic Tile, Dropped, Ceiling	.55	0.5%
Replace Smoke Detector	.53	0.5%
Maintain Wood, Solid Core, Painted, Door Locks	.50	0.5%
Repair Fire Sprinkler Head	.48	0.4%
Repair Main Switchgear, <1200 Amp.	.39	0.4%

**Task cost ($2004) per gross square foot over 50 years.*

***Percent of total M&R costs.*

Cost per Sqft. by System

Building Age	Exterior Closure	Roofing	Interior Constrution	Stairways	Interior Finishes	Conveying Systems	Plumbing	HVAC Systems	Fire Protection	Electrical	Equipment	Total per Sqft.
1	.03	.59	.03				.04	.16	.01	.28		**1.13**
2	.03	.59	.03		.00		.09	.16	.01	.28		**1.19**
3	.03	.64	.03				.04	.16	.01	.28		**1.18**
4	.03	.59	.04		.63		.09	.16	.01	.28		**1.82**
5	.03	.59	.03		.02		.05	.16	.01	.72		**1.61**
6	.03	.64	.03		.00		.09	.16	.01	.28		**1.24**
7	.03	.59	.03				.05	.16	.01	.28		**1.14**
8	.03	.59	.04		1.26		.10	.16	.01	.28		**2.47**
9	.03	.64	.03		.05		.04	.16	.01	.28		**1.23**
10	.80	.59	.08		.13		.21	3.31	.17	1.98		**7.26**
11	.03	.59	.03				.04	.16	.01	.28		**1.13**
12	.08	.64	.04		.63		.09	.17	.01	.28		**1.93**
13	.03	.59	.03				.08	.16	.01	.28		**1.17**
14	.08	2.33	.03		.00		.10	.16	.01	.28		**2.99**
15	.03	.64	.03		.07		.05	5.14	.01	1.01		**6.96**
16	.03	.59	.04		1.26		.10	.16	.01	.28		**2.47**
17	.03	.59	.03				.04	.16	.01	.28		**1.13**
18	.03	.64	.03		2.09		.09	.16	.01	.68		**3.72**
19	.03	.59	.03				.04	.16	.01	.28		**1.13**
20	.81	.59	.09		1.00		.23	.16	.14	1.77		**4.79**
21	.03	.64	.03				.05	.16	.01	.28		**1.19**
22	.03	.59	.03		.00		.09	.16	.01	.28		**1.19**
23	.03	.59	.03				.04	.16	.01	.65		**1.50**
24	.08	.64	.04		1.49		.10	.17	.01	.28		**2.81**
25	1.29	.59	.03		.11		2.86	3.31	.01	.46		**8.65**
26	.03	.59	.03		.00		.12	.16	.01	.28		**1.22**
27	.03	.64	.03		.05		.04	.16	.01	.28		**1.23**
28	.08	3.45	.04		.63		.10	.16	.01	.65		**5.11**
29	.03	.59	.03				.04	.16	.01	.28		**1.13**
30	.80	.59	.23		.18		.20	5.20	.17	3.16		**10.53**
31	.03	.64	.03				.04	.16	.01	.28		**1.18**
32	.03	.59	.03		1.26		.10	.16	.01	.28		**2.45**
33	.03	.59	.03				.04	.16	.01	.65		**1.50**
34	.03	.64	.04		.00		.09	.16	.01	.28		**1.24**
35	.13	.59	.03		.02		.38	.16	.01	.36		**1.68**
36	.08	.59	.03		2.71		.06	.17	.01	.68		**4.32**
37	.03	.64	.03				.07	.16	.01	.28		**1.21**
38	.03	.59	.04		.00		.07	.16	.01	.28		**1.17**
39	.03	.59	.03				.10	.16	.01	.28		**1.19**
40	.80	.64	.08		1.63		.19	3.31	.14	1.87		**8.66**
41	.03	.59	.03				.07	.16	.01	.65		**1.53**
42	.08	2.33	.04		.00		.07	.16	.01	.28		**2.97**
43	.03	.64	.03				.07	.16	.01	.28		**1.21**
44	.03	.59	.03		.63		.06	.16	.01	.28		**1.78**
45	.05	.59	.03		.12		.17	5.14	.01	.64		**6.73**
46	.03	.64	.04		.00		.06	.16	.01	.65		**1.58**
47	.03	.59	.03				.07	.16	.01	.28		**1.16**
48	.08	.59	.03		1.49		.07	.17	.01	.28		**2.72**
49	.03	.64	.03				.08	.16	.01	.28		**1.22**
50	2.04	.59	.09		.22		2.91	.16	.17	1.51		**7.69**
Total	8.34	36.43	1.90		17.71		10.10	32.51	1.07	27.38		135.43

50-Year Total M&R Cost Profile per Sqft.

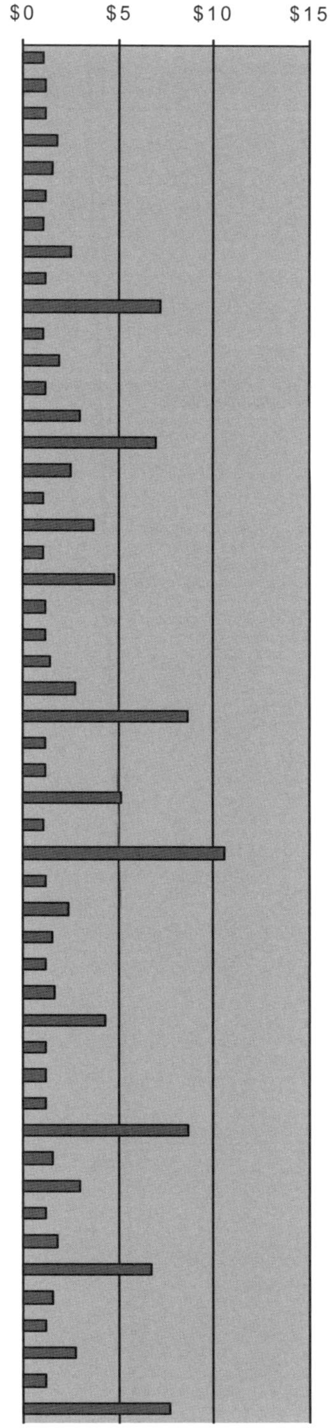

A value of "0.00" means a cost of more than $.000 but less than $.005 per gross square foot.

Childcare Center

Gross Sqft:	12,000
Height in Ft:	12
Exterior:	Clay Brick
Floor coverings:	Vinyl Tile/Carpet
HVAC:	Heat Pumps, Gas Boiler
Occupancy:	170
Replacement Cost:	$1,413,236

50-Year M&R Cost Summary

	50 Year Total	Annual Cost per Sqft	Annual cost as % replacement
PM & Minor Repair	$411,657	$0.69	0.58%
Unscheduled Maintenance	$434,678	$0.72	0.62%
Renewal & Replacement	$873,102	$1.46	1.24%
Total M&R Costs	$1,719,437	$2.87	2.43%

Components

Component	Quantity	Unit
B20 Exterior Enclosure		
Clay Brick, Exterior, 1st Floor	4400	Sq Ft
Aluminum Fixed Window, 12 sf, 1st Floor	70	Each
Steel, Painted, Exterior Door	9	Each
Aluminum Frame, Fully Glazed, Exterior Door	2	Each
B30 Roofing		
Built-up Roof	12000	Sq Ft
C10 Interior Construction		
Wood, Solid Core, Painted, Interior Door	26	Each
C30 Interior Finishes		
Sheetrock, Stippled, Interior Wall Finish	11910	Sq Ft
Concrete Block, Painted, Interior Wall Finish	600	Sq Ft
Vinyl Interior Wall Finish	3000	Sq Ft
Vinyl Tile Flooring	3600	Sq Ft
Carpet, Nylon 20 oz., Low Traffic	8400	Sq Ft
Sheetrock, Stippled, Ceiling	3600	Sq Ft
Acoustical Tile, Dropped, Ceiling	8400	Sq Ft
D20 Plumbing		
Tankless Water Closet	5	Each
Urinal, Vitreous China	3	Each
Lavatory, Vitreous China	4	Each
Sink, Stainless Steel	2	Each
Service Sink, Iron Enamel	2	Each
Drinking Fountain, Vitreous China	4	Each
Water Heater, Gas/Oil 65 Gph	1	Each
Pipe & Fittings, 3/4" Copper, Cold Water	0.7	K Ln Ft
Pipe & Fittings, 2" Copper, Cold Water	0.9	K Ln Ft
Pipe & Fittings, 3/4" Copper, Hot Water	0.3	K Ln Ft
Pipe Insulation, Cold Water	0.7	K Ln Ft
Pipe Insulation, Hot Water	0.3	K Ln Ft
Pipe & Fittings, 6" Cast Iron	0.5	K Ln Ft
Pipe & Fittings, 10" Cast Iron	0.3	K Ln Ft
Pipe & Fittings, 4" DWV PVC	0.1	K Ln Ft
Roof Drain, 4-6"	3	Each
D30 HVAC		
Pipe & Fittings, 4" Steel, Gas	0.3	K Ln Ft
Air Conditioner, Indoor DX Packaged, Air Cooled, 20 Ton	2	Each
D40 Fire Protection		
Fire Sprinkler Head	60	Each
D50 Electrical		
Main Switchgear, <1200 Amp.	1	Each
Distribution Switch, Fused, <600 V	3	Each
Fluorescent Lighting Fixture, 160w	128	Each
Wiring Device, Switch	40	Each
Receptacle, 120V, 15 Amp	48	Each
Annunciation Panel	1	Each
Fire Alarm Control Panel	1	Each
Manual Pull Station	10	Each
Smoke Detector	24	Each

Distribution of Repair Costs

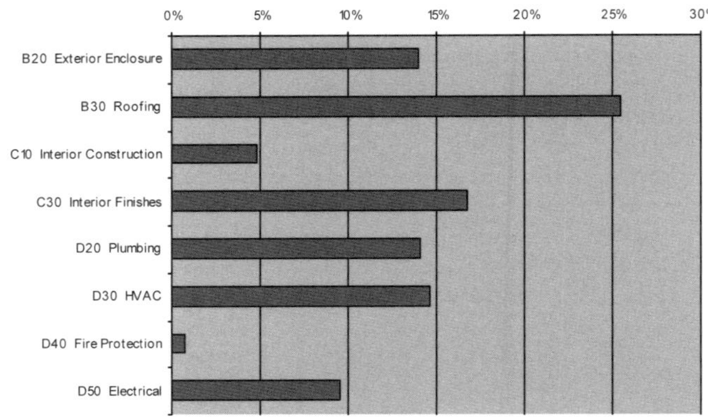

Thirty Most Costly Repair Tasks

Major Repair Task	Task Cost*	Pct.**
Maintain Built-up Roof	14.62	13.7%
Replace Carpet, Nylon 20 oz., Low Traffic	10.16	9.5%
Replace Air Conditioner, Indoor DX Packaged, Air Cooled, 20 Ton	7.13	6.7%
Refinish Sheetrock, Stippled, Interior Wall Finish	6.03	5.6%
Maintain Air Conditioner, Indoor DX Packaged, Air Cooled, 20 Ton	5.78	5.4%
Clean & Reseal Clay Brick, Exterior, 1st Floor	5.45	5.1%
Replace Pipe & Fittings, 2" Copper, Cold Water	5.08	4.7%
Repair Air Conditioner, Indoor DX Packaged, Air Cooled, 20 Ton	3.75	3.5%
Place New Membrane Over Existing, Built-up Roof	3.49	3.3%
Replace Pipe & Fittings, 3/4" Copper, Cold Water	3.48	3.2%
Remove & Replace Membrane, Built-up Roof	3.45	3.2%
Repoint (50% surface) Clay Brick, Exterior, 1st Floor	3.25	3.0%
Replace Ballast & Lamps, Fluorescent Lighting Fixture, 160w	2.94	2.7%
Replace Fluorescent Lighting Fixture, 160w	2.09	1.9%
Replace Vinyl Interior Wall Finish	2.06	1.9%
Refinish Sheetrock, Stippled, Ceiling	1.83	1.7%
Maintain Wood, Solid Core, Painted, Door Locks	1.81	1.7%
Replace Pipe & Fittings, 3/4" Copper, Hot Water	1.48	1.4%
Check Operation, Smoke Detector	1.33	1.2%
Replace Vinyl Tile Flooring	1.26	1.2%
Replace Water Heater, Gas/Oil 65 Gph	.99	0.9%
Non-Destructive Moisture Inspection	.81	0.8%
Replace Wood, Solid Core, Painted, Interior Door	.81	0.8%
Replace Solid Core, Painted, Door Locks	.80	0.8%
Clean Water Heater, Gas/Oil 65 Gph	.80	0.8%
Repair Steel, Painted, Exterior Door	.73	0.7%
Replace Smoke Detector	.70	0.7%
Repair Clay Brick, Exterior, 1st Floor	.70	0.7%
Repair Main Switchgear, <1200 Amp.	.66	0.6%
Maintain Steel, Painted, Exterior Door, Locks	.64	0.6%

Task cost ($2004) per gross square foot over 50 years.

***Percent of total M&R costs.*

Cost per Sqft. by System

Building Age	Exterior Closure	Roofing	Interior Construtio n	Stairways	Interior Finishes	Conveying Systems	Plumbing	HVAC Systems	Fire Protection	Electrical	Equipment	Total per Sqft.
1	.04	.59	.10				.08	.21	.01	.09		**1.11**
2	.04	.59	.10				.17	.21	.01	.09		**1.19**
3	.04	.64	.10				.08	.21	.01	.09		**1.16**
4	.05	.59	.13		.69		.17	.21	.01	.09		**1.92**
5	.04	.59	.10		.09		.13	.21	.01	.21		**1.37**
6	.04	.64	.10				.17	.21	.01	.09		**1.25**
7	.04	.59	.10				.10	.21	.01	.09		**1.12**
8	.05	.59	.13		.69		.31	.21	.01	.09		**2.05**
9	.04	.64	.10		.01		.08	.21	.01	.09		**1.17**
10	2.63	.59	.30		.52		.32	1.46	.17	1.38		**7.36**
11	.04	.59	.10				.08	.21	.01	.09		**1.11**
12	.08	.64	.13		.69		.17	.22	.01	.09		**2.02**
13	.04	.59	.10				.18	.21	.01	.09		**1.21**
14	.28	2.33	.10				.19	.21	.01	.09		**3.20**
15	.04	.64	.10		3.46		.09	.21	.01	.72		**5.26**
16	.05	.59	.13		.69		.31	.21	.01	.09		**2.05**
17	.04	.59	.10				.08	.21	.01	.09		**1.11**
18	.04	.64	.10		.63		.17	.21	.01	.09		**1.88**
19	.04	.59	.10				.08	.21	.01	.09		**1.11**
20	2.70	.59	.33		1.86		.36	3.56	.14	1.49		**11.03**
21	.04	.64	.10				.14	.21	.01	.09		**1.21**
22	.04	.59	.10				.17	.21	.01	.09		**1.19**
23	.04	.59	.10				.08	.21	.01	.09		**1.11**
24	.08	.64	.13		.71		.31	.22	.01	.09		**2.18**
25	2.02	.59	.10		.09		5.61	.21	.01	.40		**9.01**
26	.04	.59	.10				.22	.21	.01	.09		**1.24**
27	.04	.64	.10		.01		.09	.21	.01	.09		**1.17**
28	.29	3.45	.13		.69		.18	.21	.01	.09		**5.04**
29	.04	.59	.10				.13	.21	.01	.09		**1.15**
30	2.63	.59	.84		3.89		.31	1.56	.17	2.27		**12.25**
31	.04	.64	.10				.09	.21	.01	.09		**1.16**
32	.05	.59	.10		.69		.30	.21	.01	.09		**2.02**
33	.04	.59	.10				.09	.21	.01	.09		**1.11**
34	.04	.64	.13				.16	.21	.01	.09		**1.27**
35	.04	.59	.10		.09		.86	.21	.01	.21		**2.10**
36	.08	.59	.10		1.32		.09	.22	.01	.09		**2.49**
37	.04	.64	.10				.20	.21	.01	.09		**1.27**
38	.04	.59	.13				.10	.21	.01	.09		**1.15**
39	.04	.59	.10				.21	.21	.01	.09		**1.24**
40	2.70	.64	.30		1.86		.35	3.56	.14	1.67		**11.22**
41	.04	.59	.10				.16	.21	.01	.09		**1.18**
42	.28	2.33	.13				.11	.21	.01	.09		**3.15**
43	.04	.64	.10				.17	.21	.01	.09		**1.24**
44	.05	.59	.10		.69		.09	.21	.01	.09		**1.81**
45	.04	.59	.10		3.47		.37	.21	.01	.72		**5.49**
46	.04	.64	.13				.09	.21	.01	.09		**1.19**
47	.04	.59	.10				.16	.21	.01	.09		**1.18**
48	.08	.59	.10		.71		.22	.22	.01	.09		**2.01**
49	.04	.64	.10				.18	.21	.01	.09		**1.25**
50	4.61	.59	.33		.52		5.64	1.46	.17	1.19		**14.50**
Total	19.99	36.43	6.84		24.05		20.17	21.01	1.07	13.73		143.29

A value of "0.00" means a cost of more than $.000 but less than $.005 per gross square foot.

50-Year Total M&R Cost Profile per Sqft.

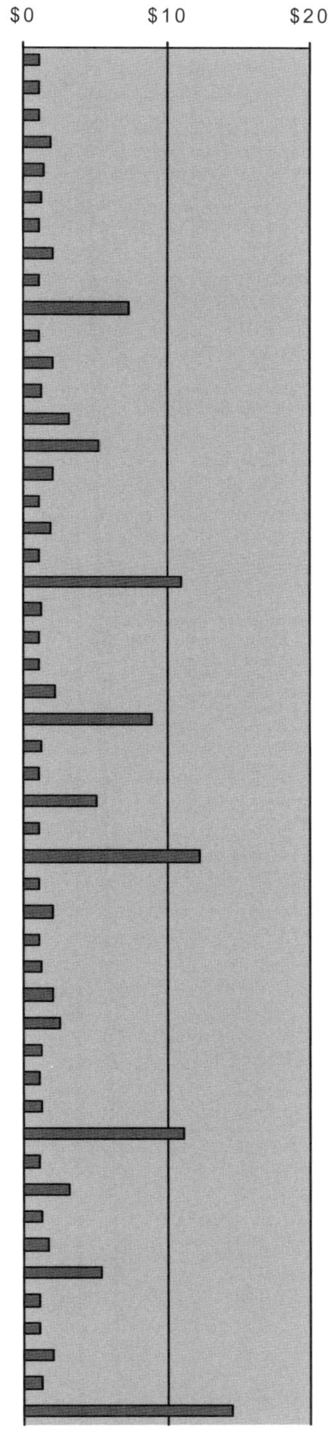

Church

Gross Sqft:	17,000
Height in Ft:	24
Exterior:	Brick Veneer
Floor coverings:	Carpet
HVAC:	Split System/Oil Heat/Wall Fin Radiation
Occupancy:	850
Replacement Cost:	$2,015,066

Components

Component	Quantity	Unit
B20 Exterior Enclosure		
Clay Brick, Exterior, 1st Floor	12312	Sq Ft
Steel Frame, Painted, Operable Window, 12 sf, 1st Floor	50	Each
Steel Frame, Painted, Operable Window, 12 sf, 2nd Floor	23	Each
Wood, Solid Core, Painted, Exterior Door	10	Each
Aluminum Frame, Fully Glazed, Exterior Door	2	Each
B30 Roofing		
Asphalt Shingle Roof	20400	Sq Ft
C10 Interior Construction		
Wood Frame, Fully Glazed, Painted, Interior Door	6	Each
Steel, Painted, Interior Door	42	Each
C20 Stairs		
Concrete Interior Steps	100	Sq Ft
C30 Interior Finishes		
Clay Brick Interior Wall Finish	860	Sq Ft
Concrete, Painted, Interior Wall Finish	10368	Sq Ft
Vinyl Tile Flooring	8500	Sq Ft
Carpet, Nylon 20 oz., High Traffic	8500	Sq Ft
Wood Finished Ceiling	20400	Sq Ft
D20 Plumbing		
Lavatory, Vitreous China	3	Each
Sink, Enameled Steel	4	Each
Drinking Fountain, Refrigerated	2	Each
Pipe & Fittings, 4" Steel	1.343	K Ln Ft
Pipe & Fittings, 3/4" Copper, Cold Water	0.48	K Ln Ft
Water Heater, Gas/Oil 65 Gph	2	Each
Pipe & Fittings, 2" Copper, Cold Water	0.766	K Ln Ft
Pipe & Fittings, 3/4" Copper, Hot Water	0.23	K Ln Ft
Pipe Insulation, Cold Water	0.902	K Ln Ft
Pipe Insulation, Hot Water	0.616	K Ln Ft
Pipe & Fittings, 6" Cast Iron	0.333	K Ln Ft
Pipe & Fittings, 10" Cast Iron	0.286	K Ln Ft
Pipe & Fittings, 4" DWV PVC	0.59	K Ln Ft
Aluminum Gutter, Downspouts, Fittings	0.54	K Ln Ft
D30 HVAC		
Pipe & Fittings, 2" Copper, Fuel Oil	0.42	K Ln Ft
Gas Boiler, 2000 Mbh	1	Each
Circulation Pump 5 HP, Hot Water	1	Each
Condenser, Air-Cooled, 20 Ton	1	Each
Condenser, Air-Cooled, 40 Ton	1	Each
Air Handler, Singlezone, 6500 Cfm	1	Each
Air Handler, Multizone, 20,000 Cfm	1	Each
Direct Digital Controls, System Points	4	Each
D40 Fire Protection		
Fire Sprinkler System	1	Each
Fire Sprinkler Head	125	Each
D50 Electrical		
Safety Switch, Fused, 400 Amp., 3 Ph.	7	Each
Main Switchgear, <1200 Amp.	1	Each
Distribution Panel Board	4	Each
Emergency Horn & Strobe	15	Each
Light Dimming Panel	1	Each
Exit Lighting Fixture, w/ Battery	15	Each
Fluorescent Lighting Fixture, 160w	85	Each
Metal Halide Lighting Fixture, High Bay, 400w	60	Each
Wiring Device, Switch	17	Each
Receptacle, 120V, 15 Amp	43	Each
Sound System	1	Each
Annunciation Panel	1	Each
Fire Alarm Bell, 6"	12	Each
Fire Alarm Control Panel	1	Each
Manual Pull Station	12	Each
Smoke Detector	25	Each

50-Year M&R Cost Summary

	50 Year Total	Annual Cost per Sqft	Annual cost as % replacement
PM & Minor Repair	$815,645	$0.96	0.81%
Unscheduled Maintenance	$822,503	$0.97	0.82%
Renewal & Replacement	$1,694,923	$1.99	1.68%
Total M&R Costs	$3,333,071	$3.92	3.31%

Distribution of Repair Costs

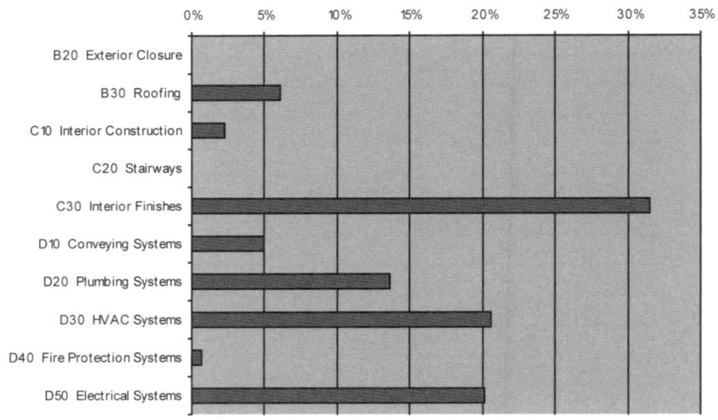

Thirty Most Costly Repair Tasks

Major Repair Task	Task Cost*	Pct.**
Replace Carpet, Nylon 20 oz., High Traffic	14.51	9.8%
Clean & Reseal Clay Brick, Exterior, 1st Floor	10.76	7.3%
Maintain Asphalt Shingle Roof	9.98	6.8%
Inspect & Test Gas Boiler, 2000 Mbh	7.70	5.2%
Repoint (50% surface) Clay Brick, Exterior, 1st Floor	6.42	4.3%
Minor Replacement, Asphalt Shingle Roof	6.34	4.3%
Fire Sprinkler System, Annual PM	5.08	3.4%
Refinish Wood Finished Ceiling	4.86	3.3%
Refinish Concrete, Painted, Interior Wall Finish	4.68	3.2%
Minor Repair, Concrete, Painted, Interior Wall Finish	3.56	2.4%
Replace Steel, Painted, Door Locks	3.16	2.1%
Replace Pipe & Fittings, 2" Copper, Cold Water	3.05	2.1%
Replace Air Handler, Multizone, 20,000 Cfm	3.04	2.1%
Replace Condenser, Air-Cooled, 40 Ton	2.90	2.0%
Replace Metal Halide Lighting Fixture, High Bay, 400w	2.88	2.0%
Remove & Replace Asphalt Shingle Roof	2.55	1.7%
Replace Safety Switch, Fused, 400 Amp., 3 Ph.	2.22	1.5%
Maintain Steel, Painted, Door Locks	2.10	1.4%
Replace Vinyl Tile Flooring	2.10	1.4%
Maintain Air Handler, Multizone, 20,000 Cfm	2.05	1.4%
Place New Over Existing, Asphalt Shingle Roof	2.00	1.4%
Replace Ballast, Metal Halide Lighting Fixture, High Bay, 400w	1.87	1.3%
Replace Pipe & Fittings, 2" Copper, Fuel Oil	1.69	1.1%
Replace Pipe & Fittings, 3/4" Copper, Cold Water	1.68	1.1%
Maintain Air Handler, Singlezone, 6500 Cfm	1.60	1.1%
Replace Gas Boiler, 2000 Mbh	1.55	1.0%
Replace Air Handler, Singlezone, 6500 Cfm	1.54	1.0%
Replace Lamp, Metal Halide Lighting Fixture, High Bay, 400w	1.49	1.0%
Replace Water Heater, Gas/Oil 65 Gph	1.40	0.9%
Repair Clay Brick, Exterior, 1st Floor	1.38	0.9%

*Task cost ($2004) per gross square foot over 50 years.

**Percent of total M&R costs.

Cost per Sqft. by System

Building Age	Exterior Closure	Roofing	Interior Constrution	Stairways	Interior Finishes	Conveying Systems	Plumbing	HVAC Systems	Fire Protection	Electrical	Equipment	Total per Sqft.
1	.03	.40	.13				.13	.47	.11	.18		**1.45**
2	.03	.40	.13		.01		.21	.48	.11	.18		**1.56**
3	.03	.40	.13				.13	.47	.11	.18		**1.45**
4	.04	.40	.16		.40		.21	.48	.11	.18		**1.99**
5	.12	1.10	.13				.20	.48	.15	.53		**2.70**
6	.03	.40	.13		.62		.21	.48	.11	.18		**2.16**
7	.12	.40	.13				.13	.67	.11	.18		**1.74**
8	.04	.40	.16		2.81		.39	.48	.11	.19		**4.58**
9	.03	.40	.13		.02		.13	.47	.11	.18		**1.47**
10	5.24	1.10	.81		.01		.31	.84	.39	1.71		**10.41**
11	.03	.40	.13				.13	.47	.11	.18		**1.45**
12	.07	.40	.16		1.01		.21	.48	.11	.18		**2.62**
13	.03	.40	.13				.28	.47	.11	.18		**1.59**
14	.12	.40	.13		.01		.22	.68	.11	.18		**1.85**
15	.12	1.10	.13	.01	1.29		.14	3.34	.15	1.01		**7.30**
16	.04	.40	.16		2.81		.39	.48	.11	.19		**4.58**
17	.03	.40	.13				.13	.47	.11	.18		**1.45**
18	.03	.40	.13		1.67		.21	.48	.11	.18		**3.21**
19	.03	.40	.13				.13	.47	.11	.18		**1.45**
20	5.30	3.10	.84		.40		.59	.59	.35	2.78		**13.96**
21	.12	.40	.13				.19	.67	.11	.18		**1.80**
22	.03	.40	.13		.01		.21	.48	.11	.18		**1.56**
23	.03	.40	.13				.13	.47	.11	.18		**1.45**
24	.07	.40	.16		3.42		.39	.48	.11	.19		**5.22**
25	4.02	1.10	.13		.05		3.71	1.57	.15	1.69		**12.43**
26	.03	.40	.13		.01		.29	.48	.11	.18		**1.63**
27	.03	.40	.13		.02		.14	.47	.11	.18		**1.48**
28	.13	.40	.16		.40		.21	.68	.11	.18		**2.28**
29	.03	.40	.13				.19	.47	.11	.18		**1.51**
30	5.24	1.10	.81	.01	1.91		.30	4.70	.39	2.10		**16.57**
31	.03	.40	.13				.14	.47	.11	.18		**1.46**
32	.04	.40	.16		2.81		.38	.48	.11	.18		**4.57**
33	.03	.40	.13				.14	.47	.11	.19		**1.47**
34	.03	.40	.13		.01		.21	.48	.11	.18		**1.55**
35	.21	1.10	.13				.29	.48	.15	.53		**2.89**
36	.07	.40	.16		2.06		.17	.48	.11	.18		**3.63**
37	.03	.40	.13				.23	.67	.11	.18		**1.75**
38	.03	.40	.13		.01		.17	.48	.11	.18		**1.52**
39	.03	.40	.13				.26	.47	.11	.18		**1.58**
40	5.71	2.55	.84		2.81		.64	.83	.35	2.89		**16.62**
41	.03	.40	.13				.17	.47	.11	.19		**1.50**
42	.03	.40	.13		.62		.17	.48	.11	.18		**2.12**
43	.03	.40	.13				.19	.47	.11	.18		**1.51**
44	.04	.40	.16		.40		.17	.68	.11	.18		**2.15**
45	.12	1.10	.13	.01	1.31		.31	3.35	.15	1.01		**7.49**
46	.03	.40	.13		.01		.17	.48	.11	.18		**1.51**
47	.12	.40	.13				.17	.47	.11	.18		**1.58**
48	.07	.40	.16		3.42		.33	.48	.11	.18		**5.15**
49	.03	.40	.13				.18	.47	.11	.19		**1.51**
50	9.14	1.10	1.01		.07		3.79	1.43	.39	2.67		**19.60**
Total	37.09	30.31	10.48	.03	30.46		18.55	37.84	7.07	24.24		196.06

50-Year Total M&R Cost Profile per Sqft.

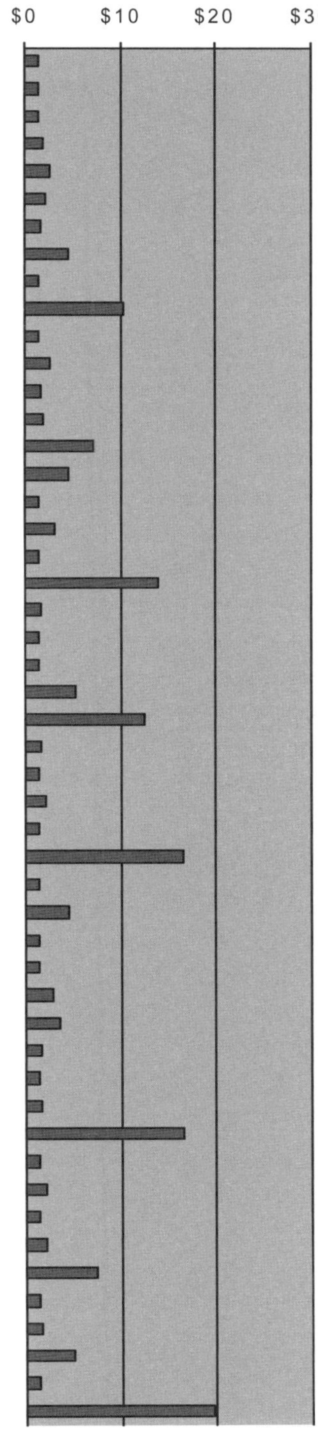

A value of "0.00" means a cost of more than $.000 but less than $.005 per gross square foot.

Club, Country

Gross Sqft:	6,000
Height in Ft:	12
Exterior:	Stone Ashlar Veneer
Floor coverings:	Carpet/Hardwood/Ceramic Tile
HVAC:	Electric Cool, Gas Heat, Multizone Unit
Occupancy:	180
Replacement Cost:	$824,059

Components

Component	Quantity	Unit
B20 Exterior Enclosure		
Stucco, Exterior, 1st Floor	2652	Sq Ft
Steel Frame, Painted, Operable Window, 12 sf, 1st Floor	95	Each
Steel, Painted, Exterior Door	4	Each
Aluminum Frame, Fully Glazed, Exterior Door	2	Each
B30 Roofing		
Asphalt Shingle Roof	7380	Sq Ft
Concrete Decking	200	Sq Ft
Wood Decking	200	Sq Ft
Wooden Steps	40	Sq Ft
C10 Interior Construction		
Aluminum, w/ Safety Glass, Plain or Anodized, Interior Door	20	Each
C30 Interior Finishes		
Sheetrock, Stippled, Interior Wall Finish	4114	Sq Ft
Concrete Block, Painted, Interior Wall Finish	2652	Sq Ft
Wallpaper Interior Wall Finish	4114	Sq Ft
Ceramic Tile Wall Finish	2057	Sq Ft
Ceramic Tile Flooring	1200	Sq Ft
Wood Parquetry Flooring	1800	Sq Ft
Carpet, Wool 45 oz., Premium	3000	Sq Ft
Sheetrock, Stippled, Ceiling	6000	Sq Ft
D20 Plumbing		
Flush Tank Water Closet	12	Each
Urinal, Vitreous China	3	Each
Lavatory, Vitreous China	15	Each
Sink, Stainless Steel	5	Each
Shower, Ceramic Tile	6	Each
Pipe & Fittings, 3/4" Copper, Hot Water	0.805	K Ln Ft
Pipe & Fittings, 3/4" Copper, Cold Water	1.415	K Ln Ft
Pipe & Fittings, 2" Copper, Cold Water	1.527	K Ln Ft
Pipe & Fittings, 4" Steel	0.887	K Ln Ft
Pipe Insulation, Cold Water	1.749	K Ln Ft
Pipe Insulation, Hot Water	0.984	K Ln Ft
Water Heater, Gas/Oil 175 Gph	2	Each
Pipe & Fittings, 6" Cast Iron	0.611	K Ln Ft
Pipe & Fittings, 10" Cast Iron	0.155	K Ln Ft
Aluminum Gutter, Downspouts, Fittings	0.34	K Ln Ft
Pipe & Fittings, 4" DWV PVC	0.213	K Ln Ft
D30 HVAC		
Air Conditioner, Rooftop, 5 Ton	1	Each
Air Conditioner, Rooftop, 20 Ton	1	Each
D40 Fire Protection		
Fire Sprinkler Head	45	Each
Kitchen Fire Suppression System (CO2)	1	Each
D50 Electrical		
Safety Switch, Fused, 400 Amp., 3 Ph.	4	Each
Main Switchgear, <1200 Amp.	1	Each
Exit Lighting Fixture, w/ Battery	6	Each
Light Dimming Panel	2	Each
Incandescent Lighting Fixture, Basic, 200w	20	Each
Fluorescent Lighting Fixture, 160w	30	Each
Wiring Device, Switch	12	Each
Grounded Fault Circuit Interruptor, 15 Amps	6	Each
Receptacle, 120V, 15 Amp	18	Each
TV Cable Outlet	1	Each
Annunciation Panel	1	Each
Fire Alarm Bell, 6"	4	Each
Fire Alarm Control Panel	1	Each
Manual Pull Station	4	Each
Smoke Detector	25	Each

50-Year M&R Cost Summary

	50 Year Total	Annual Cost per Sqft	Annual cost as % replacement
PM & Minor Repair	$265,242	$0.88	0.64%
Unscheduled Maintenance	$274,950	$0.92	0.67%
Renewal & Replacement	$997,998	$3.33	2.42%
Total M&R Costs	$1,538,190	$5.13	3.73%

Distribution of Repair Costs

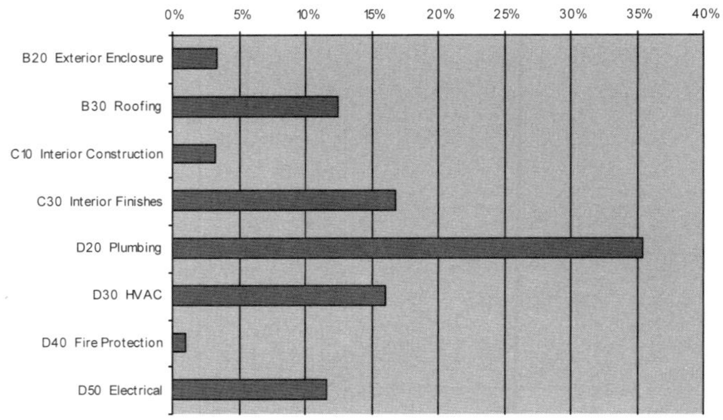

Thirty Most Costly Repair Tasks

Major Repair Task	Task Cost*	Pct.**
Replace Pipe & Fittings, 2" Copper, Cold Water	17.23	8.2%
Replace Air Conditioner, Rooftop, 20 Ton	15.20	7.2%
Replace Pipe & Fittings, 3/4" Copper, Cold Water	14.05	6.7%
Replace Carpet, Wool 45 oz., Premium	11.62	5.5%
Maintain Asphalt Shingle Roof	10.23	4.9%
Replace Water Heater, Gas/Oil 175 Gph	9.73	4.6%
Replace Pipe & Fittings, 3/4" Copper, Hot Water	7.95	3.8%
Minor Replacement, Asphalt Shingle Roof	6.50	3.1%
Refinish Sheetrock, Stippled, Ceiling	6.10	2.9%
Repair Air Conditioner, Rooftop, 20 Ton	6.09	2.9%
Maintain Air Conditioner, Rooftop, 20 Ton	5.67	2.7%
Replace Wallpaper Interior Wall Finish	5.30	2.5%
Replace Air Conditioner, Rooftop, 5 Ton	4.19	2.0%
Refinish Sheetrock, Stippled, Interior Wall Finish	4.17	2.0%
Replace Safety Switch, Fused, 400 Amp., 3 Ph.	3.59	1.7%
Refinish Stucco, Exterior, 1st Floor	3.56	1.7%
Replace Wood Parquetry Floor	3.50	1.7%
Refinish Concrete Block, Painted, Interior Wall Finish	3.39	1.6%
Clean Water Heater, Gas/Oil 175 Gph	3.20	1.5%
Maintain Aluminum, w/ Safety Glass, Plain or Anodized, Door Locks	2.84	1.3%
Check Operation, Smoke Detector	2.77	1.3%
Replace Faucet Washer & Clean Trap, Lavatory, Vitreous China	2.75	1.3%
Remove & Replace Asphalt Shingle Roof	2.62	1.2%
Replace Ceramic Tile Flooring	2.55	1.2%
Maintain Air Conditioner, Rooftop, 5 Ton	2.49	1.2%
Place New Over Existing, Asphalt Shingle Roof	2.05	1.0%
Minor Repair, Ceramic Tile Wall Finish	1.97	0.9%
Replace Pipe Insulation, Cold Water	1.94	0.9%
Repair Air Conditioner, Rooftop, 5 Ton	1.87	0.9%
Annual PM, Safety Switch, Fused, 400 Amp., 3 Ph.	1.84	0.9%

**Task cost ($2004) per gross square foot over 50 years.*

***Percent of total M&R costs.*

Cost per Sqft. by System

Building Age	Exterior Closure	Roofing	Interior Constrution	Stairways	Interior Finishes	Conveying Systems	Plumbing	HVAC Systems	Fire Protection	Electrical	Equipment	Total per Sqft.
1	.04	.41	.15				.28	.30	.03	.28		**1.49**
2	.04	.41	.15				.72	.30	.03	.28		**1.93**
3	.04	.41	.15				.28	.30	.03	.28		**1.49**
4	.05	.41	.15		1.14		.72	.30	.03	.28		**3.07**
5	.19	1.15	.15		.19		.72	.30	.03	.71		**3.43**
6	.04	.41	.15				.72	.30	.03	.28		**1.93**
7	.04	.45	.15				.35	.30	.03	.28		**1.59**
8	.05	.41	.15		1.14		2.20	.30	.03	.30		**4.57**
9	.04	.41	.15				.28	.30	.03	.28		**1.49**
10	.90	1.15	.29		1.72		1.67	2.95	.27	1.38		**10.32**
11	.04	.41	.15				.28	.30	.03	.28		**1.49**
12	.12	.41	.15		1.14		.72	.30	.03	.28		**3.14**
13	.04	.41	.15				.68	.30	.03	.28		**1.89**
14	.26	.45	.15				.78	.30	.03	.28		**2.25**
15	.19	1.15	.15		.19		.71	6.46	.03	1.79		**10.68**
16	.05	.41	.15		1.14		2.20	.30	.03	.30		**4.57**
17	.04	.41	.15				.28	.30	.03	.28		**1.49**
18	.04	.41	.15				.72	.30	.03	.28		**1.93**
19	.04	.41	.15				.28	.30	.03	.28		**1.49**
20	1.06	3.39	.29		3.92		2.29	.30	.23	2.04		**13.52**
21	.04	.45	.15				.50	.30	.03	.28		**1.75**
22	.04	.41	.15				.72	.30	.03	.28		**1.93**
23	.04	.41	.15				.28	.30	.03	.28		**1.49**
24	.12	.41	.15		1.35		2.20	.30	.03	.30		**4.84**
25	.19	1.15	.15		5.97		22.73	2.95	.03	2.64		**35.80**
26	.04	.41	.15				.89	.30	.03	.28		**2.10**
27	.04	.41	.15				.31	.30	.03	.28		**1.52**
28	.27	.45	.15		1.14		.75	.30	.03	.28		**3.37**
29	.04	.41	.15				.47	.30	.03	.28		**1.68**
30	.90	1.15	.29		1.72		2.38	6.46	.27	2.25		**15.42**
31	.04	.41	.15				.31	.30	.03	.28		**1.52**
32	.05	.41	.15		1.14		2.16	.30	.03	.28		**4.52**
33	.04	.41	.15				.31	.30	.03	.30		**1.54**
34	.04	.41	.15				.69	.30	.03	.28		**1.89**
35	.19	1.19	.15		.19		3.14	.30	.03	.71		**5.89**
36	.12	.41	.15		1.14		.38	.30	.03	.28		**2.80**
37	.04	.41	.15				.77	.30	.03	.28		**1.98**
38	.04	.41	.15				.42	.30	.03	.28		**1.63**
39	.04	.41	.15				.82	.30	.03	.28		**2.03**
40	1.06	2.83	.29		7.19		3.30	2.95	.23	2.25		**20.10**
41	.04	.41	.15				.62	.30	.03	.30		**1.84**
42	.26	.45	.15				.45	.30	.03	.28		**1.91**
43	.04	.41	.15				.62	.30	.03	.28		**1.83**
44	.05	.41	.15		1.14		.38	.30	.03	.28		**2.74**
45	.19	1.15	.15		.19		1.48	6.46	.03	1.79		**11.44**
46	.04	.41	.15				.38	.30	.03	.28		**1.59**
47	.04	.41	.15				.62	.30	.03	.28		**1.83**
48	.12	.41	.15		1.35		1.86	.30	.03	.28		**4.49**
49	.04	.45	.15				.68	.30	.03	.30		**1.95**
50	.90	1.15	.29		10.04		23.39	.30	.27	2.89		**39.21**
Total	8.48	31.92	8.32		43.12		90.84	41.22	2.64	29.82		256.37

50-Year Total M&R Cost Profile per Sqft.

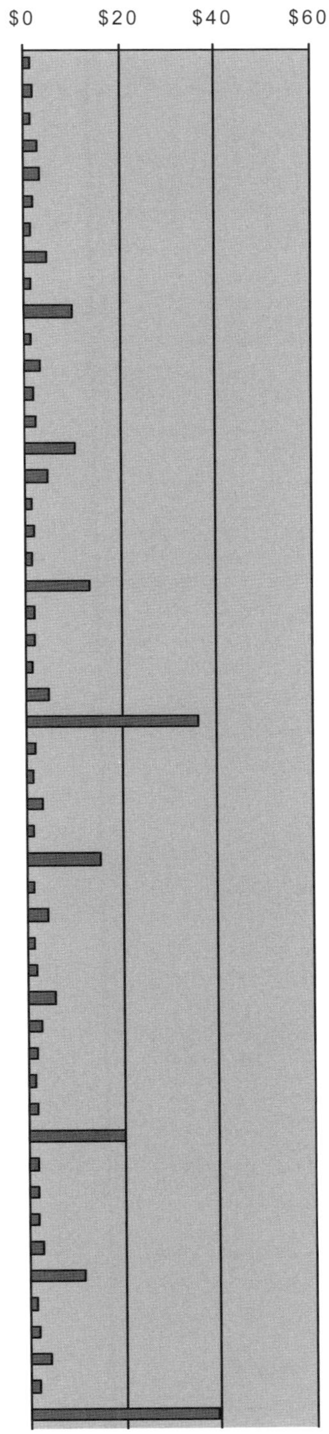

A value of "0.00" means a cost of more than $.000 but less than $.005 per gross square foot.

Club, Social

Gross Sqft:	22,000
Height in Ft:	12
Exterior:	Stone Ashlar Veneer
Floor coverings:	Carpet/Hardwood/Ceramic Tile
HVAC:	Electric Cool, Gas Heat, Multizone Unit
Occupancy:	770
Replacement Cost:	$2,148,832

Components

Component	Quantity	Unit
B20 Exterior Enclosure		
Stucco, Exterior, 1st Floor	4992	Sq Ft
Steel Frame, Painted, Operable Window, 12 sf, 1st Floor	90	Each
Steel, Painted, Exterior Door	4	Each
Aluminum Frame, Fully Glazed, Exterior Door	2	Each
B30 Roofing		
Single-Ply Modified Bituminous/Thermoplastic Roof	22000	Sq Ft
Concrete Decking	200	Sq Ft
Wood Decking	200	Sq Ft
Wooden Steps	40	Sq Ft
C10 Interior Construction		
Wood, Solid Core, Painted, Interior Door	157	Each
C30 Interior Finishes		
Sheetrock, Stippled, Interior Wall Finish	24508	Sq Ft
Concrete Block, Painted, Interior Wall Finish	4992	Sq Ft
Wallpaper Interior Wall Finish	9400	Sq Ft
Ceramic Tile Wall Finish	3770	Sq Ft
Ceramic Tile Flooring	3300	Sq Ft
Wood Parquetry Flooring	5500	Sq Ft
Carpet, Wool 45 oz., Premium	13200	Sq Ft
Acoustical Tile Ceiling	22000	Sq Ft
D20 Plumbing		
Tankless Water Closet	8	Each
Urinal, Vitreous China	3	Each
Lavatory, Vitreous China	11	Each
Sink, Stainless Steel	6	Each
Shower, Ceramic Tile	3	Each
Drinking Fountain, Refrigerated	3	Each
Pipe & Fittings, 3/4" Copper, Cold Water	1.16	K Ln Ft
Pipe & Fittings, 3/4" Copper, Hot Water	0.63	K Ln Ft
Pipe & Fittings, 2" Copper, Cold Water	1.442	K Ln Ft
Pipe & Fittings, 4" Steel	1.287	K Ln Ft
Pipe Insulation, Cold Water	1.804	K Ln Ft
Pipe Insulation, Hot Water	0.977	K Ln Ft
Water Heater, Gas/Oil 175 Gph	1	Each
Pipe & Fittings, 6" Cast Iron	0.588	K Ln Ft
Pipe & Fittings, 10" Cast Iron	0.297	K Ln Ft
Roof Drain, 2"	9	Each
Pipe & Fittings, 4" DWV PVC	0.318	K Ln Ft
D30 HVAC		
Exhaust Fan, Ceiling, 200-500 Cfm	3	Each
Air Conditioner, Rooftop, 20 Ton	4	Each
D40 Fire Protection		
Fire Sprinkler System	1	Each
Fire Sprinkler Head	160	Each
Fire Suppression System Water Pump	1	Each
D50 Electrical		
Safety Switch, Fused, 400 Amp., 3 Ph.	4	Each
Distribution Panel Board	4	Each
Main Switchgear, <1200 Amp.	1	Each
Exit Lighting Fixture, w/ Battery	10	Each
Emergency Horn & Strobe	12	Each
Light Dimming Panel	2	Each
Incandescent Lighting Fixture, Basic, 100w	20	Each
Fluorescent Lighting Fixture, 160w	110	Each
Wiring Device, Switch	22	Each
Grounded Fault Circuit Interruptor, 15 Amps	3	Each
Receptacle, 120V, 15 Amp	55	Each
TV Cable Outlet	1	Each
Annunciation Panel	1	Each
Fire Alarm Bell, 6"	4	Each
Fire Alarm Control Panel	1	Each
Manual Pull Station	4	Each
Smoke Detector	7	Each

50-Year M&R Cost Summary

	50 Year Total	Annual Cost per Sqft	Annual cost as % replacement
PM & Minor Repair	$671,448	$0.61	0.62%
Unscheduled Maintenance	$656,880	$0.60	0.61%
Renewal & Replacement	$2,278,585	$2.07	2.12%
Total M&R Costs	$3,606,913	$3.28	3.36%

Distribution of Repair Costs

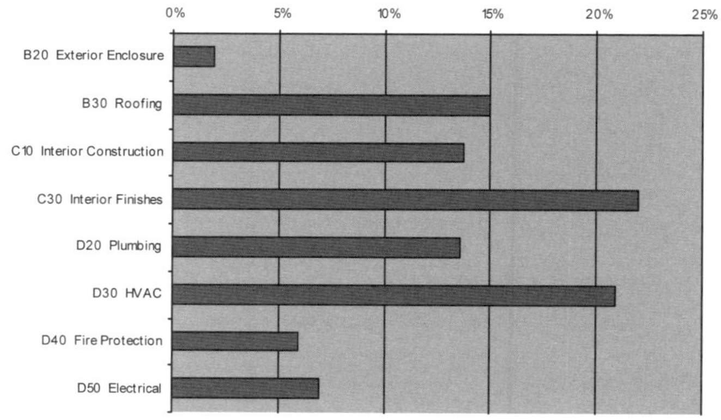

Thirty Most Costly Repair Tasks

Major Repair Task	Task Cost*	Pct.**
Replace Air Conditioner, Rooftop, 20 Ton	16.59	12.4%
Replace Carpet, Wool 45 oz., Premium	13.95	10.4%
Maintain Single-Ply Modified Bituminous/Thermoplastic Roof	7.54	5.6%
Remove & Replace Membrane, Single-Ply Modified Bituminous/Ther	7.01	5.2%
Refinish Sheetrock, Stippled, Interior Wall Finish	6.77	5.0%
Repair Air Conditioner, Rooftop, 20 Ton	6.65	5.0%
Maintain Air Conditioner, Rooftop, 20 Ton	6.18	4.6%
Maintain Wood, Solid Core, Painted, Door Locks	5.95	4.4%
Replace Pipe & Fittings, 2" Copper, Cold Water	4.44	3.3%
Fire Sprinkler System, Annual PM	3.93	2.9%
Replace Wallpaper Interior Wall Finish	3.30	2.5%
Replace Pipe & Fittings, 3/4" Copper, Cold Water	3.14	2.3%
Replace Wood Parquetry Floor	2.91	2.2%
Replace Wood, Solid Core, Painted, Interior Door	2.66	2.0%
Replace Solid Core, Painted, Door Locks	2.65	2.0%
Repair Single-Ply Modified Bituminous/Thermoplastic Roof	2.50	1.9%
Replace Ceramic Tile Flooring	1.91	1.4%
Refinish Stucco, Exterior, 1st Floor	1.83	1.4%
Minor Repair, Acoustic Tile Ceiling	1.78	1.3%
Refinish Concrete Block, Painted, Interior Wall Finish	1.74	1.3%
Replace Pipe & Fittings, 3/4" Copper, Hot Water	1.70	1.3%
Replace Ballast & Lamps, Fluorescent Lighting Fixture, 160w	1.38	1.0%
Replace Water Heater, Gas/Oil 175 Gph	1.33	1.0%
Annual PM, Fire Suppression System Water Pump Motor	1.27	0.9%
Annual PM, Fire Suppression System Water Pump	1.06	0.8%
Refinish Wood, Solid Core, Painted, Interior Door	1.06	0.8%
Minor Repair, Ceramic Tile Wall Finish	.99	0.7%
Replace Safety Switch, Fused, 400 Amp., 3 Ph.	.98	0.7%
Replace Fluorescent Lighting Fixture, 160w	.98	0.7%
Annual PM, Distribution Panel Board	.86	0.6%

**Task cost ($2004) per gross square foot over 50 years.*

***Percent of total M&R costs.*

Cost per Sqft. by System

Building Age	Exterior Closure	Roofing	Interior Constrution	Stairways	Interior Finishes	Conveying Systems	Plumbing	HVAC Systems	Fire Protection	Electrical	Equipment	Total per Sqft.
1	.01	.29	.33				.08	.23	.17	.10		**1.22**
2	.01	.29	.33				.22	.23	.17	.10		**1.35**
3	.01	.34	.33				.08	.23	.17	.10		**1.27**
4	.01	.29	.42		.71		.22	.23	.17	.10		**2.15**
5	.05	.30	.33		.14		.14	.23	.20	.23		**1.61**
6	.01	.34	.33				.22	.23	.17	.10		**1.41**
7	.01	.30	.33				.10	.23	.17	.10		**1.24**
8	.01	.29	.42		.71		.43	.23	.17	.11		**2.37**
9	.01	.34	.33		.36		.08	.23	.17	.10		**1.62**
10	.41	1.13	.99		1.09		.49	2.45	.43	.73		**7.72**
11	.01	.29	.33				.08	.23	.17	.10		**1.22**
12	.03	.34	.42		.71		.22	.23	.17	.10		**2.22**
13	.01	.29	.33				.18	.23	.17	.10		**1.32**
14	.07	.30	.33				.24	.23	.17	.10		**1.44**
15	.05	.35	.33		.14		.12	5.53	.20	.46		**7.18**
16	.01	.29	.42		.71		.43	.23	.17	.11		**2.37**
17	.01	.29	.33				.08	.23	.17	.10		**1.22**
18	.01	.34	.33		.36		.22	.23	.17	.10		**1.76**
19	.01	.29	.33				.08	.23	.17	.10		**1.22**
20	.46	3.56	1.08		2.27		.52	.29	.39	1.13		**9.70**
21	.01	.30	.33				.12	.23	.17	.10		**1.26**
22	.01	.29	.33				.22	.23	.17	.10		**1.35**
23	.01	.34	.33				.08	.23	.17	.10		**1.27**
24	.03	.29	.42		.82		.43	.23	.17	.11		**2.49**
25	.05	.30	.33		7.07		5.56	2.45	.20	.74		**16.69**
26	.01	.34	.33				.28	.23	.17	.10		**1.47**
27	.01	.29	.33		.36		.09	.23	.17	.10		**1.58**
28	.07	.30	.42		.71		.23	.23	.17	.10		**2.23**
29	.01	.34	.33				.11	.23	.17	.10		**1.30**
30	.41	1.13	2.76		1.09		.56	5.53	.43	.92		**12.84**
31	.01	.29	.33				.09	.23	.17	.10		**1.22**
32	.01	.34	.33		.71		.42	.23	.17	.10		**2.32**
33	.01	.29	.33				.09	.23	.17	.11		**1.23**
34	.01	.29	.42				.21	.23	.17	.10		**1.43**
35	.05	.36	.33		.14		.58	.23	.20	.23		**2.11**
36	.03	.29	.33		1.07		.15	.23	.17	.10		**2.37**
37	.01	.29	.33				.17	.23	.17	.10		**1.31**
38	.01	.34	.42				.16	.23	.17	.10		**1.43**
39	.01	.29	.33				.22	.23	.17	.10		**1.36**
40	.46	3.56	.99		5.00		.67	2.51	.39	1.17		**14.74**
41	.01	.29	.33				.15	.23	.17	.11		**1.29**
42	.07	.30	.42				.17	.23	.17	.10		**1.46**
43	.01	.34	.33				.15	.23	.17	.10		**1.34**
44	.01	.29	.33		.71		.15	.23	.17	.10		**2.00**
45	.05	.30	.33		.50		.35	5.53	.20	.46		**7.72**
46	.01	.34	.42				.15	.23	.17	.10		**1.42**
47	.01	.29	.33				.15	.23	.17	.10		**1.29**
48	.03	.29	.33		.82		.36	.23	.17	.10		**2.33**
49	.01	.35	.33				.17	.23	.17	.11		**1.37**
50	.41	1.13	1.08		9.93		5.75	.23	.43	1.16		**20.13**
Total	3.14	24.51	22.54		36.09		22.25	34.28	9.80	11.34		163.95

50-Year Total M&R Cost Profile per Sqft.

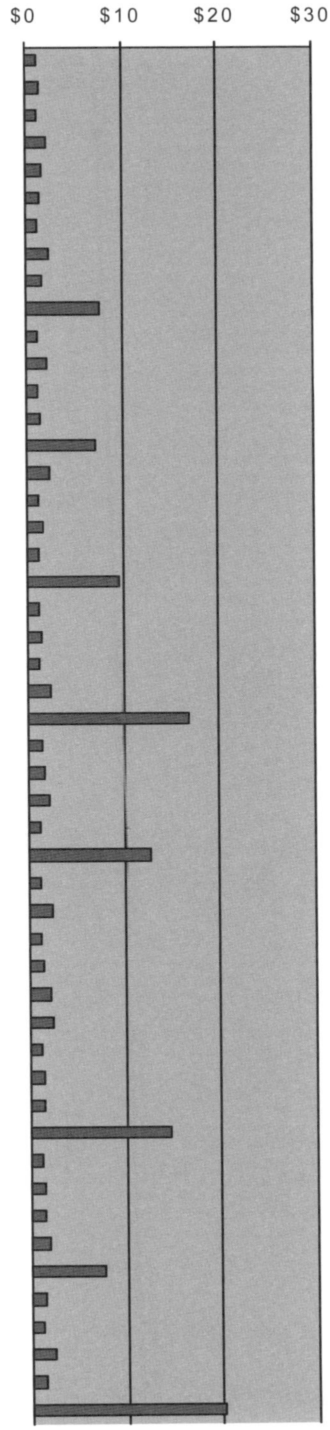

A value of "0.00" means a cost of more than $.000 but less than $.005 per gross square foot.

College Classrooms

Gross Sqft:	90,000
Height in Ft:	24
Exterior:	Decorative Concrete Block
Floor coverings:	Carpet/Vinyl Tile/Ceramic Tile
HVAC:	Electric Cool, Gas Heat, Multizone Heat
Occupancy:	4500
Replacement Cost:	$9,651,343

Components

Component	Quantity	Unit
B20 Exterior Enclosure		
Concrete Block, Exterior, 1st Floor	17160	Sq Ft
Steel Frame, Painted, Operable Window, 12 sf, 1st Floor	308	Each
Aluminum Frame, Fully Glazed, Exterior Door	6	Each
B30 Roofing		
Concrete Steps	125	Sq Ft
Concrete Decking	1000	Sq Ft
Built-up Roof	45000	Sq Ft
C10 Interior Construction		
Steel, Painted, Interior Door	450	Each
C20 Stairs		
Concrete Interior Steps	400	Sq Ft
C30 Interior Finishes		
Sheetrock, Stippled, Interior Wall Finish	102600	Sq Ft
Ceramic Tile Wall Finish	5400	Sq Ft
Ceramic Tile Flooring	4500	Sq Ft
Vinyl Tile Flooring	63000	Sq Ft
Carpet, Nylon 20 oz., High Traffic	22500	Sq Ft
Acoustical Tile Ceiling	90000	Sq Ft
D10 Conveying		
Elevator, Hydraulic, 2500 lbs, 3-5 Floor, 200 fpm	2	Each
D20 Plumbing		
Tankless Water Closet	80	Each
Urinal, Vitreous China	54	Each
Lavatory, Vitreous China	69	Each
Sink, Stainless Steel	118	Each
Drinking Fountain, Refrigerated	16	Each
Pipe & Fittings, 4" Steel	1.797	K Ln Ft
Pipe & Fittings, 3/4" Copper, Cold Water	11.74	K Ln Ft
Pipe & Fittings, 2" Copper, Cold Water	10.319	K Ln Ft
Pipe & Fittings, 3/4" Copper, Hot Water	6.2	K Ln Ft
Pipe Insulation, Cold Water	12.638	K Ln Ft
Pipe Insulation, Hot Water	6.674	K Ln Ft
Water Heater, Gas/Oil 175 Gph	2	Each
Pipe & Fittings, 6" Cast Iron	5.362	K Ln Ft
Pipe & Fittings, 10" Cast Iron	0.424	K Ln Ft
Pipe & Fittings, 4" DWV PVC	0.65	K Ln Ft
Roof Drain, 2"	24	Each
D30 HVAC		
Exhaust Fan, Ceiling, 200-500 Cfm	8	Each
Air Conditioner, Rooftop, 50 Ton	7	Each
D40 Fire Protection		
Fire Sprinkler System	2	Each
Fire Sprinkler Head	642	Each
D50 Electrical		
Safety Switch, Fused, 400 Amp., 3 Ph.	8	Each
Main Switchgear, <1200 Amp.	1	Each
Secondary Transformer, Dry, 50 KVA	3	Each
Distribution Panel Board	12	Each
Exit Lighting Fixture, w/ Battery	15	Each
Fluorescent Lighting Fixture, 160w	1296	Each
Wiring Device, Switch	90	Each
Receptacle, 120V, 15 Amp	900	Each
TV Cable Outlet	47	Each
Annunciation Panel	1	Each
Fire Alarm Bell, 6"	10	Each
Fire Alarm Control Panel	1	Each
Manual Pull Station	10	Each
Heat Detector	7	Each
Smoke Detector	100	Each
Time Control Clock	30	Each
Generator Switchgear	1	Each

50-Year M&R Cost Summary

	50 Year Total	Annual Cost per Sqft	Annual cost as % replacement
PM & Minor Repair	$2,202,939	$0.49	0.46%
Unscheduled Maintenance	$2,534,987	$0.56	0.53%
Renewal & Replacement	$8,976,565	$1.99	1.86%
Total M&R Costs	$13,714,491	$3.05	2.84%

Distribution of Repair Costs

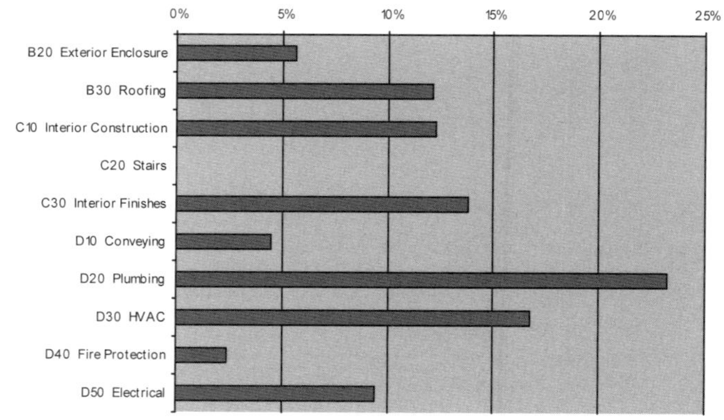

Thirty Most Costly Repair Tasks

Major Repair Task	Task Cost*	Pct.**
Replace Air Conditioner, Rooftop, 50 Ton	11.99	9.7%
Replace Pipe & Fittings, 3/4" Copper, Cold Water	7.77	6.3%
Replace Pipe & Fittings, 2" Copper, Cold Water	7.76	6.2%
Repair Air Conditioner, Rooftop, 50 Ton	7.34	5.9%
Maintain Built-up Roof	7.31	5.9%
Replace Carpet, Nylon 20 oz., High Traffic	7.25	5.8%
Refinish Sheetrock, Stippled, Interior Wall Finish	6.93	5.6%
Replace Steel, Painted, Door Locks	6.39	5.1%
Maintain Elevator, Hydraulic, 2500 lbs, 3-5 Floor, 200 fpm	5.76	4.6%
Maintain Steel, Painted, Door Locks	4.26	3.4%
Replace Pipe & Fittings, 3/4" Copper, Hot Water	4.08	3.3%
Replace Ballast & Lamps, Fluorescent Lighting Fixture, 160w	3.97	3.2%
Maintain Air Conditioner, Rooftop, 50 Ton	3.42	2.7%
Replace Vinyl Tile Flooring	2.94	2.4%
Clean & Seal Concrete Block, Exterior, 1st Floor	2.84	2.3%
Replace Fluorescent Lighting Fixture, 160w	2.82	2.3%
Fire Sprinkler System, Annual PM	1.92	1.5%
Minor Repair, Acoustic Tile Ceiling	1.78	1.4%
Place New Membrane Over Existing, Built-up Roof	1.75	1.4%
Remove & Replace Membrane, Built-up Roof	1.73	1.4%
Renovate Elevator, Hydraulic, 2500 lbs, 3-5 Floor, 200 fpm	1.04	0.8%
Replace Pipe Insulation, Cold Water	.93	0.8%
Replace Faucet Washer & Clean Trap, Lavatory, Vitreous China	.84	0.7%
Repoint (50% surface) Concrete Block, Exterior, 1st Floor	.82	0.7%
Refinish Steel, Painted, Interior Door	.78	0.6%
Replace Sink, Stainless Steel	.76	0.6%
Check Operation, Smoke Detector	.74	0.6%
Replace Faucet Washer & Clean Trap, Sink, Stainless Steel	.72	0.6%
Repair Fire Sprinkler Head	.68	0.6%
Minor Repair, Sheetrock, Stippled, Interior Wall Finish	.68	0.5%

*Task cost ($2004) per gross square foot over 50 years.

**Percent of total M&R costs.

Cost per Sqft. by System

Building Age	Exterior Closure	Roofing	Interior Constrution	Stairways	Interior Finishes	Conveying Systems	Plumbing	HVAC Systems	Fire Protection	Electrical	Equipment	Total per Sqft.
1	.00	.29	.23			.12	.05	.13	.05	.10		**.96**
2	.00	.29	.23		.01	.12	.29	.13	.05	.10		**1.21**
3	.00	.32	.23			.12	.05	.13	.05	.10		**.99**
4	.00	.29	.29		.58	.12	.29	.13	.05	.10		**1.86**
5	.03	.29	.23			.12	.09	.13	.06	.12		**1.07**
6	.00	.32	.23		.01	.12	.29	.13	.05	.10		**1.24**
7	.00	.31	.23			.12	.08	.13	.05	.10		**1.01**
8	.00	.29	.29		1.79	.12	.47	.13	.05	.10		**3.23**
9	.00	.32	.23		.38	.12	.05	.13	.05	.10		**1.37**
10	1.39	.29	1.51		.08	.12	.73	2.58	.29	1.56		**8.54**
11	.00	.29	.23			.12	.05	.13	.05	.10		**.96**
12	.02	.32	.29		.58	.12	.29	.13	.05	.10		**1.90**
13	.00	.29	.23			.12	.14	.13	.05	.10		**1.05**
14	.00	1.18	.23		.01	.12	.33	.13	.05	.10		**2.14**
15	.03	.32	.23	.01	.00	.12	.07	4.00	.06	.33		**5.17**
16	.00	.29	.29		1.79	.12	.47	.13	.05	.10		**3.23**
17	.00	.29	.23			.12	.05	.13	.05	.10		**.96**
18	.00	.32	.23		1.83	.12	.29	.13	.05	.17		**3.14**
19	.00	.29	.23			.12	.05	.13	.05	.10		**.96**
20	1.42	.36	1.57		1.14	.12	.81	.17	.25	1.93		**7.78**
21	.00	.33	.23			.12	.09	.13	.05	.10		**1.05**
22	.00	.29	.23		.01	.12	.29	.13	.05	.10		**1.21**
23	.00	.29	.23			.12	.05	.13	.05	.10		**.96**
24	.02	.32	.29		1.79	.12	.47	.13	.05	.10		**3.27**
25	.64	.29	.23			.63	10.87	2.58	.06	.45		**15.75**
26	.00	.29	.23		.01	.12	.34	.13	.05	.10		**1.26**
27	.00	.32	.23		.38	.12	.06	.13	.05	.10		**1.38**
28	.00	1.74	.29		.58	.12	.32	.13	.05	.10		**3.33**
29	.00	.29	.23			.12	.07	.13	.05	.10		**.99**
30	1.39	.29	1.51	.01	.08	.12	.72	4.00	.29	1.76		**10.16**
31	.00	.32	.23			.12	.06	.13	.05	.10		**1.00**
32	.00	.29	.29		1.79	.12	.45	.13	.05	.10		**3.22**
33	.00	.29	.23			.12	.06	.13	.05	.10		**.98**
34	.00	.32	.23		.01	.12	.28	.13	.05	.10		**1.23**
35	.03	.31	.23			.12	1.15	.13	.06	.12		**2.15**
36	.02	.29	.29		2.41	.12	.19	.13	.05	.17		**3.66**
37	.00	.32	.23			.12	.17	.13	.05	.10		**1.10**
38	.00	.29	.23		.01	.12	.21	.13	.05	.10		**1.13**
39	.00	.29	.23			.12	.22	.13	.05	.10		**1.13**
40	1.42	.39	1.57		2.34	.12	1.09	2.61	.25	2.03		**11.82**
41	.00	.29	.23			.12	.16	.13	.05	.10		**1.07**
42	.00	1.18	.23		.01	.12	.22	.13	.05	.10		**2.03**
43	.00	.32	.23			.12	.16	.13	.05	.10		**1.09**
44	.00	.29	.29		.58	.12	.19	.13	.05	.10		**1.75**
45	.03	.29	.23	.01	.38	.12	.48	4.00	.06	.33		**5.94**
46	.00	.32	.23		.01	.12	.19	.13	.05	.10		**1.13**
47	.00	.29	.23			.12	.16	.13	.05	.10		**1.07**
48	.02	.29	.29		1.79	.12	.36	.13	.05	.10		**3.14**
49	.00	.33	.23			.12	.19	.13	.05	.10		**1.15**
50	2.00	.29	1.51		.71	.63	11.21	.13	.29	1.70		**18.46**
Total	8.55	18.46	18.66	.02	21.08	6.80	35.38	25.46	3.60	14.37		152.38

A value of "0.00" means a cost of more than $.000 but less than $.005 per gross square foot.

50-Year Total M&R Cost Profile per Sqft.

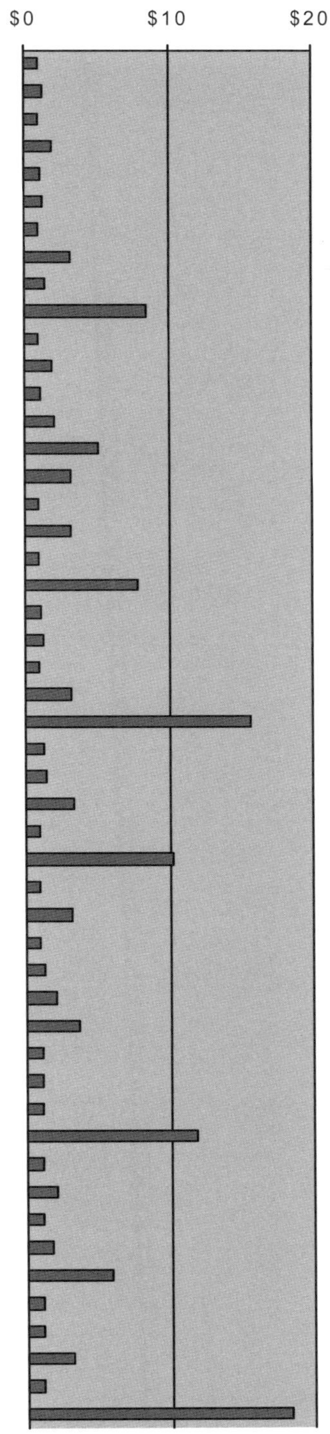

College Student Union

Gross Sqft:	25,000
Height in Ft:	24
Exterior:	Brick Veneer
Floor coverings:	Carpet/Vinyl Tile
HVAC:	Electric Cool, Gas Heat, Multizone Heat
Occupancy:	1250
Replacement Cost:	$2,663,038

50-Year M&R Cost Summary

	50 Year Total	Annual Cost per Sqft	Annual cost as % replacement
PM & Minor Repair	$705,214	$0.56	0.53%
Unscheduled Maintenance	$835,837	$0.67	0.63%
Renewal & Replacement	$2,917,814	$2.33	2.19%
Total M&R Costs	$4,458,865	$3.57	3.35%

Components

Component	Quantity	Unit
B20 Exterior Enclosure		
Clay Brick, Exterior, 1st Floor	8226	Sq Ft
Steel Frame, Painted, Operable Window, 12 sf, 1st Floor	91	Each
Aluminum Frame, Fully Glazed, Exterior Door	4	Each
B30 Roofing		
Concrete Steps	100	Sq Ft
Concrete Decking	400	Sq Ft
Built-up Roof	12500	Sq Ft
C10 Interior Construction		
Steel, Painted, Interior Door	179	Each
C20 Stairs		
Concrete Interior Steps	480	Sq Ft
C30 Interior Finishes		
Sheetrock, Stippled, Interior Wall Finish	21432	Sq Ft
Vinyl Tile Flooring	12500	Sq Ft
Carpet, Nylon 20 oz., High Traffic	12500	Sq Ft
Acoustical Tile Ceiling	25000	Sq Ft
D10 Conveying		
Elevator, Hydraulic, 2500 lbs, 3-5 Floor, 200 fpm	2	Each
D20 Plumbing		
Tankless Water Closet	13	Each
Urinal, Vitreous China	3	Each
Lavatory, Vitreous China	12	Each
Sink, Stainless Steel	6	Each
Drinking Fountain, Refrigerated	3	Each
Pipe & Fittings, 4" Steel	0.994	K Ln Ft
Pipe & Fittings, 3/4" Copper, Cold Water	1.34	K Ln Ft
Pipe & Fittings, 2" Copper, Cold Water	1.419	K Ln Ft
Pipe & Fittings, 3/4" Copper, Hot Water	0.66	K Ln Ft
Pipe Insulation, Cold Water	1.838	K Ln Ft
Pipe Insulation, Hot Water	0.934	K Ln Ft
Water Heater, Gas/Oil 175 Gph	2	Each
Pipe & Fittings, 6" Cast Iron	0.702	K Ln Ft
Pipe & Fittings, 10" Cast Iron	0.224	K Ln Ft
Pipe & Fittings, 4" DWV PVC	0.181	K Ln Ft
Roof Drain, 2"	3	Each
D30 HVAC		
Air Conditioner, Rooftop, 50 Ton	2	Each
D40 Fire Protection		
Fire Sprinkler System	1	Each
Fire Sprinkler Head	180	Each
D50 Electrical		
Safety Switch, Fused, 400 Amp., 3 Ph.	3	Each
Main Switchgear, >1200 Amp.	1	Each
Distribution Panel Board	4	Each
Emergency Horn & Strobe	15	Each
Exit Lighting Fixture, w/ Battery	10	Each
Fluorescent Lighting Fixture, 160w	375	Each
Wiring Device, Switch	50	Each
Receptacle, 120V, 15 Amp	200	Each
TV Cable Outlet	9	Each
Annunciation Panel	1	Each
Fire Alarm Bell, 6"	7	Each
Fire Alarm Control Panel	1	Each
Manual Pull Station	10	Each
Heat Detector	4	Each
Smoke Detector	25	Each

Distribution of Repair Costs

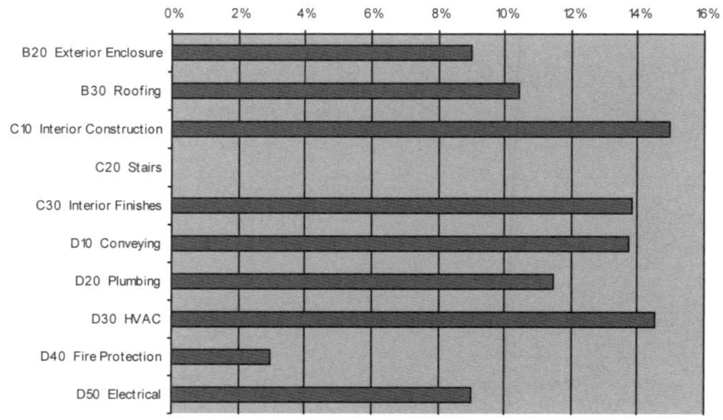

Thirty Most Costly Repair Tasks

Major Repair Task	Task Cost*	Pct.**
Maintain Elevator, Hydraulic, 2500 lbs, 3-5 Floor, 200 fpm	20.75	14.3%
Replace Carpet, Nylon 20 oz., High Traffic	14.51	10.0%
Replace Air Conditioner, Rooftop, 50 Ton	12.34	8.5%
Replace Steel, Painted, Door Locks	9.15	6.3%
Repair Air Conditioner, Rooftop, 50 Ton	7.55	5.2%
Maintain Built-up Roof	7.31	5.0%
Maintain Steel, Painted, Door Locks	6.10	4.2%
Refinish Sheetrock, Stippled, Interior Wall Finish	5.21	3.6%
Clean & Reseal Clay Brick, Exterior, 1st Floor	4.89	3.4%
Replace Ballast & Lamps, Fluorescent Lighting Fixture, 160w	4.13	2.9%
Replace Pipe & Fittings, 2" Copper, Cold Water	3.84	2.7%
Renovate Elevator, Hydraulic, 2500 lbs, 3-5 Floor, 200 fpm	3.73	2.6%
Maintain Air Conditioner, Rooftop, 50 Ton	3.51	2.4%
Fire Sprinkler System, Annual PM	3.46	2.4%
Replace Pipe & Fittings, 3/4" Copper, Cold Water	3.19	2.2%
Replace Fluorescent Lighting Fixture, 160w	2.93	2.0%
Repoint (50% surface) Clay Brick, Exterior, 1st Floor	2.92	2.0%
Replace Water Heater, Gas/Oil 175 Gph	2.34	1.6%
Replace Vinyl Tile Flooring	2.10	1.4%
Minor Repair, Acoustic Tile Ceiling	1.78	1.2%
Place New Membrane Over Existing, Built-up Roof	1.75	1.2%
Remove & Replace Membrane, Built-up Roof	1.73	1.2%
Replace Pipe & Fittings, 3/4" Copper, Hot Water	1.56	1.1%
Refinish Steel, Painted, Interior Door	1.11	0.8%
Clean Water Heater, Gas/Oil 175 Gph	.77	0.5%
Annual PM, Distribution Panel Board	.75	0.5%
Repair Fire Sprinkler Head	.69	0.5%
Check Operation, Smoke Detector	.66	0.5%
Replace Safety Switch, Fused, 400 Amp., 3 Ph.	.65	0.4%
Repair Clay Brick, Exterior, 1st Floor	.63	0.4%

**Task cost ($2004) per gross square foot over 50 years.*

***Percent of total M&R costs.*

Cost per Sqft. by System

Building Age	Exterior Closure	Roofing	Interior Construtions	Stairways	Interior Finishes	Conveying Systems	Plumbing	HVAC Systems	Fire Protection	Electrical	Equipment	Total per Sqft.
1	.01	.29	.33			.42	.07	.13	.08	.10		**1.42**
2	.01	.29	.33		.01	.42	.19	.13	.08	.10		**1.56**
3	.01	.32	.33			.42	.07	.13	.08	.10		**1.45**
4	.01	.29	.42		.45	.42	.19	.13	.08	.10		**2.08**
5	.04	.29	.33			.42	.12	.13	.11	.18		**1.61**
6	.01	.32	.33		.01	.42	.19	.13	.08	.10		**1.58**
7	.01	.32	.33			.42	.08	.13	.08	.10		**1.46**
8	.01	.29	.42		2.85	.42	.54	.13	.08	.11		**4.85**
9	.01	.32	.33		.37	.42	.07	.13	.08	.10		**1.82**
10	2.37	.29	2.16		.01	.42	.40	2.65	.34	1.68		**10.31**
11	.01	.29	.33			.42	.07	.13	.08	.10		**1.42**
12	.04	.32	.42		.45	.42	.19	.13	.08	.10		**2.14**
13	.01	.29	.33			.42	.17	.13	.08	.10		**1.52**
14	.01	1.19	.33		.01	.42	.20	.13	.08	.10		**2.47**
15	.04	.32	.33	.03		.42	.08	4.11	.11	.45		**5.88**
16	.01	.29	.42		2.85	.42	.54	.13	.08	.11		**4.85**
17	.01	.29	.33			.42	.07	.13	.08	.10		**1.42**
18	.01	.32	.33		1.42	.42	.19	.13	.08	.10		**2.99**
19	.01	.29	.33			.42	.07	.13	.08	.10		**1.42**
20	2.40	.41	2.25		.82	.42	.43	.13	.30	2.25		**9.40**
21	.01	.34	.33			.42	.12	.13	.08	.10		**1.52**
22	.01	.29	.33		.01	.42	.19	.13	.08	.10		**1.56**
23	.01	.29	.33			.42	.07	.13	.08	.10		**1.42**
24	.04	.32	.42		2.85	.42	.54	.13	.08	.11		**4.90**
25	1.81	.29	.33			2.28	5.03	2.65	.11	.60		**13.10**
26	.01	.29	.33		.01	.42	.24	.13	.08	.10		**1.60**
27	.01	.32	.33		.37	.42	.07	.13	.08	.10		**1.83**
28	.01	1.75	.42		.45	.42	.19	.13	.08	.10		**3.55**
29	.01	.29	.33			.42	.11	.13	.08	.10		**1.47**
30	2.37	.29	2.16	.03	.01	.42	.39	4.11	.34	1.83		**11.95**
31	.01	.32	.33			.42	.07	.13	.08	.10		**1.45**
32	.01	.29	.42		2.85	.42	.54	.13	.08	.10		**4.84**
33	.01	.29	.33			.42	.07	.13	.08	.11		**1.43**
34	.01	.32	.33		.01	.42	.18	.13	.08	.10		**1.57**
35	.04	.32	.33			.42	.61	.13	.11	.18		**2.13**
36	.04	.29	.42		1.85	.42	.12	.13	.08	.10		**3.45**
37	.01	.32	.33			.42	.17	.13	.08	.10		**1.55**
38	.01	.29	.33		.01	.42	.13	.13	.08	.10		**1.50**
39	.01	.29	.33			.42	.19	.13	.08	.10		**1.54**
40	2.40	.43	2.25		3.22	.42	.66	2.65	.30	2.36		**14.68**
41	.01	.29	.33			.42	.13	.13	.08	.11		**1.49**
42	.01	1.19	.33		.01	.42	.13	.13	.08	.10		**2.40**
43	.01	.32	.33			.42	.13	.13	.08	.10		**1.51**
44	.01	.29	.42		.45	.42	.12	.13	.08	.10		**2.02**
45	.04	.29	.33	.03	.37	.42	.35	4.11	.11	.45		**6.49**
46	.01	.32	.33		.01	.42	.12	.13	.08	.10		**1.51**
47	.01	.29	.33			.42	.13	.13	.08	.10		**1.49**
48	.04	.29	.42		2.85	.42	.48	.13	.08	.10		**4.81**
49	.01	.34	.33			.42	.15	.13	.08	.11		**1.55**
50	4.14	.29	2.16		.01	2.28	5.16	.13	.34	1.88		**16.39**
Total	16.06	18.61	26.71	.10	24.66	24.48	20.48	25.92	5.28	16.06		178.35

50-Year Total M&R Cost Profile per Sqft.

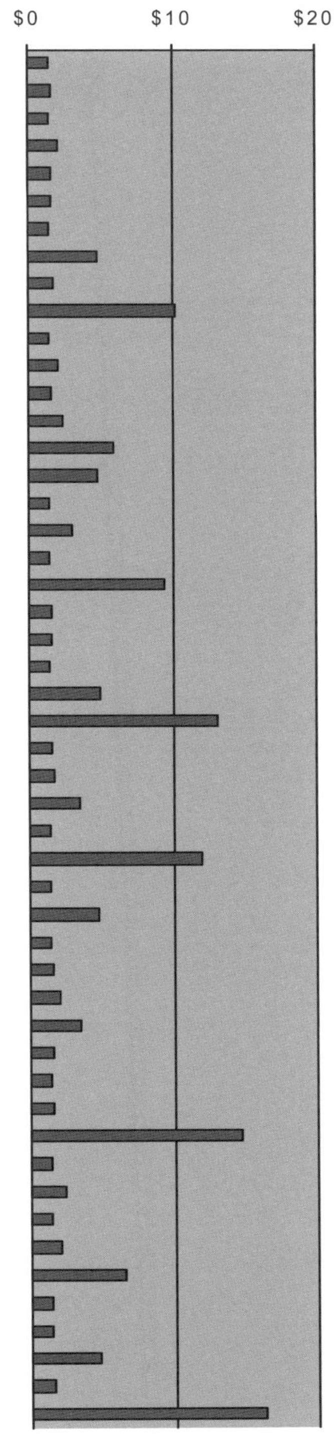

A value of "0.00" means a cost of more than $.000 but less than $.005 per gross square foot.

Community Center

Gross Sqft:	10,000
Height in Ft:	12
Exterior:	Brick Veneer
Floor coverings:	Carpet/Vinyl Tile
HVAC:	Electric Cool, Gas Heat, Multizone Heat
Occupancy:	600
Replacement Cost:	$993,655

Components

Component	Quantity	Unit
B20 Exterior Enclosure		
Clay Brick, Exterior, 1st Floor	4349	Sq Ft
Steel Frame, Painted, Operable Window, 12 sf, 1st Floor	3	Each
Aluminum Frame, Fully Glazed, Exterior Door	4	Each
B30 Roofing		
Concrete Steps	100	Sq Ft
Concrete Decking	400	Sq Ft
Built-up Roof	10000	Sq Ft
C10 Interior Construction		
Steel, Painted, Interior Door	72	Each
C30 Interior Finishes		
Sheetrock, Stippled, Interior Wall Finish	17160	Sq Ft
Vinyl Tile Flooring	5000	Sq Ft
Carpet, Nylon 20 oz., High Traffic	5000	Sq Ft
Acoustical Tile Ceiling	10000	Sq Ft
D20 Plumbing		
Tankless Water Closet	6	Each
Urinal, Vitreous China	2	Each
Lavatory, Vitreous China	7	Each
Sink, Stainless Steel	4	Each
Drinking Fountain, Refrigerated	3	Each
Pipe & Fittings, 3/4" Copper, Cold Water	0.79	K Ln Ft
Pipe & Fittings, 4" Steel	0.9	K Ln Ft
Pipe & Fittings, 2" Copper, Cold Water	0.935	K Ln Ft
Pipe & Fittings, 3/4" Copper, Hot Water	0.35	K Ln Ft
Pipe Insulation, Cold Water	1.24	K Ln Ft
Pipe Insulation, Hot Water	0.6	K Ln Ft
Water Heater, Gas/Oil 175 Gph	2	Each
Pipe & Fittings, 6" Cast Iron	0.43	K Ln Ft
Pipe & Fittings, 10" Cast Iron	0.2	K Ln Ft
Pipe & Fittings, 4" DWV PVC	0.145	K Ln Ft
Roof Drain, 2"	4	Each
Aluminum Gutter, Downspouts, Fittings	0.453	K Ln Ft
D30 HVAC		
Exhaust Fan, Ceiling, 200-500 Cfm	4	Each
Air Conditioner, Rooftop, 50 Ton	1	Each
D40 Fire Protection		
Fire Sprinkler System	1	Each
Fire Sprinkler Head	71	Each
D50 Electrical		
Safety Switch, Fused, 400 Amp., 3 Ph.	1	Each
Main Switchgear, <1200 Amp.	1	Each
Distribution Panel Board	2	Each
Emergency Horn & Strobe	5	Each
Exit Lighting Fixture, w/ Battery	4	Each
Incandescent Lighting Fixture, Basic, 100w	60	Each
Fluorescent Lighting Fixture, 160w	60	Each
Wiring Device, Switch	30	Each
Receptacle, 120V, 15 Amp	25	Each
TV Cable Outlet	1	Each
Annunciation Panel	1	Each
Fire Alarm Bell, 6"	4	Each
Fire Alarm Control Panel	1	Each
Manual Pull Station	4	Each
Smoke Detector	6	Each

50-Year M&R Cost Summary

	50 Year Total	Annual Cost per Sqft	Annual cost as % replacement
PM & Minor Repair	$436,087	$0.87	0.88%
Unscheduled Maintenance	$506,453	$1.01	1.02%
Renewal & Replacement	$1,221,714	$2.44	2.46%
Total M&R Costs	$2,164,254	$4.33	4.36%

Distribution of Repair Costs

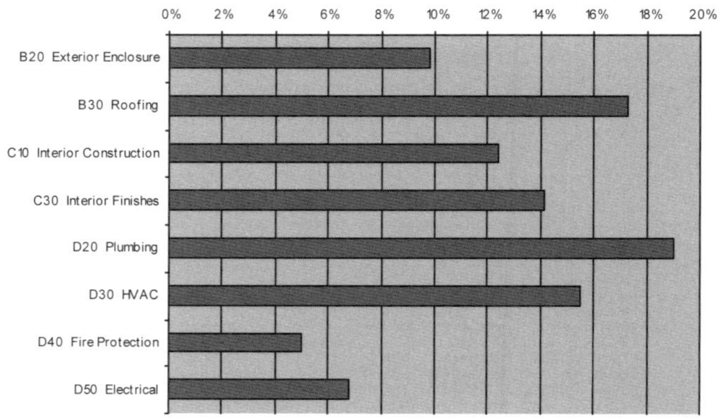

Thirty Most Costly Repair Tasks

Major Repair Task	Task Cost*	Pct.**
Replace Air Conditioner, Rooftop, 50 Ton	15.42	9.3%
Maintain Built-up Roof	14.62	8.8%
Replace Carpet, Nylon 20 oz., High Traffic	14.51	8.8%
Refinish Sheetrock, Stippled, Interior Wall Finish	10.42	6.3%
Repair Air Conditioner, Rooftop, 50 Ton	9.44	5.7%
Replace Steel, Painted, Door Locks	9.20	5.5%
Fire Sprinkler System, Annual PM	8.64	5.2%
Clean & Reseal Clay Brick, Exterior, 1st Floor	6.46	3.9%
Replace Pipe & Fittings, 2" Copper, Cold Water	6.33	3.8%
Maintain Steel, Painted, Door Locks	6.13	3.7%
Replace Water Heater, Gas/Oil 175 Gph	5.84	3.5%
Replace Pipe & Fittings, 3/4" Copper, Cold Water	4.71	2.8%
Maintain Air Conditioner, Rooftop, 50 Ton	4.39	2.6%
Repoint (50% surface) Clay Brick, Exterior, 1st Floor	3.86	2.3%
Place New Membrane Over Existing, Built-up Roof	3.49	2.1%
Remove & Replace Membrane, Built-up Roof	3.45	2.1%
Replace Vinyl Tile Flooring	2.10	1.3%
Replace Pipe & Fittings, 3/4" Copper, Hot Water	2.07	1.3%
Clean Water Heater, Gas/Oil 175 Gph	1.92	1.2%
Minor Repair, Acoustic Tile Ceiling	1.78	1.1%
Replace Ballast & Lamps, Fluorescent Lighting Fixture, 160w	1.65	1.0%
Replace Fluorescent Lighting Fixture, 160w	1.17	0.7%
Refinish Steel, Painted, Interior Door	1.12	0.7%
Replace Drinking Fountain, Refrigerated	1.08	0.6%
Replace Incandescent Lighting Fixture, Basic, 100w	1.05	0.6%
Minor Repair, Sheetrock, Stippled, Interior Wall Finish	1.02	0.6%
Annual PM, Distribution Panel Board	.94	0.6%
Repair Clay Brick, Exterior, 1st Floor	.83	0.5%
Replace Pipe Insulation, Cold Water	.82	0.5%
Non-Destructive Moisture Inspection	.81	0.5%

*Task cost ($2004) per gross square foot over 50 years.

**Percent of total M&R costs.

Cost per Sqft. by System

Building Age	Exterior Closure	Roofing	Interior Construtiion	Stairways	Interior Finishes	Conveying Systems	Plumbing	HVAC Systems	Fire Protection	Electrical	Equipment	Total per Sqft.
1	.02	.59	.33				.21	.18	.18	.12		**1.62**
2	.02	.59	.33		.01		.43	.18	.18	.12		**1.85**
3	.02	.64	.33				.21	.18	.18	.12		**1.67**
4	.02	.59	.42		.88		.43	.18	.18	.12		**2.81**
5	.02	.59	.33				.33	.18	.25	.35		**2.05**
6	.02	.64	.33		.01		.43	.18	.18	.12		**1.90**
7	.02	.64	.33				.23	.18	.18	.12		**1.70**
8	.02	.59	.42		3.29		1.31	.18	.18	.12		**6.10**
9	.02	.64	.33		.37		.21	.18	.18	.12		**2.04**
10	3.10	.59	2.17		.01		.77	3.32	.48	1.02		**11.47**
11	.02	.59	.33				.21	.18	.18	.12		**1.62**
12	.10	.64	.42		.88		.43	.18	.18	.12		**2.95**
13	.02	.59	.33				.43	.18	.18	.12		**1.84**
14	.02	2.39	.33		.01		.45	.18	.18	.12		**3.68**
15	.02	.64	.33				.23	5.16	.25	.73		**7.36**
16	.02	.59	.42		3.29		1.31	.18	.18	.12		**6.10**
17	.02	.59	.33				.21	.18	.18	.12		**1.62**
18	.02	.64	.33		1.42		.43	.18	.18	.12		**3.31**
19	.02	.59	.33				.21	.18	.18	.12		**1.62**
20	3.10	.87	2.26		1.62		1.19	.35	.44	1.80		**11.63**
21	.02	.70	.33				.32	.18	.18	.12		**1.84**
22	.02	.59	.33		.01		.43	.18	.18	.12		**1.85**
23	.02	.59	.33				.21	.18	.18	.12		**1.62**
24	.10	.64	.42		3.29		1.31	.18	.18	.12		**6.23**
25	2.36	.59	.33				7.98	3.32	.25	.73		**15.57**
26	.02	.59	.33		.01		.53	.18	.18	.12		**1.95**
27	.02	.64	.33		.37		.22	.18	.18	.12		**2.05**
28	.02	3.51	.42		.88		.44	.18	.18	.12		**5.75**
29	.02	.59	.33				.32	.18	.18	.12		**1.72**
30	3.10	.59	2.17		.01		.76	5.16	.48	1.28		**13.54**
31	.02	.64	.33				.22	.18	.18	.12		**1.68**
32	.02	.59	.42		3.29		1.30	.18	.18	.12		**6.09**
33	.02	.59	.33				.22	.18	.18	.12		**1.63**
34	.02	.64	.33		.01		.42	.18	.18	.12		**1.89**
35	.02	.64	.33				.93	.18	.25	.35		**2.71**
36	.10	.59	.42		2.29		.33	.18	.18	.12		**4.20**
37	.02	.64	.33				.40	.18	.18	.12		**1.86**
38	.02	.59	.33		.01		.34	.18	.18	.12		**1.77**
39	.02	.59	.33				.42	.18	.18	.12		**1.83**
40	3.10	.92	2.26		4.02		1.95	3.49	.44	1.92		**18.10**
41	.02	.59	.33				.31	.18	.18	.12		**1.72**
42	.02	2.39	.33		.01		.35	.18	.18	.12		**3.58**
43	.02	.64	.33				.31	.18	.18	.12		**1.77**
44	.02	.59	.42		.88		.33	.18	.18	.12		**2.72**
45	.02	.59	.33		.37		.64	5.16	.25	.73		**8.09**
46	.02	.64	.33		.01		.33	.18	.18	.12		**1.81**
47	.02	.59	.33				.31	.18	.18	.12		**1.71**
48	.10	.59	.42		3.29		1.21	.18	.18	.12		**6.08**
49	.02	.70	.33				.33	.18	.18	.12		**1.85**
50	5.44	.59	2.17		.01		8.26	.18	.48	1.16		**18.30**
Total	21.29	37.41	26.86		30.61		41.11	33.58	10.86	14.71		216.43

50-Year Total M&R Cost Profile per Sqft.

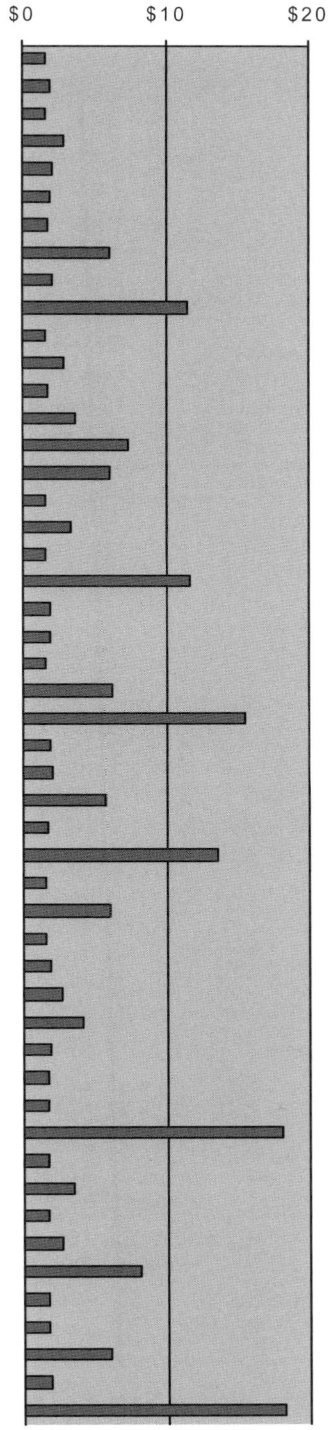

A value of "0.00" means a cost of more than $.000 but less than $.005 per gross square foot.

County Jail

Gross Sqft:	318,455
Height in Ft:	173
Exterior:	Cast Concrete
Floor coverings:	Finished Concrete/Vinyl Tile/Carpet
HVAC:	Chilled Water, Gas Boiler, Multizone Air
Occupancy:	1600
Replacement Cost:	$126,898,222

Components

Component	Quantity	Unit
B20 Exterior Enclosure		
Concrete, Painted, Exterior, 3rd+ Floor	125000	Sq Ft
Aluminum Fixed Window, 24 sf, 3rd+ Floor	910	Each
Steel, Painted, Exterior Door	36	Each
Aluminum Frame, Fully Glazed, Exterior Door	4	Each
Steel Single 12'x12', Painted, Roll-up Door	2	Each
B30 Roofing		
Built-up Roof	35700	Sq Ft
C10 Interior Construction		
Wood, Solid Core, Painted, Interior Door	83	Each
Steel, Painted, Interior Door	631	Each
C20 Stairs		
Concrete Interior Steps	4800	Sq Ft
Metal Interior Railing, Painted	1280	Ln Ft
C30 Interior Finishes		
Sheetrock, Unstippled, Interior Wall Finish	105160	Sq Ft
Concrete, Painted, Interior Wall Finish	312340	Sq Ft
Ceramic Tile Wall Finish	151280	Sq Ft
Concrete, Finished Flooring	179110	Sq Ft
Vinyl Tile Flooring	82210	Sq Ft
Carpet, Nylon 20 oz., High Traffic	38280	Sq Ft
Concrete, Finished Ceiling	104600	Sq Ft
Sheetrock, Unstippled, Ceiling	29315	Sq Ft
Acoustical Tile Ceiling	114300	Sq Ft
D10 Conveying		
Elevator, Geared, 3500 lbs, 6-15 Floor, 350 fpm	7	Each
D20 Plumbing		
Tankless Water Closet	69	Each
Urinal, Vitreous China	13	Each
Toilet & Wash Basin, Stainless Steel, Detention Fixture	806	Each
Lavatory, Vitreous China	128	Each
Service Sink, Iron Enamel	33	Each
Shower, Fiberglass	84	Each
Drinking Fountain, Refrigerated	4	Each
Booster Pump, 10 HP	3	Each
Pipe & Fittings, 3/4" Copper, Cold Water	34	K Ln Ft
Pipe & Fittings, 2" Copper, Cold Water	12	K Ln Ft
Pipe & Fittings, 3/4" Copper, Hot Water	8	K Ln Ft
Water Storage Tank, 500 Gal.	3	Each
Circulator Pump, 1 HP, Hot Water	2	Each
Water Softener	1	Each
Water Heater, Electric 1500 Gph	2	Each
Pipe & Fittings, 6" Cast Iron	27	K Ln Ft
Pipe & Fittings, 10" Cast Iron	1	K Ln Ft
Sewage Ejector Pump, 1 HP	2	Each
Pipe & Fittings, 4" DWV PVC	2	K Ln Ft
Roof Drain, 4-6"	17	Each
Sump Pump, 3 HP	2	Each
Air Compressor, Gas, 7.5 HP	2	Each
Compressed Air Dryer	2	Each
D30 HVAC		
Pipe & Fittings, 4" Steel, Gas	1	K Ln Ft
Fuel Oil Pump, 1.5 HP	1	Each
Circulation Pump 25 HP, Hot Water	2	Each
Gas Boiler, 2500 Mbh	2	Each
Chemical Feed System	5	Each
Expansion Tank, 60 Gal.	3	Each
Circulation Pump, 7.5 HP, Chiller & Condenser Water	4	Each
Circulation Pump, 25 HP, Chiller & Condenser Water	4	Each
Chiller, Centrifugal Hermetic, Bundled w/ Condenser, 300 Ton	2	Each
Cooling Tower, 300 Ton	2	Each
Air Handler, Multizone, 2500 Cfm	19	Each
Air Handler, Multizone, 10,000 Cfm	6	Each
Air Handler, Multizone, 15,000 Cfm	2	Each
Supply Fan, Roof Mounted, 5000 Cfm	24	Each
Supply Fan, Roof Mounted, 10,000 Cfm	3	Each
Fan, Axial Flow, 5000 Cfm	23	Each
Exhaust Fan, Roof Mounted, 5000 Cfm	9	Each
Exhaust Fan, Roof Mounted, 10,000 Cfm	3	Each
Exhaust Fan, Ceiling, 200-500 Cfm	12	Each
D40 Fire Protection		
Fire Sprinkler Head	3500	Each
D50 Electrical		
Primary Transformer, Dry, >500 KVA	1	Each
Secondary Transformer, Dry, 150 KVA	4	Each
Main Switchgear, <1200 Amp.	5	Each
Fluorescent Lighting Fixture, 160w	2891	Each
Metal Halide Lighting Fixture, Low Bay, 250w	75	Each
Wiring Device, Switch	2000	Each
Receptacle, 120V, 15 Amp	1109	Each
Annunciation Panel	1	Each
Fire Alarm Bell, 6"	2	Each
Fire Alarm Control Panel	23	Each

50-Year M&R Cost Summary

	50 Year Total	Annual Cost per Sqft	Annual cost as % replacement
PM & Minor Repair	$5,391,484	$0.34	0.08%
Unscheduled Maintenance	$5,209,861	$0.33	0.08%
Renewal & Replacement	$30,091,556	$1.89	0.47%
Total M&R Costs	$40,692,901	$2.56	0.64%

Distribution of Repair Costs

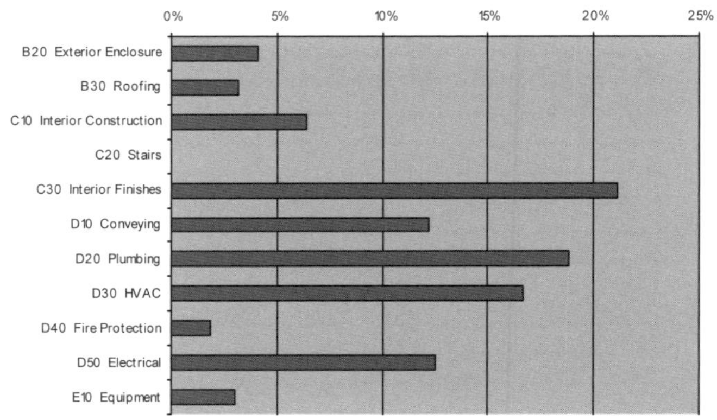

Thirty Most Costly Repair Tasks

Major Repair Task	Task Cost*	Pct.**
Maintain Elevator, Geared, 3500 lbs, 6-15 Floor, 350 fpm	12.28	11.0%
Refinish Concrete, Painted, Interior Wall Finish	7.53	6.8%
Replace Pipe & Fittings, 3/4" Copper, Cold Water	6.36	5.7%
Replace Toilet & Wash Basin, Stainless Steel, Detention Fixture	5.86	5.3%
Minor Repair, Concrete, Painted, Interior Wall Finish	5.73	5.1%
Refinish Concrete, Painted, Exterior, 3rd+ Floor	4.59	4.1%
Replace Carpet, Nylon 20 oz., High Traffic	3.49	3.1%
Renovate Elevator, Geared, 3500 lbs, 6-15 Floor, 350 fpm	3.30	3.0%
Replace Generator, Diesel, 1000kw	3.10	2.8%
Minor Repair, Ceramic Tile Wall Finish	2.73	2.5%
Replace Pipe & Fittings, 2" Copper, Cold Water	2.55	2.3%
Replace Steel, Painted, Door Locks	2.53	2.3%
Replace Ballast & Lamps, Fluorescent Lighting Fixture, 160w	2.50	2.2%
Replace Chiller, Centrifugal Hermetic, Bnd'd w/ Condenser, 300 Ton	2.36	2.1%
Refinish Sheetrock, Unstippled, Interior Wall Finish	2.01	1.8%
Repair Chiller, Centrifugal Hermetic, Bnd'd w/ Condenser, 300 Ton	2.00	1.8%
Maintain & Repair Detention Doors & Hardware	1.95	1.7%
Check Operation, Smoke Detector	1.90	1.7%
Replace Fluorescent Lighting Fixture, 160w	1.78	1.6%
Maintain Steel, Painted, Door Locks	1.69	1.5%
Maintain Air Handler, Multizone, 2500 Cfm	1.64	1.5%
Maintain Built-up Roof	1.64	1.5%
Replace Pipe & Fittings, 3/4" Copper, Hot Water	1.49	1.3%
Replace Water Heater, Electric 1500 Gph	1.20	1.1%
Replace Air Handler, Multizone, 2500 Cfm	1.11	1.0%
Replace Vinyl Tile Flooring	1.08	1.0%
Repair Fire Sprinkler Head	1.05	0.9%
Replace Smoke Detector	1.00	0.9%
Maintain Fan, Axial Flow, 5000 Cfm	1.00	0.9%
Maintain Chiller, Centrifugal Hermetic, Bnd'd w/ Condenser, 300 Ton	.90	0.8%

**Task cost ($2004) per gross square foot over 50 years.*

***Percent of total M&R costs.*

Cost per Sqft. by System

Building Age	Exterior Closure	Roofing	Interior Constrution	Stairways	Interior Finishes	Conveying Systems	Plumbing	HVAC Systems	Fire Protection	Electrical	Equipment	Total per Sqft.
1	.01	.07	.10			.25	.03	.24	.01	.10	.08	**.88**
2	.01	.07	.10		.00	.25	.10	.24	.01	.10	.08	**.96**
3	.01	.07	.10			.25	.03	.24	.01	.10	.08	**.88**
4	.01	.07	.13		.84	.25	.12	.24	.01	.10	.08	**1.84**
5	.01	.07	.10			.25	.07	.24	.01	.14	.08	**.97**
6	.01	.07	.10		.00	.25	.10	.24	.01	.10	.08	**.96**
7	.01	.07	.10	.00		.25	.04	.26	.01	.10	.08	**.91**
8	.01	.07	.13		1.42	.25	.13	.24	.01	.10	.08	**2.43**
9	.01	.07	.10		.14	.25	.03	.24	.01	.10	.08	**1.02**
10	.93	.07	.63		.55	.25	.39	1.08	.37	1.27	.08	**5.61**
11	.01	.07	.10			.25	.03	.24	.01	.10	.08	**.88**
12	.01	.07	.13		.84	.25	.12	.24	.01	.10	.08	**1.85**
13	.01	.07	.10			.25	.10	.24	.01	.10	.08	**.95**
14	.04	.26	.10	.00	.00	.25	.11	.26	.01	.10	.08	**1.22**
15	.01	.07	.10	.03	2.47	.25	.48	1.20	.01	.73	.08	**5.43**
16	.01	.07	.13		1.42	.25	.13	.24	.01	.10	.08	**2.43**
17	.01	.07	.10			.25	.03	.24	.01	.10	.08	**.88**
18	.01	.07	.10		.67	.25	.10	.24	.01	.10	.08	**1.63**
19	.01	.07	.10			.25	.03	.24	.01	.10	.08	**.88**
20	1.00	.07	.66		1.60	.25	.57	2.07	.31	1.23	.08	**7.84**
21	.01	.07	.10	.00		.25	.04	.26	.01	.10	.08	**.91**
22	.01	.07	.10		.00	.25	.10	.24	.01	.10	.08	**.96**
23	.01	.07	.10			.25	.03	.24	.01	.10	.08	**.88**
24	.01	.07	.13		1.42	.25	.13	.24	.01	.10	.08	**2.44**
25	.01	.07	.10		.53	1.89	5.60	.33	.01	1.94	.08	**10.55**
26	.01	.07	.10		.00	.25	.17	.24	.01	.10	.08	**1.02**
27	.01	.07	.10		.14	.25	.03	.24	.01	.10	.08	**1.02**
28	.05	.39	.13	.00	.84	.25	.11	.26	.01	.10	.08	**2.22**
29	.01	.07	.10			.25	.04	.24	.01	.10	.08	**.89**
30	.93	.07	.70	.03	3.02	.25	.80	2.13	.37	1.79	.08	**10.15**
31	.01	.07	.10			.25	.03	.24	.01	.10	.08	**.88**
32	.01	.07	.13		1.42	.25	.11	.24	.01	.10	.08	**2.41**
33	.01	.07	.10			.25	.04	.24	.01	.10	.08	**.89**
34	.01	.07	.11		.00	.25	.10	.24	.01	.10	.08	**.97**
35	.02	.07	.10	.00		.25	.40	.24	.01	.14	.08	**1.31**
36	.01	.07	.13		1.51	.25	.05	.24	.01	.10	.08	**2.44**
37	.01	.07	.10			.25	.10	.26	.01	.10	.08	**.97**
38	.01	.07	.11		.00	.25	.06	.24	.01	.10	.08	**.91**
39	.01	.07	.10			.25	.15	.24	.01	.10	.08	**.99**
40	1.00	.07	.66		2.18	.25	6.20	2.16	.31	1.49	.08	**14.38**
41	.01	.07	.10			.25	.10	.24	.01	.10	.08	**.95**
42	.04	.26	.11	.00	.00	.25	.06	.24	.01	.10	.08	**1.15**
43	.01	.07	.10			.25	.09	.24	.01	.10	.08	**.94**
44	.01	.07	.13		.84	.25	.05	.26	.01	.10	.08	**1.79**
45	.01	.07	.10	.03	2.61	.25	.68	1.20	.01	.73	.08	**5.77**
46	.01	.07	.11		.00	.25	.05	.24	.01	.10	.08	**.91**
47	.01	.07	.10			.25	.08	.24	.01	.10	.08	**.93**
48	.01	.07	.13		1.42	.25	.05	.24	.01	.10	.08	**2.35**
49	.01	.07	.10	.00		.25	.11	.24	.01	.10	.08	**.96**
50	.93	.07	.64		1.08	1.89	5.79	1.00	.37	2.55	.08	**14.39**
Total	5.24	4.08	8.22	.10	27.03	15.58	24.06	21.27	2.35	15.98	3.87	127.78

A value of "0.00" means a cost of more than $.000 but less than $.005 per gross square foot.

50-Year Total M&R Cost Profile per Sqft.

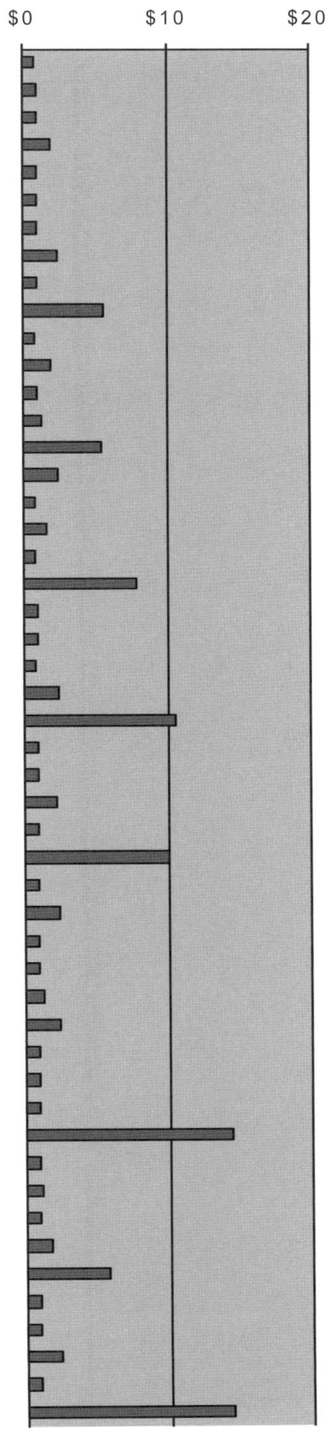

Components

Manual Pull Station	32	Each
Heat Detector	10	Each
Smoke Detector	911	Each
Generator, Diesel, 1000kw	2	Each
E10 Equipment		
Detention Doors & Hardware	831	Each

Court House, 1 Story

Gross Sqft:	30,000
Height in Ft:	14
Exterior:	Limestone Panels
Floor coverings:	Carpet/Hardwood/Terrazzo Tile
HVAC:	Electric Cool, Gas Heat, Multizone Unit
Occupancy:	1050
Replacement Cost:	$4,068,162

Components

Component	Quantity	Unit
B20 Exterior Enclosure		
Concrete Block, Exterior, 1st Floor	8621	Sq Ft
Steel Frame, Painted, Operable Window, 12 sf, 1st Floor	111	Each
Wood, Solid Core, Painted, Exterior Door	7	Each
B30 Roofing		
Built-up Roof	30000	Sq Ft
Concrete Decking	1000	Sq Ft
Concrete Steps	200	Sq Ft
C10 Interior Construction		
Wood, Solid Core, Painted, Interior Door	300	Each
C30 Interior Finishes		
Sheetrock, Stippled, Interior Wall Finish	29400	Sq Ft
Wallpaper Interior Wall Finish	4200	Sq Ft
Vinyl Interior Wall Finish	8400	Sq Ft
Terrazzo Flooring	6000	Sq Ft
Wood Parquetry Flooring	18000	Sq Ft
Carpet, Nylon 20 oz., High Traffic	6000	Sq Ft
Plaster Ceiling	30000	Sq Ft
D20 Plumbing		
Tankless Water Closet	14	Each
Urinal, Vitreous China	6	Each
Lavatory, Vitreous China	22	Each
Sink, Stainless Steel	8	Each
Drinking Fountain, Refrigerated	10	Each
Pipe & Fittings, 3/4" Copper, Cold Water	2.02	K Ln Ft
Pipe & Fittings, 3/4" Copper, Hot Water	0.94	K Ln Ft
Pipe & Fittings, 2" Copper, Cold Water	2.511	K Ln Ft
Pipe & Fittings, 4" Steel	1.486	K Ln Ft
Pipe Insulation, Cold Water	2.762	K Ln Ft
Pipe Insulation, Hot Water	1.336	K Ln Ft
Water Heater, Gas/Oil 175 Gph	2	Each
Pipe & Fittings, 6" Cast Iron	0.943	K Ln Ft
Pipe & Fittings, 10" Cast Iron	0.346	K Ln Ft
Roof Drain, 2"	8	Each
Pipe & Fittings, 4" DWV PVC	0.434	K Ln Ft
D30 HVAC		
Air Conditioner, Rooftop, 50 Ton	2	Each
D40 Fire Protection		
Fire Sprinkler Head	215	Each
D50 Electrical		
Safety Switch, Fused, 400 Amp., 3 Ph.	2	Each
Distribution Panel Board	3	Each
Main Switchgear, >1200 Amp.	1	Each
Exit Lighting Fixture, w/ Battery	7	Each
Emergency Horn & Strobe	10	Each
Fluorescent Lighting Fixture, 160w	300	Each
Wiring Device, Switch	60	Each
Grounded Fault Circuit Interruptor, 15 Amps	4	Each
Receptacle, 120V, 15 Amp	150	Each
TV Cable Outlet	1	Each
Annunciation Panel	1	Each
Fire Alarm Bell, 6"	7	Each
Fire Alarm Control Panel	1	Each
Manual Pull Station	4	Each
Heat Detector	10	Each
Smoke Detector	20	Each
Generator Switchgear	1	Each

50-Year M&R Cost Summary

	50 Year Total	Annual Cost per Sqft	Annual cost as % replacement
PM & Minor Repair	$1,148,047	$0.77	0.56%
Unscheduled Maintenance	$1,358,524	$0.91	0.67%
Renewal & Replacement	$2,993,203	$2.00	1.47%
Total M&R Costs	$5,499,774	$3.67	2.70%

Distribution of Repair Costs

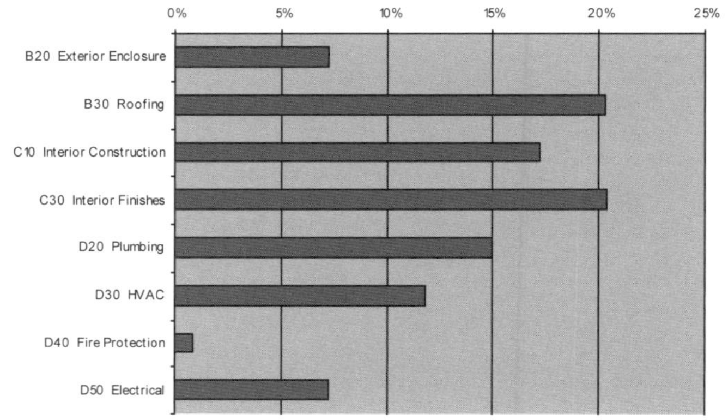

Thirty Most Costly Repair Tasks

Major Repair Task	Task Cost*	Pct.**
Maintain Built-up Roof	14.62	10.6%
Replace Air Conditioner, Rooftop, 50 Ton	10.28	7.4%
Maintain Wood, Solid Core, Painted, Door Locks	8.34	6.0%
Replace Wood Parquetry Floor	6.99	5.1%
Repair Air Conditioner, Rooftop, 50 Ton	6.30	4.6%
Refinish Plaster Ceiling	6.10	4.4%
Refinish Sheetrock, Stippled, Interior Wall Finish	5.95	4.3%
Replace Carpet, Nylon 20 oz., High Traffic	5.80	4.2%
Replace Pipe & Fittings, 2" Copper, Cold Water	5.67	4.1%
Minor Repair, Plaster Ceiling	4.95	3.6%
Clean & Seal Concrete Block, Exterior, 1st Floor	4.29	3.1%
Replace Pipe & Fittings, 3/4" Copper, Cold Water	4.01	2.9%
Replace Wood, Solid Core, Painted, Interior Door	3.73	2.7%
Replace Solid Core, Painted, Door Locks	3.71	2.7%
Place New Membrane Over Existing, Built-up Roof	3.49	2.5%
Remove & Replace Membrane, Built-up Roof	3.45	2.5%
Maintain Air Conditioner, Rooftop, 50 Ton	2.93	2.1%
Replace Ballast & Lamps, Fluorescent Lighting Fixture, 160w	2.76	2.0%
Replace Vinyl Interior Wall Finish	2.31	1.7%
Replace Fluorescent Lighting Fixture, 160w	1.96	1.4%
Replace Water Heater, Gas/Oil 175 Gph	1.95	1.4%
Replace Pipe & Fittings, 3/4" Copper, Hot Water	1.86	1.3%
Sand & Refinish Wood Parquetry Flooring	1.81	1.3%
Refinish Wood, Solid Core, Painted, Interior Door	1.48	1.1%
Repoint (50% surface) Concrete Block, Exterior, 1st Floor	1.24	0.9%
Replace Drinking Fountain, Refrigerated	1.20	0.9%
Replace Wallpaper Interior Wall Finish	1.08	0.8%
Annual PM, Generator Switchgear	.90	0.7%
Replace Coolant & Adjust Drinking Fountain, Refrigerated	.84	0.6%
Non-Destructive Moisture Inspection	.81	0.6%

**Task cost ($2004) per gross square foot over 50 years.*

***Percent of total M&R costs.*

Cost per Sqft. by System

Building Age	Exterior Closure	Roofing	Interior Constrution	Stairways	Interior Finishes	Conveying Systems	Plumbing	HVAC Systems	Fire Protection	Electrical	Equipment	Total per Sqft.
1	.01	.59	.46				.08	.11	.01	.11		**1.36**
2	.01	.59	.46		.01		.31	.11	.01	.11		**1.59**
3	.01	.64	.46				.08	.11	.01	.11		**1.41**
4	.01	.59	.58		1.01		.31	.11	.01	.11		**2.72**
5	.04	.59	.46		.11		.13	.11	.01	.17		**1.62**
6	.01	.64	.46		.01		.31	.11	.01	.11		**1.64**
7	.05	.63	.46				.10	.11	.01	.11		**1.47**
8	.01	.59	.58		1.97		.61	.11	.01	.11		**3.99**
9	.01	.64	.46				.08	.11	.01	.11		**1.41**
10	2.08	.59	1.39		1.14		.62	2.21	.24	1.18		**9.44**
11	.01	.59	.46				.08	.11	.01	.11		**1.36**
12	.01	.64	.58		2.38		.31	.11	.01	.11		**4.14**
13	.01	.59	.46				.19	.11	.01	.11		**1.47**
14	.05	2.38	.46		.01		.33	.11	.01	.11		**3.44**
15	.04	.64	.46		.22		.10	3.43	.01	.38		**5.27**
16	.01	.59	.58		1.97		.61	.11	.01	.11		**3.99**
17	.01	.59	.46				.08	.11	.01	.11		**1.36**
18	.01	.64	.46		.01		.31	.11	.01	.33		**1.86**
19	.01	.59	.46				.08	.11	.01	.11		**1.36**
20	2.12	.81	1.51		2.56		.67	.11	.20	1.54		**9.52**
21	.05	.68	.46				.13	.11	.01	.11		**1.55**
22	.01	.59	.46		.01		.31	.11	.01	.11		**1.59**
23	.01	.59	.46				.08	.11	.01	.11		**1.36**
24	.01	.64	.58		3.34		.61	.11	.01	.11		**5.41**
25	.96	.59	.46		.11		6.67	2.21	.01	.43		**11.44**
26	.01	.59	.46		.01		.36	.11	.01	.11		**1.65**
27	.01	.64	.46				.09	.11	.01	.11		**1.42**
28	.05	3.50	.58		1.01		.32	.11	.01	.11		**5.69**
29	.01	.59	.46				.12	.11	.01	.11		**1.40**
30	2.08	.59	3.86		1.24		.61	3.43	.24	1.30		**13.35**
31	.01	.64	.46				.09	.11	.01	.11		**1.42**
32	.01	.59	.46		1.97		.60	.11	.01	.11		**3.85**
33	.01	.59	.46				.09	.11	.01	.11		**1.37**
34	.01	.64	.58		.01		.30	.11	.01	.11		**1.76**
35	.08	.63	.46		.11		.73	.11	.01	.17		**2.30**
36	.01	.59	.46		2.38		.21	.11	.01	.33		**4.09**
37	.01	.64	.46				.21	.11	.01	.11		**1.54**
38	.01	.59	.58		.01		.22	.11	.01	.11		**1.63**
39	.01	.59	.46				.25	.11	.01	.11		**1.53**
40	2.28	.86	1.39		10.06		.81	2.21	.20	1.63		**19.44**
41	.01	.59	.46				.18	.11	.01	.11		**1.46**
42	.01	2.38	.58		.01		.23	.11	.01	.11		**3.43**
43	.01	.64	.46				.18	.11	.01	.11		**1.51**
44	.01	.59	.46		1.01		.21	.11	.01	.11		**2.50**
45	.04	.59	.46		.22		.42	3.43	.01	.38		**5.54**
46	.01	.64	.58		.01		.21	.11	.01	.11		**1.67**
47	.05	.59	.46				.18	.11	.01	.11		**1.50**
48	.01	.59	.46		3.34		.51	.11	.01	.11		**5.13**
49	.01	.68	.46				.20	.11	.01	.11		**1.58**
50	3.00	.59	1.51		1.14		6.96	.11	.24	1.26		**14.80**
Total	13.36	37.21	31.58		37.34		27.42	21.60	1.53	13.28		183.33

50-Year Total M&R Cost Profile per Sqft.

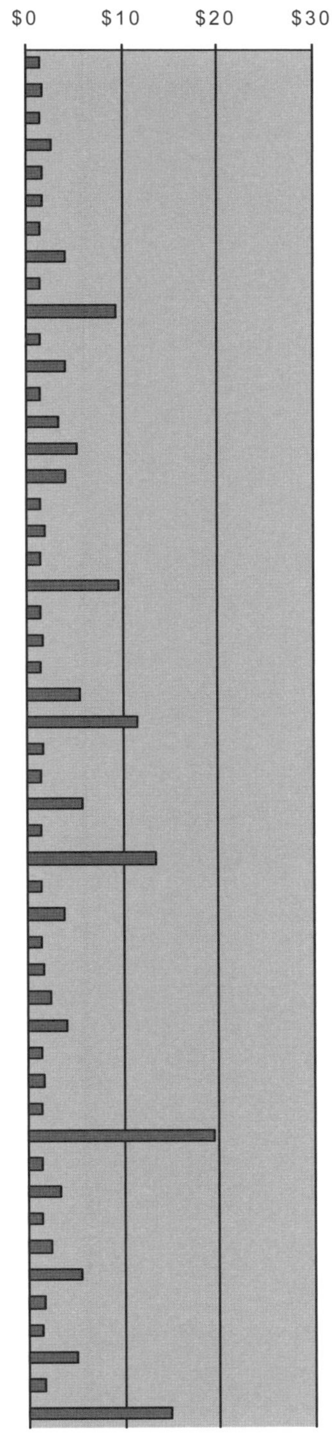

A value of "0.00" means a cost of more than $.000 but less than $.005 per gross square foot.

Department Store

Gross Sqft:	94,000
Height in Ft:	16
Exterior:	Clay Brick
Floor coverings:	Finished Concrete/Terrazzo/Carpet
HVAC:	Single Zone Rooftop Units
Occupancy:	1500
Replacement Cost:	$6,502,987

Components

Component	Quantity	Unit
B20 Exterior Enclosure		
Clay Brick, Exterior, 1st Floor	17642	Sq Ft
Aluminum Fixed Window, 24 sf, 1st Floor	84	Each
Steel w/ Safety Glass, Painted, Exterior Door	4	Each
Steel, Painted, Exterior Door	8	Each
Aluminum Frame, Fully Glazed, Exterior Door	10	Each
Steel Single 12'x12', Painted, Roll-up Door	4	Each
B30 Roofing		
Built-up Roof	94000	Sq Ft
C10 Interior Construction		
Wood, Solid Core, Painted, Interior Door	150	Each
Steel, Painted, Interior Door	6	Each
C30 Interior Finishes		
Sheetrock, Unstippled, Interior Wall Finish	56800	Sq Ft
Concrete Block, Painted, Interior Wall Finish	2200	Sq Ft
Concrete, Finished Flooring	4700	Sq Ft
Terrazzo Flooring	51700	Sq Ft
Carpet, Nylon 20 oz., High Traffic	37600	Sq Ft
Sheetrock, Unstippled, Ceiling	18800	Sq Ft
Acoustical Tile, Dropped, Ceiling	75200	Sq Ft
D20 Plumbing		
Tankless Water Closet	24	Each
Urinal, Vitreous China	6	Each
Lavatory, Vitreous China	24	Each
Service Sink, Iron Enamel	3	Each
Drinking Fountain, Refrigerated	3	Each
Water Heater, Gas/Oil 65 Gph	3	Each
Pipe & Fittings, 3/4" Copper, Cold Water	1.9	K Ln Ft
Pipe & Fittings, 2" Copper, Cold Water	2.2	K Ln Ft
Pipe & Fittings, 3/4" Copper, Hot Water	0.9	K Ln Ft
Pipe Insulation, Cold Water	1.9	K Ln Ft
Pipe Insulation, Hot Water	0.9	K Ln Ft
Pipe & Fittings, 6" Cast Iron	1.4	K Ln Ft
Pipe & Fittings, 10" Cast Iron	0.7	K Ln Ft
Pipe & Fittings, 4" DWV PVC	0.9	K Ln Ft
Roof Drain, 4-6"	27	Each
D30 HVAC		
Pipe & Fittings, 4" Steel, Gas	0.7	K Ln Ft
Air Conditioner, Rooftop, 50 Ton	6	Each
D40 Fire Protection		
Fire Sprinkler Head	470	Each
D50 Electrical		
Main Switchgear, >1200 Amp.	1	Each
Distribution Switch, Fused, <600 V	16	Each
Fluorescent Lighting Fixture, 160w	1470	Each
Wiring Device, Switch	104	Each
Receptacle, 120V, 15 Amp	415	Each
Annunciation Panel	1	Each
Fire Alarm Bell, 6"	20	Each
Fire Alarm Control Panel	1	Each
Manual Pull Station	20	Each
Smoke Detector	190	Each

50-Year M&R Cost Summary

	50 Year Total	Annual Cost per Sqft	Annual cost as % replacement
PM & Minor Repair	$2,389,583	$0.51	0.73%
Unscheduled Maintenance	$2,511,374	$0.53	0.77%
Renewal & Replacement	$5,815,407	$1.24	1.79%
Total M&R Costs	$10,716,364	$2.28	3.30%

Distribution of Repair Costs

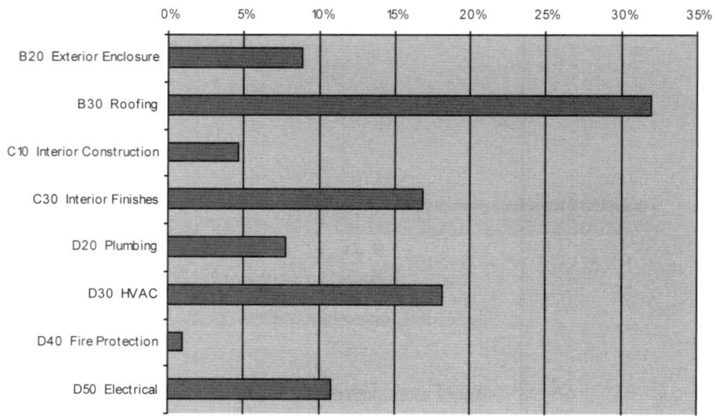

Thirty Most Costly Repair Tasks

Major Repair Task	Task Cost*	Pct.**
Maintain Built-up Roof	14.62	16.7%
Replace Carpet, Nylon 20 oz., High Traffic	11.61	13.3%
Replace Air Conditioner, Rooftop, 50 Ton	9.84	11.3%
Repair Air Conditioner, Rooftop, 50 Ton	6.03	6.9%
Replace Ballast & Lamps, Fluorescent Lighting Fixture, 160w	4.31	4.9%
Refinish Sheetrock, Unstippled, Interior Wall Finish	3.67	4.2%
Place New Membrane Over Existing, Built-up Roof	3.49	4.0%
Remove & Replace Membrane, Built-up Roof	3.45	4.0%
Replace Fluorescent Lighting Fixture, 160w	3.06	3.5%
Maintain Air Conditioner, Rooftop, 50 Ton	2.80	3.2%
Clean & Reseal Clay Brick, Exterior, 1st Floor	2.79	3.2%
Repoint (50% surface) Clay Brick, Exterior, 1st Floor	1.66	1.9%
Replace Pipe & Fittings, 2" Copper, Cold Water	1.58	1.8%
Check Operation, Smoke Detector	1.34	1.5%
Maintain Wood, Solid Core, Painted, Door Locks	1.33	1.5%
Refinish Sheetrock, Unstippled, Ceiling	1.21	1.4%
Replace Pipe & Fittings, 3/4" Copper, Cold Water	1.20	1.4%
Minor Repair, Terrazzo Flooring	.88	1.0%
Non-Destructive Moisture Inspection	.81	0.9%
Replace Smoke Detector	.71	0.8%
Replace Wood, Solid Core, Painted, Interior Door	.60	0.7%
Replace Solid Core, Painted, Door Locks	.59	0.7%
Replace Pipe & Fittings, 3/4" Copper, Hot Water	.57	0.6%
Maintain Roof Drain, 4-6"	.52	0.6%
Minor Repair, Acoustic Tile, Dropped, Ceiling	.50	0.6%
Repair Fire Sprinkler Head	.48	0.5%
Replace Distribution Switch, Fused, <600 V	.39	0.4%
Replace Water Heater, Gas/Oil 65 Gph	.38	0.4%
Minor Repair, Sheetrock, Unstippled, Interior Wall Finish	.36	0.4%
Repair Clay Brick, Exterior, 1st Floor	.36	0.4%

**Task cost ($2004) per gross square foot over 50 years.*

***Percent of total M&R costs.*

Cost per Sqft. by System

Building Age	Exterior Closure	Roofing	Interior Constrution	Stairways	Interior Finishes	Conveying Systems	Plumbing	HVAC Systems	Fire Protection	Electrical	Equipment	Total per Sqft.
1	.02	.59	.08				.06	.10	.01	.06		**.91**
2	.02	.59	.08		.01		.10	.10	.01	.06		**.96**
3	.02	.64	.08				.06	.10	.01	.06		**.96**
4	.02	.59	.10		.43		.10	.10	.01	.06		**1.41**
5	.02	.59	.08				.08	.10	.01	.07		**.94**
6	.02	.64	.08		.01		.10	.10	.01	.06		**1.01**
7	.02	.59	.08				.06	.10	.01	.06		**.91**
8	.02	.59	.10		2.36		.15	.10	.01	.06		**3.38**
9	.02	.64	.08				.06	.10	.01	.06		**.96**
10	1.36	.59	.24		.11		.19	2.11	.17	1.61		**6.37**
11	.02	.59	.08				.06	.10	.01	.06		**.91**
12	.04	.64	.10		.43		.10	.11	.01	.06		**1.48**
13	.02	.59	.08				.09	.10	.01	.06		**.94**
14	.06	2.33	.08		.01		.11	.10	.01	.06		**2.76**
15	.02	.64	.08		.31		.06	3.28	.01	.33		**4.72**
16	.02	.59	.10		2.36		.15	.10	.01	.06		**3.38**
17	.02	.59	.08				.06	.10	.01	.06		**.91**
18	.02	.64	.08		.01		.10	.10	.01	.06		**1.01**
19	.02	.59	.08				.06	.10	.01	.06		**.91**
20	1.38	.59	.26		.95		.20	.10	.14	1.79		**5.41**
21	.02	.64	.08				.08	.10	.01	.06		**.98**
22	.02	.59	.08		.01		.10	.10	.01	.06		**.96**
23	.02	.59	.08				.06	.10	.01	.06		**.91**
24	.04	.64	.10		2.37		.15	.11	.01	.06		**3.47**
25	1.04	.59	.08		.03		1.92	2.11	.01	.17		**5.95**
26	.02	.59	.08		.01		.12	.10	.01	.06		**.98**
27	.02	.64	.08				.06	.10	.01	.06		**.96**
28	.06	3.45	.10		.43		.11	.10	.01	.06		**4.32**
29	.02	.59	.08				.07	.10	.01	.06		**.92**
30	1.36	.59	.64		.42		.18	3.31	.17	2.15		**8.82**
31	.02	.64	.08				.06	.10	.01	.06		**.96**
32	.02	.59	.08		2.36		.15	.10	.01	.06		**3.36**
33	.02	.59	.08				.06	.10	.01	.06		**.91**
34	.02	.64	.10		.01		.10	.10	.01	.06		**1.03**
35	.11	.59	.08				.34	.10	.01	.07		**1.29**
36	.04	.59	.08		.43		.07	.11	.01	.06		**1.38**
37	.02	.64	.08				.11	.10	.01	.06		**1.01**
38	.02	.59	.10		.01		.07	.10	.01	.06		**.95**
39	.02	.59	.08				.11	.10	.01	.06		**.96**
40	1.37	.64	.24		2.87		.17	2.11	.14	1.89		**9.43**
41	.02	.59	.08				.09	.10	.01	.06		**.94**
42	.06	2.33	.10		.01		.07	.10	.01	.06		**2.74**
43	.02	.64	.08				.09	.10	.01	.06		**.99**
44	.02	.59	.08		.43		.07	.10	.01	.06		**1.35**
45	.03	.59	.08		.31		.19	3.28	.01	.33		**4.81**
46	.02	.64	.10		.01		.07	.10	.01	.06		**1.00**
47	.02	.59	.08				.09	.10	.01	.06		**.94**
48	.04	.59	.08		2.37		.11	.11	.01	.06		**3.36**
49	.02	.64	.08				.10	.10	.01	.06		**1.00**
50	2.36	.59	.26		.15		1.95	.10	.17	1.51		**7.08**
Total	10.11	36.42	5.28		19.24		8.86	20.73	1.07	12.28		114.00

A value of "0.00" means a cost of more than $.000 but less than $.005 per gross square foot.

50-Year Total M&R Cost Profile per Sqft.

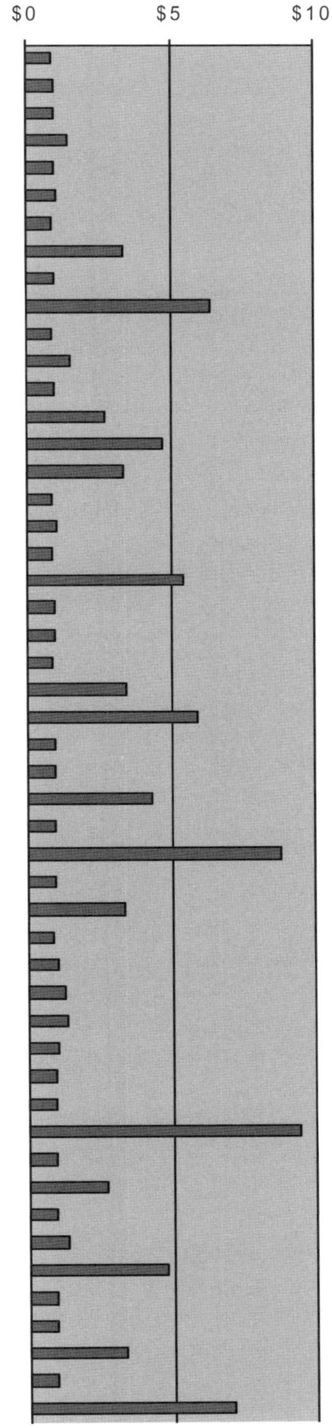

Dormitory, 50 Room

Gross Sqft:	25,000
Height in Ft:	36
Exterior:	Clay Brick
Floor coverings:	Finished Concrete/Ceramic Tile/Carpet
HVAC:	Chilled Water, Gas Boiler, Fan Coil Units
Occupancy:	100
Replacement Cost:	$2,157,174

Components

Component	Quantity	Unit
B20 Exterior Enclosure		
Clay Brick, Exterior, 1st Floor	3305	Sq Ft
Clay Brick, Exterior, 2nd Floor	3300	Sq Ft
Clay Brick, Exterior, 3rd+ Floor	3300	Sq Ft
Aluminum Operable Window, 12 sf, 1st Floor	90	Each
Aluminum Operable Window, 12 sf, 2nd Floor	90	Each
Aluminum Operable Window, 12 sf, 3rd+ Floor	90	Each
Steel, Painted, Exterior Door	8	Each
Aluminum Frame, Fully Glazed, Exterior Door	2	Each
B30 Roofing		
Single-Ply Modified Bituminous/Thermoplastic Roof	8340	Sq Ft
C10 Interior Construction		
Wood, Solid Core, Painted, Interior Door	84	Each
Wood, Hollow Core, Painted, Interior Door	111	Each
C20 Stairs		
Concrete Interior Steps	450	Sq Ft
Metal Interior Railing, Painted	120	Ln Ft
C30 Interior Finishes		
Sheetrock, Unstippled, Interior Wall Finish	60490	Sq Ft
Concrete Block, Painted, Interior Wall Finish	2780	Sq Ft
Concrete, Finished Flooring	1250	Sq Ft
Ceramic Tile Flooring	2000	Sq Ft
Carpet, Nylon 20 oz., Low Traffic	21750	Sq Ft
Sheetrock, Unstippled, Ceiling	25000	Sq Ft
D10 Conveying		
Elevator, Hydraulic, 2500 lbs, 3-5 Floor, 200 fpm	1	Each
D20 Plumbing		
Tankless Water Closet	14	Each
Lavatory, Vitreous China	12	Each
Service Sink, Iron Enamel	3	Each
Shower, Fiberglass	16	Each
Drinking Fountain, Vitreous China	3	Each
Pipe & Fittings, 3/4" Copper, Cold Water	1.6	K Ln Ft
Pipe & Fittings, 2" Copper, Cold Water	2	K Ln Ft
Pipe & Fittings, 3/4" Copper, Hot Water	1	K Ln Ft
Pipe Insulation, Cold Water	1.9	K Ln Ft
Pipe Insulation, Hot Water	1.4	K Ln Ft
Circulator Pump, 1/2 HP, Hot Water	2	Each
Water Heater, Gas/Oil 175 Gph	2	Each
Pipe & Fittings, 6" Cast Iron	0.6	K Ln Ft
Pipe & Fittings, 10" Cast Iron	0.2	K Ln Ft
Pipe & Fittings, 4" DWV PVC	0.2	K Ln Ft
Roof Drain, 4-6"	2	Each
D30 HVAC		
Pipe & Fittings, 4" Steel, Gas	0.9	K Ln Ft
Gas Boiler, 1000 Mbh	1	Each
Chemical Feed System	1	Each
Circulation Pump 5 HP, Hot Water	1	Each
Expansion Tank, 60 Gal.	2	Each
Circulation Pump, 5 HP, Chiller & Condenser Water	2	Each
Chiller, Reciprocal Air-Cooled Hermetic, 60 Ton	1	Each
Condenser, Air-Cooled, 60 Ton	1	Each
Fan Coil, Two-Pipe, 400 Cfm	65	Each
D40 Fire Protection		
Fire Sprinkler Head	125	Each
D50 Electrical		
Main Switchgear, <1200 Amp.	1	Each
Distribution Switch, Fused, <600 V	3	Each
Motor Starter, 5-20 HP, <600 V	3	Each
Motor Starter, 21-50 HP, <600 V	2	Each
Fluorescent Lighting Fixture, 160w	235	Each
Wiring Device, Switch	100	Each
Receptacle, 120V, 15 Amp	250	Each
Fire Alarm Bell, 6"	6	Each
Fire Alarm Control Panel	1	Each
Manual Pull Station	6	Each
Smoke Detector	60	Each

50-Year M&R Cost Summary

	50 Year Total	Annual Cost per Sqft	Annual cost as % replacement
PM & Minor Repair	$1,108,757	$0.89	1.03%
Unscheduled Maintenance	$1,133,844	$0.91	1.05%
Renewal & Replacement	$2,991,755	$2.39	2.77%
Total M&R Costs	$5,234,356	$4.19	4.85%

Distribution of Repair Costs

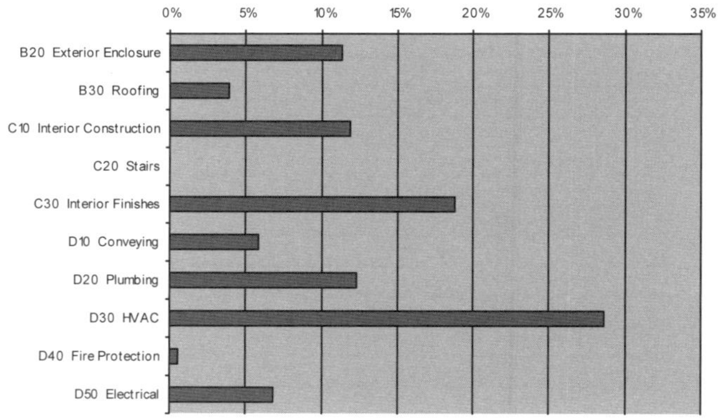

Thirty Most Costly Repair Tasks

Major Repair Task	Task Cost*	Pct.**
Refinish Sheetrock, Unstippled, Interior Wall Finish	14.70	9.0%
Replace Carpet, Nylon 20 oz., Low Traffic	12.62	7.7%
Maintain Fan Coil, Two-Pipe, 400 Cfm	11.94	7.3%
Replace Fan Coil, Two-Pipe, 400 Cfm	10.40	6.3%
Maintain Elevator, Hydraulic, 2500 lbs, 3-5 Floor, 200 fpm	10.38	6.3%
Refinish Sheetrock, Unstippled, Ceiling	6.07	3.7%
Replace Pipe & Fittings, 2" Copper, Cold Water	5.42	3.3%
Replace Chiller, Reciprocal Air Cooled Hermetic, 60 Ton	4.42	2.7%
Maintain Chiller, Reciprocal Air Cooled Hermetic, 60 Ton	4.00	2.4%
Replace Pipe & Fittings, 3/4" Copper, Cold Water	3.81	2.3%
Maintain Wood, Hollow Core, Painted, Door Locks	3.70	2.3%
Clean & Reseal Clay Brick, Exterior, 3rd+ Floor	2.94	1.8%
Repair Chiller, Reciprocal Air Cooled Hermetic, 60 Ton	2.84	1.7%
Maintain Wood, Solid Core, Painted, Door Locks	2.80	1.7%
Inspect & Test Gas Boiler, 1000 Mbh	2.62	1.6%
Replace Ballast & Lamps, Fluorescent Lighting Fixture, 160w	2.59	1.6%
Maintain Single-Ply Modified Bituminous/Thermoplastic Roof	2.51	1.5%
Clean & Reseal Clay Brick, Exterior, 2nd Floor	2.45	1.5%
Replace Pipe & Fittings, 3/4" Copper, Hot Water	2.37	1.4%
Remove & Replace Membrane, Single-Ply Modified Bituminous/Ther	2.34	1.4%
Replace Water Heater, Gas/Oil 175 Gph	2.34	1.4%
Replace Wood, Hollow Core, Painted, Door Locks	2.24	1.4%
Replace Condenser, Air-Cooled, 60 Ton	2.17	1.3%
Repair Fan Coil, Two-Pipe, 400 Cfm	2.02	1.2%
Clean & Reseal Clay Brick, Exterior, 1st Floor	1.96	1.2%
Renovate Elevator, Hydraulic, 2500 lbs, 3-5 Floor, 200 fpm	1.86	1.1%
Replace Fluorescent Lighting Fixture, 160w	1.84	1.1%
Check Operation, Smoke Detector	1.59	1.0%
Minor Repair, Sheetrock, Unstippled, Interior Wall Finish	1.44	0.9%
Replace Wood, Hollow Core, Painted, Interior Door	1.33	0.8%

**Task cost ($2004) per gross square foot over 50 years.*

***Percent of total M&R costs.*

Cost per Sqft. by System

Building Age	Exterior Closure	Roofing	Interior Constrution	Stairways	Interior Finishes	Conveying Systems	Plumbing	HVAC Systems	Fire Protection	Electrical	Equipment	Total per Sqft.
1	.02	.10	.36			.21	.07	.74	.01	.11		**1.60**
2	.02	.10	.36			.21	.18	.75	.01	.11		**1.73**
3	.02	.11	.36			.21	.07	.74	.01	.11		**1.62**
4	.02	.10	.45		1.80	.21	.18	.75	.01	.11		**3.63**
5	.02	.10	.36		.02	.21	.23	.75	.01	.23		**1.92**
6	.02	.11	.36			.21	.18	.75	.01	.11		**1.74**
7	.02	.10	.36	.00		.21	.07	.82	.01	.11		**1.70**
8	.02	.10	.45		1.80	.21	.53	.75	.01	.11		**3.98**
9	.02	.11	.36			.21	.07	.74	.01	.11		**1.62**
10	3.56	.37	1.23		.02	.21	.80	2.47	.17	1.28		**10.11**
11	.02	.10	.36			.21	.07	.74	.01	.11		**1.60**
12	.04	.11	.45		1.80	.21	.18	.77	.01	.11		**3.69**
13	.02	.10	.36			.21	.14	.74	.01	.11		**1.68**
14	.12	.10	.36	.00		.21	.19	.83	.01	.11		**1.92**
15	.02	.11	.36	.03	4.23	.21	.19	4.74	.01	.66		**10.56**
16	.02	.10	.45		1.80	.21	.53	.75	.01	.11		**3.98**
17	.02	.10	.36			.21	.07	.74	.01	.11		**1.60**
18	.02	.11	.36			.21	.18	.75	.01	.25		**1.88**
19	.02	.10	.36			.21	.07	.74	.01	.11		**1.60**
20	3.72	1.17	1.33		3.64	.21	.94	2.82	.14	1.47		**15.44**
21	.02	.10	.36	.00		.21	.11	.82	.01	.11		**1.73**
22	.02	.10	.36			.21	.18	.75	.01	.11		**1.73**
23	.02	.11	.36			.21	.07	.74	.01	.17		**1.68**
24	.04	.10	.45		1.85	.21	.53	.77	.01	.11		**4.07**
25	2.24	.10	.36		.06	1.14	6.53	1.51	.01	.36		**12.30**
26	.02	.11	.36			.21	.20	.75	.01	.11		**1.76**
27	.02	.10	.36			.21	.08	.74	.01	.11		**1.61**
28	.13	.10	.45	.00	1.80	.21	.18	.83	.01	.17		**3.88**
29	.02	.11	.36			.21	.11	.74	.01	.11		**1.67**
30	3.56	.37	2.67	.03	4.23	.21	.79	6.57	.17	1.74		**20.34**
31	.02	.10	.36			.21	.08	.74	.01	.11		**1.61**
32	.02	.11	.36		1.80	.21	.52	.75	.01	.11		**3.89**
33	.02	.10	.36			.21	.08	.74	.01	.17		**1.67**
34	.02	.10	.45			.21	.17	.75	.01	.11		**1.81**
35	.02	.11	.36	.00	.02	.21	.82	.75	.01	.17		**2.47**
36	.04	.10	.36		1.80	.21	.10	.77	.01	.25		**3.63**
37	.02	.10	.36			.21	.19	.82	.01	.11		**1.81**
38	.02	.11	.45			.21	.11	.75	.01	.11		**1.77**
39	.02	.10	.36			.21	.18	.74	.01	.11		**1.71**
40	3.72	1.17	1.23		3.64	.21	1.02	3.58	.14	1.66		**16.37**
41	.02	.10	.36			.21	.15	.74	.01	.17		**1.75**
42	.12	.10	.45	.00		.21	.11	.75	.01	.11		**1.85**
43	.02	.11	.36			.21	.16	.74	.01	.11		**1.71**
44	.02	.10	.36		1.80	.21	.10	.83	.01	.11		**3.53**
45	.02	.10	.36	.03	4.23	.21	.51	4.74	.01	.60		**10.79**
46	.02	.11	.45			.21	.10	.75	.01	.17		**1.81**
47	.02	.10	.36			.21	.15	.74	.01	.11		**1.69**
48	.04	.10	.36		1.85	.21	.44	.77	.01	.11		**3.88**
49	.02	.11	.36	.00		.21	.16	.74	.01	.11		**1.72**
50	5.78	.37	1.33		1.08	1.14	6.91	1.75	.17	1.04		**19.57**
Total	23.76	8.10	24.89	.12	39.30	12.24	25.76	59.94	1.07	14.20		209.37

50-Year Total M&R Cost Profile per Sqft.

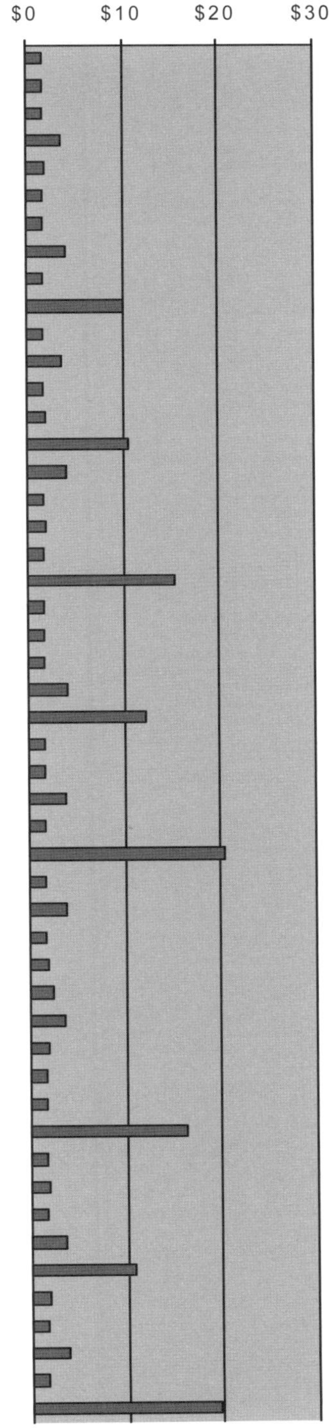

A value of "0.00" means a cost of more than $.000 but less than $.005 per gross square foot.

Elementary School

Gross Sqft:	47,000
Height in Ft:	12
Exterior:	Cast Concrete
Floor coverings:	Finished Concrete/Tile/Carpet
HVAC:	Chilled Air, Gas Boiler, Multizone Air
Occupancy:	600
Replacement Cost:	$3,426,265

Components

Component	Quantity	Unit
B20 Exterior Enclosure		
Concrete, Painted, Exterior, 1st Floor	7770	Sq Ft
Aluminum Fixed Window, 12 sf, 1st Floor	220	Each
Steel w/ Safety Glass, Painted, Exterior Door	8	Each
Aluminum Frame, Fully Glazed, Exterior Door	4	Each
B30 Roofing		
Built-up Roof	47000	Sq Ft
C10 Interior Construction		
Wood, Solid Core, Painted, Interior Door	16	Each
Wood, Solid Core w/ Safety Glass, Painted, Interior Door	52	Each
C30 Interior Finishes		
Plaster Interior Wall Finish	50650	Sq Ft
Concrete Block, Painted, Interior Wall Finish	2350	Sq Ft
Concrete, Finished Flooring	2350	Sq Ft
Quarry Tile Flooring	11750	Sq Ft
Vinyl Tile Flooring	28200	Sq Ft
Carpet, Nylon 20 oz., Low Traffic	4700	Sq Ft
Plaster Ceiling	47000	Sq Ft
D20 Plumbing		
Tankless Water Closet	22	Each
Urinal, Vitreous China	12	Each
Lavatory, Vitreous China	18	Each
Service Sink, Iron Enamel	4	Each
Drinking Fountain, Vitreous China	18	Each
Pipe & Fittings, 3/4" Copper, Cold Water	2.5	K Ln Ft
Pipe & Fittings, 2" Copper, Cold Water	3	K Ln Ft
Pipe & Fittings, 3/4" Copper, Hot Water	0.7	K Ln Ft
Pipe Insulation, Cold Water	3	K Ln Ft
Pipe Insulation, Hot Water	1	K Ln Ft
Circulator Pump, 1/12 HP, Hot Water	1	Each
Water Heater, Gas/Oil 130 Gph	1	Each
Pipe & Fittings, 6" Cast Iron	1.5	K Ln Ft
Pipe & Fittings, 10" Cast Iron	0.5	K Ln Ft
Pipe & Fittings, 4" DWV PVC	0.4	K Ln Ft
Roof Drain, 4-6"	14	Each
D30 HVAC		
Pipe & Fittings, 4" Steel, Gas	1.8	K Ln Ft
Gas Boiler, 2500 Mbh	1	Each
Chemical Feed System	1	Each
Circulation Pump 5 HP, Hot Water	1	Each
Expansion Tank, 60 Gal.	2	Each
Circulation Pump, 5 HP, Chiller & Condenser Water	3	Each
Chiller, Reciprocal Water-Cooled Hermetic, 165 Ton	1	Each
Cooling Tower, 200 Ton	1	Each
Air Handler, Multizone, 15,000 Cfm	1	Each
Air Handler, Multizone, 50,000 Cfm	1	Each
D40 Fire Protection		
Fire Sprinkler Head	235	Each
D50 Electrical		
Secondary Transformer, Dry, 150 KVA	3	Each
Main Switchgear, <1200 Amp.	1	Each
Distribution Switch, Fused, <600 V	5	Each
Motor Starter, 5-20 HP, <600 V	4	Each
Motor Starter, 21-50 HP, <600 V	5	Each
Motor Starter, 50-75 HP, <600 V	1	Each
Fluorescent Lighting Fixture, 160w	881	Each
Wiring Device, Switch	70	Each
Receptacle, 120V, 15 Amp	270	Each
Annunciation Panel	1	Each
Fire Alarm Bell, 6"	6	Each
Fire Alarm Control Panel	1	Each
Smoke Detector	94	Each

50-Year M&R Cost Summary

	50 Year Total	Annual Cost per Sqft	Annual cost as % replacement
PM & Minor Repair	$1,418,705	$0.60	0.83%
Unscheduled Maintenance	$1,330,028	$0.57	0.78%
Renewal & Replacement	$4,009,555	$1.71	2.34%
Total M&R Costs	$6,758,288	$2.88	3.94%

Distribution of Repair Costs

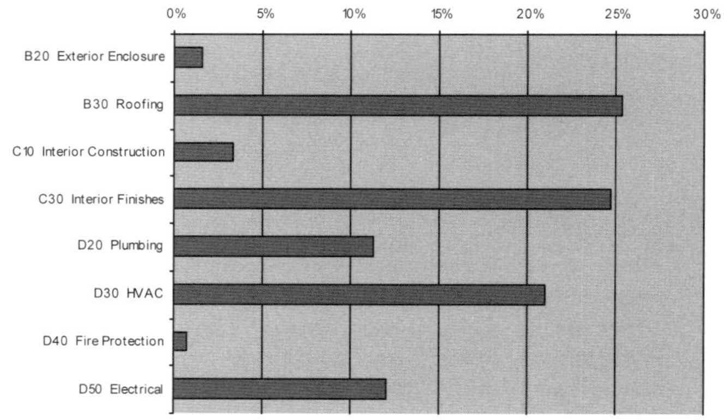

Thirty Most Costly Repair Tasks

Major Repair Task	Task Cost*	Pct.**
Maintain Built-up Roof	14.62	12.7%
Replace Quarry Tile Flooring	10.59	9.2%
Refinish Plaster Interior Wall Finish	6.55	5.7%
Refinish Plaster Ceiling	6.10	5.3%
Replace Ballast & Lamps, Fluorescent Lighting Fixture, 160w	5.17	4.5%
Minor Repair, Plaster Ceiling	4.95	4.3%
Replace Chiller, Reciprocal Water-Cooled Hermetic, 165 Ton	4.51	3.9%
Replace Pipe & Fittings, 2" Copper, Cold Water	4.32	3.7%
Replace Fluorescent Lighting Fixture, 160w	3.67	3.2%
Place New Membrane Over Existing, Built-up Roof	3.49	3.0%
Remove & Replace Membrane, Built-up Roof	3.45	3.0%
Replace Air Handler, Multizone, 50,000 Cfm	3.27	2.8%
Replace Pipe & Fittings, 3/4" Copper, Cold Water	3.17	2.7%
Inspect & Test Gas Boiler, 2500 Mbh	2.78	2.4%
Replace Vinyl Tile Flooring	2.52	2.2%
Repair Chiller, Reciprocal Water-Cooled Hermetic, 165 Ton	2.14	1.9%
Maintain Chiller, Reciprocal Water-Cooled Hermetic, 165 Ton	2.12	1.8%
Minor Repair, Plaster Interior Wall Finish	1.58	1.4%
Replace Carpet, Nylon 20 oz., Low Traffic	1.45	1.3%
Check Operation, Smoke Detector	1.33	1.2%
Refinish Concrete, Painted, Exterior, 1st Floor	1.30	1.1%
Replace Cooling Tower, 200 Ton	1.18	1.0%
Maintain Cooling Tower, 200 Ton	1.16	1.0%
Replace Air Handler, Multizone, 15,000 Cfm	1.04	0.9%
Maintain Air Handler, Multizone, 50,000 Cfm	.95	0.8%
Maintain Circulation Pump, 5 HP, Chiller & Condenser Water	.94	0.8%
Maintain Wood, Solid Core w/ Safety Glass, Painted, Door Locks	.92	0.8%
Replace Pipe & Fittings, 3/4" Copper, Hot Water	.88	0.8%
Non-Destructive Moisture Inspection	.81	0.7%
Maintain Air Handler, Multizone, 15,000 Cfm	.74	0.6%

**Task cost ($2004) per gross square foot over 50 years.*

***Percent of total M&R costs.*

Cost per Sqft. by System

Building Age	Exterior Closure	Roofing	Interior Constrution	Stairways	Interior Finishes	Conveying Systems	Plumbing	HVAC Systems	Fire Protection	Electrical	Equipment	Total per Sqft.
1	.01	.59	.07				.05	.32	.01	.10		**1.14**
2	.01	.59	.07				.13	.33	.01	.10		**1.23**
3	.01	.64	.07				.05	.32	.01	.10		**1.19**
4	.01	.59	.08		1.09		.13	.33	.01	.10		**2.33**
5	.01	.59	.07		.00		.08	.33	.01	.20		**1.28**
6	.01	.64	.07				.13	.33	.01	.10		**1.28**
7	.01	.59	.07				.07	.39	.01	.10		**1.23**
8	.01	.59	.08		1.09		.23	.33	.01	.10		**2.43**
9	.01	.64	.07		.02		.05	.32	.01	.10		**1.22**
10	.27	.59	.24		.00		.28	1.22	.17	2.03		**4.79**
11	.01	.59	.07				.05	.32	.01	.10		**1.14**
12	.03	.64	.08		2.98		.13	.35	.01	.10		**4.32**
13	.01	.59	.07				.10	.32	.01	.10		**1.19**
14	.07	2.33	.07				.15	.40	.01	.10		**3.12**
15	.01	.64	.07		.55		.07	2.22	.01	.47		**4.03**
16	.01	.59	.08		1.09		.23	.33	.01	.10		**2.43**
17	.01	.59	.07				.05	.32	.01	.10		**1.14**
18	.01	.64	.07		1.26		.13	.33	.01	.32		**2.76**
19	.01	.59	.07				.05	.32	.01	.10		**1.14**
20	.32	.59	.25		1.09		.31	2.51	.14	2.20		**7.41**
21	.01	.64	.07				.08	.39	.01	.10		**1.29**
22	.01	.59	.07				.13	.33	.01	.10		**1.23**
23	.01	.59	.07				.05	.32	.01	.17		**1.21**
24	.03	.64	.08		3.01		.23	.35	.01	.10		**4.44**
25	.01	.59	.07		.04		4.68	.50	.01	.23		**6.11**
26	.01	.59	.07				.16	.33	.01	.10		**1.26**
27	.01	.64	.07		.02		.06	.32	.01	.10		**1.22**
28	.07	3.45	.08		1.09		.15	.40	.01	.17		**5.40**
29	.01	.59	.07				.07	.32	.01	.10		**1.16**
30	.27	.59	.32		.55		.27	3.65	.17	2.69		**8.51**
31	.01	.64	.07				.06	.32	.01	.10		**1.20**
32	.01	.59	.08		1.09		.22	.33	.01	.10		**2.42**
33	.01	.59	.07				.06	.32	.01	.17		**1.21**
34	.01	.64	.07				.13	.33	.01	.10		**1.28**
35	.01	.59	.07		.00		.88	.33	.01	.13		**2.01**
36	.03	.59	.08		4.24		.05	.35	.01	.32		**5.66**
37	.01	.64	.07				.14	.39	.01	.10		**1.35**
38	.01	.59	.07				.06	.33	.01	.10		**1.16**
39	.01	.59	.07				.17	.32	.01	.10		**1.26**
40	.32	.64	.59		1.09		.21	2.68	.14	2.30		**7.97**
41	.01	.59	.07				.13	.32	.01	.17		**1.29**
42	.07	2.33	.07				.07	.33	.01	.10		**2.97**
43	.01	.64	.07				.14	.32	.01	.10		**1.28**
44	.01	.59	.08		1.09		.05	.40	.01	.10		**2.32**
45	.01	.59	.07		.57		.32	2.22	.01	.40		**4.19**
46	.01	.64	.07				.05	.33	.01	.17		**1.27**
47	.01	.59	.07				.13	.32	.01	.10		**1.22**
48	.03	.59	.08		3.01		.14	.35	.01	.10		**4.30**
49	.01	.64	.07				.15	.32	.01	.10		**1.29**
50	.27	.59	.24		10.63		4.68	1.07	.17	1.87		**19.52**
Total	2.20	36.43	4.79		35.59		16.19	30.20	1.07	17.33		143.79

50-Year Total M&R Cost Profile per Sqft.

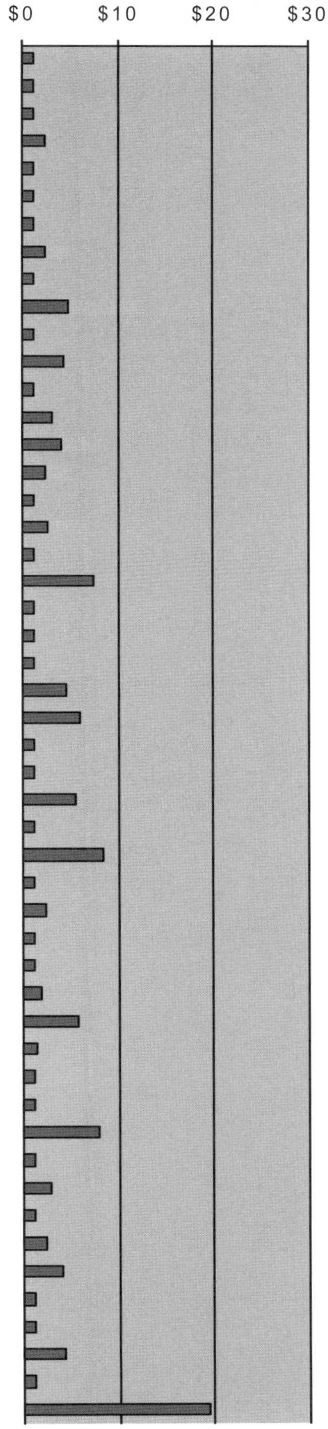

A value of "0.00" means a cost of more than $.000 but less than $.005 per gross square foot.

Fire Station

Gross Sqft:	6,000
Height in Ft:	14
Exterior:	Brick Veneer
Floor coverings:	Finished Concrete
HVAC:	Multizone Unit
Occupancy:	150
Replacement Cost:	$627,169

50-Year M&R Cost Summary

	50 Year Total	Annual Cost per Sqft	Annual cost as % replacement
PM & Minor Repair	$342,804	$1.14	1.09%
Unscheduled Maintenance	$378,638	$1.26	1.21%
Renewal & Replacement	$1,053,760	$3.51	3.36%
Total M&R Costs	$1,775,202	$5.92	5.66%

Components

Component	Quantity	Unit
B20 Exterior Enclosure		
Clay Brick, Exterior, 1st Floor	3360	Sq Ft
Aluminum Operable Window, 12 sf, 1st Floor	14	Each
Steel, Painted, Exterior Door	3	Each
Aluminum Frame, Fully Glazed, Exterior Door	1	Each
Steel 20'x20', Painted, Roll-up Door	5	Each
B30 Roofing		
Single-Ply Modified Bituminous/Thermoplastic Roof	6000	Sq Ft
C10 Interior Construction		
Steel, Painted, Interior Door	12	Each
C30 Interior Finishes		
Concrete Block, Painted, Interior Wall Finish	13242	Sq Ft
Concrete, Unfinished Flooring	3000	Sq Ft
Vinyl Tile Flooring	3000	Sq Ft
Acoustical Tile Ceiling	3000	Sq Ft
Metal Ceiling, Painted	3000	Sq Ft
D20 Plumbing		
Tankless Water Closet	5	Each
Urinal, Vitreous China	2	Each
Lavatory, Vitreous China	7	Each
Sink, Stainless Steel	4	Each
Shower, Ceramic Tile	4	Each
Drinking Fountain, Refrigerated	2	Each
Pipe & Fittings, 3/4" Copper, Hot Water	0.59	K Ln Ft
Pipe & Fittings, 3/4" Copper, Cold Water	0.96	K Ln Ft
Pipe & Fittings, 2" Copper, Cold Water	1.235	K Ln Ft
Pipe & Fittings, 4" Steel	0.82	K Ln Ft
Pipe Insulation, Cold Water	1.37	K Ln Ft
Pipe Insulation, Hot Water	0.845	K Ln Ft
Water Heater, Gas/Oil 175 Gph	2	Each
Pipe & Fittings, 6" Cast Iron	0.427	K Ln Ft
Pipe & Fittings, 10" Cast Iron	0.155	K Ln Ft
Roof Drain, 2"	3	Each
Pipe & Fittings, 4" DWV PVC	0.173	K Ln Ft
D30 HVAC		
Exhaust Fan, Ceiling, 200-500 Cfm	2	Each
Exhaust Fan, Roof Mounted, 5000 Cfm	2	Each
Air Conditioner, Rooftop, 20 Ton	2	Each
D40 Fire Protection		
Fire Sprinkler System	1	Each
Fire Sprinkler Head	45	Each
D50 Electrical		
Safety Switch, Fused, 400 Amp., 3 Ph.	1	Each
Distribution Panel Board	1	Each
Main Switchgear, <1200 Amp.	1	Each
Exit Lighting Fixture, w/ Battery	5	Each
Emergency Horn & Strobe	4	Each
Fluorescent Lighting Fixture, 160w	30	Each
Wiring Device, Switch	6	Each
Grounded Fault Circuit Interrupter, 15 Amps	2	Each
Receptacle, 120V, 15 Amp	15	Each
TV Cable Outlet	1	Each
Annunciation Panel	1	Each
Fire Alarm Bell, 6"	4	Each
Fire Alarm Control Panel	1	Each
Manual Pull Station	4	Each
Smoke Detector	4	Each
Generator Switchgear	1	Each

Distribution of Repair Costs

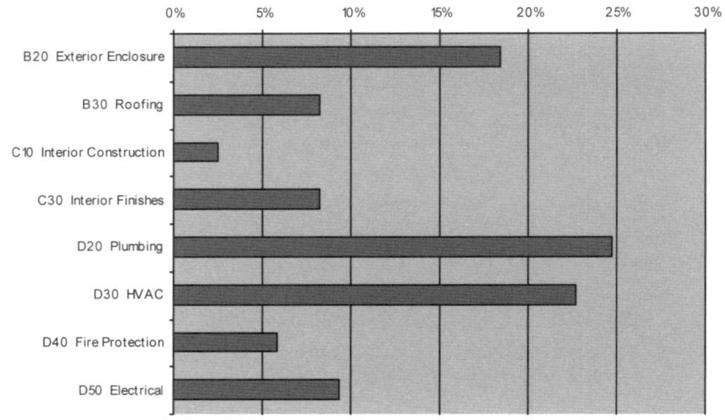

Thirty Most Costly Repair Tasks

Major Repair Task	Task Cost*	Pct.**
Replace Air Conditioner, Rooftop, 20 Ton	30.41	13.1%
Refinish Concrete Block, Painted, Interior Wall Finish	16.94	7.3%
Fire Sprinkler System, Annual PM	14.40	6.2%
Replace Pipe & Fittings, 2" Copper, Cold Water	13.93	6.0%
Repair Air Conditioner, Rooftop, 20 Ton	12.18	5.2%
Maintain Air Conditioner, Rooftop, 20 Ton	11.33	4.9%
Replace Water Heater, Gas/Oil 175 Gph	9.73	4.2%
Replace Pipe & Fittings, 3/4" Copper, Cold Water	9.53	4.1%
Clean & Reseal Clay Brick, Exterior, 1st Floor	8.32	3.6%
Maintain Single-Ply Modified Bituminous/Thermoplastic Roof	7.54	3.2%
Remove & Replace Membrane, Single-Ply Modified Bituminous/Ther	7.01	3.0%
Maintain Steel 20'x20', Painted, Roll-up Door	6.44	2.8%
Replace Pipe & Fittings, 3/4" Copper, Hot Water	5.82	2.5%
Replace Steel 20'x20', Painted, Roll-up Door	5.56	2.4%
Repoint (50% surface) Clay Brick, Exterior, 1st Floor	4.97	2.1%
Annual PM, Generator Switchgear	4.52	1.9%
Clean Water Heater, Gas/Oil 175 Gph	3.20	1.4%
Replace Generator Switchgear	2.56	1.1%
Replace Steel, Painted, Door Locks	2.56	1.1%
Repair Single-Ply Modified Bituminous/Thermoplastic Roof	2.50	1.1%
Repair Steel 20'x20', Painted, Roll-up Door	2.43	1.0%
Replace Vinyl Tile Flooring	2.10	0.9%
Refinish Steel 20'x20', Painted, Roll-up Door	2.07	0.9%
Maintain Steel, Painted, Door Locks	1.70	0.7%
Maintain Exhaust Fan, Roof Mounted, 5000 Cfm	1.60	0.7%
Replace Pipe Insulation, Cold Water	1.52	0.7%
Replace Ballast & Lamps, Fluorescent Lighting Fixture, 160w	1.38	0.6%
Minor Repair, Concrete Block, Painted, Interior Wall Finish	1.34	0.6%
Repair Main Switchgear, <1200 Amp.	1.31	0.6%
Replace Faucet Washer & Clean Trap, Lavatory, Vitreous China	1.28	0.6%

**Task cost ($2004) per gross square foot over 50 years.*

***Percent of total M&R costs.*

Cost per Sqft. by System

Building Age	Exterior Closure	Roofing	Interior Construction	Stairways	Interior Finishes	Conveying Systems	Plumbing	HVAC Systems	Fire Protection	Electrical	Equipment	Total per Sqft.
1	.36	.29	.09				.27	.49	.30	.33		**2.13**
2	.36	.29	.09				.63	.49	.30	.33		**2.49**
3	.36	.34	.09				.27	.49	.30	.33		**2.18**
4	.37	.29	.12		1.41		.63	.49	.30	.33		**3.93**
5	.59	.29	.09		.09		.58	.49	.42	.62		**3.16**
6	.36	.34	.09				.63	.49	.30	.33		**2.54**
7	.36	.29	.09				.31	.49	.30	.33		**2.16**
8	.37	.29	.12		1.56		2.10	.49	.30	.33		**5.55**
9	.36	.34	.09		.20		.27	.49	.30	.33		**2.38**
10	5.16	1.12	.60		.09		1.43	4.65	.66	1.16		**14.86**
11	.36	.29	.09				.27	.49	.30	.33		**2.13**
12	.40	.34	.12		1.41		.63	.49	.30	.33		**4.01**
13	.36	.29	.09				.63	.49	.30	.33		**2.49**
14	.52	.29	.09				.66	.49	.30	.33		**2.68**
15	.59	.34	.09		.27		.42	10.21	.42	1.08		**13.42**
16	.37	.29	.12		1.56		2.10	.49	.30	.33		**5.55**
17	.36	.29	.09				.27	.49	.30	.33		**2.13**
18	.36	.34	.09		1.23		.63	.49	.30	1.43		**4.87**
19	.36	.29	.09				.27	.49	.30	.33		**2.13**
20	5.19	3.50	.63		1.50		1.55	1.08	.61	1.74		**15.81**
21	.36	.29	.09				.47	.49	.30	.33		**2.32**
22	.36	.29	.09				.63	.49	.30	.33		**2.49**
23	.36	.34	.09				.27	.49	.30	.33		**2.18**
24	.40	.29	.12		2.60		2.10	.49	.30	.33		**6.63**
25	3.61	.29	.09		.09		17.31	4.55	.42	1.12		**27.48**
26	.36	.34	.09				.77	.49	.30	.33		**2.68**
27	.36	.29	.09		.20		.30	.49	.30	.33		**2.36**
28	.53	.29	.12		1.41		.64	.49	.30	.33		**4.10**
29	.36	.34	.09				.46	.49	.30	.33		**2.37**
30	5.16	1.12	.60		.27		1.80	10.31	.66	1.55		**21.46**
31	.36	.29	.09				.30	.49	.30	.33		**2.16**
32	.37	.34	.12		1.56		2.07	.49	.30	.33		**5.57**
33	.36	.29	.09				.30	.49	.30	.33		**2.16**
34	.36	.29	.09				.60	.49	.30	.33		**2.46**
35	5.82	.34	.09		.09		1.50	.49	.42	.62		**9.36**
36	.40	.29	.12		2.64		.46	.49	.30	1.43		**6.12**
37	.36	.29	.09				.60	.49	.30	.33		**2.46**
38	.36	.34	.09				.49	.49	.30	.33		**2.40**
39	.36	.29	.09				.61	.49	.30	.33		**2.47**
40	4.59	3.50	.63		1.65		2.78	5.14	.61	1.82		**20.72**
41	.36	.29	.09				.44	.49	.30	.33		**2.30**
42	.52	.29	.09				.49	.49	.30	.33		**2.51**
43	.36	.34	.09				.44	.49	.30	.33		**2.35**
44	.37	.29	.12		1.41		.46	.49	.30	.33		**3.76**
45	1.20	.29	.09		.47		1.11	10.21	.42	1.08		**14.86**
46	.36	.34	.09				.46	.49	.30	.33		**2.37**
47	.36	.29	.09				.44	.49	.30	.33		**2.30**
48	.40	.29	.12		2.60		1.93	.49	.30	.33		**6.45**
49	.36	.34	.09				.48	.49	.30	.33		**2.39**
50	7.57	1.12	.60		.09		17.91	.58	.66	1.51		**30.04**
Total	54.55	24.28	7.47		24.39		73.17	67.18	17.18	27.65		295.87

50-Year Total M&R Cost Profile per Sqft.

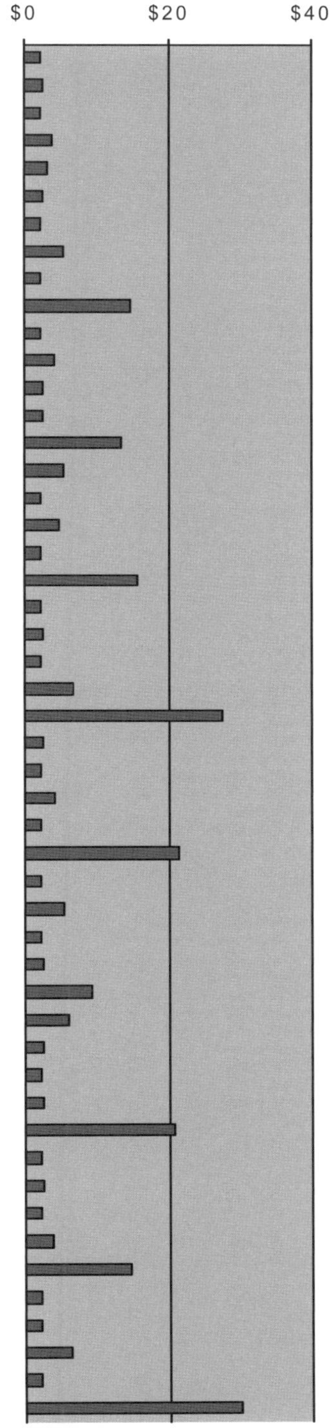

A value of "0.00" means a cost of more than $.000 but less than $.005 per gross square foot.

Funeral Home

Gross Sqft: 10,000
Height in Ft: 10
Exterior: Redwood Siding
Floor coverings: Carpet/Ceramic Tile
HVAC: Electric Cool, Gas Heat, Multizone Unit
Occupancy: 700
Replacement Cost: $893,983

Components

Component	Quantity	Unit
B20 Exterior Enclosure		
Aluminum Siding, Exterior, 1st Floor	3825	Sq Ft
Steel Frame, Painted, Operable Window, 12 sf, 1st Floor	38	Each
Wood, Solid Core, Painted, Exterior Door	4	Each
B30 Roofing		
Built-up Roof	14140	Sq Ft
Concrete Decking	200	Sq Ft
Concrete Steps	100	Sq Ft
C10 Interior Construction		
Wood Frame, Fully Glazed, Painted, Interior Door	67	Each
C30 Interior Finishes		
Sheetrock, Stippled, Interior Wall Finish	3335	Sq Ft
Fabric Interior Wall Finish	3335	Sq Ft
Wallpaper Interior Wall Finish	6670	Sq Ft
Ceramic Tile Flooring	3000	Sq Ft
Carpet, Wool 45 oz., Premium	7000	Sq Ft
Acoustical Tile Ceiling	10000	Sq Ft
D20 Plumbing		
Flush Tank Water Closet	4	Each
Urinal, Vitreous China	2	Each
Lavatory, Vitreous China	4	Each
Sink, Stainless Steel	6	Each
Drinking Fountain, Refrigerated	2	Each
Pipe & Fittings, 3/4" Copper, Cold Water	0.73	K Ln Ft
Pipe & Fittings, 3/4" Copper, Hot Water	0.39	K Ln Ft
Pipe & Fittings, 2" Copper, Cold Water	0.86	K Ln Ft
Pipe & Fittings, 4" Steel	0.88	K Ln Ft
Pipe Insulation, Cold Water	1.17	K Ln Ft
Pipe Insulation, Hot Water	0.63	K Ln Ft
Water Heater, Gas/Oil 175 Gph	2	Each
Pipe & Fittings, 6" Cast Iron	0.36	K Ln Ft
Pipe & Fittings, 10" Cast Iron	0.2	K Ln Ft
Aluminum Gutter, Downspouts, Fittings	0.425	K Ln Ft
Pipe & Fittings, 4" DWV PVC	0.116	K Ln Ft
D30 HVAC		
Exhaust Fan, Ceiling, 200-500 Cfm	2	Each
Exhaust Fan, Roof Mounted, 1000 Cfm	2	Each
Air Conditioner, Rooftop, 50 Ton	1	Each
D40 Fire Protection		
Fire Sprinkler Head	71	Each
D50 Electrical		
Safety Switch, Fused, 400 Amp., 3 Ph.	1	Each
Distribution Panel Board	2	Each
Main Switchgear, <1200 Amp.	1	Each
Exit Lighting Fixture, w/ Battery	6	Each
Emergency Horn & Strobe	4	Each
Fluorescent Lighting Fixture, 160w	50	Each
Wiring Device, Switch	10	Each
Grounded Fault Circuit Interruptor, 15 Amps	2	Each
Receptacle, 120V, 15 Amp	25	Each
TV Cable Outlet	1	Each
Annunciation Panel	1	Each
Fire Alarm Bell, 6"	4	Each
Fire Alarm Control Panel	1	Each
Manual Pull Station	4	Each
Heat Detector	1	Each
Smoke Detector	4	Each

50-Year M&R Cost Summary

	50 Year Total	Annual Cost per Sqft	Annual cost as % replacement
PM & Minor Repair	$424,448	$0.85	0.95%
Unscheduled Maintenance	$448,627	$0.90	1.00%
Renewal & Replacement	$1,119,741	$2.24	2.51%
Total M&R Costs	$1,992,816	$3.99	4.46%

Distribution of Repair Costs

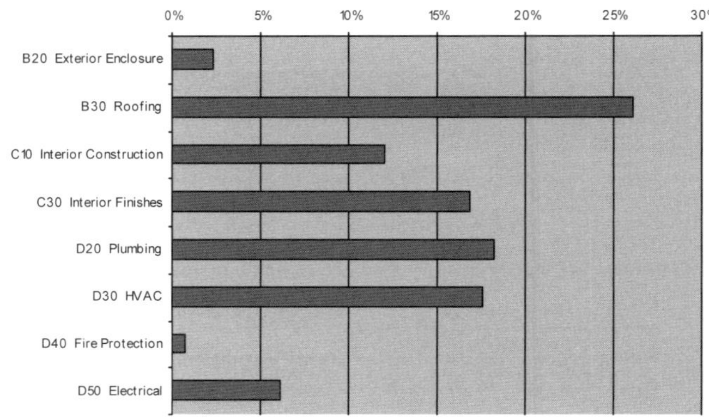

Thirty Most Costly Repair Tasks

Major Repair Task	Task Cost*	Pct.**
Maintain Built-up Roof	20.67	13.4%
Replace Carpet, Wool 45 oz., Premium	16.27	10.5%
Replace Air Conditioner, Rooftop, 50 Ton	15.42	10.0%
Repair Air Conditioner, Rooftop, 50 Ton	9.44	6.1%
Replace Water Heater, Gas/Oil 175 Gph	5.84	3.8%
Replace Pipe & Fittings, 2" Copper, Cold Water	5.82	3.8%
Maintain Wood Frame, Fully Glazed, Painted, Door Locks	5.59	3.6%
Replace Wallpaper Interior Wall Finish	5.16	3.3%
Place New Membrane Over Existing, Built-up Roof	4.94	3.2%
Replace Wood Frame, Fully Glazed, Painted, Interior Door	4.93	3.2%
Remove & Replace Membrane, Built-up Roof	4.88	3.2%
Maintain Air Conditioner, Rooftop, 50 Ton	4.39	2.8%
Replace Pipe & Fittings, 3/4" Copper, Cold Water	4.35	2.8%
Replace Ceramic Tile Flooring	3.82	2.5%
Replace Wood Frame, Fully Glazed, Painted, Door Locks	3.38	2.2%
Replace Fabric Interior Wall Finish	2.76	1.8%
Replace Pipe & Fittings, 3/4" Copper, Hot Water	2.31	1.5%
Refinish Sheetrock, Stippled, Interior Wall Finish	2.03	1.3%
Clean Water Heater, Gas/Oil 175 Gph	1.92	1.2%
Minor Repair, Acoustic Tile Ceiling	1.78	1.2%
Replace Aluminum Siding, Exterior, 1st Floor	1.69	1.1%
Replace Ballast & Lamps, Fluorescent Lighting Fixture, 160w	1.38	0.9%
Non-Destructive Moisture Inspection	1.15	0.7%
Replace Fluorescent Lighting Fixture, 160w	.98	0.6%
Maintain Exhaust Fan, Roof Mounted, 1000 Cfm	.96	0.6%
Annual PM, Distribution Panel Board	.94	0.6%
Repair Main Switchgear, <1200 Amp.	.79	0.5%
Replace Pipe Insulation, Cold Water	.78	0.5%
Replace Aluminum Gutter, Downspouts, Fittings	.75	0.5%
Install New Gasket & Bolts, Pipe & Fittings, 4" Steel	.74	0.5%

**Task cost ($2004) per gross square foot over 50 years.*

***Percent of total M&R costs.*

Cost per Sqft. by System

Building Age	Exterior Closure	Roofing	Interior Construtiion	Stairways	Interior Finishes	Conveying Systems	Plumbing	HVAC Systems	Fire Protection	Electrical	Equipment	Total per Sqft.
1	.02	.83	.31				.16	.20	.01	.11		**1.64**
2	.02	.83	.31				.33	.20	.01	.11		**1.80**
3	.02	.90	.31				.16	.20	.01	.11		**1.71**
4	.03	.83	.35		.17		.33	.20	.01	.11		**2.02**
5	.05	.83	.31		.27		.30	.20	.01	.29		**2.25**
6	.02	.90	.31				.33	.20	.01	.11		**1.88**
7	.08	.86	.31				.18	.20	.01	.11		**1.75**
8	.03	.83	.35		.17		1.22	.20	.01	.11		**2.92**
9	.02	.90	.31		.36		.16	.20	.01	.11		**2.07**
10	.05	.83	1.15		1.63		.55	3.39	.24	.81		**8.64**
11	.02	.83	.31				.16	.20	.01	.11		**1.64**
12	.03	.90	.35		.17		.33	.20	.01	.11		**2.09**
13	.02	.83	.31				.38	.20	.01	.11		**1.85**
14	.08	3.33	.31				.34	.20	.01	.11		**4.38**
15	.05	.90	.31		.28		.23	5.18	.01	.58		**7.55**
16	.03	.83	.35		.17		1.22	.20	.01	.11		**2.92**
17	.02	.83	.31				.16	.20	.01	.11		**1.64**
18	.02	.90	.31		.36		.33	.20	.01	.11		**2.23**
19	.02	.83	.31				.16	.20	.01	.11		**1.64**
20	.60	1.00	1.19		1.94		.94	.40	.20	1.21		**7.47**
21	.08	.93	.31				.27	.20	.01	.11		**1.92**
22	.02	.83	.31				.33	.20	.01	.11		**1.80**
23	.02	.83	.31				.16	.20	.01	.11		**1.64**
24	.03	.90	.35		.17		1.22	.20	.01	.11		**2.99**
25	.05	.83	.31		8.36		7.67	3.35	.01	.61		**21.18**
26	.02	.83	.31				.42	.20	.01	.11		**1.90**
27	.02	.90	.31		.36		.18	.20	.01	.11		**2.08**
28	.09	4.92	.35		.17		.33	.20	.01	.11		**6.18**
29	.02	.83	.31				.27	.20	.01	.11		**1.75**
30	.05	.83	1.15		1.64		.56	5.22	.24	1.05		**10.74**
31	.02	.90	.31				.18	.20	.01	.11		**1.72**
32	.03	.83	.35		.17		1.21	.20	.01	.11		**2.90**
33	.02	.83	.31				.18	.20	.01	.11		**1.65**
34	.02	.90	.31				.32	.20	.01	.11		**1.86**
35	2.16	.86	.31		.27		.74	.20	.01	.29		**4.83**
36	.03	.83	.35		.52		.27	.20	.01	.11		**2.32**
37	.02	.90	.31				.32	.20	.01	.11		**1.87**
38	.02	.83	.31				.28	.20	.01	.11		**1.75**
39	.02	.83	.31				.33	.20	.01	.11		**1.81**
40	.38	1.07	1.19		1.94		1.89	3.55	.20	1.27		**11.48**
41	.02	.83	.31				.23	.20	.01	.11		**1.70**
42	.02	3.33	.31				.28	.20	.01	.11		**4.26**
43	.02	.90	.31				.23	.20	.01	.11		**1.77**
44	.03	.83	.35		.17		.27	.20	.01	.11		**1.96**
45	.05	.83	.31		.63		.45	5.18	.01	.58		**8.05**
46	.02	.90	.31				.27	.20	.01	.11		**1.81**
47	.08	.83	.31				.23	.20	.01	.11		**1.76**
48	.03	.83	.35		.17		1.16	.20	.01	.11		**2.85**
49	.02	.93	.31				.24	.20	.01	.11		**1.83**
50	.05	.83	4.97		13.54		7.94	.24	.24	1.01		**28.81**
Total	4.56	52.09	23.94		33.61		36.40	35.06	1.51	12.10		199.28

50-Year Total M&R Cost Profile per Sqft.

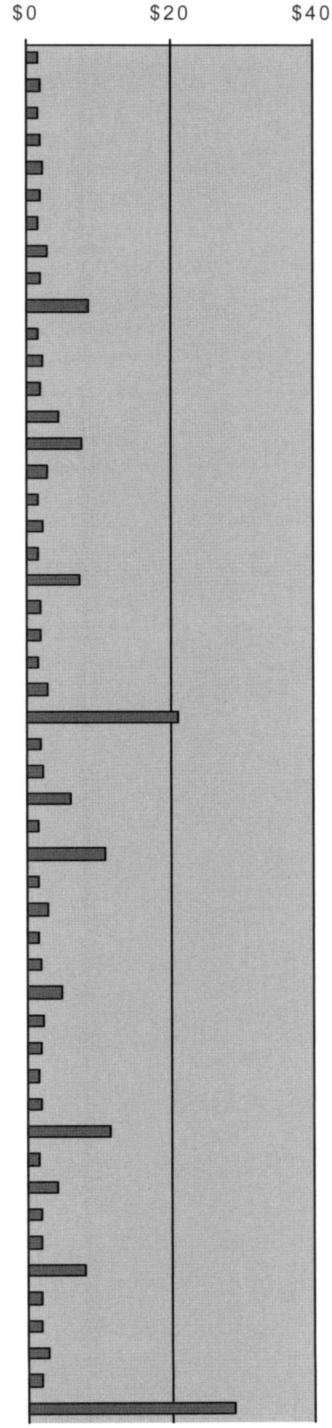

A value of "0.00" means a cost of more than \$.000 but less than \$.005 per gross square foot.

Garage, Auto Sales

Gross Sqft:	21,000
Height in Ft:	14
Exterior:	Metal Panel
Floor coverings:	Finished Concrete/Vinyl Tile
HVAC:	Electric Cool, Gas Heat< Singlezone Air
Occupancy:	840
Replacement Cost:	$1,500,604

Components

Component	Quantity	Unit
B20 Exterior Enclosure		
Stucco, Exterior, 1st Floor	4165	Sq Ft
Aluminum Operable Window, 12 sf, 1st Floor	18	Each
Steel, Painted, Exterior Door	3	Each
Aluminum Frame, Fully Glazed, Exterior Door	3	Each
Steel 20'x20', Painted, Roll-up Door	12	Each
B30 Roofing		
Concrete Decking	200	Sq Ft
Built-up Roof	21000	Sq Ft
C10 Interior Construction		
Steel, Painted, Interior Door	75	Each
C30 Interior Finishes		
Sheetrock, Stippled, Interior Wall Finish	21000	Sq Ft
Concrete, Finished Flooring	10500	Sq Ft
Vinyl Tile Flooring	10500	Sq Ft
Metal Ceiling, Painted	10500	Sq Ft
Acoustical Tile Ceiling	10500	Sq Ft
D20 Plumbing		
Tankless Water Closet	2	Each
Urinal, Vitreous China	2	Each
Lavatory, Vitreous China	5	Each
Sink, Stainless Steel	4	Each
Drinking Fountain, Refrigerated	2	Each
Pipe & Fittings, 3/4" Copper, Cold Water	0.63	K Ln Ft
Pipe & Fittings, 4" Steel	1.215	K Ln Ft
Pipe & Fittings, 2" Copper, Cold Water	0.934	K Ln Ft
Pipe & Fittings, 3/4" Copper, Hot Water	0.35	K Ln Ft
Pipe Insulation, Cold Water	1.038	K Ln Ft
Pipe Insulation, Hot Water	0.668	K Ln Ft
Water Heater, Gas/Oil 175 Gph	2	Each
Pipe & Fittings, 6" Cast Iron	0.333	K Ln Ft
Pipe & Fittings, 10" Cast Iron	0.29	K Ln Ft
Pipe & Fittings, 4" DWV PVC	0.17	K Ln Ft
Roof Drain, 2"	7	Each
D30 HVAC		
House Furnace, Gas/Oil, 100 Mbh	2	Each
Exhaust Fan, Roof Mounted, 1000 Cfm	2	Each
Exhaust Fan, Ceiling, 200-500 Cfm	2	Each
Air Conditioner, Rooftop, 50 Ton	1	Each
D40 Fire Protection		
Fire Sprinkler Head	150	Each
D50 Electrical		
Safety Switch, Fused, 400 Amp., 3 Ph.	2	Each
Main Switchgear, <1200 Amp.	1	Each
Distribution Panel Board	2	Each
Emergency Horn & Strobe	5	Each
Grounded Fault Circuit Interruptor, 15 Amps	2	Each
Exit Lighting Fixture, w/ Battery	3	Each
Fluorescent Lighting Fixture, 160w	210	Each
Wiring Device, Switch	18	Each
Receptacle, 120V, 15 Amp	26	Each
TV Cable Outlet	1	Each
Annunciation Panel	1	Each
Fire Alarm Bell, 6"	2	Each
Fire Alarm Control Panel	1	Each
Manual Pull Station	2	Each
Heat Detector	4	Each
Smoke Detector	8	Each

50-Year M&R Cost Summary

	50 Year Total	Annual Cost per Sqft	Annual cost as % replacement
PM & Minor Repair	$668,989	$0.64	0.89%
Unscheduled Maintenance	$745,407	$0.71	0.99%
Renewal & Replacement	$1,394,939	$1.33	1.86%
Total M&R Costs	$2,809,335	$2.68	3.74%

Distribution of Repair Costs

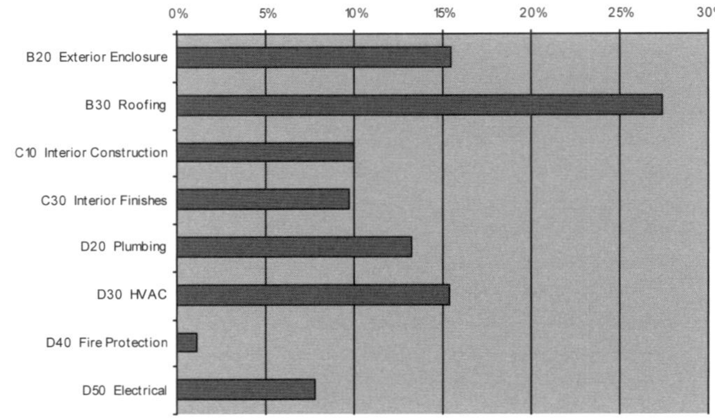

Thirty Most Costly Repair Tasks

Major Repair Task	Task Cost*	Pct.**
Maintain Built-up Roof	14.62	14.9%
Replace Air Conditioner, Rooftop, 50 Ton	7.34	7.5%
Refinish Sheetrock, Stippled, Interior Wall Finish	6.07	6.2%
Replace Steel, Painted, Door Locks	4.56	4.6%
Repair Air Conditioner, Rooftop, 50 Ton	4.50	4.6%
Maintain Steel 20'x20', Painted, Roll-up Door	4.41	4.5%
Replace Steel 20'x20', Painted, Roll-up Door	3.81	3.9%
Place New Membrane Over Existing, Built-up Roof	3.49	3.6%
Remove & Replace Membrane, Built-up Roof	3.45	3.5%
Maintain Steel, Painted, Door Locks	3.04	3.1%
Replace Pipe & Fittings, 2" Copper, Cold Water	3.01	3.1%
Replace Water Heater, Gas/Oil 175 Gph	2.78	2.8%
Replace Ballast & Lamps, Fluorescent Lighting Fixture, 160w	2.76	2.8%
Replace Vinyl Tile Flooring	2.10	2.1%
Maintain Air Conditioner, Rooftop, 50 Ton	2.09	2.1%
Replace Fluorescent Lighting Fixture, 160w	1.96	2.0%
Replace Pipe & Fittings, 3/4" Copper, Cold Water	1.79	1.8%
Maintain House Furnace, Gas/Oil, 100 Mbh	1.73	1.8%
Repair Steel 20'x20', Painted, Roll-up Door	1.66	1.7%
Refinish Stucco, Exterior, 1st Floor	1.60	1.6%
Refinish Steel 20'x20', Painted, Roll-up Door	1.42	1.4%
Replace Pipe & Fittings, 3/4" Copper, Hot Water	.99	1.0%
Clean Water Heater, Gas/Oil 175 Gph	.92	0.9%
Minor Repair, Acoustic Tile Ceiling	.89	0.9%
Refinish Metal Ceiling, Painted	.88	0.9%
Non-Destructive Moisture Inspection	.81	0.8%
Repair Fire Sprinkler Head	.69	0.7%
Refinish Concrete, Finished Flooring	.68	0.7%
Maintain Roof Drain, 2"	.61	0.6%
Minor Repair, Metal Ceiling, Painted	.61	0.6%

**Task cost ($2004) per gross square foot over 50 years.*

***Percent of total M&R costs.*

Cost per Sqft. by System

Building Age	Exterior Closure	Roofing	Interior Constrution	Stairways	Interior Finishes	Conveying Systems	Plumbing	HVAC Systems	Fire Protection	Electrical	Equipment	Total per Sqft.
1	.24	.59	.16				.10	.16	.01	.07		**1.32**
2	.24	.59	.16				.18	.16	.01	.07		**1.40**
3	.24	.64	.16				.10	.16	.01	.07		**1.38**
4	.24	.59	.21		.51		.18	.16	.01	.07		**1.96**
5	.40	.59	.16		.09		.15	.16	.01	.15		**1.70**
6	.24	.64	.16				.18	.16	.01	.07		**1.45**
7	.24	.60	.16				.10	.16	.01	.07		**1.34**
8	.24	.59	.21		.65		.59	.16	.01	.07		**2.53**
9	.24	.64	.16		.20		.10	.16	.01	.07		**1.57**
10	1.13	.59	1.08		.09		.28	1.75	.24	1.11		**6.26**
11	.24	.59	.16				.10	.16	.01	.07		**1.32**
12	.27	.64	.21		.51		.18	.16	.01	.07		**2.04**
13	.24	.59	.16				.21	.16	.01	.07		**1.44**
14	.29	2.34	.16				.18	.16	.01	.07		**3.21**
15	.40	.64	.16		.27		.11	2.65	.01	.31		**4.55**
16	.24	.59	.21		.65		.59	.16	.01	.07		**2.53**
17	.24	.59	.16				.10	.16	.01	.07		**1.32**
18	.24	.64	.16		1.23		.18	.16	.01	.07		**2.68**
19	.24	.59	.16				.10	.16	.01	.07		**1.32**
20	1.14	.64	1.12		1.02		.30	.26	.20	1.31		**6.00**
21	.24	.65	.16				.15	.16	.01	.07		**1.44**
22	.24	.59	.16				.18	.16	.01	.07		**1.40**
23	.24	.59	.16				.10	.16	.01	.07		**1.32**
24	.27	.64	.21		.65		.59	.16	.01	.07		**2.61**
25	.40	.59	.16		.43		3.61	1.74	.01	.44		**7.37**
26	.24	.59	.16				.23	.16	.01	.07		**1.46**
27	.24	.64	.16		.20		.10	.16	.01	.07		**1.58**
28	.29	3.46	.21		.51		.18	.16	.01	.07		**4.88**
29	.24	.59	.16				.15	.16	.01	.07		**1.38**
30	1.13	.59	1.08		.27		.27	2.66	.24	1.23		**7.48**
31	.24	.64	.16				.10	.16	.01	.07		**1.38**
32	.24	.59	.21		.65		.59	.16	.01	.07		**2.52**
33	.24	.59	.16				.10	.16	.01	.07		**1.33**
34	.24	.64	.16				.17	.16	.01	.07		**1.45**
35	3.99	.60	.16		.09		.30	.16	.01	.15		**5.45**
36	.27	.59	.21		1.73		.14	.16	.01	.07		**3.18**
37	.24	.64	.16				.18	.16	.01	.07		**1.46**
38	.24	.59	.16				.15	.16	.01	.07		**1.37**
39	.24	.59	.16				.20	.16	.01	.07		**1.43**
40	.73	.69	1.12		1.17		.73	1.83	.20	1.35		**7.83**
41	.24	.59	.16				.13	.16	.01	.07		**1.36**
42	.29	2.34	.16				.15	.16	.01	.07		**3.18**
43	.24	.64	.16				.13	.16	.01	.07		**1.41**
44	.24	.59	.21		.51		.14	.16	.01	.07		**1.92**
45	.81	.59	.16		.47		.25	2.65	.01	.31		**5.25**
46	.24	.64	.16				.14	.16	.01	.07		**1.42**
47	.24	.59	.16				.13	.16	.01	.07		**1.36**
48	.27	.59	.21		.65		.56	.16	.01	.07		**2.52**
49	.24	.65	.16				.14	.16	.01	.07		**1.43**
50	.72	.59	1.08		.43		3.71	.18	.24	1.31		**8.25**
Total	20.69	36.61	13.33		12.98		17.70	20.54	1.53	10.40		133.78

50-Year Total M&R Cost Profile per Sqft.

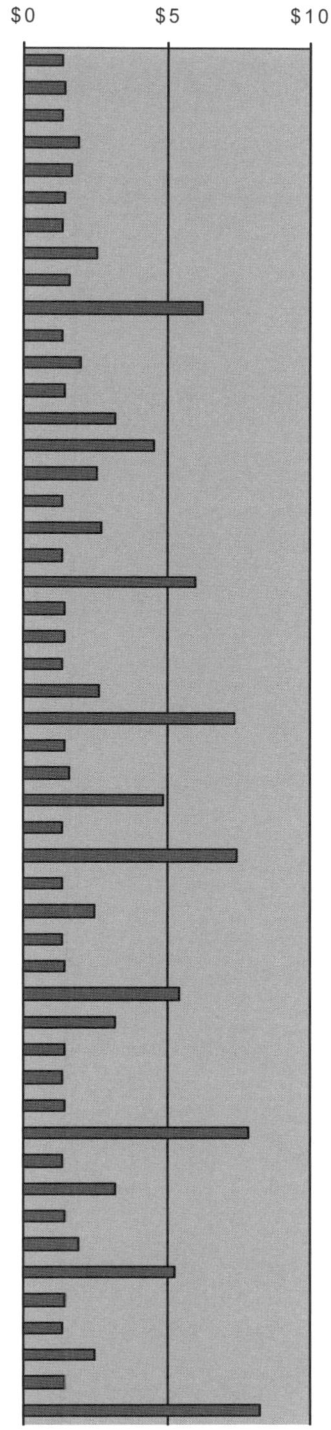

A value of "0.00" means a cost of more than $.000 but less than $.005 per gross square foot.

Garage, Parking

Gross Sqft:	110,808
Height in Ft:	40
Exterior:	Brick Veneer
Floor coverings:	Unfinished concrete
HVAC:	N/A
Occupancy:	0
Replacement Cost:	$3,709,301

Components

Component	Quantity	Unit
B20 Exterior Enclosure		
Clay Brick, Exterior, 1st Floor	1950	Sq Ft
Clay Brick, Exterior, 2nd Floor	1950	Sq Ft
Clay Brick, Exterior, 3rd+ Floor	3900	Sq Ft
Concrete, Painted, Exterior, 1st Floor	13710	Sq Ft
Concrete, Painted, Exterior, 2nd Floor	13710	Sq Ft
Concrete, Painted, Exterior, 3rd+ Floor	22710	Sq Ft
Aluminum Fixed Window, 24 sf, 1st Floor	4	Each
Aluminum Fixed Window, 24 sf, 2nd Floor	4	Each
Aluminum Fixed Window, 24 sf, 3rd+ Floor	8	Each
Steel, Painted, Exterior Door	6	Each
B30 Roofing		
Metal Roof	900	Sq Ft
C20 Stairs		
Concrete Interior Steps	640	Sq Ft
Metal Interior Railing, Painted	4776	Ln Ft
C30 Interior Finishes		
Concrete, Finished Ceiling	83106	Sq Ft
D10 Conveying		
Elevator, Hydraulic, 2500 lbs, 3-5 Floor, 200 fpm	2	Each
D20 Plumbing		
Service Sink, Iron Enamel	1	Each
Pipe & Fittings, 2" Copper, Cold Water	1.045	K Ln Ft
Water Heater, Electric 52 Gal.	1	Each
Pipe & Fittings, 6" Cast Iron	0.568	K Ln Ft
Pipe & Fittings, 10" Cast Iron	0.975	K Ln Ft
Roof Drain, 4-6"	8	Each
Pipe & Fittings, 4" DWV PVC	0.641	K Ln Ft
D40 Fire Protection		
Fire Standpipe System	1	Each
Fire Sprinkler System	1	Each
Fire Sprinkler Head	8	Each
Fire Suppression System Water Pump	1	Each
D50 Electrical		
Distribution Panel Board	4	Each
Secondary Transformer, Dry, 50 KVA	1	Each
Main Switchgear, <1200 Amp.	1	Each
Exit Lighting Fixture, w/ Battery	48	Each
Emergency Horn & Strobe	32	Each
Metal Halide Lighting Fixture, Low Bay, 250w	222	Each
Metal Halide Lighting Fixture, High Bay, 400w	12	Each
Wiring Device, Switch	8	Each
Receptacle, 120v, 20 Amp	16	Each
Fire Alarm Bell, 6"	8	Each
Manual Pull Station	8	Each
G20 Site Improvements		
Parking control equipment, barrier gate & controller	2	Each
Parking control equipment, card reader	2	Each
Parking stall	1966	Each

50-Year M&R Cost Summary

	50 Year Total	Annual Cost per Sqft	Annual cost as % replacement
PM & Minor Repair	$417,751	$0.08	0.23%
Unscheduled Maintenance	$342,199	$0.06	0.18%
Renewal & Replacement	$2,144,799	$0.39	1.16%
Total M&R Costs	$2,904,749	$0.52	1.57%

Distribution of Repair Costs

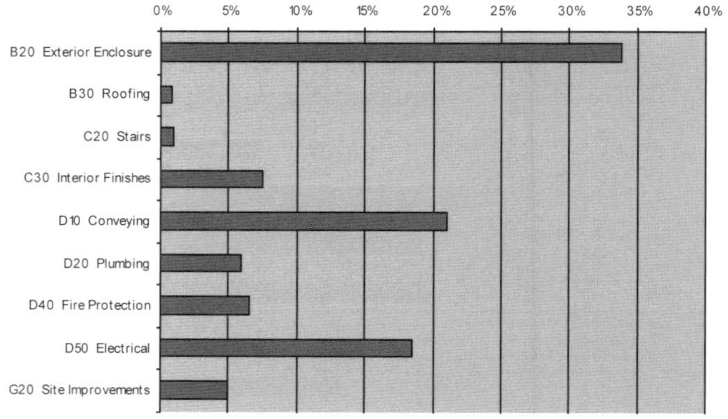

Thirty Most Costly Repair Tasks

Major Repair Task	Task Cost*	Pct.**
Maintain Elevator, Hydraulic, 2500 lbs, 3-5 Floor, 200 fpm	4.68	20.2%
Refinish Concrete, Painted, Exterior, 3rd+ Floor	2.40	10.4%
Replace Metal Halide Lighting Fixture, Low Bay, 250w	1.42	6.1%
Refinish Concrete, Painted, Exterior, 2nd Floor	1.21	5.2%
Refinish Concrete, Painted, Exterior, 1st Floor	.98	4.2%
Minor Repair, Concrete, Finished Ceiling	.94	4.1%
Replace Ballast, Metal Halide Lighting Fixture, Low Bay, 250w	.89	3.9%
Renovate Elevator, Hydraulic, 2500 lbs, 3-5 Floor, 200 fpm	.84	3.6%
Replace Lamp, Metal Halide Lighting Fixture, Low Bay, 250w	.80	3.5%
Clean & Reseal Clay Brick, Exterior, 3rd+ Floor	.78	3.4%
Fire Sprinkler System, Annual PM	.78	3.4%
Paint parking stall	.69	3.0%
Refinish Concrete, Finished Ceiling	.65	2.8%
Replace Pipe & Fittings, 2" Copper, Cold Water	.64	2.8%
Replace barrier gate and controller	.39	1.7%
Finish Repaired Concrete, Finished Ceiling	.36	1.6%
Repoint (50% surface) Clay Brick, Exterior, 3rd+ Floor	.34	1.5%
Clean & Reseal Clay Brick, Exterior, 2nd Floor	.33	1.4%
Clean & Reseal Clay Brick, Exterior, 1st Floor	.26	1.1%
Annual PM, Fire Suppression System Water Pump Motor	.25	1.1%
Replace parking control equipment, card reader	.22	0.9%
Annual PM, Fire Suppression System Water Pump	.21	0.9%
Annual PM, Emergency Horn & Strobe	.20	0.8%
Replace Exit Lighting Fixture, w/ Battery	.19	0.8%
Refinish Metal Interior Railing, Painted	.18	0.8%
Annual PM, Distribution Panel Board	.17	0.7%
Annual PM, Fire Suppression System Water Pump Testing	.17	0.7%
Repoint (50% Surface) Clay Brick, Exterior, 2nd Floor	.16	0.7%
Repoint (50% surface) Clay Brick, Exterior, 1st Floor	.16	0.7%
Replace Lamp, Exit Lighting Fixture, w/ Battery	.14	0.6%

*Task cost ($2004) per gross square foot over 50 years.

**Percent of total M&R costs.

Cost per Sqft. by System

Building Age	Exterior Closure	Roofing	Interior Constrution	Stairways	Interior Finishes	Conveying Systems	Plumbing	HVAC Systems	Fire Protection	Electrical	Equipment	Total per Sqft.
1	.00	.00				.09	.01		.03	.02		**.16**
2	.00	.00				.09	.01		.03	.02		**.16**
3	.00	.00				.09	.01		.03	.02		**.16**
4	.00	.00				.09	.01		.03	.02		**.16**
5	.00	.01				.09	.01		.04	.16		**.32**
6	.00	.00				.09	.01		.03	.02		**.16**
7	.00	.00		.03		.09	.01		.03	.02		**.18**
8	.00	.00				.09	.02		.03	.02		**.16**
9	.00	.00				.09	.01		.03	.02		**.16**
10	1.58	.01				.09	.01		.04	.49		**2.23**
11	.00	.00				.09	.01		.03	.02		**.16**
12	.00	.00				.09	.01		.03	.02		**.16**
13	.00	.00				.09	.03		.03	.02		**.17**
14	.02	.00		.03		.09	.01		.03	.02		**.20**
15	.00	.01		.02	.44	.09	.02		.04	.17		**.79**
16	.00	.00				.09	.01		.03	.02		**.16**
17	.00	.00				.09	.01		.03	.02		**.16**
18	.00	.00				.09	.01		.03	.02		**.16**
19	.00	.00				.09	.01		.03	.02		**.16**
20	1.59	.01				.09	.02		.04	.95		**2.71**
21	.00	.00		.03		.09	.01		.03	.02		**.18**
22	.00	.00				.09	.01		.03	.02		**.16**
23	.00	.00				.09	.02		.03	.02		**.16**
24	.00	.00				.09	.01		.03	.02		**.16**
25	.40	.01			.32	.51	.41		.04	.17		**1.87**
26	.00	.00				.09	.03		.03	.02		**.17**
27	.00	.00				.09	.01		.03	.02		**.16**
28	.02	.00		.03		.09	.01		.03	.02		**.20**
29	.00	.00				.09	.01		.03	.02		**.16**
30	1.58	.12		.02	.44	.09	.02		.04	.51		**2.83**
31	.00	.00				.09	.01		.03	.02		**.16**
32	.00	.00				.09	.01		.03	.02		**.16**
33	.00	.00				.09	.01		.03	.02		**.16**
34	.00	.00				.09	.01		.03	.02		**.16**
35	.00	.01		.03		.09	.02		.04	.16		**.35**
36	.00	.00				.09	.01		.03	.02		**.16**
37	.00	.00				.09	.01		.03	.02		**.16**
38	.00	.00				.09	.02		.03	.02		**.16**
39	.00	.00				.09	.03		.03	.02		**.17**
40	1.59	.01				.09	.02		.04	.96		**2.71**
41	.00	.00				.09	.01		.03	.02		**.16**
42	.02	.00		.03		.09	.01		.03	.02		**.20**
43	.00	.00				.09	.01		.03	.02		**.16**
44	.00	.00				.09	.01		.03	.02		**.16**
45	.00	.01		.02	.44	.09	.02		.04	.17		**.79**
46	.00	.00				.09	.01		.03	.02		**.16**
47	.00	.00				.09	.01		.03	.02		**.16**
48	.00	.00				.09	.01		.03	.02		**.16**
49	.00	.00		.03		.09	.01		.03	.02		**.18**
50	1.98	.01			.32	.51	.41		.04	.48		**3.77**
Total	8.88	.23		.24	1.95	5.52	1.56		1.70	4.83		24.92

50-Year Total M&R Cost Profile per Sqft.

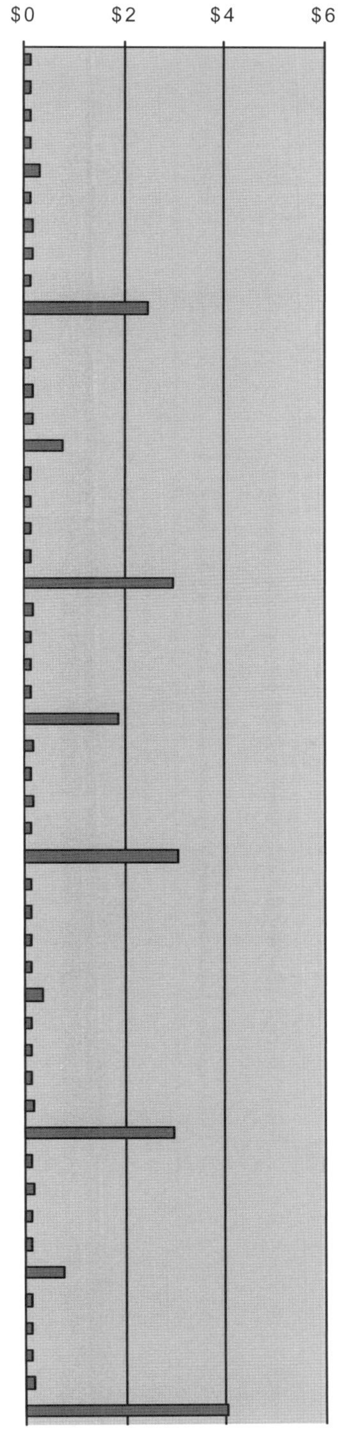

A value of "0.00" means a cost of more than $.000 but less than $.005 per gross square foot.

Garage, Service Station

Gross Sqft:	1,400
Height in Ft:	10
Exterior:	Brick Veneer
Floor coverings:	Finished Concrete/Vinyl Tile
HVAC:	Spilt System, Gas Heat
Occupancy:	14
Replacement Cost:	$169,882

Components

Component	Quantity	Unit
B20 Exterior Enclosure		
Clay Brick, Exterior, 1st Floor	918	Sq Ft
Aluminum Fixed Window, 12 sf, 1st Floor	2	Each
Steel, Painted, Exterior Door	2	Each
Aluminum Frame, Fully Glazed, Exterior Door	1	Each
Steel Single 12'x12', Painted, Roll-up Door	2	Each
B30 Roofing		
Asphalt Shingle Roof	1400	Sq Ft
C10 Interior Construction		
Steel, Painted, Interior Door	2	Each
C30 Interior Finishes		
Concrete Block, Painted, Interior Wall Finish	2038	Sq Ft
Concrete, Finished Flooring	910	Sq Ft
Vinyl Tile Flooring	490	Sq Ft
Acoustical Tile Ceiling	490	Sq Ft
Metal Ceiling, Painted	910	Sq Ft
D20 Plumbing		
Tankless Water Closet	3	Each
Lavatory, Vitreous China	3	Each
Service Sink, Iron Enamel	4	Each
Drinking Fountain, Refrigerated	2	Each
Pipe & Fittings, 3/4" Copper, Hot Water	0.23	K Ln Ft
Pipe & Fittings, 3/4" Copper, Cold Water	0.44	K Ln Ft
Pipe & Fittings, 2" Copper, Cold Water	0.517	K Ln Ft
Pipe & Fittings, 4" Steel	0.346	K Ln Ft
Pipe Insulation, Cold Water	0.614	K Ln Ft
Pipe Insulation, Hot Water	0.327	K Ln Ft
Water Heater, Gas/Oil 75 Gph	1	Each
Pipe & Fittings, 6" Cast Iron	0.148	K Ln Ft
Pipe & Fittings, 10" Cast Iron	0.077	K Ln Ft
D30 HVAC		
House Furnace, Gas/Oil, 100 Mbh	2	Each
Condensor, Air Cooled, 7.5 Ton	1	Each
Exhaust Fan, Roof Mounted, 5000 Cfm	3	Each
D40 Fire Protection		
Fire Sprinkler Head	10	Each
D50 Electrical		
Safety Switch, Fused, 400 Amp., 3 Ph.	2	Each
Distribution Panel Board	1	Each
Main Switchgear, <1200 Amp.	1	Each
Exit Lighting Fixture, w/ Battery	3	Each
Fluorescent Lighting Fixture, 160w	14	Each
Wiring Device, Switch	6	Each
Grounded Fault Circuit Interruptor, 15 Amps	2	Each
Receptacle, 120V, 15 Amp	11	Each
Annunciation Panel	1	Each
Manual Pull Station	2	Each
Smoke Detector	2	Each

50-Year M&R Cost Summary

	50 Year Total	Annual Cost per Sqft	Annual cost as % replacement
PM & Minor Repair	$148,289	$2.12	1.75%
Unscheduled Maintenance	$151,845	$2.17	1.79%
Renewal & Replacement	$255,248	$3.65	3.01%
Total M&R Costs	$555,382	$7.93	6.54%

Distribution of Repair Costs

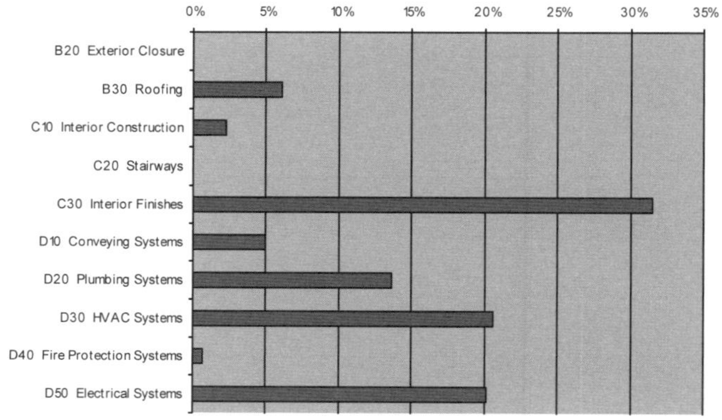

Thirty Most Costly Repair Tasks

Major Repair Task	Task Cost*	Pct.**
Maintain House Furnace, Gas/Oil, 100 Mbh	25.98	9.0%
Replace Pipe & Fittings, 2" Copper, Cold Water	25.00	8.7%
Replace Pipe & Fittings, 3/4" Copper, Cold Water	18.73	6.5%
Refinish Concrete Block, Painted, Interior Wall Finish	11.17	3.9%
Maintain Exhaust Fan, Roof Mounted, 5000 Cfm	10.29	3.6%
Clean & Reseal Clay Brick, Exterior, 1st Floor	9.74	3.4%
Replace Pipe & Fittings, 3/4" Copper, Hot Water	9.73	3.4%
Replace Condenser, Air-Cooled, 7.5 Ton	9.55	3.3%
Replace Water Heater, Gas/Oil 75 Gph	8.48	2.9%
Maintain Asphalt Shingle Roof	8.33	2.9%
Replace House Furnace, Gas/Oil, 100 Mbh	8.04	2.8%
Replace Safety Switch, Fused, 400 Amp., 3 Ph.	7.69	2.7%
Clean Water Heater, Gas/Oil 75 Gph	6.88	2.4%
Maintain Steel Single 12'x12', Painted, Roll-up Door	6.83	2.4%
Replace Exhaust Fan, Roof Mounted, 5000 Cfm	6.52	2.3%
Repoint (50% surface) Clay Brick, Exterior, 1st Floor	5.82	2.0%
Maintain Condenser, Air-Cooled, 7.5 Ton	5.71	2.0%
Repair Main Switchgear, <1200 Amp.	5.63	2.0%
Minor Replacement, Asphalt Shingle Roof	5.28	1.8%
Replace Drinking Fountain, Refrigerated	5.13	1.8%
Annual PM, Safety Switch, Fused, 400 Amp., 3 Ph.	3.94	1.4%
Replace Main Switchgear, <1200 Amp.	3.94	1.4%
Replace Coolant & Adjust Drinking Fountain, Refrigerated	3.61	1.3%
Repair House Furnace, Gas/Oil, 100 Mbh	3.38	1.2%
Annual PM, Distribution Panel Board	3.36	1.2%
Replace Steel Single 12'x12', Painted, Roll-up Door	3.18	1.1%
Replace Pipe Insulation, Cold Water	2.92	1.0%
Replace Ballast & Lamps, Fluorescent Lighting Fixture, 160w	2.76	1.0%
Inspect & Maintain Annunciation Panel	2.52	0.9%
Replace Faucet Washer & Clean Trap, Lavatory, Vitreous China	2.37	0.8%

**Task cost ($2004) per gross square foot over 50 years.*

***Percent of total M&R costs.*

Cost per Sqft. by System

Building Age	Exterior Closure	Roofing	Interior Construtiion	Stairways	Interior Finishes	Conveying Systems	Plumbing	HVAC Systems	Fire Protection	Electrical	Equipment	Total per Sqft.
1	.44	.33	.07				.42	1.57	.01	.50		**3.34**
2	.44	.33	.07				1.37	1.57	.01	.50		**4.30**
3	.44	.33	.07				.42	1.57	.01	.50		**3.34**
4	.46	33	.08		.93		1.37	1.57	.01	.50		**5.26**
5	.57	.92	.07		.12		.84	1.57	.01	1.61		**5.71**
6	.44	.33	.07				1.37	1.57	.01	.50		**4.30**
7	.44	.33	.07				.46	1.57	.01	.50		**3.38**
8	.46	.33	.08		1.12		2.54	1.57	.01	.54		**6.66**
9	.44	.33	.07		.14		.42	1.57	.01	.50		**3.48**
10	5.56	.92	.43		.12		2.84	3.97	.24	2.75		**16.81**
11	.44	.33	.07				.42	1.57	.01	.50		**3.34**
12	.61	.33	.08		.93		1.37	1.57	.01	.50		**5.41**
13	.44	.33	.07				1.11	1.57	.01	.50		**4.03**
14	.91	.33	.07				1.42	1.57	.01	.50		**4.80**
15	.57	.92	.07		.35		.51	6.24	.01	2.20		**10.86**
16	.46	.33	.08		1.12		2.54	1.57	.01	.54		**6.66**
17	.44	.33	.07				.42	1.57	.01	.50		**3.34**
18	.44	.33	.07		.86		1.37	1.57	.01	.50		**5.16**
19	.44	.33	.07				.42	1.57	.01	.50		**3.34**
20	5.59	2.59	.45		1.05		3.06	4.46	.20	4.17		**21.56**
21	.44	.33	.07				.80	1.57	.01	.50		**3.72**
22	.44	.33	.07				1.37	1.57	.01	.50		**4.30**
23	.44	.33	.07				.42	1.57	.01	.50		**3.34**
24	.61	.33	.08		1.81		2.54	1.57	.01	.54		**7.49**
25	4.11	.92	.07		.56		31.16	3.35	.01	5.52		**45.69**
26	.44	.33	.07				1.63	1.57	.01	.50		**4.55**
27	.44	.33	.07		.14		.46	1.57	.01	.50		**3.52**
28	.93	.33	.08		.93		1.37	1.57	.01	.50		**5.73**
29	.44	.33	.07				.80	1.57	.01	.50		**3.72**
30	5.56	.92	.43		.35		2.79	6.86	.24	3.12		**20.26**
31	.44	.33	.07				.46	1.57	.01	.50		**3.39**
32	.46	.33	.08		1.12		2.49	1.57	.01	.50		**6.58**
33	.44	.33	.07				.46	1.57	.01	.54		**3.42**
34	.44	.33	.07				1.33	1.57	.01	.50		**4.25**
35	3.40	.92	.07		.12		3.70	1.57	.01	1.61		**11.39**
36	.61	.33	.08		1.79		.89	1.57	.01	.50		**5.79**
37	.44	.33	.07				1.24	1.57	.01	.50		**4.16**
38	.44	.33	.07				.94	1.57	.01	.50		**3.86**
39	.44	.33	.07				1.20	1.57	.01	.50		**4.12**
40	5.25	2.13	.45		1.24		2.53	6.23	.20	4.39		**22.41**
41	.44	.33	.07				.90	1.57	.01	.54		**3.86**
42	.91	.33	.07				.93	1.57	.01	.50		**4.32**
43	.44	.33	.07				1.07	1.57	.01	.50		**3.99**
44	.46	.33	.08		.93		.89	1.57	.01	.50		**4.78**
45	.92	.92	.07		.49		2.37	6.24	.01	2.20		**13.20**
46	.44	.33	.07				.89	1.57	.01	.50		**3.81**
47	.44	.33	.07				.90	1.57	.01	.50		**3.82**
48	.61	.33	.08		1.81		1.89	1.57	.01	.50		**6.80**
49	.44	.33	.07				.94	1.57	.01	.54		**3.90**
50	8.75	.92	.43		.56		32.19	2.19	.24	6.22		**51.49**
Total	60.20	25.25	5.33		18.59		126.22	105.56	1.53	54.02		396.70

50-Year Total M&R Cost Profile per Sqft.

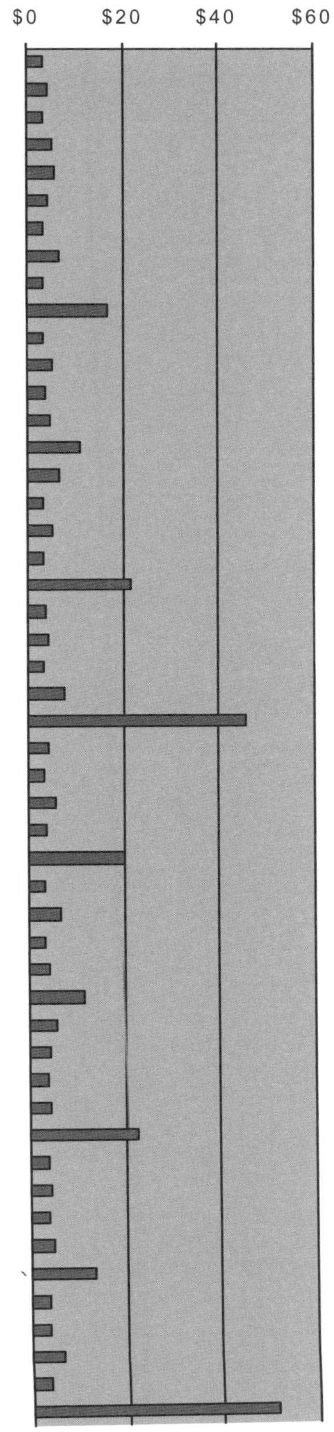

A value of "0.00" means a cost of more than $.000 but less than $.005 per gross square foot.

Gymnasium

Gross Sqft:	40,000
Height in Ft:	24
Exterior:	Clay Brick
Floor coverings:	Finished Conc/Maple/Rubber Tile/Carp
HVAC:	Single Zone Rooftop Units
Occupancy:	560
Replacement Cost:	$3,818,650

Components

Component	Quantity	Unit
B20 Exterior Enclosure		
Clay Brick, Exterior, 1st Floor	5480	Sq Ft
Clay Brick, Exterior, 2nd Floor	6130	Sq Ft
Aluminum Fixed Window, 12 sf, 1st Floor	80	Each
Aluminum Fixed Window, 12 sf, 2nd Floor	84	Each
Steel, Painted, Exterior Door	14	Each
Aluminum Frame, Fully Glazed, Exterior Door	2	Each
B30 Roofing		
Built-up Roof	40000	Sq Ft
C10 Interior Construction		
Steel, Painted, Interior Door	30	Each
Aluminum, w/ Safety Glass, Plain or Anodized, Interior Door	12	Each
C30 Interior Finishes		
Sheetrock, Unstippled, Interior Wall Finish	27840	Sq Ft
Concrete Block, Painted, Interior Wall Finish	960	Sq Ft
Ceramic Tile Wall Finish	2000	Sq Ft
Concrete, Finished Flooring	2000	Sq Ft
Ceramic Tile Flooring	6000	Sq Ft
Maple Sports Floor	10000	Sq Ft
Rubber Tile Flooring	12000	Sq Ft
Carpet, Nylon 20 oz., Low Traffic	10000	Sq Ft
Sheetrock, Unstippled, Ceiling	15000	Sq Ft
Metal Ceiling, Painted	25000	Sq Ft
D20 Plumbing		
Tankless Water Closet	11	Each
Lavatory, Vitreous China	13	Each
Service Sink, Iron Enamel	2	Each
Shower, Ceramic Tile	27	Each
Drinking Fountain, Vitreous China	6	Each
Pipe & Fittings, 3/4" Copper, Hot Water	1.3	K Ln Ft
Pipe & Fittings, 3/4" Copper, Cold Water	2.2	K Ln Ft
Pipe & Fittings, 2" Copper, Cold Water	2.9	K Ln Ft
Pipe Insulation, Cold Water	2.2	K Ln Ft
Pipe Insulation, Hot Water	1.3	K Ln Ft
Circulator Pump 1/6 HP, Cold Water	2	Each
Circulator Pump, 1/2 HP, Hot Water	1	Each
Water Heater, Gas/Oil 850 Gph	3	Each
Water Storage Tank, 2500 Gal.	1	Each
Pipe & Fittings, 6" Cast Iron	1.1	K Ln Ft
Pipe & Fittings, 10" Cast Iron	0.4	K Ln Ft
Roof Drain, 4-6"	12	Each
D30 HVAC		
Pipe & Fittings, 4" Steel, Gas	0.4	K Ln Ft
Air Conditioner, Rooftop, 20 Ton	1	Each
Air Conditioner, Rooftop, 50 Ton	3	Each
D40 Fire Protection		
Fire Sprinkler Head	200	Each
D50 Electrical		
Distribution Switch, Fused, <600 V	11	Each
Secondary Transformer, Dry, 150 KVA	3	Each
Main Switchgear, <1200 Amp.	1	Each
Fluorescent Lighting Fixture, 160w	150	Each
Metal Halide Lighting Fixture, High Bay, 400w	20	Each
Wiring Device, Switch	40	Each
Receptacle, 120V, 15 Amp	160	Each
Annunciation Panel	1	Each
Fire Alarm Control Panel	1	Each
Smoke Detector	80	Each

50-Year M&R Cost Summary

	50 Year Total	Annual Cost per Sqft	Annual cost as % replacement
PM & Minor Repair	$1,522,717	$0.76	0.80%
Unscheduled Maintenance	$1,571,168	$0.79	0.82%
Renewal & Replacement	$3,331,158	$1.67	1.74%
Total M&R Costs	$6,425,043	$3.21	3.37%

Distribution of Repair Costs

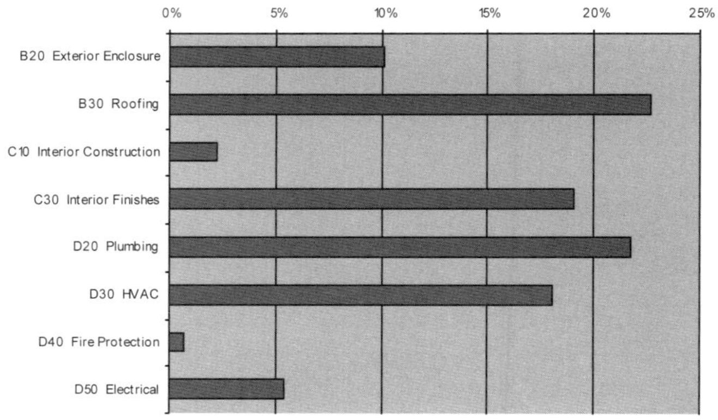

Thirty Most Costly Repair Tasks

Major Repair Task	Task Cost*	Pct.**
Maintain Built-up Roof	14.62	12.0%
Replace Air Conditioner, Rooftop, 50 Ton	11.57	9.5%
Repair Air Conditioner, Rooftop, 50 Ton	7.08	5.8%
Replace Rubber Tile Flooring	5.00	4.1%
Replace Pipe & Fittings, 2" Copper, Cold Water	4.91	4.0%
Screen & Recoat Maple Sports Floor	4.73	3.9%
Replace Water Heater, Gas/Oil 850 Gph	4.68	3.9%
Refinish Sheetrock, Unstippled, Interior Wall Finish	4.23	3.5%
Replace Carpet, Nylon 20 oz., Low Traffic	3.63	3.0%
Clean & Service Water Heater, Gas/Oil 850 Gph	3.51	2.9%
Place New Membrane Over Existing, Built-up Roof	3.49	2.9%
Remove & Replace Membrane, Built-up Roof	3.45	2.8%
Maintain Air Conditioner, Rooftop, 50 Ton	3.29	2.7%
Replace Pipe & Fittings, 3/4" Copper, Cold Water	3.28	2.7%
Clean & Reseal Clay Brick, Exterior, 2nd Floor	2.85	2.3%
Replace Air Conditioner, Rooftop, 20 Ton	2.28	1.9%
Refinish Sheetrock, Unstippled, Ceiling	2.28	1.9%
Replace Maple Sports Floor	2.09	1.7%
Clean & Reseal Clay Brick, Exterior, 1st Floor	2.04	1.7%
Replace Pipe & Fittings, 3/4" Copper, Hot Water	1.92	1.6%
Replace Ceramic Tile Flooring	1.91	1.6%
Repoint (50% Surface) Clay Brick, Exterior, 2nd Floor	1.42	1.2%
Check Operation, Smoke Detector	1.33	1.1%
Repoint (50% surface) Clay Brick, Exterior, 1st Floor	1.22	1.0%
Refinish Metal Ceiling, Painted	1.11	0.9%
Replace Ballast & Lamps, Fluorescent Lighting Fixture, 160w	1.03	0.9%
Replace Steel, Painted, Door Locks	.96	0.8%
Repair Air Conditioner, Rooftop, 20 Ton	.91	0.8%
Replace Valve Set, Shower, Ceramic Tile	.87	0.7%
Maintain Air Conditioner, Rooftop, 20 Ton	.85	0.7%

**Task cost ($2004) per gross square foot over 50 years.*

***Percent of total M&R costs.*

Cost per Sqft. by System

Building Age	Exterior Closure	Roofing	Interior Constrution	Stairways	Interior Finishes	Conveying Systems	Plumbing	HVAC Systems	Fire Protection	Electrical	Equipment	Total per Sqft.
1	.02	.59	.05		.13		.04	.15	.01	.06		**1.05**
2	.02	.59	.05		.13		.68	.15	.01	.06		**1.68**
3	.02	.64	.05		.13		.04	.15	.01	.06		**1.10**
4	.02	.59	.06		.69		.68	.15	.01	.06		**2.25**
5	.02	.59	.05		.25		.20	.15	.01	.13		**1.39**
6	.02	.64	.05		.13		.68	.15	.01	.06		**1.73**
7	.02	.59	.05		.13		.05	.15	.01	.06		**1.06**
8	.02	.59	.06		.88		.97	.15	.01	.06		**2.73**
9	.02	.64	.05		.19		.04	.15	.01	.06		**1.16**
10	2.36	.59	.25		.50		1.14	2.82	.17	.67		**8.49**
11	.02	.59	.05		.13		.04	.15	.01	.06		**1.05**
12	.03	.64	.06		.69		.68	.16	.01	.06		**2.32**
13	.02	.59	.05		.13		.12	.15	.01	.06		**1.12**
14	.13	2.33	.05		.13		.68	.15	.01	.06		**3.55**
15	.02	.64	.05		1.48		.17	4.62	.01	.39		**7.36**
16	.02	.59	.06		.88		.97	.15	.01	.06		**2.73**
17	.02	.59	.05		.13		.04	.15	.01	.06		**1.05**
18	.02	.64	.05		2.63		.68	.15	.01	.06		**4.24**
19	.02	.59	.05		.13		.04	.15	.01	.06		**1.05**
20	2.41	.59	.26		1.64		1.22	.15	.14	.85		**7.25**
21	.02	.64	.05		.13		.08	.15	.01	.06		**1.14**
22	.02	.59	.05		.13		.68	.15	.01	.06		**1.68**
23	.02	.59	.05		.13		.04	.15	.01	.06		**1.05**
24	.03	.64	.06		.89		.97	.16	.01	.06		**2.81**
25	1.61	.59	.05		.28		5.63	2.82	.01	.22		**11.20**
26	.02	.59	.05		.13		.70	.15	.01	.06		**1.70**
27	.02	.64	.05		.19		.05	.15	.01	.06		**1.17**
28	.14	3.45	.06		.69		.68	.15	.01	.06		**5.23**
29	.02	.59	.05		.13		.08	.15	.01	.06		**1.09**
30	2.36	.59	.25		1.72		1.54	4.66	.17	1.84		**13.12**
31	.02	.64	.05		.13		.05	.15	.01	.06		**1.11**
32	.02	.59	.06		.88		.96	.15	.01	.06		**2.72**
33	.02	.59	.05		.13		.05	.15	.01	.06		**1.06**
34	.02	.64	.05		.13		.67	.15	.01	.06		**1.73**
35	.02	.59	.05		.25		.57	.15	.01	.13		**1.76**
36	.03	.59	.06		3.19		.62	.16	.01	.06		**4.71**
37	.02	.64	.05		.13		.14	.15	.01	.06		**1.19**
38	.02	.59	.05		.13		.63	.15	.01	.06		**1.63**
39	.02	.59	.05		.13		.13	.15	.01	.06		**1.14**
40	2.41	.64	.26		3.60		1.37	2.82	.14	.93		**12.16**
41	.02	.59	.05		.13		.10	.15	.01	.06		**1.11**
42	.13	2.33	.05		.13		.62	.15	.01	.06		**3.49**
43	.02	.64	.05		.13		.11	.15	.01	.06		**1.16**
44	.02	.59	.06		.69		.62	.15	.01	.06		**2.19**
45	.02	.59	.05		1.53		.38	4.62	.01	.39		**7.58**
46	.02	.64	.05		.13		.62	.15	.01	.06		**1.67**
47	.02	.59	.05		.13		.10	.15	.01	.06		**1.11**
48	.03	.59	.06		.89		.90	.16	.01	.06		**2.70**
49	.02	.64	.05		.13		.11	.15	.01	.06		**1.17**
50	3.95	.59	.25		2.44		6.56	.15	.17	.58		**14.69**
Total	16.28	36.42	3.55		30.65		34.92	29.03	1.07	8.71		160.63

50-Year Total M&R Cost Profile per Sqft.

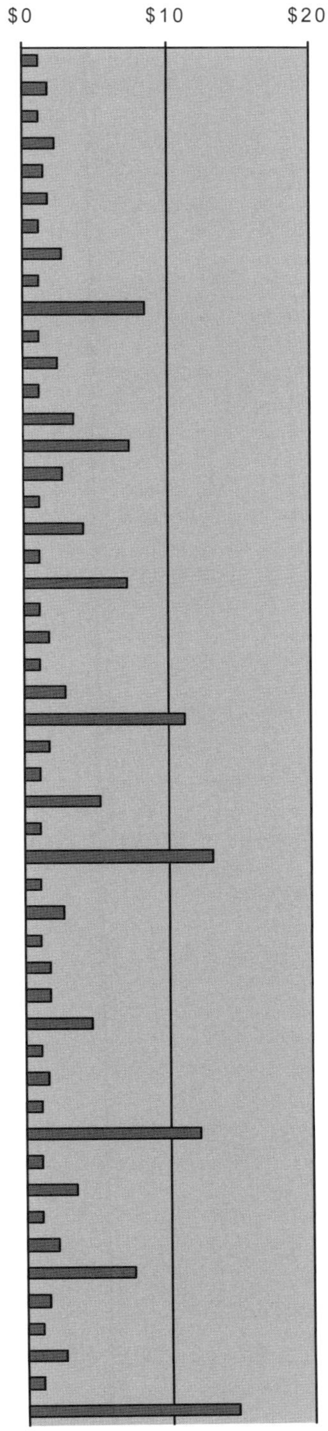

A value of "0.00" means a cost of more than $.000 but less than $.005 per gross square foot.

Hockey Rink

Gross Sqft:	30,000
Height in Ft:	24
Exterior:	Concrete Brick
Floor coverings:	Finished Concrete/Vinyl Sheet
HVAC:	Pkgd DX AC, Gas Boiler, Unit Heaters
Occupancy:	900
Replacement Cost:	$3,362,789

Components

Component	Quantity	Unit
B20 Exterior Enclosure		
Concrete Block, Exterior, 1st Floor	16872	Sq Ft
Aluminum Fixed Window, 12 sf, 1st Floor	9	Each
Steel w/ Safety Glass, Painted, Exterior Door	6	Each
Aluminum Frame, Fully Glazed, Exterior Door	2	Each
B30 Roofing		
Built-up Roof	30000	Sq Ft
C10 Interior Construction		
Wood, Hollow Core, Painted, Interior Door	12	Each
C30 Interior Finishes		
Concrete, Painted, Interior Wall Finish	10320	Sq Ft
Concrete, Unfinished Flooring	3000	Sq Ft
Vinyl Tile Flooring	12000	Sq Ft
Metal Ceiling, Painted	20000	Sq Ft
Acoustical Tile Ceiling	10000	Sq Ft
D20 Plumbing		
Tankless Water Closet	10	Each
Urinal, Vitreous China	7	Each
Lavatory, Vitreous China	8	Each
Sink, Stainless Steel	2	Each
Shower, Ceramic Tile	17	Each
Drinking Fountain, Refrigerated	3	Each
Pipe & Fittings, 3/4" Copper, Cold Water	1.57	K Ln Ft
Pipe & Fittings, 4" Steel	1.482	K Ln Ft
Pipe & Fittings, 2" Copper, Cold Water	1.92	K Ln Ft
Pipe & Fittings, 3/4" Copper, Hot Water	0.82	K Ln Ft
Pipe Insulation, Cold Water	2.31	K Ln Ft
Pipe Insulation, Hot Water	1.214	K Ln Ft
Water Heater, Gas/Oil 75 Gph	3	Each
Pipe & Fittings, 6" Cast Iron	0.871	K Ln Ft
Pipe & Fittings, 10" Cast Iron	0.346	K Ln Ft
D30 HVAC		
Gas Boiler, 1000 Mbh	2	Each
Circulation Pump 5 HP, Hot Water	2	Each
Exhaust Fan, Ceiling, 200-500 Cfm	6	Each
Air Conditioner, Indoor DX Packaged, Air Cooled, 20 Ton	1	Each
Air Conditioner, Indoor DX Packaged, Air Cooled, 50 Ton	1	Each
Terminal Reheat Unit, 10,000 Cfm	3	Each
D50 Electrical		
Safety Switch, Fused, 400 Amp., 3 Ph.	2	Each
Main Switchgear, <1200 Amp.	1	Each
Secondary Transformer, Dry, 50 KVA	1	Each
Distribution Panel Board	6	Each
Grounded Fault Circuit Interruptor, 15 Amps	5	Each
Emergency Horn & Strobe	12	Each
Exit Lighting Fixture, w/ Battery	14	Each
Fluorescent Lighting Fixture, 160w	60	Each
Metal Halide Lighting Fixture, High Bay, 400w	300	Each
Wiring Device, Switch	10	Each
Receptacle, 120V, 15 Amp	38	Each
Annunciation Panel	1	Each
Fire Alarm Bell, 6"	6	Each
Fire Alarm Control Panel	1	Each
Manual Pull Station	6	Each
Smoke Detector	18	Each
Time Control Clock	1	Each
Generator Switchgear	4	Each

50-Year M&R Cost Summary

	50 Year Total	Annual Cost per Sqft	Annual cost as % replacement
PM & Minor Repair	$1,397,295	$0.93	0.83%
Unscheduled Maintenance	$1,303,346	$0.87	0.78%
Renewal & Replacement	$1,965,710	$1.31	1.17%
Total M&R Costs	$4,666,351	$3.11	2.78%

Distribution of Repair Costs

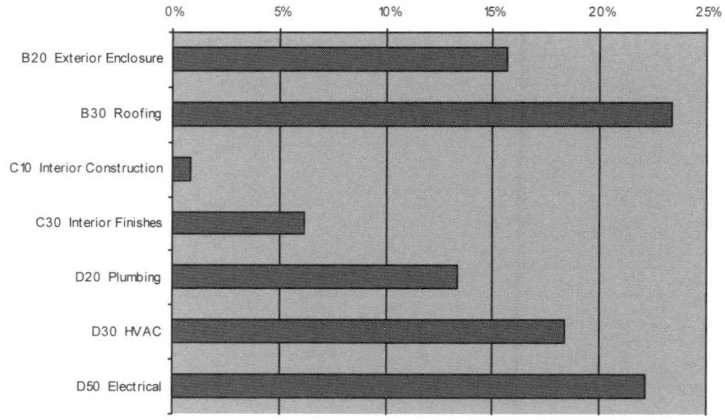

Thirty Most Costly Repair Tasks

Major Repair Task	Task Cost*	Pct.**
Maintain Built-up Roof	14.62	13.0%
Clean & Seal Concrete Block, Exterior, 1st Floor	8.39	7.5%
Replace Metal Halide Lighting Fixture, High Bay, 400w	8.16	7.3%
Replace Ballast, Metal Halide Lighting Fixture, High Bay, 400w	5.31	4.7%
Inspect & Test Gas Boiler, 1000 Mbh	4.36	3.9%
Replace Pipe & Fittings, 2" Copper, Cold Water	4.33	3.9%
Replace Lamp, Metal Halide Lighting Fixture, High Bay, 400w	4.22	3.8%
Annual PM, Generator Switchgear	3.62	3.2%
Place New Membrane Over Existing, Built-up Roof	3.49	3.1%
Remove & Replace Membrane, Built-up Roof	3.45	3.1%
Replace Air Conditioner, Indoor DX Packaged, Air Cooled, 50 Ton	3.43	3.1%
Replace Pipe & Fittings, 3/4" Copper, Cold Water	3.12	2.8%
Refinish Concrete, Painted, Interior Wall Finish	2.64	2.4%
Maintain Terminal Reheat Unit, 10,000 Cfm	2.62	2.3%
Repoint (50% surface) Concrete Block, Exterior, 1st Floor	2.42	2.2%
Replace Generator Switchgear	2.05	1.8%
Minor Repair, Concrete, Painted, Interior Wall Finish	2.01	1.8%
Repair Air Conditioner, Indoor DX Packaged, Air Cooled, 50 Ton	1.72	1.5%
Replace Vinyl Tile Flooring	1.68	1.5%
Replace Pipe & Fittings, 3/4" Copper, Hot Water	1.62	1.4%
Maintain Air Conditioner, Indoor DX Packaged, Air Cooled, 50 Ton	1.49	1.3%
Replace Air Conditioner, Indoor DX Packaged, Air Cooled, 20 Ton	1.43	1.3%
Replace Gas Boiler, 1000 Mbh	1.37	1.2%
Replace Water Heater, Gas/Oil 75 Gph	1.19	1.1%
Repair Concrete Block, Exterior, 1st Floor	1.18	1.1%
Refinish Metal Ceiling, Painted	1.18	1.1%
Maintain Air Conditioner, Indoor DX Packaged, Air Cooled, 20 Ton	1.16	1.0%
Clean Water Heater, Gas/Oil 75 Gph	.96	0.9%
Annual PM, Distribution Panel Board	.94	0.8%
Replace Terminal Reheat Unit, 10,000 Cfm	.83	0.7%

**Task cost ($2004) per gross square foot over 50 years.*

***Percent of total M&R costs.*

Cost per Sqft. by System

Building Age	Exterior Closure	Roofing	Interior Constrution	Stairways	Interior Finishes	Conveying Systems	Plumbing	HVAC Systems	Fire Protection	Electrical	Equipment	Total per Sqft.
1	.01	.59	.02				.06	.36		.23		**1.26**
2	.01	.59	.02				.17	.37		.23		**1.38**
3	.01	.64	.02				.06	.36		.23		**1.31**
4	.01	.59	.02		.22		.17	.37		.23		**1.61**
5	.01	.59	.02		.12		.21	.37		.87		**2.18**
6	.01	.64	.02				.17	.37		.23		**1.43**
7	.01	.59	.02				.07	.50		.23		**1.41**
8	.01	.59	.02		.42		.31	.37		.23		**1.95**
9	.01	.64	.02		.13		.06	.36		.23		**1.44**
10	4.00	.59	.07		.12		.57	1.21		2.87		**9.42**
11	.01	.59	.02				.06	.36		.23		**1.26**
12	.03	.64	.02		.22		.17	.37		.23		**1.67**
13	.01	.59	.02				.18	.36		.23		**1.38**
14	.08	2.33	.02				.18	.51		.23		**3.34**
15	.01	.64	.02		.88		.16	.68		1.02		**3.41**
16	.01	.59	.02		.42		.31	.37		.23		**1.95**
17	.01	.59	.02				.06	.36		.23		**1.26**
18	.01	.64	.02		.96		.17	.37		1.11		**3.27**
19	.01	.59	.02				.06	.36		.23		**1.26**
20	4.00	.59	.07		.34		.61	2.79		4.84		**13.25**
21	.01	.64	.02				.12	.50		.23		**1.51**
22	.01	.59	.02				.17	.37		.23		**1.38**
23	.01	.59	.02				.06	.36		.23		**1.26**
24	.03	.64	.02		.42		.31	.37		.23		**2.01**
25	1.81	.59	.02		.12		5.41	.37		1.08		**9.40**
26	.01	.59	.02				.22	.37		.23		**1.44**
27	.01	.64	.02		.13		.07	.36		.23		**1.45**
28	.08	3.45	.02		.22		.17	.51		.23		**4.68**
29	.01	.59	.02				.11	.36		.23		**1.32**
30	4.00	.59	.12		.88		.90	2.71		3.07		**12.28**
31	.01	.64	.02				.07	.36		.23		**1.32**
32	.01	.59	.02		.42		.30	.37		.23		**1.93**
33	.01	.59	.02				.07	.36		.23		**1.27**
34	.01	.64	.02				.16	.37		.23		**1.43**
35	.01	.59	.02		.12		.56	.37		.87		**2.53**
36	.03	.59	.02		1.18		.12	.37		1.11		**3.42**
37	.01	.64	.02				.15	.50		.23		**1.54**
38	.01	.59	.02				.13	.37		.23		**1.35**
39	.01	.59	.02				.17	.36		.23		**1.37**
40	4.00	.64	.07		.54		.62	2.80		4.88		**13.55**
41	.01	.59	.02				.10	.36		.23		**1.30**
42	.08	2.33	.02				.14	.37		.23		**3.17**
43	.01	.64	.02				.10	.36		.23		**1.35**
44	.01	.59	.02		.22		.12	.51		.23		**1.70**
45	.01	.59	.02		1.02		.36	.68		1.02		**3.69**
46	.01	.64	.02				.12	.37		.23		**1.39**
47	.01	.59	.02				.10	.36		.23		**1.30**
48	.03	.59	.02		.42		.27	.37		.23		**1.92**
49	.01	.64	.02				.11	.36		.23		**1.37**
50	5.80	.59	.07		.12		5.70	1.20		3.00		**16.48**
Total	24.39	36.42	1.29		9.62		20.77	28.61		34.44		155.55

A value of "0.00" means a cost of more than $.000 but less than $.005 per gross square foot.

50-Year Total M&R Cost Profile per Sqft.

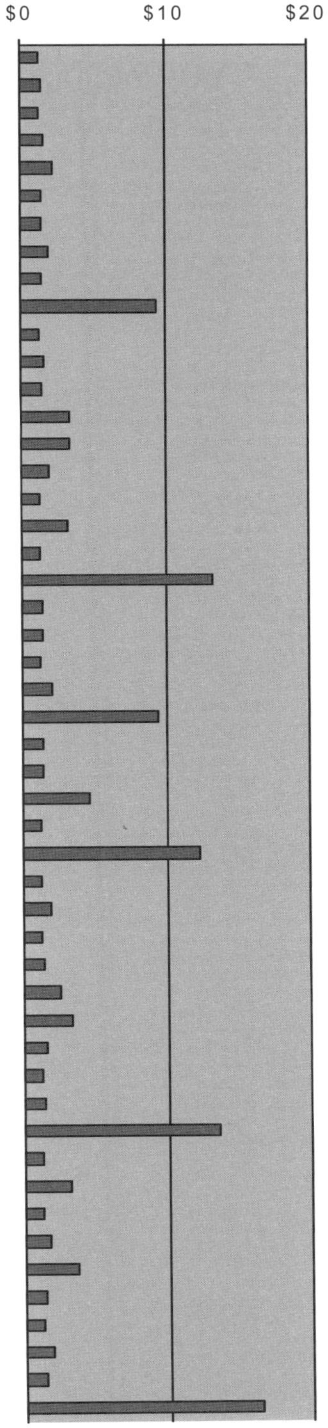

Hospital, General

Gross Sqft:	125,000
Height in Ft:	36
Exterior:	Cast Concrete
Floor coverings:	Finished Concrete/Tile/Carpet
HVAC:	Chilled Water, Gas Boiler, Fan Coil Units
Occupancy:	1500
Replacement Cost:	$15,963,874

50-Year M&R Cost Summary

	50 Year Total	Annual Cost per Sqft	Annual cost as % replacement
PM & Minor Repair	$5,361,497	$0.86	0.67%
Unscheduled Maintenance	$5,363,400	$0.86	0.67%
Renewal & Replacement	$15,845,214	$2.54	1.99%
Total M&R Costs	$26,570,111	$4.25	3.33%

Components

B20 Exterior Enclosure		
Concrete, Painted, Exterior, 1st Floor	8840	Sq Ft
Concrete, Painted, Exterior, 2nd Floor	8840	Sq Ft
Concrete, Painted, Exterior, 3rd+ Floor	8840	Sq Ft
Aluminum Fixed Window, 12 sf, 1st Floor	60	Each
Aluminum Fixed Window, 12 sf, 2nd Floor	90	Each
Aluminum Fixed Window, 12 sf, 3rd+ Floor	90	Each
Steel, Painted, Exterior Door	8	Each
Aluminum Frame, Fully Glazed, Exterior Door	6	Each
Aluminum Frame, Fully Glazed, Sliding Exterior Door	2	Each
B30 Roofing		
Built-up Roof	41700	Sq Ft
C10 Interior Construction		
Wood, Solid Core, Painted, Interior Door	1317	Each
Steel, Painted, Interior Door	50	Each
Aluminum, w/ Safety Glass, Plain or Anodized, Interior Door	8	Each
Wood Lead-Lined Medical Interior Door	4	Each
C30 Interior Finishes		
Plaster Interior Wall Finish	188180	Sq Ft
Vinyl Interior Wall Finish	111120	Sq Ft
Concrete, Finished Flooring	6250	Sq Ft
Ceramic Tile Flooring	31250	Sq Ft
Vinyl Tile Flooring	62500	Sq Ft
Carpet, Nylon 20 oz., Low Traffic	25000	Sq Ft
Plaster Ceiling	93750	Sq Ft
Acoustical Tile Ceiling	31250	Sq Ft
D10 Conveying		
Elevator, Hydraulic, 2500 lbs, 3-5 Floor, 200 fpm	2	Each
D20 Plumbing		
Tankless Water Closet	112	Each
Flush Tank Water Closet	75	Each
Urinal, Vitreous China	8	Each
Lavatory, Vitreous China	150	Each
Sink, Enameled Steel	70	Each
Sink, Stainless Steel	24	Each
Service Sink, Iron Enamel	6	Each
Shower, Fiberglass	75	Each
Bath Tub, Enameled Steel	5	Each
Drinking Fountain, Refrigerated	6	Each
Pipe & Fittings, 3/4" Copper, Cold Water	16.5	K Ln Ft
Pipe & Fittings, 3/4" Copper, Hot Water	10.3	K Ln Ft
Pipe & Fittings, 2" Copper, Cold Water	15.2	K Ln Ft
Pipe Insulation, Cold Water	17	K Ln Ft
Pipe Insulation, Hot Water	11.1	K Ln Ft
Circulator Pump, 1/2 HP, Hot Water	4	Each
Water Heater, Gas/Oil 275 Gph	1	Each
Water Heater, Gas/Oil 575 Gph	2	Each
Water Storage Tank, 1000 Gal.	1	Each
Pipe & Fittings, 6" Cast Iron	6.3	K Ln Ft
Pipe & Fittings, 10" Cast Iron	0.5	K Ln Ft
Pipe & Fittings, 4" DWV PVC	0.9	K Ln Ft
Pipe & Fittings, 3/4" Copper, Oxygen	2	K Ln Ft
D30 HVAC		
Pipe & Fittings, 4" Steel, Gas	6.6	K Ln Ft
Chemical Feed System	1	Each
Circulation Pump 25 HP, Hot Water	2	Each
Expansion Tank, 100 Gal.	2	Each
Gas Boiler, 2000 Mbh	2	Each
Circulation Pump, 75 HP, Chiller & Condenser Water	2	Each
Chiller, Centrifugal Hermetic, Bundled w/ Condenser, 200 Ton	2	Each
Cooling Tower, 200 Ton	2	Each
Fan Coil, Two-Pipe, 400 Cfm	390	Each
D40 Fire Protection		
Fire Sprinkler Head	625	Each
D50 Electrical		
Distribution Switch, Fused, <600 V	16	Each
Primary Transformer, Dry, >500 KVA	1	Each
Secondary Transformer, Dry, 150 KVA	8	Each
Main Switchgear, >1200 Amp.	1	Each
Motor Starter, 21-50 HP, <600 V	9	Each
Motor Starter, 50-75 HP, <600 V	2	Each
Incandescent Lighting Fixture, Basic, 100w	75	Each
Fluorescent Lighting Fixture, 160w	2344	Each
Wiring Device, Switch	700	Each
Receptacle, 120V, 15 Amp	1000	Each
Annunciation Panel	1	Each
Fire Alarm Control Panel	1	Each
Manual Pull Station	100	Each
Smoke Detector	500	Each
Public Address Speaker	1	Each

Distribution of Repair Costs

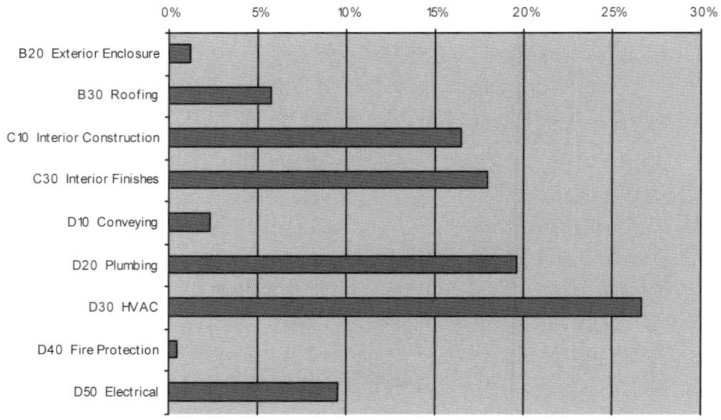

Thirty Most Costly Repair Tasks

Major Repair Task	Task Cost*	Pct.**
Maintain Fan Coil, Two-Pipe, 400 Cfm	14.32	8.4%
Replace Fan Coil, Two-Pipe, 400 Cfm	12.48	7.4%
Refinish Plaster Interior Wall Finish	9.15	5.4%
Maintain Wood, Solid Core, Painted, Door Locks	8.79	5.2%
Replace Pipe & Fittings, 2" Copper, Cold Water	8.23	4.9%
Replace Pipe & Fittings, 3/4" Copper, Cold Water	7.87	4.6%
Replace Vinyl Interior Wall Finish	7.34	4.3%
Replace Ballast & Lamps, Fluorescent Lighting Fixture, 160w	5.17	3.0%
Replace Pipe & Fittings, 3/4" Copper, Hot Water	4.88	2.9%
Maintain Built-up Roof	4.88	2.9%
Refinish Plaster Ceiling	4.57	2.7%
Maintain Elevator, Hydraulic, 2500 lbs, 3-5 Floor, 200 fpm	4.15	2.4%
Replace Chiller, Centrifugal Hermetic, Bnd'd w/ Condenser, 200 Ton	4.08	2.4%
Replace Wood, Solid Core, Painted, Interior Door	3.93	2.3%
Replace Solid Core, Painted, Door Locks	3.91	2.3%
Minor Repair, Plaster Ceiling	3.71	2.2%
Replace Fluorescent Lighting Fixture, 160w	3.67	2.2%
Replace Ceramic Tile Flooring	3.18	1.9%
Replace Carpet, Nylon 20 oz., Low Traffic	2.90	1.7%
Check Operation, Smoke Detector	2.66	1.6%
Repair Fan Coil, Two-Pipe, 400 Cfm	2.42	1.4%
Maintain Chiller, Centrifugal Hermetic, Bnd'd w/ Condenser, 200 Ton	2.29	1.3%
Minor Repair, Plaster Interior Wall Finish	2.20	1.3%
Replace Vinyl Tile Flooring	2.10	1.2%
Inspect & Test Gas Boiler, 2000 Mbh	2.09	1.2%
Repair Chiller, Centrifugal Hermetic, Bnd'd w/ Condenser, 200 Ton	1.79	1.1%
Refinish Wood, Solid Core, Painted, Interior Door	1.56	0.9%
Replace Smoke Detector	1.40	0.8%
Replace Faucet Washer & Clean Trap, Lavatory, Vitreous China	1.32	0.8%
Replace Pipe & Fittings, 3/4" Copper, Oxygen	1.22	0.7%

**Task cost ($2004) per gross square foot over 50 years.*

***Percent of total M&R costs.*

Cost per Sqft. by System

Building Age	Exterior Closure	Roofing	Interior Constrution	Stairways	Interior Finishes	Conveying Systems	Plumbing	HVAC Systems	Fire Protection	Electrical	Equipment	Total per Sqft.
1	.00	.20	.51			.08	.02	.69	.01	.13		**1.63**
2	.00	.20	.51			.08	.39	.69	.01	.13		**2.00**
3	.00	.21	.51			.08	.02	.69	.01	.13		**1.64**
4	.01	.20	.64		1.14	.08	.40	.69	.01	.13		**3.29**
5	.00	.20	.51		.26	.08	.18	.70	.01	.17		**2.10**
6	.00	.21	.51			.08	.39	.69	.01	.13		**2.02**
7	.00	.20	.51			.08	.04	.74	.01	.13		**1.71**
8	.01	.20	.64		1.14	.08	.52	.69	.01	.13		**3.41**
9	.00	.21	.51		.11	.08	.02	.69	.01	.13		**1.76**
10	.42	.20	1.60		1.47	.08	1.16	2.16	.17	2.22		**9.47**
11	.00	.20	.51			.08	.02	.69	.01	.13		**1.63**
12	.02	.21	.64		2.91	.08	.40	.72	.01	.13		**5.12**
13	.00	.20	.51			.08	.10	.69	.01	.13		**1.71**
14	.03	.78	.51			.08	.41	.74	.01	.13		**2.69**
15	.00	.21	.51		1.24	.08	.21	5.00	.01	.78		**8.05**
16	.01	.20	.64		1.14	.08	.52	.69	.01	.13		**3.41**
17	.00	.20	.51			.08	.02	.69	.01	.13		**1.63**
18	.00	.21	.51		1.14	.08	.39	.69	.01	.26		**3.30**
19	.00	.20	.51			.08	.02	.69	.01	.13		**1.63**
20	.44	.20	1.73		2.62	.08	1.35	2.66	.14	2.32		**11.53**
21	.00	.21	.51			.08	.05	.74	.01	.13		**1.74**
22	.00	.20	.51			.08	.39	.69	.01	.13		**2.00**
23	.00	.20	.51			.08	.02	.69	.01	.16		**1.65**
24	.02	.21	.64		2.91	.08	.52	.72	.01	.13		**5.24**
25	.00	.20	.51		.29	.46	12.17	1.56	.01	.45		**15.65**
26	.00	.20	.51			.08	.41	.69	.01	.13		**2.02**
27	.00	.21	.51		.11	.08	.03	.69	.01	.13		**1.78**
28	.03	1.15	.64		1.14	.08	.41	.74	.01	.16		**4.36**
29	.00	.20	.51			.08	.05	.69	.01	.13		**1.66**
30	.42	.20	4.20		2.46	.08	1.18	6.17	.17	3.43		**18.31**
31	.00	.21	.51			.08	.03	.69	.01	.13		**1.66**
32	.01	.20	.51		1.14	.08	.50	.69	.01	.13		**3.27**
33	.00	.20	.51			.08	.03	.69	.01	.16		**1.67**
34	.00	.21	.64			.08	.37	.69	.01	.13		**2.13**
35	.00	.20	.51		.26	.08	1.79	.70	.01	.15		**3.68**
36	.02	.20	.52		4.05	.08	.19	.72	.01	.26		**6.04**
37	.00	.21	.51			.08	.23	.74	.01	.13		**1.91**
38	.00	.20	.64			.08	.22	.69	.01	.13		**1.96**
39	.00	.20	.51			.08	.27	.69	.01	.13		**1.88**
40	.44	.21	1.60		2.62	.08	.87	3.52	.14	2.62		**12.12**
41	.00	.20	.51			.08	.22	.69	.01	.16		**1.86**
42	.03	.78	.64			.08	.21	.69	.01	.13		**2.56**
43	.00	.21	.51			.08	.27	.69	.01	.13		**1.90**
44	.01	.20	.51		1.14	.08	.19	.74	.01	.13		**3.00**
45	.00	.20	.51		1.35	.08	.79	5.00	.01	.75		**8.70**
46	.00	.21	.64			.08	.19	.69	.01	.16		**1.97**
47	.00	.20	.51			.08	.23	.69	.01	.13		**1.84**
48	.02	.20	.51		2.91	.08	.27	.72	.01	.13		**4.85**
49	.00	.21	.51			.08	.25	.69	.01	.13		**1.87**
50	.42	.20	1.73		4.69	.46	12.69	1.31	.17	1.89		**23.54**
Total	2.48	12.15	35.07		38.25	4.90	41.63	56.71	1.07	20.31		212.56

50-Year Total M&R Cost Profile per Sqft.

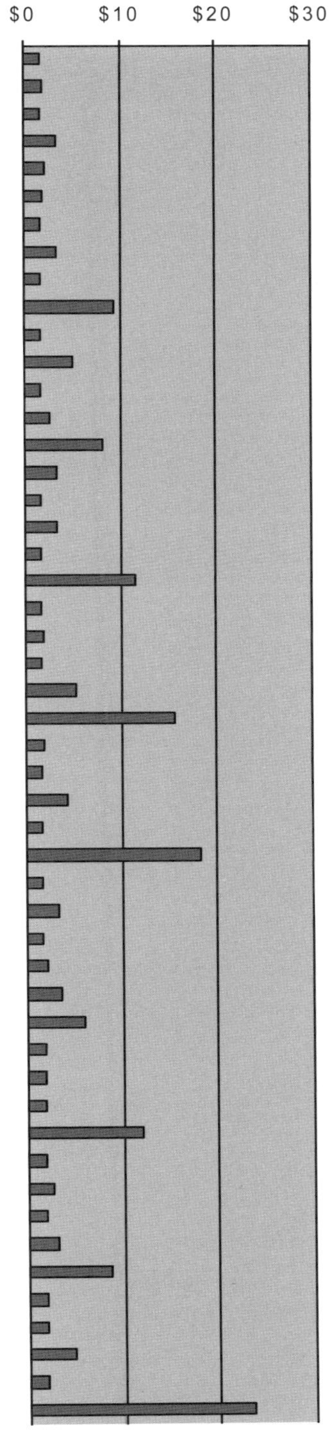

A value of "0.00" means a cost of more than $.000 but less than $.005 per gross square foot.

Research Hospital

Gross Sqft:	540,200
Height in Ft:	60
Exterior:	Clay Brick
Floor coverings:	Vinyl Tile/Terrazzo, Carpet
HVAC:	Chilled Water, Gas Boiler, Fan Coil Units
Occupancy:	6000
Replacement Cost:	$150,422,976

Components

Component	Quantity	Unit
B20 Exterior Enclosure		
Clay Brick, Exterior, 1st Floor	94650	Sq Ft
Clay Brick, Exterior, 2nd Floor	31550	Sq Ft
Clay Brick, Exterior, 3rd+ Floor	31550	Sq Ft
Aluminum Fixed Window, 12 sf, 1st Floor	416	Each
Aluminum Fixed Window, 12 sf, 2nd Floor	416	Each
Aluminum Fixed Window, 12 sf, 3rd+ Floor	1248	Each
Steel, Painted, Exterior Door	10	Each
Aluminum Frame, Fully Glazed, Exterior Door	45	Each
Steel Single 12'x12', Painted, Roll-up Door	12	Each
B30 Roofing		
Single-Ply Modified Bituminous/Thermoplastic Roof	129325	Sq Ft
C10 Interior Construction		
Wood, Solid Core, Painted, Interior Door	955	Each
Steel, Painted, Interior Door	580	Each
Aluminum Frame, Fully Glazed, Interior Door	265	Each
C30 Interior Finishes		
Sheetrock, Unstippled, Interior Wall Finish	884300	Sq Ft
Concrete, Painted, Interior Wall Finish	25400	Sq Ft
Fabric Interior Wall Finish	6300	Sq Ft
Vinyl Interior Wall Finish	7600	Sq Ft
Ceramic Tile Flooring	14300	Sq Ft
Terrazzo Flooring	119107	Sq Ft
Wood Parquetry Flooring	5400	Sq Ft
Vinyl Tile Flooring	225400	Sq Ft
Rubber Tile Flooring	2500	Sq Ft
Vinyl Sheet Flooring	16050	Sq Ft
Carpet, Nylon 20 oz., Low Traffic	79800	Sq Ft
Concrete, Finished Ceiling	14800	Sq Ft
Sheetrock, Unstippled, Ceiling	132650	Sq Ft
Acoustical Tile Ceiling	273000	Sq Ft
D10 Conveying		
Elevator, Hydraulic, 6000 lbs, 3-5 Floor, 150 fpm	2	Each
Elevator, Geared, 3000 lbs, 3-5 Floor, 300 fpm	4	Each
Platform Lifter, 5 Ton	1	Each
Wheel Chair Lift	1	Each
Dock Ramp Leveler	5	Each
D20 Plumbing		
Flush Tank Water Closet	81	Each
Urinal, Vitreous China	21	Each
Lavatory, Vitreous China	106	Each
Sink, Enameled Steel	375	Each
Sink, Stainless Steel	511	Each
Service Sink, Iron Enamel	35	Each
Shower, Fiberglass	163	Each
Drinking Fountain, Refrigerated	12	Each
Pipe & Fittings, 3/4" Copper, Hot Water	39	K Ln Ft
Pipe & Fittings, 3/4" Copper, Cold Water	56	K Ln Ft
Pipe & Fittings, 2" Copper, Cold Water	50	K Ln Ft
Pipe Insulation, Hot Water	95	K Ln Ft
Booster Pump, 10 HP	1	Each
Steam Converter, Domestic Hot Water, 10 Gpm	1	Each
Pipe & Fittings, 6" Cast Iron	4	K Ln Ft
Pipe & Fittings, 10" Cast Iron	1	K Ln Ft
Backflow Preventer, 2"	16	Each
Backflow Preventer, 4"	4	Each
Backflow Preventer, 8"	3	Each
Roof Drain, 4-6"	80	Each
Pipe & Fittings, 4" DWV PVC	17	K Ln Ft
Pipe & Fittings, 3/4" Copper, Oxygen	69	K Ln Ft
Air Compressor, Gas, 7.5 HP	1	Each
Anesthesia Vacuum Pump, <16 HP	2	Each
Medical Gas Compressor/Dryer	5	Each
Medical Gas Outlet	2245	Each
Reverse Osmosis/Deionized Water System	2	Each
D30 HVAC		
Leak Detection, Panel & Float Assembly	2	Each
Fuel Oil Pump, 1.5 HP	5	Each
Oil Storage Tank, 25,000 Gal.	2	Each
Pipe & Fittings, 2" Copper, Fuel Oil	5	K Ln Ft
Pipe & Fittings, 4" Steel, Gas	16.4	K Ln Ft
Pressure Reducer Valve, 2"	4	Each
Boiler Blowoff System	1	Each
Chemical Feed System	4	Each
Deaerator	1	Each
Circulation Pump 5 HP, Hot Water	17	Each
Circulation Pump 7.5 HP, Hot Water	4	Each
Circulation Pump 25 HP, Hot Water	7	Each
Circulation Pump 75 HP, Hot Water	4	Each
Circulation Pump 100 HP, Hot Water	4	Each
Circulation Pump 150 HP, Hot Water	2	Each
Expansion Tank, 250 Gal.	3	Each
Surge Tank, 1000 Gal.	1	Each

50-Year M&R Cost Summary

	50 Year Total	Annual Cost per Sqft	Annual cost as % replacement
PM & Minor Repair	$25,785,990	$0.95	0.34%
Unscheduled Maintenance	$25,628,867	$0.95	0.34%
Renewal & Replacement	$63,686,202	$2.36	0.85%
Total M&R Costs	$115,101,059	$4.26	1.53%

Distribution of Repair Costs

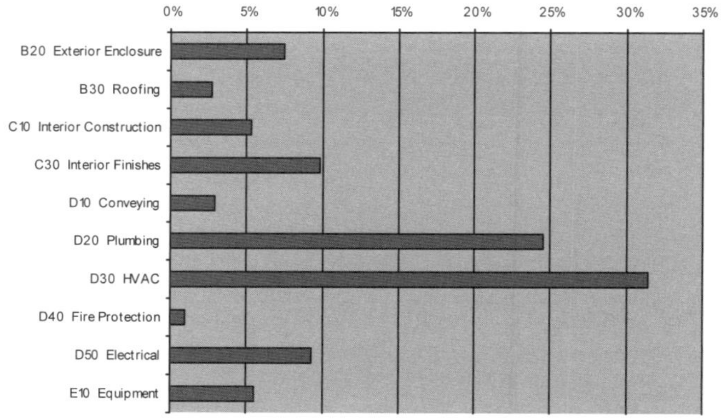

Thirty Most Costly Repair Tasks

Major Repair Task	Task Cost*	Pct.**
Annual PM, Variable Air Volume Box	12.15	7.3%
Refinish Sheetrock, Unstippled, Interior Wall Finish	9.94	6.0%
Replace Pipe & Fittings, 3/4" Copper, Oxygen	9.73	5.9%
Replace Chiller, Centrifugal Hermetic, Bnd'd w/ Condenser, 1000 To	6.30	3.8%
Replace Pipe & Fittings, 2" Copper, Cold Water	6.27	3.8%
Replace Pipe & Fittings, 3/4" Copper, Cold Water	6.18	3.7%
Replace Pipe & Fittings, 3/4" Copper, Hot Water	4.28	2.6%
Replace Laboratory Exhaust Hood	4.24	2.6%
Annual PM, Distribution Panel Board	3.07	1.9%
Annual PM, Medical Gas Outlet	3.02	1.8%
Replace Air Handler, Singlezone, 100,000 Cfm	2.87	1.7%
Repair Chiller, Centrifugal Hermetic, Bnd'd w/ Condenser, 1000 Ton	2.86	1.7%
Annual PM, Laboratory Exhaust Hood	2.77	1.7%
Replace Generator, Diesel, 1000kw	2.74	1.7%
Annual PM, Humidifier, Steam, Duct w/ Controls	2.74	1.7%
Clean & Reseal Clay Brick, Exterior, 1st Floor	2.60	1.6%
Maintain Elevator, Hydraulic, 6000 lbs, 3-5 Floor, 150 fpm	2.33	1.4%
Maintain Elevator, Geared, 3000 lbs, 3-5 Floor, 300 fpm	2.25	1.4%
Replace Ballast & Lamps, Fluorescent Lighting Fixture, 160w	2.18	1.3%
Replace Carpet, Nylon 20 oz., Low Traffic	2.14	1.3%
Replace Gas Boiler, Steam, 24,000 Mbh	1.99	1.2%
Maintain Exhaust Fan, Roof Mounted, 10,000 Cfm	1.89	1.1%
Replace Medical Gas Outlet	1.88	1.1%
Replace Humidifier, Steam, Duct w/ Controls	1.86	1.1%
Maintain Single-Ply Modified Bituminous/Thermoplastic Roof	1.80	1.1%
Replace Vinyl Tile Flooring	1.75	1.1%
Remove & Replace Membrane, Single-Ply Modified Bituminous/Ther	1.68	1.0%
Repoint (50% surface) Clay Brick, Exterior, 1st Floor	1.55	0.9%
Replace Fluorescent Lighting Fixture, 160w	1.55	0.9%
Replace Receptacle	1.51	0.9%

**Task cost ($2004) per gross square foot over 50 years.*

***Percent of total M&R costs.*

Cost per Sqft. by System

Building Age	Exterior Closure	Roofing	Interior Constrution	Stairways	Interior Finishes	Conveying Systems	Plumbing	HVAC Systems	Fire Protection	Electrical	Equipment	Total per Sqft.
1	.01	.07	.15			.10	.27	.79	.02	.15	.13	**1.70**
2	.01	.07	.15			.10	.45	.79	.02	.15	.13	**1.88**
3	.01	.08	.15			.10	.27	.82	.02	.15	.13	**1.74**
4	.01	.07	.19		.98	.10	.46	.79	.02	.15	.13	**2.91**
5	.01	.07	.15		.01	.10	.32	.81	.02	.18	.13	**1.81**
6	.01	.08	.15			.10	.45	.83	.02	.15	.13	**1.94**
7	.01	.07	.15			.10	.27	.95	.02	.15	.13	**1.86**
8	.01	.07	.19		.98	.10	.56	.79	.02	.15	.13	**3.01**
9	.01	.08	.15		.20	.10	.27	.82	.02	.15	.13	**1.94**
10	2.41	.27	.72		.05	.10	1.01	2.10	.31	1.30	.35	**8.63**
11	.01	.07	.15			.10	.27	.79	.02	.15	.13	**1.70**
12	.03	.08	.19		.98	.10	.46	.82	.02	.15	.13	**2.97**
13	.01	.07	.15			.10	.35	.82	.02	.15	.13	**1.81**
14	.02	.07	.15			.10	.46	.96	.02	.15	.13	**2.06**
15	.01	.08	.15		.96	.10	.36	3.81	.02	.33	1.43	**7.25**
16	.01	.07	.19		.98	.10	.56	.82	.02	.15	.13	**3.04**
17	.01	.07	.15			.10	.27	.79	.02	.15	.13	**1.70**
18	.01	.08	.15		1.57	.10	.45	.80	.02	.28	.13	**3.60**
19	.01	.07	.15			.10	.27	.82	.02	.15	.13	**1.73**
20	2.46	.84	.75		1.93	.10	1.15	5.71	.26	2.33	.35	**15.88**
21	.01	.07	.15			.10	.27	.95	.02	.15	.13	**1.86**
22	.01	.07	.15			.10	.45	.79	.02	.15	.13	**1.88**
23	.01	.08	.15			.10	.27	.82	.02	.17	.13	**1.75**
24	.03	.07	.19		.98	.10	.56	.82	.02	.15	.13	**3.06**
25	1.62	.07	.15		.02	.62	15.01	1.35	.02	1.73	.13	**20.73**
26	.01	.08	.15			.10	.42	.82	.02	.15	.13	**1.90**
27	.01	.07	.15		.20	.10	.30	.79	.02	.15	.13	**1.93**
28	.02	.07	.19		.98	.10	.42	.97	.02	.17	.13	**3.07**
29	.01	.08	.15			.10	.31	.82	.02	.15	.13	**1.78**
30	2.41	.27	1.15		1.00	.10	1.02	7.16	.31	1.89	1.66	**16.97**
31	.01	.07	.15			.10	.30	.79	.02	.15	.13	**1.73**
32	.01	.08	.17		.98	.10	.52	.79	.02	.15	.13	**2.96**
33	.01	.07	.15			.10	.31	.82	.02	.17	.13	**1.78**
34	.01	.07	.17			.10	.42	.79	.02	.15	.13	**1.87**
35	.06	.08	.15		.01	.10	.81	.81	.02	.17	.13	**2.35**
36	.03	.07	.17		2.55	.10	.35	.85	.02	.28	.13	**4.55**
37	.01	.07	.15			.10	.38	.95	.02	.15	.13	**1.97**
38	.01	.08	.17			.10	.42	.79	.02	.15	.13	**1.89**
39	.01	.07	.15			.10	.38	.82	.02	.15	.13	**1.84**
40	2.45	.84	.73		2.04	.10	1.29	5.93	.26	2.43	.35	**16.43**
41	.01	.07	.15			.10	.38	.79	.02	.17	.13	**1.82**
42	.02	.07	.17			.10	.35	.80	.02	.15	.13	**1.82**
43	.01	.08	.15			.10	.42	.82	.02	.15	.13	**1.89**
44	.01	.07	.17		.98	.10	.35	.96	.02	.15	.13	**2.94**
45	.02	.07	.15		1.16	.10	.62	3.82	.02	.32	1.43	**7.70**
46	.01	.08	.17			.10	.35	.82	.02	.17	.13	**1.85**
47	.01	.07	.15			.10	.37	.79	.02	.15	.13	**1.80**
48	.03	.07	.17		.98	.10	.40	.82	.02	.15	.13	**2.88**
49	.01	.08	.15			.10	.38	.82	.02	.15	.13	**1.86**
50	4.02	.27	.74		.40	.62	15.51	2.21	.31	2.64	.35	**27.07**
Total	15.96	5.81	11.34		20.95	6.16	52.17	67.03	2.15	19.82	11.67	213.07

50-Year Total M&R Cost Profile per Sqft.

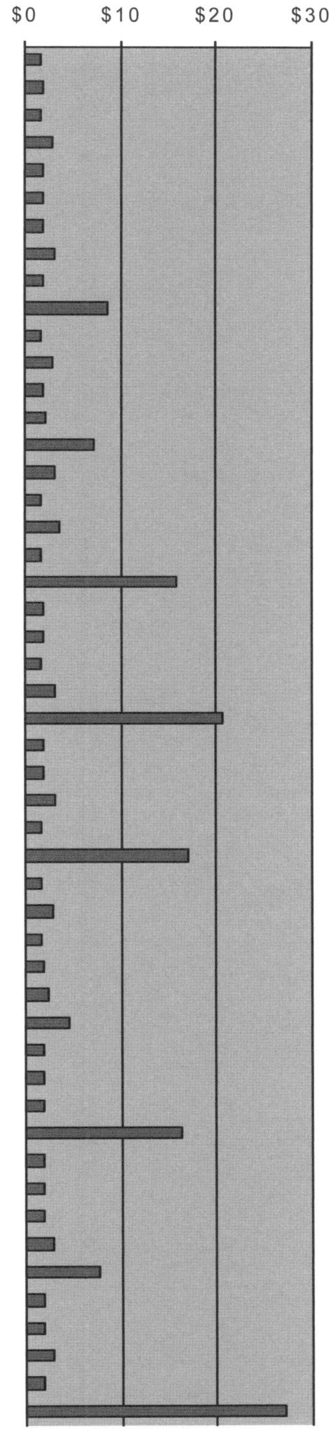

A value of "0.00" means a cost of more than $.000 but less than $.005 per gross square foot.

Components

Gas Boiler, Steam, 24,000 Mbh	3	Each
Heat Exchanger, Steam-to-Water, 130 Gph	3	Each
Steam Converter, 200 Gph	1	Each
Steam Trap, F&T, 2"	50	Each
Circulation Pump, 25 HP, Chiller & Condenser Water	4	Each
Cooling Tower, 50 Ton	1	Each
Cooling Tower, 100 Ton	3	Each
Cooling Tower, 1000 Ton	3	Each
Chiller, Centrifugal Hermetic, Bundled w/ Condenser, 1000 Ton	4	Each
Variable Air Volume Box	1950	Each
Humidifer, Steam, Duct w/ Controls	115	Each
Air Handler, Singlezone, 50,000 Cfm	4	Each
Air Handler, Singlezone, 100,000 Cfm	5	Each
Air Handler, Singlezone, 150,000 Cfm	1	Each
Exhaust Fan, Ceiling, 200-500 Cfm	17	Each
Exhaust Fan, Roof Mounted, 10,000 Cfm	65	Each
Exhaust Fan, Roof Mounted, 20,000 Cfm	32	Each
Fan Coil, Two-Pipe, 400 Cfm	51	Each
Secondary Coil, 30" x 72"	50	Each
Terminal Reheat Unit, 25,000 Cfm	8	Each
Direct Digital Controls, System Points	31	Each
D40 Fire Protection		
Fire Sprinkler Head	5035	Each
Fire Suppression System Water Pump	1	Each
D50 Electrical		
Transfer Switch, Auto, 600 V, 400 Amp.	8	Each
Transfer Switch, Auto, 600 V, 800 Amp.	7	Each
Distribution Panel Board	352	Each
Secondary Transformer, Dry, 150 KVA	63	Each
Main Switchgear, >1200 Amp.	1	Each
Main Switchgear, <1200 Amp.	1	Each
Motor Control Center	8	Each
Exit Lighting Fixture, w/ Battery	80	Each
Incandescent Lighting Fixture, Basic, 100w	250	Each
Fluorescent Lighting Fixture, 160w	4266	Each
High Pressure Sodium Lighting Fixture, 250w	385	Each
Wiring Device, Switch	1300	Each
Receptacle, 120V, 15 Amp	12900	Each
Annunciation Panel	1	Each
Fire Alarm Control Panel	1	Each
Manual Pull Station	121	Each
Smoke Detector	100	Each
Time Control Clock	149	Each
Generator, Diesel, 30kw	2	Each
Generator, Diesel, 1000kw	3	Each
E10 Equipment		
Commercial Walk-In Freezer/Cooler	6	Each
Dust Collection System	1	Each
Mortuary Unit Cooler	1	Each
Laboratory Exhaust Hood	404	Each

Laundromat

Gross Sqft:	3,000
Height in Ft:	12
Exterior:	Decorative Concrete Block
Floor coverings:	Vinyl Tile
HVAC:	Single Zone Rooftop Unit
Occupancy:	60
Replacement Cost:	$364,953

Components

Component	Qty	Unit
B20 Exterior Enclosure		
Clay Brick, Exterior, 1st Floor	2628	Sq Ft
Aluminum Operable Window, 24 sf, 1st Floor	9	Each
Aluminum Frame, Fully Glazed, Exterior Door	1	Each
B30 Roofing		
Built-up Roof	3000	Sq Ft
C10 Interior Construction		
Wood, Solid Core w/ Safety Glass, Painted, Interior Door	12	Each
C30 Interior Finishes		
Sheetrock, Stippled, Interior Wall Finish	2880	Sq Ft
Vinyl Tile Flooring	3000	Sq Ft
Acoustical Tile Ceiling	3000	Sq Ft
D20 Plumbing		
Tankless Water Closet	3	Each
Lavatory, Vitreous China	3	Each
Pipe & Fittings, 3/4" Copper, Cold Water	0.1	K Ln Ft
Pipe & Fittings, 3/4" Copper, Hot Water	0.03	K Ln Ft
Pipe & Fittings, 2" Copper, Cold Water	0.172	K Ln Ft
Pipe & Fittings, 4" Steel	0.486	K Ln Ft
Pipe Insulation, Cold Water	0.344	K Ln Ft
Pipe Insulation, Hot Water	0.164	K Ln Ft
Circulator Pump, 1/12 HP, Hot Water	1	Each
Water Heater, Gas/Oil 175 Gph	1	Each
Water Storage Tank, 2500 Gal.	1	Each
Pipe & Fittings, 6" Cast Iron	0.109	K Ln Ft
Pipe & Fittings, 10" Cast Iron	0.11	K Ln Ft
Aluminum Gutter, Downspouts, Fittings	0.055	K Ln Ft
Roof Drain, 2"	2	Each
Pipe & Fittings, 4" DWV PVC	0.024	K Ln Ft
D30 HVAC		
Exhaust Fan, Ceiling, 200-500 Cfm	2	Each
Air Conditioner, Rooftop, 5 Ton	2	Each
D40 Fire Protection		
Fire Sprinkler Head	24	Each
D50 Electrical		
Safety Switch, Fused, 400 Amp., 3 Ph.	2	Each
Distribution Panel Board	1	Each
Main Switchgear, >1200 Amp.	1	Each
Exit Lighting Fixture, w/ Battery	2	Each
Emergency Horn & Strobe	2	Each
Fluorescent Lighting Fixture, 160w	30	Each
Wiring Device, Switch	6	Each
Grounded Fault Circuit Interruptor, 15 Amps	2	Each
Receptacle, 120V, 15 Amp	8	Each
Receptacle, 120v, 20 Amp	15	Each
Receptacle, 208v, 3 phase	15	Each
Annunciation Panel	1	Each
Fire Alarm Bell, 6"	1	Each
Manual Pull Station	1	Each
Heat Detector	4	Each
Smoke Detector	2	Each
Generator Switchgear	1	Each

50-Year M&R Cost Summary

	50 Year Total	Annual Cost per Sqft	Annual cost as % replacement
PM & Minor Repair	$230,738	$1.54	1.26%
Unscheduled Maintenance	$258,748	$1.72	1.42%
Renewal & Replacement	$318,755	$2.13	1.75%
Total M&R Costs	$808,241	$5.39	4.43%

Distribution of Repair Costs

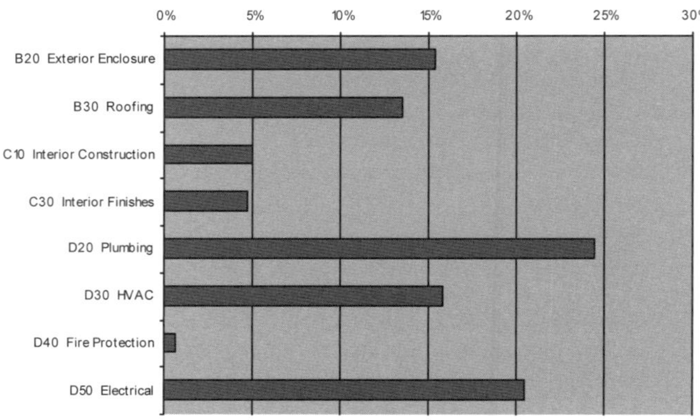

Thirty Most Costly Repair Tasks

Major Repair Task	Task Cost*	Pct.**
Replace Air Conditioner, Rooftop, 5 Ton	16.74	9.1%
Maintain Built-up Roof	14.62	8.0%
Clean & Reseal Clay Brick, Exterior, 1st Floor	13.01	7.1%
Maintain Air Conditioner, Rooftop, 5 Ton	9.95	5.4%
Replace Water Heater, Gas/Oil 175 Gph	9.73	5.3%
Annual PM, Generator Switchgear	9.04	4.9%
Repoint (50% surface) Clay Brick, Exterior, 1st Floor	7.77	4.2%
Repair Air Conditioner, Rooftop, 5 Ton	7.49	4.1%
Refinish Sheetrock, Stippled, Interior Wall Finish	5.83	3.2%
Clean & Scrape Water Storage Tank, 2500 Gal.	5.50	3.0%
Replace Generator Switchgear	5.13	2.8%
Replace Vinyl Tile Flooring	4.20	2.3%
Replace Pipe & Fittings, 2" Copper, Cold Water	3.88	2.1%
Replace Safety Switch, Fused, 400 Amp., 3 Ph.	3.59	2.0%
Place New Membrane Over Existing, Built-up Roof	3.49	1.9%
Remove & Replace Membrane, Built-up Roof	3.45	1.9%
Maintain Wood, Solid Core w/ Safety Glass, Painted, Door Locks	3.33	1.8%
Clean Water Heater, Gas/Oil 175 Gph	3.21	1.8%
Replace Main Switchgear, >1200 Amp.	3.02	1.6%
Maintain Water Storage Tank, 2500 Gal.	2.96	1.6%
Replace Ballast & Lamps, Fluorescent Lighting Fixture, 160w	2.76	1.5%
Repair Main Switchgear, >1200 Amp.	2.63	1.4%
Replace Water Storage Tank, 2500 Gal.	2.08	1.1%
Replace Wood, Solid Core w/ Safety Glass, Painted, Door Locks	2.02	1.1%
Replace Pipe & Fittings, 3/4" Copper, Cold Water	1.99	1.1%
Replace Fluorescent Lighting Fixture, 160w	1.96	1.1%
Replace Wood, Solid Core w/ Safety Glass, Painted, Interior Door	1.94	1.1%
Annual PM, Safety Switch, Fused, 400 Amp., 3 Ph.	1.84	1.0%
Minor Repair, Acoustic Tile Ceiling	1.78	1.0%
Repair Clay Brick, Exterior, 1st Floor	1.67	0.9%

**Task cost ($2004) per gross square foot over 50 years.*

***Percent of total M&R costs.*

Cost per Sqft. by System

Building Age	Exterior Closure	Roofing	Interior Constrution	Stairways	Interior Finishes	Conveying Systems	Plumbing	HVAC Systems	Fire Protection	Electrical	Equipment	Total per Sqft.
1	.01	.59	.18				.46	.38	.01	.65		**2.27**
2	.01	.59	.18				1.20	.38	.01	.65		**3.01**
3	.01	.64	.18				.46	.38	.01	.65		**2.32**
4	.01	.59	.22		.49		1.20	.38	.01	.65		**3.54**
5	.01	.59	.18				.70	.38	.01	1.15		**3.01**
6	.01	.64	.18				1.20	.38	.01	.65		**3.06**
7	.01	.59	.18				.48	.38	.01	.65		**2.29**
8	.01	.59	.22		.49		2.62	.38	.01	.66		**4.98**
9	.01	.64	.18		.39		.46	.38	.01	.65		**2.72**
10	6.21	.59	.69				1.62	2.87	.27	2.20		**14.44**
11	.01	.59	.18				.46	.38	.01	.65		**2.27**
12	.08	.64	.22		.49		1.20	.38	.01	.65		**3.66**
13	.01	.59	.18				.79	.38	.01	.65		**2.61**
14	.01	2.33	.18				1.22	.38	.01	.65		**4.78**
15	.01	.64	.18				.54	5.61	.01	1.44		**8.43**
16	.01	.59	.22		.49		2.62	.38	.01	.66		**4.98**
17	.01	.59	.18				.46	.38	.01	.65		**2.27**
18	.01	.64	.18		2.46		1.20	.38	.01	2.85		**7.73**
19	.01	.59	.18				.46	.38	.01	.65		**2.27**
20	6.27	.59	.73		.90		1.79	.66	.22	4.03		**15.18**
21	.01	.64	.18				.63	.38	.01	.65		**2.50**
22	.01	.59	.18				1.20	.38	.01	.65		**3.01**
23	.01	.59	.18				.46	.38	.01	.65		**2.27**
24	.08	.64	.22		.49		2.62	.38	.01	.66		**5.10**
25	4.73	.59	.18				5.36	2.87	.01	3.00		**16.75**
26	.01	.59	.18				1.36	.38	.01	.65		**3.18**
27	.01	.64	.18		.39		.47	.38	.01	.65		**2.73**
28	.01	3.45	.22		.49		1.20	.38	.01	.65		**6.41**
29	.01	.59	.18				.62	.38	.01	.65		**2.44**
30	6.21	.59	.69				1.60	5.61	.27	2.36		**17.33**
31	.01	.64	.18				.47	.38	.01	.65		**2.33**
32	.01	.59	.22		.49		2.61	.38	.01	.65		**4.95**
33	.01	.59	.18				.47	.38	.01	.66		**2.30**
34	.01	.64	.18				1.19	.38	.01	.65		**3.05**
35	.01	.59	.18				1.47	.38	.01	1.15		**3.79**
36	.08	.59	.22		2.94		1.06	.38	.01	2.85		**8.13**
37	.01	.64	.18				.75	.38	.01	.65		**2.61**
38	.01	.59	.18				1.07	.38	.01	.65		**2.88**
39	.01	.59	.18				.76	.38	.01	.65		**2.58**
40	6.27	.64	1.98		.90		2.89	3.16	.22	4.16		**20.21**
41	.01	.59	.18				.59	.38	.01	.66		**2.42**
42	.01	2.33	.18				1.08	.38	.01	.65		**4.64**
43	.01	.64	.18				.59	.38	.01	.65		**2.46**
44	.01	.59	.22		.49		1.06	.38	.01	.65		**3.41**
45	.01	.59	.18		.39		1.11	5.61	.01	1.44		**9.34**
46	.01	.64	.18				1.06	.38	.01	.65		**2.93**
47	.01	.59	.18				.59	.38	.01	.65		**2.41**
48	.08	.59	.22		.49		2.49	.38	.01	.65		**4.90**
49	.01	.64	.18				.61	.38	.01	.66		**2.49**
50	10.93	.59	.69				7.37	.38	.27	3.79		**24.01**
Total	41.53	36.42	13.45		12.75		65.89	42.61	1.71	55.05		269.41

A value of "0.00" means a cost of more than $.000 but less than $.005 per gross square foot.

50-Year Total M&R Cost Profile per Sqft.

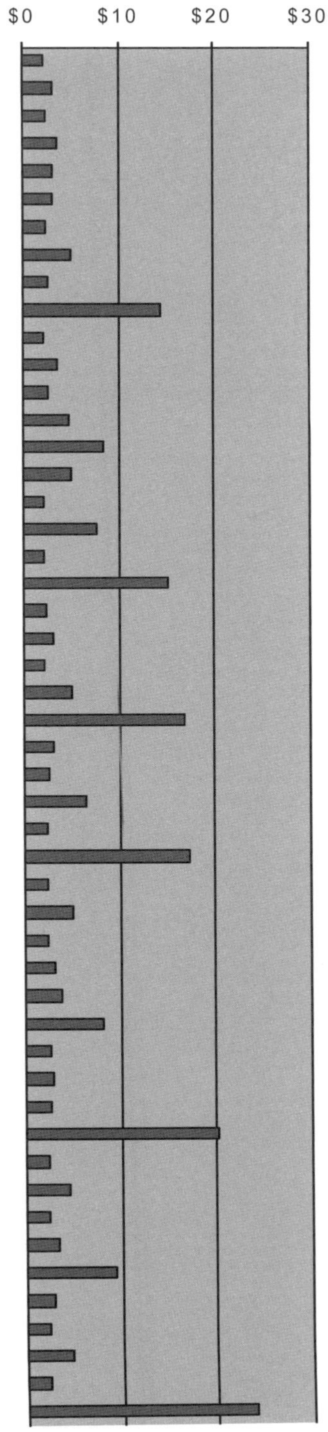

Light Manufacturing Plant

Gross Sqft:	45,000
Height in Ft:	20
Exterior:	Clay Brick
Floor coverings:	Finished Concrete/Vinyl Tile/Carpet
HVAC:	Single Zone Rooftop Units
Occupancy:	400
Replacement Cost:	$2,684,893

50-Year M&R Cost Summary

	50 Year Total	Annual Cost per Sqft	Annual cost as % replacement
PM & Minor Repair	$1,283,714	$0.57	0.96%
Unscheduled Maintenance	$1,343,372	$0.60	1.00%
Renewal & Replacement	$2,979,791	$1.32	2.22%
Total M&R Costs	$5,606,877	$2.49	4.18%

Components

Component	Quantity	Unit
B20 Exterior Enclosure		
Clay Brick, Exterior, 1st Floor	12460	Sq Ft
Steel Frame, Painted, Operable Window, 12 sf, 2nd Floor	330	Each
Aluminum Fixed Window, 12 sf, 1st Floor	72	Each
Steel, Painted, Exterior Door	6	Each
Steel Single 12'x12', Painted, Roll-up Door	4	Each
B30 Roofing		
Built-up Roof	45000	Sq Ft
C10 Interior Construction		
Wood, Solid Core, Painted, Interior Door	12	Each
Steel, Painted, Interior Door	6	Each
C30 Interior Finishes		
Sheetrock, Unstippled, Interior Wall Finish	3263	Sq Ft
Concrete Block, Painted, Interior Wall Finish	12600	Sq Ft
Concrete, Finished Flooring	39250	Sq Ft
Vinyl Tile Flooring	1750	Sq Ft
Carpet, Nylon 20 oz., Low Traffic	4000	Sq Ft
Metal Ceiling, Painted	39250	Sq Ft
Acoustical Tile, Dropped, Ceiling	5750	Sq Ft
D20 Plumbing		
Tankless Water Closet	17	Each
Urinal, Vitreous China	7	Each
Lavatory, Vitreous China	31	Each
Service Sink, Iron Enamel	3	Each
Drinking Fountain, Refrigerated	6	Each
Pipe & Fittings, 3/4" Copper, Cold Water	2.1	K Ln Ft
Pipe & Fittings, 2" Copper, Cold Water	2.6	K Ln Ft
Pipe & Fittings, 3/4" Copper, Hot Water	1.1	K Ln Ft
Pipe Insulation, Cold Water	2.1	K Ln Ft
Circulator Pump, 1/12 HP, Hot Water	1	Each
Water Heater, Gas/Oil 175 Gph	1	Each
Pipe & Fittings, 6" Cast Iron	1.2	K Ln Ft
Pipe & Fittings, 10" Cast Iron	0.5	K Ln Ft
Pipe & Fittings, 4" DWV PVC	0.6	K Ln Ft
Roof Drain, 4-6"	14	Each
D30 HVAC		
Pipe & Fittings, 4" Steel, Gas	0.5	K Ln Ft
Air Conditioner, Rooftop, 20 Ton	1	Each
Air Conditioner, Rooftop, 50 Ton	3	Each
D40 Fire Protection		
Fire Sprinkler Head	263	Each
D50 Electrical		
Main Switchgear, <1200 Amp.	1	Each
Distribution Switch, Fused, <600 V	8	Each
Fluorescent Lighting Fixture, 160w	600	Each
Wiring Device, Switch	60	Each
Receptacle, 120V, 15 Amp	240	Each
Fire Alarm Bell, 6"	45	Each
Fire Alarm Control Panel	1	Each
Manual Pull Station	45	Each
Smoke Detector	160	Each

Distribution of Repair Costs

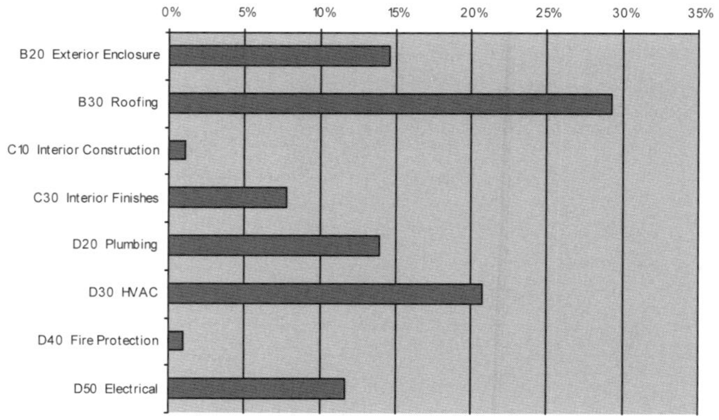

Thirty Most Costly Repair Tasks

Major Repair Task	Task Cost*	Pct.**
Maintain Built-up Roof	14.62	15.4%
Replace Air Conditioner, Rooftop, 50 Ton	10.28	10.9%
Repair Air Conditioner, Rooftop, 50 Ton	6.30	6.6%
Clean & Reseal Clay Brick, Exterior, 1st Floor	4.11	4.3%
Replace Pipe & Fittings, 2" Copper, Cold Water	3.91	4.1%
Replace Ballast & Lamps, Fluorescent Lighting Fixture, 160w	3.67	3.9%
Place New Membrane Over Existing, Built-up Roof	3.49	3.7%
Remove & Replace Membrane, Built-up Roof	3.45	3.6%
Refinish Steel Frame, Painted, Operable Window, 12 sf, 2nd Floor	3.26	3.4%
Maintain Air Conditioner, Rooftop, 50 Ton	2.93	3.1%
Replace Pipe & Fittings, 3/4" Copper, Cold Water	2.78	2.9%
Replace Fluorescent Lighting Fixture, 160w	2.61	2.8%
Repoint (50% surface) Clay Brick, Exterior, 1st Floor	2.46	2.6%
Check Operation, Smoke Detector	2.36	2.5%
Refinish Concrete Block, Painted, Interior Wall Finish	2.15	2.3%
Replace Air Conditioner, Rooftop, 20 Ton	2.03	2.1%
Refinish Metal Ceiling, Painted	1.54	1.6%
Replace Pipe & Fittings, 3/4" Copper, Hot Water	1.45	1.5%
Replace Carpet, Nylon 20 oz., Low Traffic	1.29	1.4%
Replace Smoke Detector	1.25	1.3%
Refinish Concrete, Finished Flooring	1.19	1.3%
Minor Repair, Metal Ceiling, Painted	1.06	1.1%
Minor Repair, Concrete, Finished Flooring	.96	1.0%
Non-Destructive Moisture Inspection	.81	0.9%
Repair Air Conditioner, Rooftop, 20 Ton	.81	0.9%
Replace Faucet Washer & Clean Trap, Lavatory, Vitreous China	.76	0.8%
Maintain Air Conditioner, Rooftop, 20 Ton	.76	0.8%
Replace Water Heater, Gas/Oil 175 Gph	.65	0.7%
Maintain Roof Drain, 4-6"	.57	0.6%
Repair Fire Sprinkler Head	.56	0.6%

**Task cost ($2004) per gross square foot over 50 years.*

***Percent of total M&R costs.*

Cost per Sqft. by System

Building Age	Exterior Closure	Roofing	Interior Construtiion	Stairways	Interior Finishes	Conveying Systems	Plumbing	HVAC Systems	Fire Protection	Electrical	Equipment	Total per Sqft.
1	.03	.59	.02				.06	.13	.01	.10		**.93**
2	.03	.59	.02				.19	.13	.01	.10		**1.06**
3	.03	.64	.02				.06	.13	.01	.10		**.98**
4	.03	.59	.02		.22		.19	.13	.01	.10		**1.29**
5	.36	.59	.02		.16		.08	.13	.01	.13		**1.48**
6	.03	.64	.02				.19	.13	.01	.10		**1.12**
7	.03	.59	.02				.08	.13	.01	.10		**.95**
8	.03	.59	.02		.47		.29	.13	.01	.10		**1.64**
9	.03	.64	.02		.00		.06	.13	.01	.10		**.98**
10	2.34	.59	.08		.17		.39	2.50	.19	1.55		**7.82**
11	.03	.59	.02				.06	.13	.01	.10		**.93**
12	.03	.64	.02		.22		.19	.14	.01	.10		**1.35**
13	.03	.59	.02				.11	.13	.01	.10		**.98**
14	.07	2.33	.02				.21	.13	.01	.10		**2.87**
15	.36	.64	.02		.90		.07	4.10	.01	.60		**6.70**
16	.03	.59	.02		.47		.29	.13	.01	.10		**1.64**
17	.03	.59	.02				.06	.13	.01	.10		**.93**
18	.03	.64	.02		.08		.19	.13	.01	.10		**1.20**
19	.03	.59	.02				.06	.13	.01	.10		**.93**
20	2.45	.59	.08		.42		.43	.13	.16	1.71		**5.97**
21	.03	.64	.02				.09	.13	.01	.10		**1.01**
22	.03	.59	.02				.19	.13	.01	.10		**1.06**
23	.03	.59	.02				.06	.13	.01	.10		**.93**
24	.03	.64	.02		.61		.29	.14	.01	.10		**1.84**
25	1.85	.59	.02		.75		4.43	2.50	.01	.32		**10.48**
26	.03	.59	.02				.22	.13	.01	.10		**1.09**
27	.03	.64	.02		.00		.06	.13	.01	.10		**.98**
28	.07	3.45	.02		.22		.21	.13	.01	.10		**4.22**
29	.03	.59	.02				.07	.13	.01	.10		**.94**
30	2.34	.59	.14		.92		.39	4.15	.19	2.23		**10.96**
31	.03	.64	.02				.06	.13	.01	.10		**.98**
32	.03	.59	.02		.47		.29	.13	.01	.10		**1.64**
33	.03	.59	.02				.06	.13	.01	.10		**.93**
34	.03	.64	.02				.19	.13	.01	.10		**1.12**
35	.54	.59	.02		.16		.64	.13	.01	.13		**2.21**
36	.03	.59	.02		.30		.10	.14	.01	.10		**1.28**
37	.03	.64	.02				.16	.13	.01	.10		**1.08**
38	.03	.59	.02				.11	.13	.01	.10		**.99**
39	.03	.59	.02				.18	.13	.01	.10		**1.05**
40	2.43	.64	.08		.68		.32	2.50	.16	1.90		**8.70**
41	.03	.59	.02				.15	.13	.01	.10		**1.02**
42	.07	2.33	.02				.12	.13	.01	.10		**2.79**
43	.03	.64	.02				.15	.13	.01	.10		**1.08**
44	.03	.59	.02		.22		.10	.13	.01	.10		**1.20**
45	.38	.59	.02		.91		.34	4.10	.01	.60		**6.93**
46	.03	.64	.02				.10	.13	.01	.10		**1.03**
47	.03	.59	.02				.15	.13	.01	.10		**1.02**
48	.03	.59	.02		.61		.20	.14	.01	.10		**1.69**
49	.03	.64	.02				.17	.13	.01	.10		**1.09**
50	3.81	.59	.08		.77		4.54	.13	.19	1.36		**11.47**
Total	18.12	36.42	1.34		9.71		17.35	25.82	1.25	14.58		124.60

A value of "0.00" means a cost of more than $.000 but less than $.005 per gross square foot.

50-Year Total M&R Cost Profile per Sqft.

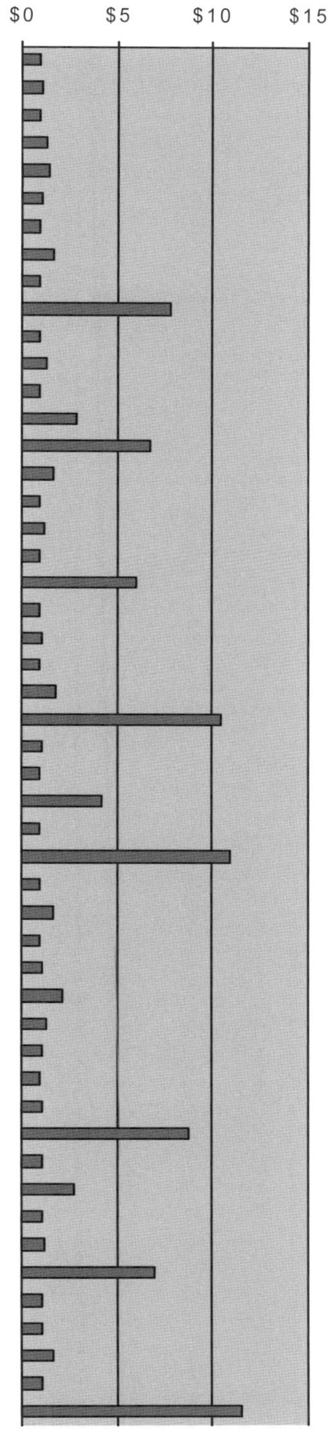

Medical Clinic

Gross Sqft:	13,000
Height in Ft:	12
Exterior:	Cast Concrete
Floor coverings:	Finished Concrete/Linoleum/Carpet
HVAC:	Packaged DX Air Conditioning
Occupancy:	180
Replacement Cost:	$2,284,251

Components

Component	Quantity	Unit
B20 Exterior Enclosure		
Concrete, Painted, Exterior, 1st Floor	4648	Sq Ft
Aluminum Fixed Window, 12 sf, 1st Floor	69	Each
Steel w/ Safety Glass, Painted, Exterior Door	2	Each
Steel, Painted, Exterior Door	2	Each
Aluminum Frame, Fully Glazed, Exterior Door	2	Each
B30 Roofing		
Built-up Roof	13000	Sq Ft
C10 Interior Construction		
Wood, Solid Core, Painted, Interior Door	215	Each
C30 Interior Finishes		
Sheetrock, Unstippled, Interior Wall Finish	31420	Sq Ft
Fabric Interior Wall Finish	10840	Sq Ft
Ceramic Tile Wall Finish	4800	Sq Ft
Concrete, Finished Flooring	650	Sq Ft
Linoleum Flooring	9100	Sq Ft
Carpet, Nylon 20 oz., High Traffic	3250	Sq Ft
Acoustical Tile, Dropped, Ceiling	13000	Sq Ft
D20 Plumbing		
Tankless Water Closet	7	Each
Urinal, Vitreous China	3	Each
Lavatory, Vitreous China	7	Each
Sink, Stainless Steel	37	Each
Service Sink, Iron Enamel	2	Each
Drinking Fountain, Refrigerated	1	Each
Pipe & Fittings, 3/4" Copper, Hot Water	1.6	K Ln Ft
Pipe & Fittings, 3/4" Copper, Cold Water	2	K Ln Ft
Pipe & Fittings, 2" Copper, Cold Water	1.9	K Ln Ft
Pipe Insulation, Cold Water	2	K Ln Ft
Pipe Insulation, Hot Water	1.6	K Ln Ft
Circulator Pump, 1/12 HP, Hot Water	1	Each
Water Heater, Gas/Oil 65 Gph	1	Each
Pipe & Fittings, 6" Cast Iron	0.5	K Ln Ft
Pipe & Fittings, 10" Cast Iron	0.3	K Ln Ft
Roof Drain, 4-6"	4	Each
Pipe & Fittings, 4" DWV PVC	0.1	K Ln Ft
D30 HVAC		
Pipe & Fittings, 4" Steel, Gas	0.3	K Ln Ft
Air Conditioner, Indoor DX Packaged, Air Cooled, 20 Ton	2	Each
D40 Fire Protection		
Fire Sprinkler Head	65	Each
D50 Electrical		
Distribution Switch, Fused, <600 V	3	Each
Main Switchgear, <1200 Amp.	1	Each
Fluorescent Lighting Fixture, 160w	204	Each
Wiring Device, Switch	130	Each
Receptacle, 120V, 15 Amp	260	Each
Fire Alarm Bell, 6"	2	Each
Manual Pull Station	4	Each
Smoke Detector	30	Each

50-Year M&R Cost Summary

	50 Year Total	Annual Cost per Sqft	Annual cost as % replacement
PM & Minor Repair	$547,859	$0.84	0.48%
Unscheduled Maintenance	$663,626	$1.02	0.58%
Renewal & Replacement	$1,619,761	$2.49	1.42%
Total M&R Costs	$2,831,246	$4.36	2.48%

Distribution of Repair Costs

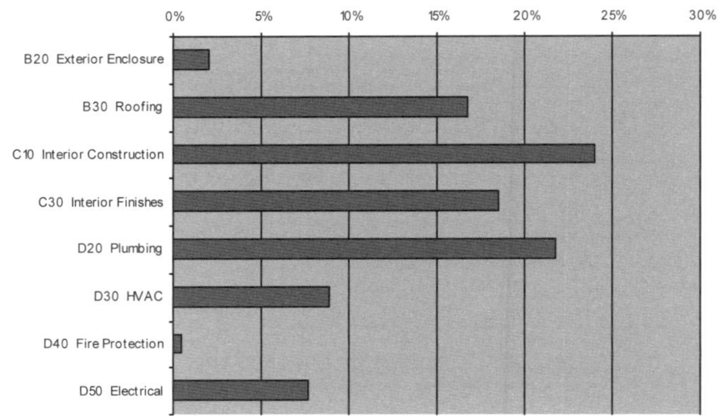

Thirty Most Costly Repair Tasks

Major Repair Task	Task Cost*	Pct.**
Refinish Sheetrock, Unstippled, Interior Wall Finish	14.68	8.8%
Maintain Built-up Roof	14.62	8.8%
Maintain Wood, Solid Core, Painted, Door Locks	13.80	8.3%
Replace Pipe & Fittings, 2" Copper, Cold Water	9.89	5.9%
Replace Pipe & Fittings, 3/4" Copper, Cold Water	9.17	5.5%
Replace Pipe & Fittings, 3/4" Copper, Hot Water	7.29	4.4%
Replace Carpet, Nylon 20 oz., High Traffic	7.25	4.4%
Replace Fabric Interior Wall Finish	6.90	4.1%
Replace Air Conditioner, Indoor DX Packaged, Air Cooled, 20 Ton	6.58	3.9%
Replace Wood, Solid Core, Painted, Interior Door	6.17	3.7%
Replace Solid Core, Painted, Door Locks	6.14	3.7%
Maintain Air Conditioner, Indoor DX Packaged, Air Cooled, 20 Ton	5.34	3.2%
Replace Linoleum Flooring	5.31	3.2%
Replace Ballast & Lamps, Fluorescent Lighting Fixture, 160w	4.32	2.6%
Place New Membrane Over Existing, Built-up Roof	3.49	2.1%
Repair Air Conditioner, Indoor DX Packaged, Air Cooled, 20 Ton	3.46	2.1%
Remove & Replace Membrane, Built-up Roof	3.45	2.1%
Replace Fluorescent Lighting Fixture, 160w	3.07	1.8%
Refinish Concrete, Painted, Exterior, 1st Floor	2.82	1.7%
Refinish Wood, Solid Core, Painted, Interior Door	2.45	1.5%
Minor Repair, Ceramic Tile Wall Finish	2.13	1.3%
Replace Sink, Stainless Steel	1.64	1.0%
Replace Faucet Washer & Clean Trap, Sink, Stainless Steel	1.57	0.9%
Check Operation, Smoke Detector	1.53	0.9%
Minor Repair, Sheetrock, Unstippled, Interior Wall Finish	1.44	0.9%
Replace Receptacle	1.26	0.8%
Replace Valve Set, Sink, Stainless Steel	1.15	0.7%
Replace Pipe Insulation, Cold Water	1.02	0.6%
Replace Wiring Device, Switch	.96	0.6%
Replace Water Heater, Gas/Oil 65 Gph	.91	0.5%

**Task cost ($2004) per gross square foot over 50 years.*

***Percent of total M&R costs.*

Cost per Sqft. by System

Building Age	Exterior Closure	Roofing	Interior Constrution	Stairways	Interior Finishes	Conveying Systems	Plumbing	HVAC Systems	Fire Protection	Electrical	Equipment	Total per Sqft.
1	.02	.59	.76				.09	.19	.01	.07		**1.72**
2	.02	.59	.76		.01		.38	.19	.01	.07		**2.02**
3	.02	.64	.76				.09	.19	.01	.07		**1.77**
4	.02	.59	.96		1.23		.38	.19	.01	.07		**3.46**
5	.02	.59	.76		.18		.17	.19	.01	.18		**2.09**
6	.02	.64	.76		.01		.38	.19	.01	.07		**2.08**
7	.02	.59	.76				.10	.19	.01	.07		**1.74**
8	.02	.59	.96		2.43		.67	.19	.01	.07		**4.94**
9	.02	.64	.76		.05		.09	.19	.01	.07		**1.82**
10	.58	.59	2.29		1.94		.92	1.35	.17	1.97		**9.80**
11	.02	.59	.76				.09	.19	.01	.07		**1.72**
12	.06	.64	.96		1.23		.38	.21	.01	.07		**3.55**
13	.02	.59	.76				.21	.19	.01	.07		**1.84**
14	.12	2.33	.76		.01		.40	.19	.01	.07		**3.89**
15	.02	.64	.76		.20		.13	.19	.01	.73		**2.67**
16	.02	.59	.96		2.43		.67	.19	.01	.07		**4.94**
17	.02	.59	.76				.09	.19	.01	.07		**1.72**
18	.02	.64	.76		2.66		.38	.19	.01	.07		**4.73**
19	.02	.59	.76				.09	.19	.01	.07		**1.72**
20	.64	.59	2.50		4.20		1.01	3.29	.14	2.45		**14.82**
21	.02	.64	.76				.14	.19	.01	.07		**1.83**
22	.02	.59	.76		.01		.38	.19	.01	.07		**2.02**
23	.02	.59	.76				.09	.19	.01	.07		**1.72**
24	.06	.64	.96		2.43		.67	.21	.01	.07		**5.04**
25	.02	.59	.76		.21		14.50	.19	.01	.52		**16.80**
26	.02	.59	.76		.01		.41	.19	.01	.07		**2.05**
27	.02	.64	.76		.05		.11	.19	.01	.07		**1.85**
28	.12	3.45	.96		1.23		.38	.19	.01	.07		**6.42**
29	.02	.59	.76				.15	.19	.01	.07		**1.78**
30	.58	.59	6.39		1.95		.89	1.44	.17	2.70		**14.71**
31	.02	.64	.76				.11	.19	.01	.07		**1.79**
32	.02	.59	.76		2.43		.65	.19	.01	.07		**4.72**
33	.02	.59	.76				.11	.19	.01	.07		**1.74**
34	.02	.64	.96		.01		.36	.19	.01	.07		**2.26**
35	.02	.59	.76		.18		.83	.19	.01	.18		**2.75**
36	.06	.59	.76		3.88		.28	.21	.01	.07		**5.85**
37	.02	.64	.76				.22	.19	.01	.07		**1.91**
38	.02	.59	.96		.01		.32	.19	.01	.07		**2.17**
39	.02	.59	.76				.24	.19	.01	.07		**1.87**
40	.64	.64	2.29		5.40		2.01	3.29	.14	2.80		**17.20**
41	.02	.59	.76				.19	.19	.01	.07		**1.82**
42	.12	2.33	.96		.01		.30	.19	.01	.07		**3.99**
43	.02	.64	.76				.19	.19	.01	.07		**1.88**
44	.02	.59	.76		1.23		.28	.19	.01	.07		**3.15**
45	.02	.59	.76		.25		.51	.19	.01	.73		**3.05**
46	.02	.64	.96		.01		.28	.19	.01	.07		**2.18**
47	.02	.59	.76				.19	.19	.01	.07		**1.82**
48	.06	.59	.76		2.43		.56	.21	.01	.07		**4.68**
49	.02	.64	.76				.20	.19	.01	.07		**1.89**
50	.58	.59	2.50		1.97		15.04	1.35	.17	1.62		**23.81**
Total	4.39	36.43	52.23		40.27		47.31	19.40	1.07	16.70		217.79

50-Year Total M&R Cost Profile per Sqft.

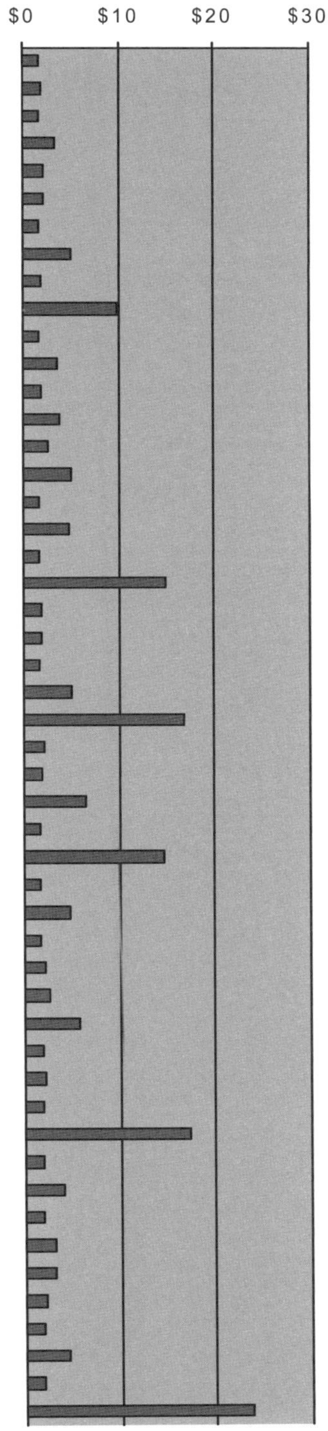

A value of "0.00" means a cost of more than $.000 but less than $.005 per gross square foot.

Motel, 18 Units

Gross Sqft:	8,000
Height in Ft:	9
Exterior:	Brick Veneer
Floor coverings:	Carpet/Ceramic Tile
HVAC:	Electric Cooling & Heating
Occupancy:	240
Replacement Cost:	$778,158

Components

Component	Quantity	Unit
B20 Exterior Enclosure		
Clay Brick, Exterior, 1st Floor	3456	Sq Ft
Aluminum Operable Window, 24 sf, 1st Floor	72	Each
Steel w/ Safety Glass, Painted, Exterior Door	22	Each
B30 Roofing		
Asphalt Shingle Roof	8000	Sq Ft
Concrete Decking	4800	Sq Ft
C10 Interior Construction		
Wood, Hollow Core, Painted, Interior Door	27	Each
C30 Interior Finishes		
Sheetrock, Stippled, Interior Wall Finish	14385	Sq Ft
Concrete, Painted, Interior Wall Finish	3024	Sq Ft
Ceramic Tile Flooring	800	Sq Ft
Carpet, Nylon 20 oz., Low Traffic	6800	Sq Ft
Sheetrock, Unstippled, Ceiling	8000	Sq Ft
D20 Plumbing		
Flush Tank Water Closet	23	Each
Lavatory, Vitreous China	23	Each
Bathtub & Shower Enclosure, Fiberglass	18	Each
Pipe & Fittings, 3/4" Copper, Cold Water	1.96	K Ln Ft
Pipe & Fittings, 3/4" Copper, Hot Water	1.23	K Ln Ft
Pipe & Fittings, 2" Copper, Cold Water	2.238	K Ln Ft
Pipe & Fittings, 4" Steel	0.752	K Ln Ft
Pipe Insulation, Cold Water	2.336	K Ln Ft
Pipe Insulation, Hot Water	1.427	K Ln Ft
Circulator Pump, 1/12 HP, Hot Water	1	Each
Water Heater, Gas/Oil 175 Gph	2	Each
Water Storage Tank, 2500 Gal.	1	Each
Pipe & Fittings, 6" Cast Iron	0.797	K Ln Ft
Pipe & Fittings, 10" Cast Iron	0.179	K Ln Ft
Aluminum Gutter, Downspouts, Fittings	0.048	K Ln Ft
Roof Drain, 4-6"	0.048	Each
Pipe & Fittings, 4" DWV PVC	0.042	K Ln Ft
D30 HVAC		
Exhaust Fan, Ceiling, 200-500 Cfm	18	Each
Air Conditioner, Window, 2 Ton	22	Each
D40 Fire Protection		
Fire Sprinkler System	1	Each
Fire Sprinkler Head	60	Each
D50 Electrical		
Distribution Panel Board	3	Each
Main Switchgear, <1200 Amp.	1	Each
Exit Lighting Fixture, w/ Battery	22	Each
Emergency Horn & Strobe	20	Each
Incandescent Lighting Fixture, Basic, 100w	40	Each
Wiring Device, Switch	40	Each
Grounded Fault Circuit Interruptor, 15 Amps	20	Each
Receptacle, 120V, 15 Amp	80	Each
TV Cable Outlet	18	Each
Fire Alarm Control Panel	1	Each
Smoke Detector	34	Each

50-Year M&R Cost Summary

	50 Year Total	Annual Cost per Sqft	Annual cost as % replacement
PM & Minor Repair	$571,224	$1.43	1.47%
Unscheduled Maintenance	$595,256	$1.49	1.53%
Renewal & Replacement	$1,424,016	$3.56	3.66%
Total M&R Costs	$2,590,496	$6.48	6.66%

Distribution of Repair Costs

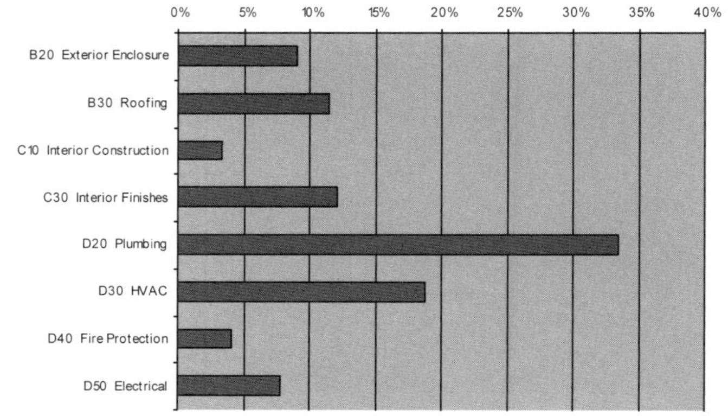

Thirty Most Costly Repair Tasks

Major Repair Task	Task Cost*	Pct.**
Replace Air Conditioner, Window, 2 Ton	24.66	9.9%
Replace Pipe & Fittings, 2" Copper, Cold Water	18.94	7.6%
Maintain Air Conditioner, Window, 2 Ton	17.68	7.1%
Replace Pipe & Fittings, 3/4" Copper, Cold Water	14.60	5.9%
Replace Carpet, Nylon 20 oz., Low Traffic	12.33	4.9%
Refinish Sheetrock, Stippled, Interior Wall Finish	10.92	4.4%
Fire Sprinkler System, Annual PM	10.80	4.3%
Replace Pipe & Fittings, 3/4" Copper, Hot Water	9.11	3.7%
Maintain Asphalt Shingle Roof	8.32	3.3%
Replace Water Heater, Gas/Oil 175 Gph	7.30	2.9%
Clean & Reseal Clay Brick, Exterior, 1st Floor	6.42	2.6%
Refinish Sheetrock, Unstippled, Ceiling	6.08	2.4%
Replace Bathtub & Shower Enclosure, Fiberglass	5.75	2.3%
Minor Replacement, Asphalt Shingle Roof	5.29	2.1%
Repoint (50% surface) Clay Brick, Exterior, 1st Floor	3.83	1.5%
Repair Concrete Decking	3.23	1.3%
Replace Faucet Washer & Clean Trap, Lavatory, Vitreous China	3.17	1.3%
Refinish Concrete, Painted, Interior Wall Finish	2.90	1.2%
Check Operation, Smoke Detector	2.83	1.1%
Maintain Wood, Hollow Core, Painted, Door Locks	2.82	1.1%
Repair Diverter Valve, Bathtub & Shower Enclosure, Fiberglass	2.71	1.1%
Repair Steel w/ Safety Glass, Painted, Exterior Door	2.69	1.1%
Refinish Concrete Decking	2.64	1.1%
Maintain Fan & Motor, Exhaust Fan, Ceiling 200-500 Cfm	2.55	1.0%
Clean Water Heater, Gas/Oil 175 Gph	2.40	1.0%
Maintain Steel w/ Safety Glass, Painted, Exterior Door, Locks	2.34	0.9%
Minor Repair, Concrete, Painted, Interior Wall Finish	2.21	0.9%
Replace Valve Set, Bathtub & Shower Enclosure, Fiberglass	2.18	0.9%
Remove & Replace Asphalt Shingle Roof	2.13	0.9%
Replace Flush Tank Water Closet	2.09	0.8%

**Task cost ($2004) per gross square foot over 50 years.*

***Percent of total M&R costs.*

Cost per Sqft. by System

Building Age	Exterior Closure	Roofing	Interior Constrution	Stairways	Interior Finishes	Conveying Systems	Plumbing	HVAC Systems	Fire Protection	Electrical	Equipment	Total per Sqft.
1	.12	.33	.16				.22	.75	.23	.28		**2.08**
2	.12	.33	.16				.95	.75	.23	.28		**2.81**
3	.12	.33	.16				.22	.75	.23	.28		**2.08**
4	.15	.33	.20		1.66		1.50	.75	.23	.28		**5.10**
5	.12	.92	.16		.02		.87	.75	.31	.64		**3.80**
6	.12	.33	.16				.95	.75	.23	.28		**2.81**
7	.12	1.04	.16				.28	.75	.23	.28		**2.85**
8	.15	.33	.20		1.66		2.57	.75	.23	.28		**6.17**
9	.12	.33	.16				.22	.75	.23	.28		**2.08**
10	3.18	.92	.58		.02		2.54	5.03	.55	.97		**13.79**
11	.12	.33	.16				.22	.75	.23	.28		**2.08**
12	.15	.33	.20		1.66		1.50	.75	.23	.28		**5.10**
13	.12	.33	.16				.55	.75	.23	.28		**2.40**
14	1.02	1.04	.16				1.01	.75	.23	.28		**4.48**
15	.12	.92	.16		4.92		.97	.75	.31	1.40		**9.53**
16	.15	.33	.20		1.66		2.57	.75	.23	.28		**6.17**
17	.12	.33	.16				.22	.75	.23	.28		**2.08**
18	.12	.33	.16				.95	.75	.23	.28		**2.81**
19	.12	.33	.16				.22	.75	.23	.28		**2.08**
20	3.38	6.00	.62		3.23		4.47	5.97	.51	2.72		**26.91**
21	.12	1.04	.16				.40	.75	.23	.28		**2.97**
22	.12	.33	.16				.95	.75	.23	.28		**2.81**
23	.12	.33	.16				.22	.75	.23	.28		**2.08**
24	.15	.33	.20		1.66		2.57	.75	.23	.28		**6.17**
25	2.45	.92	.16		.02		24.52	.75	.31	.90		**30.03**
26	.12	.33	.16				1.07	.75	.23	.28		**2.93**
27	.12	.33	.16				.26	.75	.23	.28		**2.12**
28	1.05	1.04	.20		1.66		1.52	.75	.23	.28		**6.73**
29	.12	.33	.16				.38	.75	.23	.28		**2.23**
30	3.18	.92	1.05		4.92		2.71	5.03	.55	1.47		**19.81**
31	.12	.33	.16				.26	.75	.23	.28		**2.12**
32	.15	.33	.16		1.66		2.54	.75	.23	.28		**6.09**
33	.12	.33	.16				.26	.75	.23	.28		**2.12**
34	.12	.33	.20				.91	.75	.23	.28		**2.81**
35	.12	1.63	.16		.02		3.83	.75	.31	.64		**7.47**
36	.15	.33	.16		1.66		1.11	.75	.23	.28		**4.67**
37	.12	.33	.16				.73	.75	.23	.28		**2.59**
38	.12	.33	.20				.61	.75	.23	.28		**2.51**
39	.12	.33	.16				.77	.75	.23	.28		**2.63**
40	3.38	5.54	.58		3.23		4.67	5.97	.51	2.97		**26.86**
41	.12	.33	.16				.61	.75	.23	.28		**2.47**
42	1.02	1.04	.20				.62	.75	.23	.28		**4.13**
43	.12	.33	.16				.61	.75	.23	.28		**2.47**
44	.15	.33	.16		1.66		1.11	.75	.23	.28		**4.67**
45	.12	.92	.16		4.92		1.72	.75	.31	1.40		**10.29**
46	.12	.33	.20				.56	.75	.23	.28		**2.46**
47	.12	.33	.16				.61	.75	.23	.28		**2.47**
48	.15	.33	.16		1.66		2.19	.75	.23	.28		**5.74**
49	.12	1.04	.16				.66	.75	.23	.28		**3.24**
50	5.50	.92	.62		1.30		26.33	5.03	.55	.71		**40.97**
Total	29.29	37.08	10.85		39.18		108.31	60.80	13.28	25.01		323.81

50-Year Total M&R Cost Profile per Sqft.

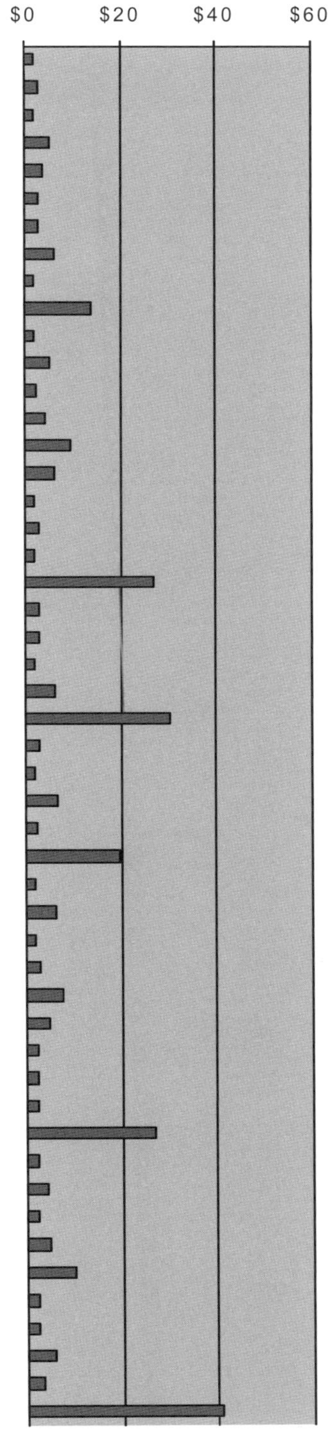

A value of "0.00" means a cost of more than $.000 but less than $.005 per gross square foot.

Motel, 40 Units

Gross Sqft:	18,000
Height in Ft:	20
Exterior:	Stucco
Floor coverings:	Finished Concrete/Quarry Tile/Carpet
HVAC:	Heat Pumps
Occupancy:	85
Replacement Cost:	$1,963,380

Components

Component	Quantity	Unit
B20 Exterior Enclosure		
Stucco, Exterior, 1st Floor	3230	Sq Ft
Stucco, Exterior, 2nd Floor	3230	Sq Ft
Aluminum Operable Window, 24 sf, 1st Floor	20	Each
Aluminum Operable Window, 24 sf, 2nd Floor	20	Each
Aluminum Fixed Window, 24 sf, 1st Floor	8	Each
Steel, Painted, Exterior Door	2	Each
Aluminum Frame, Fully Glazed, Exterior Door	2	Each
Aluminum Frame, Fully Glazed, Sliding Exterior Door	2	Each
B30 Roofing		
Single-Ply Modified Bituminous/Thermoplastic Roof	9000	Sq Ft
C10 Interior Construction		
Wood Frame, Painted, Fully Glazed, Sliding Interior Door	42	Each
Wood, Solid Core, Painted, Interior Door	90	Each
C20 Stairs		
Masonry Interior Steps, Unpainted	600	Sq Ft
Metal Interior Railing, Painted	160	Ln Ft
C30 Interior Finishes		
Sheetrock, Unstippled, Interior Wall Finish	5000	Sq Ft
Concrete Block, Painted, Interior Wall Finish	2000	Sq Ft
Vinyl Interior Wall Finish	39450	Sq Ft
Concrete, Finished Flooring	900	Sq Ft
Ceramic Tile Flooring	2000	Sq Ft
Quarry Tile Flooring	2250	Sq Ft
Carpet, Nylon 20 oz., Low Traffic	9700	Sq Ft
Carpet, Nylon 20 oz., High Traffic	3150	Sq Ft
Sheetrock, Unstippled, Ceiling	8000	Sq Ft
Textured, Spray, Ceiling	10000	Sq Ft
D20 Plumbing		
Flush Tank Water Closet	41	Each
Lavatory, Vitreous China	41	Each
Service Sink, Iron Enamel	2	Each
Bathtub & Shower Enclosure, Fiberglass	40	Each
Drinking Fountain, Refrigerated	1	Each
Pipe & Fittings, 3/4" Copper, Cold Water	3.9	K Ln Ft
Pipe & Fittings, 3/4" Copper, Hot Water	2.5	K Ln Ft
Pipe & Fittings, 2" Copper, Cold Water	4.5	K Ln Ft
Pipe Insulation, Cold Water	3.9	K Ln Ft
Pipe Insulation, Hot Water	2.5	K Ln Ft
Circulator Pump 1/6 HP, Cold Water	1	Each
Circulator Pump, 1/2 HP, Hot Water	1	Each
Water Heater, Gas/Oil 275 Gph	1	Each
Water Storage Tank, 250 Gal.	1	Each
Pipe & Fittings, 6" Cast Iron	1.4	K Ln Ft
Pipe & Fittings, 10" Cast Iron	0.2	K Ln Ft
Roof Drain, 4-6"	3	Each
Pipe & Fittings, 4" DWV PVC	0.2	K Ln Ft
D30 HVAC		
Pipe & Fittings, 4" Steel, Gas	0.2	K Ln Ft
Heat Pump, 1 Ton	40	Each
Heat Pump, 5 Ton	2	Each
D40 Fire Protection		
Fire Sprinkler Head	90	Each
D50 Electrical		
Distribution Switch, Fused, <600 V	3	Each
Main Switchgear, <1200 Amp.	1	Each
Incandescent Lighting Fixture, Basic, 100w	80	Each
Fluorescent Lighting Fixture, 160w	75	Each
Wiring Device, Switch	175	Each
Receptacle, 120V, 15 Amp	250	Each
Fire Alarm Bell, 6"	8	Each
Fire Alarm Control Panel	1	Each
Manual Pull Station	8	Each
Smoke Detector	60	Each

50-Year M&R Cost Summary

	50 Year Total	Annual Cost per Sqft	Annual cost as % replacement
PM & Minor Repair	$1,071,059	$1.19	1.09%
Unscheduled Maintenance	$978,752	$1.09	1.00%
Renewal & Replacement	$2,716,310	$3.02	2.77%
Total M&R Costs	$4,766,121	$5.30	4.86%

Distribution of Repair Costs

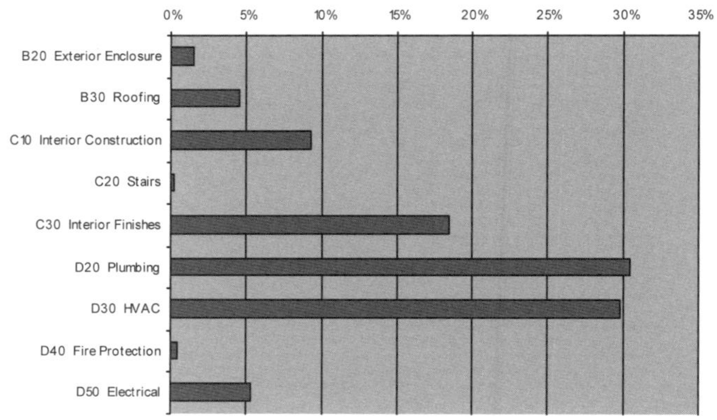

Thirty Most Costly Repair Tasks

Major Repair Task	Task Cost*	Pct.**
Maintain Heat Pump, 1 Ton	31.74	15.1%
Replace Vinyl Interior Wall Finish	18.09	8.6%
Replace Pipe & Fittings, 2" Copper, Cold Water	16.92	8.0%
Replace Pipe & Fittings, 3/4" Copper, Cold Water	12.91	6.1%
Repair Heat Pump, 1 Ton	11.67	5.5%
Replace Heat Pump, 1 Ton	8.34	4.0%
Replace Pipe & Fittings, 3/4" Copper, Hot Water	8.23	3.9%
Replace Carpet, Nylon 20 oz., Low Traffic	7.82	3.7%
Replace Bathtub & Shower Enclosure, Fiberglass	5.68	2.7%
Replace Quarry Tile Flooring	5.29	2.5%
Replace Carpet, Nylon 20 oz., High Traffic	5.08	2.4%
Maintain Wood, Solid Core, Painted, Door Locks	4.17	2.0%
Maintain Single-Ply Modified Bituminous/Thermoplastic Roof	3.77	1.8%
Remove & Replace Membrane, Single-Ply Modified Bituminous/Ther	3.50	1.7%
Refinish Sheetrock, Unstippled, Ceiling	2.70	1.3%
Repair Diverter Valve, Bathtub & Shower Enclosure, Fiberglass	2.67	1.3%
Replace Faucet Washer & Clean Trap, Lavatory, Vitreous China	2.51	1.2%
Minor Repair, Vinyl Interior Wall Finish	2.32	1.1%
Check Operation, Smoke Detector	2.21	1.1%
Replace Valve Set, Bathtub & Shower Enclosure, Fiberglass	2.16	1.0%
Replace Wood Frame, Painted, Fully Glazed, Sliding Interior Door	1.95	0.9%
Maintain Wood Frame, Painted, Fully Glazed, Sliding Door Locks	1.95	0.9%
Replace Wood, Solid Core, Painted, Interior Door	1.87	0.9%
Replace Solid Core, Painted, Door Locks	1.86	0.9%
Replace Water Heater, Gas/Oil 275 Gph	1.81	0.9%
Refinish Stucco, Exterior, 2nd Floor	1.79	0.8%
Refinish Sheetrock, Unstippled, Interior Wall Finish	1.69	0.8%
Replace Flush Tank Water Closet	1.66	0.8%
Maintain Heat Pump, 5 Ton	1.59	0.8%
Refinish Stucco, Exterior, 1st Floor	1.44	0.7%

**Task cost ($2004) per gross square foot over 50 years.*

***Percent of total M&R costs.*

Cost per Sqft. by System

Building Age	Exterior Closure	Roofing	Interior Construtiion	Stairways	Interior Finishes	Conveying Systems	Plumbing	HVAC Systems	Fire Protection	Electrical	Equipment	Total per Sqft.
1	.01	.15	.34				.08	1.18	.01	.10		**1.86**
2	.01	.15	.34		.00		.53	1.18	.01	.10		**2.32**
3	.01	.17	.34				.08	1.18	.01	.10		**1.88**
4	.01	.15	.43		.44		1.08	1.18	.01	.10		**3.39**
5	.01	.15	.34		.73		.59	1.18	.01	.24		**3.24**
6	.01	.17	.34		.00		.53	1.18	.01	.10		**2.34**
7	.01	.15	.34	.01			.12	1.18	.01	.10		**1.91**
8	.01	.15	.43		1.28		1.35	1.18	.01	.10		**4.50**
9	.01	.17	.34				.08	1.18	.01	.10		**1.88**
10	.66	.56	1.10		3.74		2.10	5.39	.17	1.00		**14.71**
11	.01	.15	.34				.08	1.18	.01	.10		**1.86**
12	.04	.17	.43		.44		1.08	1.18	.01	.10		**3.45**
13	.01	.15	.34				.21	1.18	.01	.10		**1.99**
14	.05	.15	.34	.01	.00		.58	1.18	.01	.10		**2.40**
15	.01	.17	.34	.00	3.37		.73	1.18	.01	.97		**6.78**
16	.01	.15	.43		1.28		1.35	1.18	.01	.10		**4.50**
17	.01	.15	.34				.08	1.18	.01	.10		**1.86**
18	.01	.17	.34		.00		.53	1.18	.01	.10		**2.34**
19	.01	.15	.34				.08	1.18	.01	.10		**1.86**
20	.71	1.75	1.19		4.89		3.98	4.85	.14	1.53		**19.04**
21	.01	.15	.34	.01			.15	1.18	.01	.10		**1.93**
22	.01	.15	.34		.00		.53	1.18	.01	.10		**2.32**
23	.01	.17	.34				.08	1.18	.01	.10		**1.88**
24	.04	.15	.43		1.34		1.35	1.18	.01	.10		**4.60**
25	.01	.15	.34		.77		21.14	1.18	.01	.61		**24.19**
26	.01	.17	.34		.00		.57	1.18	.01	.10		**2.38**
27	.01	.15	.34	.44			.11	1.18	.01	.10		**2.32**
28	.05	.15	.43	.01	.44		1.09	1.18	.01	.10		**3.45**
29	.01	.17	.34				.13	1.18	.01	.10		**1.94**
30	.66	.56	2.34	.00	6.38		2.24	5.44	.17	1.74		**19.51**
31	.01	.15	.34				.11	1.18	.01	.10		**1.88**
32	.01	.17	.37		1.28		1.32	1.18	.01	.10		**4.44**
33	.01	.15	.34				.11	1.18	.01	.10		**1.88**
34	.01	.15	.40		.00		.50	1.18	.01	.10		**2.35**
35	.01	.17	.34	.01	.73		3.05	1.18	.01	.24		**5.72**
36	.04	.15	.37		.44		.76	1.18	.01	.10		**3.05**
37	.01	.15	.34				.42	1.18	.01	.10		**2.19**
38	.01	.17	.40		.00		.26	1.18	.01	.10		**2.13**
39	.01	.15	.34				.46	1.18	.01	.10		**2.24**
40	.71	1.75	2.68		5.74		3.54	4.85	.14	1.90		**21.31**
41	.01	.15	.34				.39	1.18	.01	.10		**2.17**
42	.05	.15	.40	.01	.00		.26	1.18	.01	.10		**2.15**
43	.01	.17	.34				.40	1.18	.01	.10		**2.20**
44	.01	.15	.37		.44		.76	1.18	.01	.10		**3.01**
45	.01	.15	.34	.00	3.37		1.30	1.18	.01	.97		**7.32**
46	.01	.17	.40		.00		.22	1.18	.01	.10		**2.09**
47	.01	.15	.34				.39	1.18	.01	.10		**2.17**
48	.04	.15	.37		1.34		1.03	1.18	.01	.10		**4.22**
49	.01	.17	.34	.01			.43	1.18	.01	.10		**2.24**
50	.66	.56	1.16		10.48		22.35	5.39	.17	.62		**41.40**
Total	4.07	12.14	24.55	.48	48.97		80.70	78.84	1.07	13.95		264.78

50-Year Total M&R Cost Profile per Sqft.

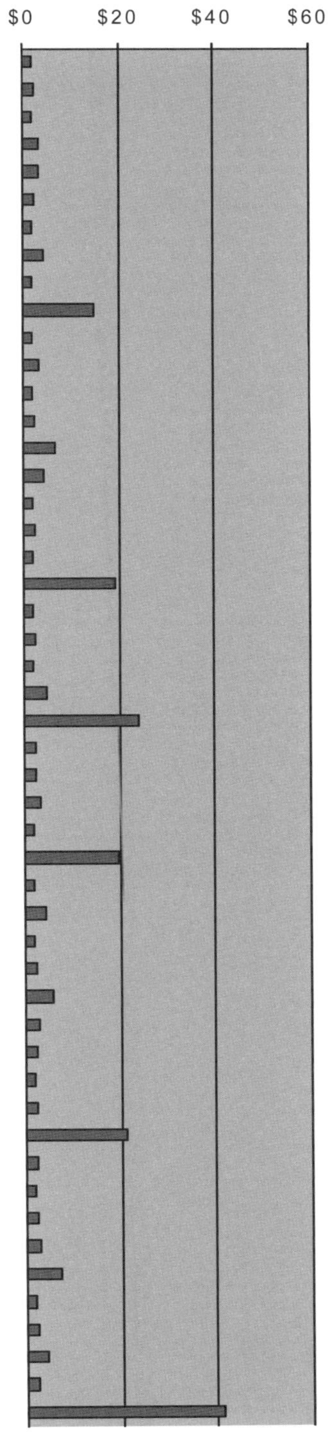

A value of "0.00" means a cost of more than $.000 but less than $.005 per gross square foot.

Movie Theater

Gross Sqft:	10,000
Height in Ft:	20
Exterior:	Clay Brick
Floor coverings:	Finished Concrete/Carpet
HVAC:	Single Zone Rooftop Units
Occupancy:	900
Replacement Cost:	$1,097,944

Components

Component	Quantity	Unit
B20 Exterior Enclosure		
Clay Brick, Painted, Exterior, 1st Floor	7620	Sq Ft
Aluminum Fixed Window, 24 sf, 1st Floor	16	Each
Steel, Painted, Exterior Door	7	Each
Aluminum Frame, Fully Glazed, Exterior Door	4	Each
B30 Roofing		
Built-up Roof	10000	Sq Ft
C10 Interior Construction		
Steel, Painted, Interior Door	50	Each
C20 Stairs		
Metal Interior Steps, Painted	150	Sq Ft
Metal Interior Railing, Painted	40	Ln Ft
C30 Interior Finishes		
Sheetrock, Unstippled, Interior Wall Finish	19470	Sq Ft
Concrete Block, Painted, Interior Wall Finish	1000	Sq Ft
Acoustical Interior Wall Tile	7160	Sq Ft
Concrete, Finished Flooring	8500	Sq Ft
Carpet, Nylon 20 oz., High Traffic	1500	Sq Ft
Sheetrock, Unstippled, Ceiling	500	Sq Ft
Acoustical Tile Ceiling	9500	Sq Ft
D20 Plumbing		
Tankless Water Closet	8	Each
Urinal, Vitreous China	3	Each
Lavatory, Vitreous China	6	Each
Sink, Stainless Steel	1	Each
Service Sink, Iron Enamel	1	Each
Drinking Fountain, Refrigerated	1	Each
Pipe & Fittings, 3/4" Copper, Cold Water	0.7	K Ln Ft
Water Heater, Gas/Oil 65 Gph	1	Each
Pipe & Fittings, 2" Copper, Cold Water	0.7	K Ln Ft
Pipe & Fittings, 3/4" Copper, Hot Water	0.3	K Ln Ft
Pipe Insulation, Cold Water	0.7	K Ln Ft
Pipe Insulation, Hot Water	0.3	K Ln Ft
Pipe & Fittings, 6" Cast Iron	0.6	K Ln Ft
Pipe & Fittings, 10" Cast Iron	0.2	K Ln Ft
Pipe & Fittings, 4" DWV PVC	0.2	K Ln Ft
Roof Drain, 4-6"	3	Each
D30 HVAC		
Pipe & Fittings, 4" Steel, Gas	0.2	K Ln Ft
Air Conditioner, Rooftop, 20 Ton	4	Each
D40 Fire Protection		
Fire Sprinkler Head	50	Each
D50 Electrical		
Main Switchgear, <1200 Amp.	1	Each
Distribution Switch, Fused, <600 V	3	Each
Incandescent Lighting Fixture, Basic, 100w	42	Each
Fluorescent Lighting Fixture, 160w	10	Each
Wiring Device, Switch	10	Each
Receptacle, 120V, 15 Amp	40	Each
Fire Alarm Control Panel	1	Each
Smoke Detector	20	Each

50-Year M&R Cost Summary

	50 Year Total	Annual Cost per Sqft	Annual cost as % replacement
PM & Minor Repair	$399,409	$0.80	0.73%
Unscheduled Maintenance	$399,503	$0.80	0.73%
Renewal & Replacement	$1,314,041	$2.63	2.39%
Total M&R Costs	$2,112,953	$4.23	3.85%

Distribution of Repair Costs

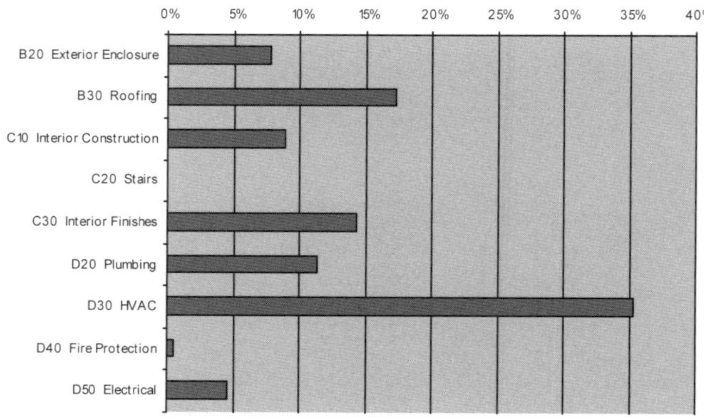

Thirty Most Costly Repair Tasks

Major Repair Task	Task Cost*	Pct.**
Replace Air Conditioner, Rooftop, 20 Ton	36.49	21.3%
Repair Air Conditioner, Rooftop, 20 Ton	14.62	8.5%
Maintain Built-up Roof	14.62	8.5%
Maintain Air Conditioner, Rooftop, 20 Ton	13.60	7.9%
Refinish Sheetrock, Unstippled, Interior Wall Finish	11.83	6.9%
Repoint (50% surface) Clay Brick, Painted, Exterior, 1st Floor	6.98	4.1%
Refinish Acoustical Interior Wall Tile	6.68	3.9%
Replace Steel, Painted, Door Locks	6.39	3.7%
Replace Pipe & Fittings, 2" Copper, Cold Water	4.74	2.8%
Refinish Clay Brick, Painted, Exterior, 1st Floor	4.56	2.7%
Replace Carpet, Nylon 20 oz., High Traffic	4.35	2.5%
Maintain Steel, Painted, Door Locks	4.26	2.5%
Replace Pipe & Fittings, 3/4" Copper, Cold Water	4.17	2.4%
Place New Membrane Over Existing, Built-up Roof	3.49	2.0%
Remove & Replace Membrane, Built-up Roof	3.45	2.0%
Replace Pipe & Fittings, 3/4" Copper, Hot Water	1.78	1.0%
Minor Repair, Acoustic Tile Ceiling	1.69	1.0%
Repair Clay Brick, Painted, Exterior, 1st Floor	1.45	0.8%
Check Operation, Smoke Detector	1.33	0.8%
Replace Water Heater, Gas/Oil 65 Gph	1.19	0.7%
Minor Repair, Sheetrock, Unstippled, Interior Wall Finish	1.16	0.7%
Refinish Concrete, Finished Flooring	1.16	0.7%
Clean Water Heater, Gas/Oil 65 Gph	.96	0.6%
Minor Repair, Concrete, Finished Flooring	.93	0.5%
Non-Destructive Moisture Inspection	.81	0.5%
Repair Main Switchgear, <1200 Amp.	.79	0.5%
Refinish Steel, Painted, Interior Door	.78	0.5%
Refinish Concrete Block, Painted, Interior Wall Finish	.77	0.4%
Replace Incandescent Lighting Fixture, Basic, 100w	.74	0.4%
Replace Smoke Detector	.70	0.4%

**Task cost ($2004) per gross square foot over 50 years.*

***Percent of total M&R costs.*

Cost per Sqft. by System

Building Age	Exterior Closure	Roofing	Interior Construtiion	Stairways	Interior Finishes	Conveying Systems	Plumbing	HVAC Systems	Fire Protection	Electrical	Equipment	Total per Sqft.
1	.05	.59	.23				.11	.50	.01	.08		**1.55**
2	.05	.59	.23		.00		.23	.50	.01	.08		**1.68**
3	.05	.64	.23				.11	.50	.01	.08		**1.61**
4	.06	.59	.29		1.64		.23	.50	.01	.08		**3.39**
5	.05	.59	.23				.17	.50	.01	.28		**1.81**
6	.05	.64	.23		.00		.23	.50	.01	.08		**1.73**
7	.05	.59	.23	.00			.13	.50	.01	.08		**1.58**
8	.06	.59	.29		2.36		.38	.50	.01	.08		**4.26**
9	.05	.64	.23	.01	.34		.11	.50	.01	.08		**1.95**
10	.96	.59	1.51		.00		.49	5.37	.17	.46		**9.54**
11	.05	.59	.23				.11	.50	.01	.08		**1.55**
12	.14	.64	.29		1.64		.23	.51	.01	.08		**3.53**
13	.05	.59	.23				.23	.50	.01	.08		**1.67**
14	.28	2.33	.23	.00	.00		.26	.50	.01	.08		**3.68**
15	.05	.64	.23	.01	.31		.12	12.16	.01	.66		**14.18**
16	.06	.59	.29		2.36		.38	.50	.01	.08		**4.26**
17	.05	.59	.23				.11	.50	.01	.08		**1.55**
18	.05	.64	.23	.01	.34		.23	.50	.01	.08		**2.08**
19	.05	.59	.23				.11	.50	.01	.08		**1.55**
20	.99	.59	1.57		2.51		.53	.50	.14	.95		**7.78**
21	.05	.64	.23	.00			.18	.50	.01	.08		**1.68**
22	.05	.59	.23		.00		.23	.50	.01	.08		**1.68**
23	.05	.59	.23				.11	.50	.01	.08		**1.55**
24	.14	.64	.29		2.67		.38	.51	.01	.08		**4.72**
25	4.26	.59	.23		.58		6.08	5.37	.01	.36		**17.48**
26	.05	.59	.23		.00		.29	.50	.01	.08		**1.74**
27	.05	.64	.23	.01	.34		.11	.50	.01	.08		**1.96**
28	.28	3.45	.29	.00	1.64		.25	.50	.01	.08		**6.50**
29	.05	.59	.23				.16	.50	.01	.08		**1.61**
30	.96	.59	1.51	.01	.32		.48	12.25	.17	1.44		**17.71**
31	.05	.64	.23				.11	.50	.01	.08		**1.61**
32	.06	.59	.29		2.36		.38	.50	.01	.08		**4.25**
33	.05	.59	.23				.11	.50	.01	.08		**1.56**
34	.05	.64	.23		.00		.23	.50	.01	.08		**1.73**
35	.05	.59	.23	.00			.98	.50	.01	.28		**2.62**
36	.14	.59	.29	.01	1.97		.15	.51	.01	.08		**3.74**
37	.05	.64	.23				.24	.50	.01	.08		**1.74**
38	.05	.59	.23		.00		.16	.50	.01	.08		**1.61**
39	.05	.59	.23				.26	.50	.01	.08		**1.70**
40	.99	.64	1.57		3.23		.44	5.37	.14	1.03		**13.42**
41	.05	.59	.23				.19	.50	.01	.08		**1.64**
42	.28	2.33	.23	.00	.00		.17	.50	.01	.08		**3.60**
43	.05	.64	.23				.20	.50	.01	.08		**1.70**
44	.06	.59	.29		1.64		.15	.50	.01	.08		**3.30**
45	.05	.59	.23	.02	.65		.49	12.16	.01	.66		**14.85**
46	.05	.64	.23		.00		.15	.50	.01	.08		**1.65**
47	.05	.59	.23				.19	.50	.01	.08		**1.64**
48	.14	.59	.29		2.67		.29	.51	.01	.08		**4.58**
49	.05	.64	.23	.00			.22	.50	.01	.08		**1.72**
50	5.18	.59	1.51		.58		6.17	.50	.17	.37		**15.06**
Total	16.52	36.43	18.66	.09	30.17		24.05	74.68	1.07	9.64		211.30

A value of "0.00" means a cost of more than $.000 but less than $.005 per gross square foot.

50-Year Total M&R Cost Profile per Sqft.

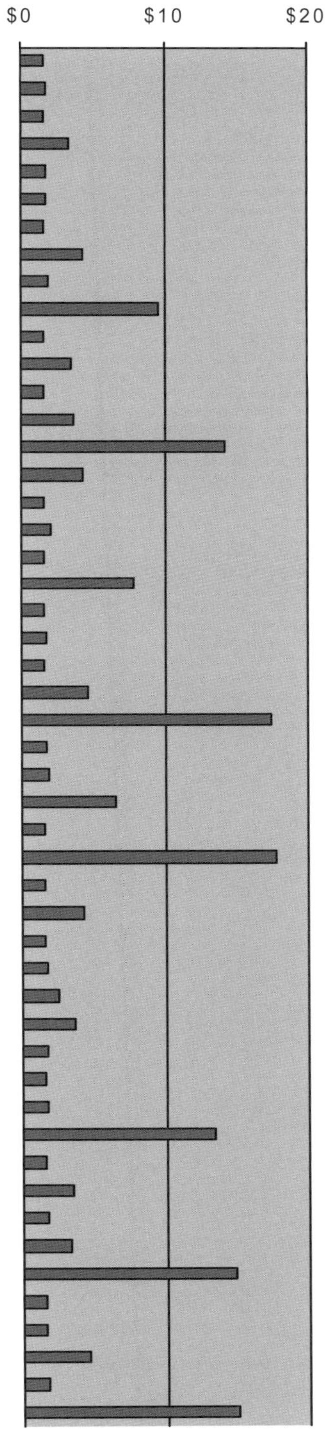

Office Building, 2 Story

Gross Sqft:	83,000 Benchmark Bldg
Height in Ft:	24
Exterior:	Clay Brick
Floor coverings:	Terrazzo/Vinyl Tile/Carpet
HVAC:	Chilled Water, Gas Boiler, Multizone Air
Occupancy:	415
Replacement Cost:	$7,816,035

Components

Component	Quantity	Unit
B20 Exterior Enclosure		
Clay Brick, Exterior, 1st Floor	6890	Sq Ft
Clay Brick, Exterior, 2nd Floor	6890	Sq Ft
Aluminum Fixed Window, 12 sf, 1st Floor	206	Each
Aluminum Fixed Window, 12 sf, 2nd Floor	230	Each
Steel w/ Safety Glass, Painted, Exterior Door	6	Each
Steel, Painted, Exterior Door	11	Each
B30 Roofing		
Single-Ply Modified Bituminous/Thermoplastic Roof	41500	Sq Ft
C10 Interior Construction		
Wood, Solid Core, Painted, Interior Door	184	Each
Steel, Painted, Interior Door	8	Each
C20 Stairs		
Concrete Interior Steps	600	Sq Ft
Metal Interior Railing, Painted	160	Ln Ft
C30 Interior Finishes		
Sheetrock, Unstippled, Interior Wall Finish	72710	Sq Ft
Concrete Block, Painted, Interior Wall Finish	7470	Sq Ft
Fabric Interior Wall Finish	13280	Sq Ft
Concrete, Finished Flooring	7470	Sq Ft
Terrazzo Flooring	11205	Sq Ft
Vinyl Tile Flooring	2075	Sq Ft
Carpet, Nylon 20 oz., Low Traffic	62250	Sq Ft
Sheetrock, Unstippled, Ceiling	7470	Sq Ft
Acoustical Tile, Dropped, Ceiling	75530	Sq Ft
D10 Conveying		
Elevator, Hydraulic, 2500 lbs, 3-5 Floor, 200 fpm	1	Each
D20 Plumbing		
Tankless Water Closet	36	Each
Urinal, Vitreous China	12	Each
Lavatory, Vitreous China	40	Each
Service Sink, Iron Enamel	2	Each
Drinking Fountain, Refrigerated	8	Each
Pipe & Fittings, 3/4" Copper, Cold Water	3.1	K Ln Ft
Pipe & Fittings, 2" Copper, Cold Water	3.2	K Ln Ft
Pipe & Fittings, 3/4" Copper, Hot Water	1.3	K Ln Ft
Pipe Insulation, Cold Water	5.7	K Ln Ft
Circulator Pump, 1/6 HP, Hot Water	1	Each
Water Heater, Gas/Oil 75 Gph	1	Each
Pipe & Fittings, 6" Cast Iron	2	K Ln Ft
Pipe & Fittings, 10" Cast Iron	0.5	K Ln Ft
Pipe & Fittings, 4" DWV PVC	0.6	K Ln Ft
Roof Drain, 4-6"	12	Each
D30 HVAC		
Pipe & Fittings, 4" Steel, Gas	1.8	K Ln Ft
Circulation Pump 10 HP, Hot Water	2	Each
Gas Boiler, 2500 Mbh	1	Each
Chemical Feed System	1	Each
Expansion Tank, 100 Gal.	2	Each
Circulation Pump, 100 HP, Chiller & Condenser Water	2	Each
Chiller, Centrifugal Hermetic, Bundled w/ Condenser, 250 Ton	1	Each
Cooling Tower, 250 Ton	1	Each
Air Handler, Multizone, 25,000 Cfm	4	Each
D40 Fire Protection		
Fire Sprinkler Head	415	Each
D50 Electrical		
Secondary Transformer, Dry, 150 KVA	6	Each
Main Switchgear, <1200 Amp.	1	Each
Distribution Switch, Fused, <600 V	16	Each
Motor Starter, 5-20 HP, <600 V	6	Each
Motor Starter, 21-50 HP, <600 V	10	Each
Motor Starter, 50-75 HP, <600 V	1	Each
Incandescent Lighting Fixture, Basic, 100w	130	Each
Fluorescent Lighting Fixture, 160w	1167	Each
Wiring Device, Switch	664	Each
Receptacle, 120V, 15 Amp	1660	Each
Fire Alarm Bell, 6"	20	Each
Fire Alarm Control Panel	1	Each
Manual Pull Station	20	Each
Smoke Detector	170	Each

50-Year M&R Cost Summary

	50 Year Total	Annual Cost per Sqft	Annual cost as % replacement
PM & Minor Repair	$1,621,705	$0.39	0.41%
Unscheduled Maintenance	$1,617,483	$0.39	0.41%
Renewal & Replacement	$5,603,054	$1.35	1.43%
Total M&R Costs	$8,842,242	$2.13	2.26%

Distribution of Repair Costs

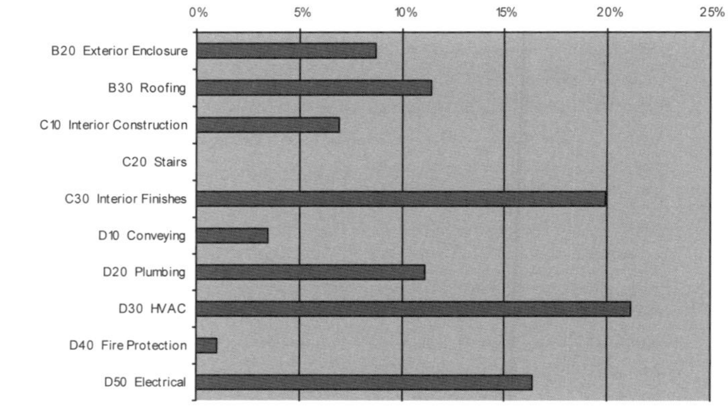

Thirty Most Costly Repair Tasks

Major Repair Task	Task Cost*	Pct.**
Replace Carpet, Nylon 20 oz., Low Traffic	10.88	12.5%
Refinish Sheetrock, Unstippled, Interior Wall Finish	5.32	6.1%
Replace Ballast & Lamps, Fluorescent Lighting Fixture, 160w	3.87	4.5%
Replace Air Handler, Multizone, 25,000 Cfm	3.83	4.4%
Maintain Single-Ply Modified Bituminous/Thermoplastic Roof	3.77	4.3%
Remove & Replace Membrane, Single-Ply Modified Bituminous/Ther	3.50	4.0%
Replace Chiller, Centrifugal Hermetic, Bnd'd w/ Condenser, 250 Ton	3.43	3.9%
Maintain Elevator, Hydraulic, 2500 lbs, 3-5 Floor, 200 fpm	3.13	3.6%
Replace Fluorescent Lighting Fixture, 160w	2.75	3.2%
Replace Pipe & Fittings, 2" Copper, Cold Water	2.61	3.0%
Replace Pipe & Fittings, 3/4" Copper, Cold Water	2.23	2.6%
Maintain Air Handler, Multizone, 25,000 Cfm	1.87	2.1%
Maintain Wood, Solid Core, Painted, Door Locks	1.85	2.1%
Maintain Chiller, Centrifugal Hermetic, Bnd'd w/ Condenser, 250 Ton	1.72	2.0%
Inspect & Test Gas Boiler, 2500 Mbh	1.58	1.8%
Clean & Reseal Clay Brick, Exterior, 2nd Floor	1.54	1.8%
Repair Chiller, Centrifugal Hermetic, Bnd'd w/ Condenser, 250 Ton	1.50	1.7%
Replace Circulation Pump, 100 HP, Chiller & Condenser Water	1.36	1.6%
Check Operation, Smoke Detector	1.36	1.6%
Replace Fabric Interior Wall Finish	1.32	1.5%
Replace Receptacle	1.26	1.4%
Repair Single-Ply Modified Bituminous/Thermoplastic Roof	1.25	1.4%
Clean & Reseal Clay Brick, Exterior, 1st Floor	1.23	1.4%
Replace Pipe & Fittings, 3/4" Copper, Hot Water	.93	1.1%
Replace Wood, Solid Core, Painted, Interior Door	.83	1.0%
Replace Solid Core, Painted, Door Locks	.82	0.9%
Replace Cooling Tower, 250 Ton	.80	0.9%
Repoint (50% Surface) Clay Brick, Exterior, 2nd Floor	.77	0.9%
Replace Wiring Device, Switch	.77	0.9%
Repoint (50% surface) Clay Brick, Exterior, 1st Floor	.74	0.8%

**Task cost ($2004) per gross square foot over 50 years.*

***Percent of total M&R costs.*

Cost per Sqft. by System

Building Age	Exterior Closure	Roofing	Interior Constrution	Stairways	Interior Finishes	Conveying Systems	Plumbing	HVAC Systems	Fire Protection	Electrical	Equipment	Total per Sqft.
1	.01	.15	.11			.06	.03	.21	.01	.10		**.66**
2	.01	.15	.11			.06	.12	.21	.01	.10		**.76**
3	.01	.17	.11			.06	.03	.21	.01	.10		**.69**
4	.01	.15	.13		.55	.06	.12	.21	.01	.10		**1.34**
5	.01	.15	.11		.06	.06	.05	.22	.01	.19		**.85**
6	.01	.17	.11			.06	.12	.21	.01	.10		**.78**
7	.01	.15	.11	.00		.06	.04	.25	.01	.10		**.72**
8	.01	.15	.13		.55	.06	.14	.21	.01	.10		**1.36**
9	.01	.17	.11		.00	.06	.03	.21	.01	.10		**.69**
10	1.34	.56	.34		.40	.06	.30	.84	.17	1.79		**5.79**
11	.01	.15	.11			.06	.03	.21	.01	.10		**.66**
12	.01	.17	.13		.55	.06	.12	.23	.01	.10		**1.38**
13	.01	.15	.11			.06	.06	.21	.01	.10		**.70**
14	.08	.15	.11	.00		.06	.14	.25	.01	.10		**.88**
15	.01	.17	.11	.01	3.77	.06	.04	2.22	.01	.67		**7.07**
16	.01	.15	.13		.55	.06	.14	.21	.01	.10		**1.36**
17	.01	.15	.11			.06	.03	.21	.01	.10		**.66**
18	.01	.17	.11		.05	.06	.12	.21	.01	.30		**1.04**
19	.01	.15	.11			.06	.03	.21	.01	.10		**.66**
20	1.40	1.75	.37		1.39	.06	.33	1.88	.14	2.31		**9.62**
21	.01	.15	.11	.00		.06	.05	.25	.01	.10		**.73**
22	.01	.15	.11			.06	.12	.21	.01	.10		**.76**
23	.01	.17	.11			.06	.03	.21	.01	.16		**.75**
24	.01	.15	.13		.59	.06	.14	.23	.01	.10		**1.41**
25	.92	.15	.11		.12	.34	3.25	.34	.01	.41		**5.64**
26	.01	.17	.11			.06	.14	.21	.01	.10		**.81**
27	.01	.15	.11		.00	.06	.03	.21	.01	.10		**.67**
28	.08	.15	.13	.00	.55	.06	.14	.25	.01	.16		**1.52**
29	.01	.17	.11			.06	.03	.21	.01	.10		**.70**
30	1.34	.56	.89	.01	4.11	.06	.30	3.13	.17	2.71		**13.27**
31	.01	.15	.11			.06	.03	.21	.01	.10		**.66**
32	.01	.17	.11		.55	.06	.14	.21	.01	.10		**1.35**
33	.01	.15	.11			.06	.03	.21	.01	.16		**.73**
34	.01	.15	.13			.06	.12	.21	.01	.10		**.79**
35	.01	.17	.11	.00	.06	.06	.54	.22	.01	.13		**1.31**
36	.01	.15	.11		.60	.06	.06	.23	.01	.30		**1.52**
37	.01	.15	.11			.06	.10	.25	.01	.10		**.77**
38	.01	.17	.13			.06	.07	.21	.01	.10		**.76**
39	.01	.15	.11			.06	.11	.21	.01	.10		**.75**
40	1.40	1.75	.34		1.39	.06	.17	1.99	.14	2.59		**9.83**
41	.01	.15	.11			.06	.09	.21	.01	.16		**.79**
42	.08	.15	.13	.00		.06	.08	.21	.01	.10		**.81**
43	.01	.17	.11			.06	.09	.21	.01	.10		**.75**
44	.01	.15	.11		.55	.06	.06	.25	.01	.10		**1.29**
45	.01	.15	.11	.01	3.77	.06	.25	2.22	.01	.61		**7.19**
46	.01	.17	.13			.06	.06	.21	.01	.16		**.81**
47	.01	.15	.11			.06	.09	.21	.01	.10		**.72**
48	.01	.15	.11		.59	.06	.08	.23	.01	.10		**1.32**
49	.01	.17	.11	.00		.06	.10	.21	.01	.10		**.77**
50	2.25	.56	.36		.46	.34	3.33	.75	.17	1.44		**9.67**
Total	9.26	12.14	7.36	.05	21.17	3.69	11.85	22.54	1.07	17.40		106.53

50-Year Total M&R Cost Profile per Sqft.

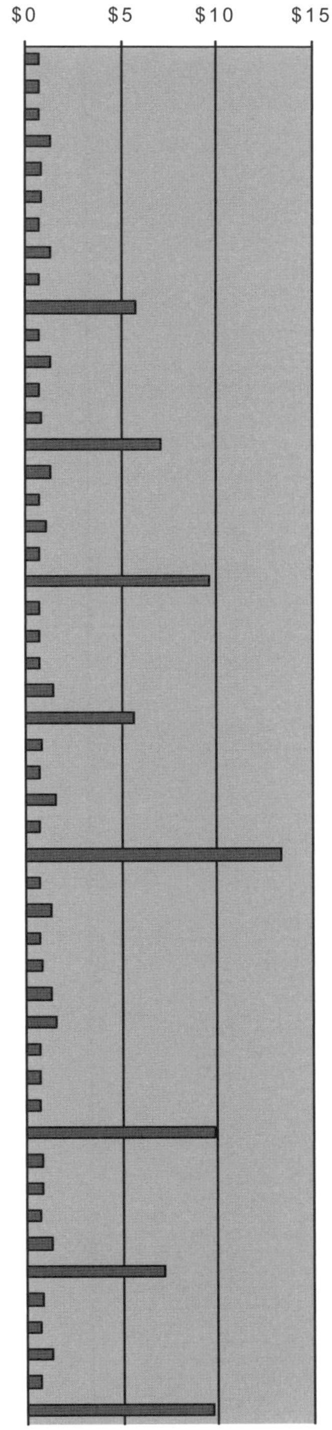

A value of "0.00" means a cost of more than $.000 but less than $.005 per gross square foot.

Office Building, 15 Story

Gross Sqft:	250,000
Height in Ft:	180
Exterior:	Cast Concrete
Floor coverings:	Vinyl Tile/Carpet
HVAC:	Chilled Water, Gas Boiler, Multizone Air
Occupancy:	1875
Replacement Cost:	$30,492,533

Components

Component	Quantity	Unit
B20 Exterior Enclosure		
Concrete, Painted, Exterior, 1st Floor	4940	Sq Ft
Concrete, Painted, Exterior, 2nd Floor	4940	Sq Ft
Concrete, Painted, Exterior, 3rd+ Floor	64100	Sq Ft
Aluminum Fixed Window, 24 sf, 1st Floor	52	Each
Aluminum Fixed Window, 24 sf, 2nd Floor	60	Each
Aluminum Fixed Window, 24 sf, 3rd+ Floor	678	Each
Aluminum Frame, Fully Glazed, Exterior Door	4	Each
Aluminum & Glass Revolving Door	4	Each
B30 Roofing		
Built-up Roof	16670	Sq Ft
C10 Interior Construction		
Wood, Solid Core, Painted, Interior Door	90	Each
Wood, Hollow Core, Painted, Interior Door	195	Each
Steel, Painted, Interior Door	30	Each
C20 Stairs		
Concrete Interior Steps	4500	Sq Ft
Metal Interior Railing, Painted	1200	Ln Ft
C30 Interior Finishes		
Sheetrock, Unstippled, Interior Wall Finish	160675	Sq Ft
Concrete Block, Painted, Interior Wall Finish	22500	Sq Ft
Concrete, Finished Flooring	22500	Sq Ft
Vinyl Tile Flooring	40000	Sq Ft
Carpet, Nylon 20 oz., Low Traffic	187500	Sq Ft
Sheetrock, Unstippled, Ceiling	62500	Sq Ft
Acoustical Tile, Dropped, Ceiling	187500	Sq Ft
D10 Conveying		
Elevator, Geared, 3500 lbs, 6-15 Floor, 350 fpm	5	Each
D20 Plumbing		
Tankless Water Closet	70	Each
Urinal, Vitreous China	28	Each
Lavatory, Vitreous China	84	Each
Service Sink, Iron Enamel	15	Each
Drinking Fountain, Refrigerated	30	Each
Booster Pump, 10 HP	2	Each
Pipe & Fittings, 3/4" Copper, Cold Water	7.3	K Ln Ft
Pipe & Fittings, 2" Copper, Cold Water	7.5	K Ln Ft
Pipe & Fittings, 3/4" Copper, Hot Water	3.1	K Ln Ft
Pipe Insulation, Cold Water	11.8	K Ln Ft
Water Storage Tank, 250 Gal.	1	Each
Circulator Pump, 1/2 HP, Hot Water	2	Each
Water Heater, Gas/Oil 275 Gph	1	Each
Pipe & Fittings, 6" Cast Iron	4	K Ln Ft
Pipe & Fittings, 10" Cast Iron	0.3	K Ln Ft
Sewage Ejector Pump, 1 HP	1	Each
Pipe & Fittings, 4" DWV PVC	1.8	K Ln Ft
Roof Drain, 4-6"	5	Each
D30 HVAC		
Pipe & Fittings, 4" Steel, Gas	1.8	K Ln Ft
Circulation Pump 75 HP, Hot Water	4	Each
Gas Boiler, 2000 Mbh	3	Each
Chemical Feed System	1	Each
Expansion Tank, 250 Gal.	2	Each
Chiller, Centrifugal Hermetic, Bundled w/ Condenser, 250 Ton	3	Each
Cooling Tower, 250 Ton	1	Each
Cooling Tower, 500 Ton	1	Each
Air Handler, Multizone, 10,000 Cfm	30	Each
D40 Fire Protection		
Fire Sprinkler Head	1250	Each
D50 Electrical		
Primary Transformer, Dry, >500 KVA	1	Each
Secondary Transformer, Dry, 150 KVA	16	Each
Main Switchgear, >1200 Amp.	1	Each
Distribution Switch, Fused, <600 V	32	Each
Motor Starter, 5-20 HP, <600 V	32	Each
Motor Starter, 21-50 HP, <600 V	2	Each
Motor Starter, 50-75 HP, <600 V	9	Each
Fluorescent Lighting Fixture, 160w	3906	Each
Wiring Device, Switch	1250	Each
Receptacle, 120V, 15 Amp	3125	Each
Annunciation Panel	1	Each
Fire Alarm Control Panel	1	Each
Manual Pull Station	62	Each
Smoke Detector	500	Each

50-Year M&R Cost Summary

	50 Year Total	Annual Cost per Sqft	Annual cost as % replacement
PM & Minor Repair	$3,606,462	$0.29	0.24%
Unscheduled Maintenance	$2,981,572	$0.24	0.20%
Renewal & Replacement	$17,829,976	$1.43	1.17%
Total M&R Costs	$24,418,010	$1.95	1.60%

Distribution of Repair Costs

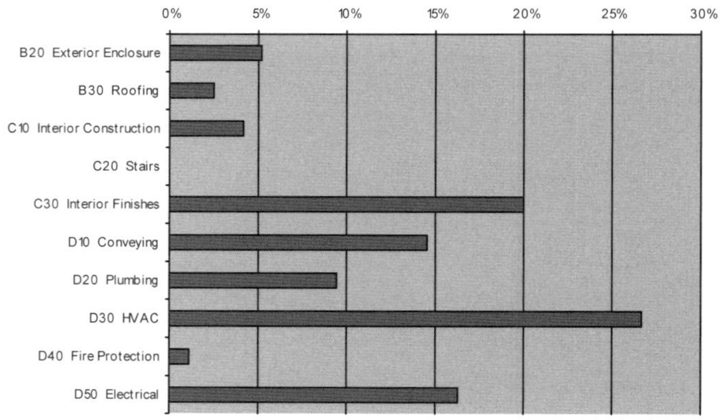

Thirty Most Costly Repair Tasks

Major Repair Task	Task Cost*	Pct.**
Maintain Elevator, Geared, 3500 lbs, 6-15 Floor, 350 fpm	11.17	13.0%
Replace Carpet, Nylon 20 oz., Low Traffic	10.88	12.7%
Replace Air Handler, Multizone, 10,000 Cfm	4.90	5.7%
Maintain Air Handler, Multizone, 10,000 Cfm	4.55	5.3%
Replace Ballast & Lamps, Fluorescent Lighting Fixture, 160w	4.31	5.0%
Refinish Sheetrock, Unstippled, Interior Wall Finish	3.90	4.6%
Replace Chiller, Centrifugal Hermetic, Bnd'd w/ Condenser, 250 Ton	3.42	4.0%
Replace Fluorescent Lighting Fixture, 160w	3.06	3.6%
Renovate Elevator, Geared, 3500 lbs, 6-15 Floor, 350 fpm	3.00	3.5%
Refinish Concrete, Painted, Exterior, 3rd+ Floor	3.00	3.5%
Replace Pipe & Fittings, 2" Copper, Cold Water	2.03	2.4%
Replace Pipe & Fittings, 3/4" Copper, Cold Water	1.74	2.0%
Maintain Chiller, Centrifugal Hermetic, Bnd'd w/ Condenser, 250 Ton	1.71	2.0%
Inspect & Test Gas Boiler, 2000 Mbh	1.57	1.8%
Refinish Sheetrock, Unstippled, Ceiling	1.52	1.8%
Repair Chiller, Centrifugal Hermetic, Bnd'd w/ Condenser, 250 Ton	1.49	1.7%
Replace Aluminum & Glass Revolving Door	1.35	1.6%
Check Operation, Smoke Detector	1.33	1.5%
Maintain Built-up Roof	.97	1.1%
Replace Circulation Pump 75 HP, Hot Water	.83	1.0%
Replace Receptacle	.79	0.9%
Replace Pipe & Fittings, 3/4" Copper, Hot Water	.73	0.9%
Replace Smoke Detector	.70	0.8%
Refinish Concrete Block, Painted, Interior Wall Finish	.69	0.8%
Replace Vinyl Tile Flooring	.67	0.8%
Maintain Wood, Hollow Core, Painted, Door Locks	.65	0.8%
Inspect & Clean Motor Starter, 5-20 HP, <600 V	.58	0.7%
Replace Wiring Device, Switch	.48	0.6%
Repair Fire Sprinkler Head	.48	0.6%
Minor Repair, Acoustic Tile, Dropped, Ceiling	.47	0.5%

**Task cost ($2004) per gross square foot over 50 years.*

***Percent of total M&R costs.*

Cost per Sqft. by System

Building Age	Exterior Closure	Roofing	Interior Constrution	Stairways	Interior Finishes	Conveying Systems	Plumbing	HVAC Systems	Fire Protection	Electrical	Equipment	Total per Sqft.
1	.00	.04	.06			.22	.01	.28	.01	.09		**.72**
2	.00	.04	.06			.22	.10	.29	.01	.09		**.80**
3	.00	.04	.06			.22	.01	.28	.01	.09		**.72**
4	.00	.04	.07		.51	.22	.10	.29	.01	.09		**1.33**
5	.00	.04	.06		.02	.22	.03	.29	.01	.15		**.82**
6	.00	.04	.06			.22	.10	.29	.01	.09		**.81**
7	.00	.04	.06	.00		.22	.02	.33	.01	.09		**.77**
8	.00	.04	.07		.51	.22	.12	.29	.01	.09		**1.35**
9	.00	.04	.06		.01	.22	.01	.28	.01	.09		**.73**
10	.67	.04	.22		.11	.22	.25	.92	.17	1.80		**4.41**
11	.00	.04	.06			.22	.01	.28	.01	.09		**.72**
12	.01	.04	.07		.51	.22	.10	.29	.01	.09		**1.34**
13	.00	.04	.06			.22	.03	.28	.01	.09		**.74**
14	.00	.16	.06	.00		.22	.11	.33	.01	.09		**.97**
15	.45	.04	.06	.03	3.66	.22	.03	2.28	.01	.51		**7.29**
16	.00	.04	.07		.51	.22	.12	.29	.01	.09		**1.35**
17	.00	.04	.06			.22	.01	.28	.01	.09		**.72**
18	.00	.04	.06		.34	.22	.10	.29	.01	.24		**1.29**
19	.00	.04	.06			.22	.01	.28	.01	.09		**.72**
20	.75	.04	.24		1.09	.22	.34	1.94	.14	2.04		**6.80**
21	.00	.04	.06	.00		.22	.03	.33	.01	.09		**.78**
22	.00	.04	.06			.22	.10	.29	.01	.09		**.80**
23	.00	.04	.06			.22	.01	.28	.01	.14		**.77**
24	.01	.04	.07		.55	.22	.12	.29	.01	.09		**1.41**
25	.00	.04	.06		.08	1.72	2.49	.42	.01	.30		**5.12**
26	.00	.04	.06			.22	.11	.29	.01	.09		**.82**
27	.00	.04	.06		.01	.22	.01	.28	.01	.09		**.73**
28	.00	.23	.07	.00	.51	.22	.11	.33	.01	.14		**1.63**
29	.00	.04	.06			.22	.02	.28	.01	.09		**.72**
30	1.12	.04	.42	.03	3.75	.22	.25	3.06	.17	2.68		**11.74**
31	.00	.04	.06			.22	.01	.28	.01	.09		**.72**
32	.00	.04	.06		.51	.22	.12	.29	.01	.09		**1.34**
33	.00	.04	.06			.22	.01	.28	.01	.14		**.77**
34	.00	.04	.07			.22	.10	.29	.01	.09		**.82**
35	.00	.04	.06	.00	.02	.22	.38	.29	.01	.09		**1.12**
36	.01	.04	.06		.85	.22	.05	.29	.01	.24		**1.76**
37	.00	.04	.06			.22	.06	.33	.01	.09		**.81**
38	.00	.04	.07			.22	.06	.29	.01	.09		**.78**
39	.00	.04	.06			.22	.07	.28	.01	.09		**.77**
40	.75	.04	.22		1.09	.22	.22	2.07	.14	2.25		**7.00**
41	.00	.04	.06			.22	.06	.28	.01	.14		**.82**
42	.00	.16	.07	.00		.22	.07	.29	.01	.09		**.90**
43	.00	.04	.06			.22	.06	.28	.01	.09		**.77**
44	.00	.04	.06		.51	.22	.05	.33	.01	.09		**1.31**
45	.45	.04	.06	.03	3.67	.22	.17	2.28	.01	.46		**7.38**
46	.00	.04	.07			.22	.05	.29	.01	.14		**.83**
47	.00	.04	.06			.22	.06	.28	.01	.09		**.76**
48	.01	.04	.06		.55	.22	.07	.29	.01	.09		**1.34**
49	.00	.04	.06	.00		.22	.07	.28	.01	.09		**.78**
50	.67	.04	.23		.18	1.72	2.60	.81	.17	1.54		**7.96**
Total	5.05	2.43	4.09	.12	19.55	14.17	9.27	26.05	1.07	15.88		97.67

50-Year Total M&R Cost Profile per Sqft.

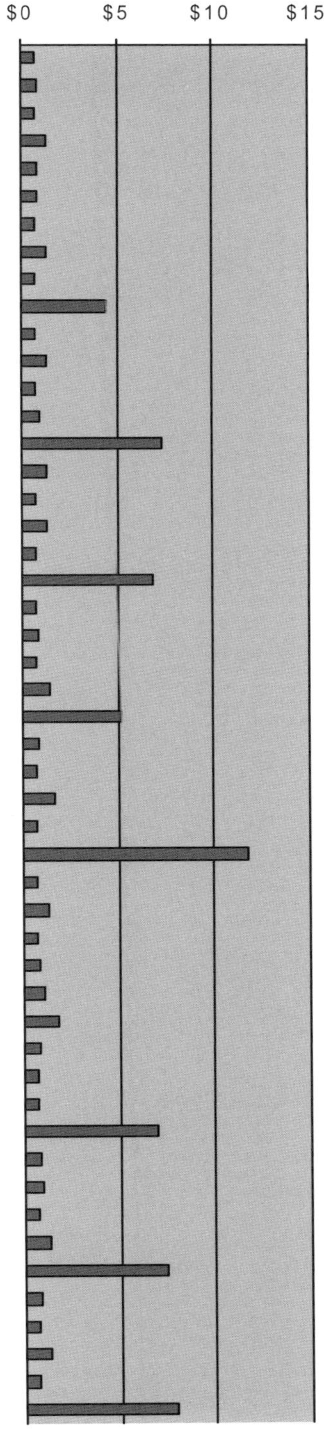

A value of "0.00" means a cost of more than $.000 but less than $.005 per gross square foot.

Office Park

Gross Sqft:	65,000
Height in Ft:	12
Exterior:	Cast Concrete
Floor coverings:	Vinyl Tile/Carpet/Quarry Tile
HVAC:	Single Zone Rooftop Units
Occupancy:	520
Replacement Cost:	$3,159,728

Components

Component	Quantity	Unit
B20 Exterior Enclosure		
Concrete, Painted, Exterior, 1st Floor	8010	Sq Ft
Aluminum Fixed Window, 12 sf, 1st Floor	358	Each
Steel w/ Safety Glass, Painted, Exterior Door	10	Each
Aluminum Frame, Fully Glazed, Exterior Door	4	Each
B30 Roofing		
Built-up Roof	65000	Sq Ft
C10 Interior Construction		
Wood, Hollow Core, Painted, Interior Door	40	Each
Steel, Painted, Interior Door	15	Each
C30 Interior Finishes		
Sheetrock, Unstippled, Interior Wall Finish	9120	Sq Ft
Vinyl Interior Wall Finish	9750	Sq Ft
Concrete, Finished Flooring	3250	Sq Ft
Quarry Tile Flooring	8125	Sq Ft
Vinyl Tile Flooring	1625	Sq Ft
Carpet, Nylon 20 oz., High Traffic	52000	Sq Ft
Sheetrock, Unstippled, Ceiling	3250	Sq Ft
Acoustical Tile, Dropped, Ceiling	61750	Sq Ft
D20 Plumbing		
Tankless Water Closet	20	Each
Urinal, Vitreous China	8	Each
Lavatory, Vitreous China	24	Each
Service Sink, Iron Enamel	4	Each
Drinking Fountain, Refrigerated	8	Each
Pipe & Fittings, 3/4" Copper, Cold Water	2.1	K Ln Ft
Pipe & Fittings, 2" Copper, Cold Water	2.5	K Ln Ft
Pipe & Fittings, 3/4" Copper, Hot Water	0.9	K Ln Ft
Pipe Insulation, Cold Water	2.1	K Ln Ft
Pipe Insulation, Hot Water	0.9	K Ln Ft
Circulator Pump, 1/12 HP, Hot Water	1	Each
Water Heater, Gas/Oil 75 Gph	1	Each
Pipe & Fittings, 6" Cast Iron	1.3	K Ln Ft
Pipe & Fittings, 10" Cast Iron	0.6	K Ln Ft
Pipe & Fittings, 4" DWV PVC	0.5	K Ln Ft
Roof Drain, 4-6"	19	Each
D30 HVAC		
Pipe & Fittings, 4" Steel, Gas	0.6	K Ln Ft
Air Conditioner, Rooftop, 50 Ton	4	Each
D40 Fire Protection		
Fire Sprinkler Head	325	Each
D50 Electrical		
Main Switchgear, <1200 Amp.	1	Each
Distribution Switch, Fused, <600 V	8	Each
Fluorescent Lighting Fixture, 160w	1056	Each
Wiring Device, Switch	580	Each
Receptacle, 120V, 15 Amp	240	Each
Fire Alarm Bell, 6"	10	Each
Fire Alarm Control Panel	1	Each
Manual Pull Station	10	Each
Smoke Detector	130	Each

50-Year M&R Cost Summary

	50 Year Total	Annual Cost per Sqft	Annual cost as % replacement
PM & Minor Repair	$1,442,854	$0.44	0.91%
Unscheduled Maintenance	$1,411,113	$0.43	0.89%
Renewal & Replacement	$4,976,001	$1.53	3.15%
Total M&R Costs	$7,829,968	$2.41	4.96%

Distribution of Repair Costs

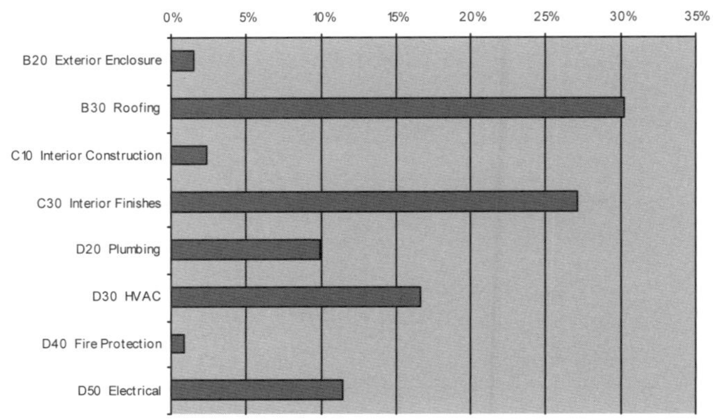

Thirty Most Costly Repair Tasks

Major Repair Task	Task Cost*	Pct.**
Replace Carpet, Nylon 20 oz., High Traffic	23.22	23.5%
Maintain Built-up Roof	14.62	14.8%
Replace Air Conditioner, Rooftop, 50 Ton	9.49	9.6%
Repair Air Conditioner, Rooftop, 50 Ton	5.81	5.9%
Replace Quarry Tile Flooring	5.29	5.4%
Replace Ballast & Lamps, Fluorescent Lighting Fixture, 160w	4.48	4.5%
Place New Membrane Over Existing, Built-up Roof	3.49	3.5%
Remove & Replace Membrane, Built-up Roof	3.45	3.5%
Replace Fluorescent Lighting Fixture, 160w	3.18	3.2%
Maintain Air Conditioner, Rooftop, 50 Ton	2.70	2.7%
Replace Pipe & Fittings, 2" Copper, Cold Water	2.60	2.6%
Replace Pipe & Fittings, 3/4" Copper, Cold Water	1.92	1.9%
Check Operation, Smoke Detector	1.33	1.3%
Replace Vinyl Interior Wall Finish	1.24	1.3%
Refinish Concrete, Painted, Exterior, 1st Floor	.97	1.0%
Replace Wiring Device, Switch	.86	0.9%
Refinish Sheetrock, Unstippled, Interior Wall Finish	.85	0.9%
Replace Pipe & Fittings, 3/4" Copper, Hot Water	.82	0.8%
Non-Destructive Moisture Inspection	.81	0.8%
Repair Wiring Device, Switch	.70	0.7%
Replace Smoke Detector	.70	0.7%
Minor Repair, Acoustic Tile, Dropped, Ceiling	.59	0.6%
Maintain Roof Drain, 4-6"	.53	0.5%
Maintain Wood, Hollow Core, Painted, Door Locks	.51	0.5%
Repair Fire Sprinkler Head	.48	0.5%
Replace Drinking Fountain, Refrigerated	.44	0.4%
Replace Faucet Washer & Clean Trap, Lavatory, Vitreous China	.41	0.4%
Minor Repair, Carpet, Nylon 20 oz., High Traffic	.34	0.3%
Replace Coolant & Adjust Drinking Fountain, Refrigerated	.31	0.3%
Replace Wood, Hollow Core, Painted, Door Locks	.31	0.3%

Task cost ($2004) per gross square foot over 50 years.

***Percent of total M&R costs.*

Cost per Sqft. by System

Building Age	Exterior Closure	Roofing	Interior Constrution	Stairways	Interior Finishes	Conveying Systems	Plumbing	HVAC Systems	Fire Protection	Electrical	Equipment	Total per Sqft.
1	.01	.59	.04				.05	.10	.01	.06		**.84**
2	.01	.59	.04		.02		.14	.10	.01	.06		**.96**
3	.01	.64	.04				.05	.10	.01	.06		**.89**
4	.01	.59	.05		.12		.14	.10	.01	.06		**1.07**
5	.01	.59	.04		.04		.06	.10	.01	.08		**.92**
6	.01	.64	.04		.02		.14	.10	.01	.06		**1.01**
7	.01	.59	.04				.06	.10	.01	.06		**.85**
8	.01	.59	.05		3.97		.16	.10	.01	.06		**4.94**
9	.01	.64	.04		.00		.05	.10	.01	.06		**.89**
10	.20	.59	.18		.39		.29	2.04	.17	1.87		**5.72**
11	.01	.59	.04				.05	.10	.01	.06		**.84**
12	.02	.64	.05		.12		.14	.10	.01	.06		**1.14**
13	.01	.59	.04				.08	.10	.01	.06		**.88**
14	.06	2.33	.04		.02		.15	.10	.01	.06		**2.76**
15	.01	.64	.04		.09		.06	3.16	.01	.58		**4.58**
16	.01	.59	.05		3.97		.16	.10	.01	.06		**4.94**
17	.01	.59	.04				.05	.10	.01	.06		**.84**
18	.01	.64	.04		.08		.14	.10	.01	.06		**1.06**
19	.01	.59	.04				.05	.10	.01	.06		**.84**
20	.26	.59	.19		.58		.31	.10	.14	1.82		**3.98**
21	.01	.64	.04				.07	.10	.01	.06		**.91**
22	.01	.59	.04		.02		.14	.10	.01	.06		**.96**
23	.01	.59	.04				.05	.10	.01	.06		**.84**
24	.02	.64	.05		3.97		.16	.10	.01	.06		**5.01**
25	.01	.59	.04		.08		2.99	2.04	.01	.38		**6.12**
26	.01	.59	.04		.02		.16	.10	.01	.06		**.98**
27	.01	.64	.04		.00		.05	.10	.01	.06		**.90**
28	.06	3.45	.05		.12		.15	.10	.01	.06		**3.99**
29	.01	.59	.04				.06	.10	.01	.06		**.85**
30	.20	.59	.26		.43		.29	3.20	.17	2.36		**7.49**
31	.01	.64	.04				.05	.10	.01	.06		**.90**
32	.01	.59	.04		3.97		.16	.10	.01	.06		**4.93**
33	.01	.59	.04				.05	.10	.01	.06		**.85**
34	.01	.64	.05		.02		.14	.10	.01	.06		**1.01**
35	.01	.59	.04		.04		.45	.10	.01	.08		**1.31**
36	.02	.59	.04		.17		.09	.10	.01	.06		**1.08**
37	.01	.64	.04				.11	.10	.01	.06		**.95**
38	.01	.59	.05		.02		.09	.10	.01	.06		**.92**
39	.01	.59	.04				.12	.10	.01	.06		**.92**
40	.26	.64	.18		4.43		.22	2.04	.14	2.12		**10.01**
41	.01	.59	.04				.10	.10	.01	.06		**.89**
42	.06	2.33	.05		.02		.10	.10	.01	.06		**2.72**
43	.01	.64	.04				.10	.10	.01	.06		**.95**
44	.01	.59	.04		.12		.09	.10	.01	.06		**1.01**
45	.01	.59	.04		.09		.23	3.16	.01	.58		**4.70**
46	.01	.64	.05		.02		.09	.10	.01	.06		**.96**
47	.01	.59	.04				.10	.10	.01	.06		**.89**
48	.02	.59	.04		3.97		.11	.10	.01	.06		**4.89**
49	.01	.64	.04				.11	.10	.01	.06		**.96**
50	.20	.59	.18		5.72		3.08	.10	.17	1.57		**11.61**
Total	1.76	36.42	2.84		32.64		11.96	20.00	1.07	13.76		120.46

50-Year Total M&R Cost Profile per Sqft.

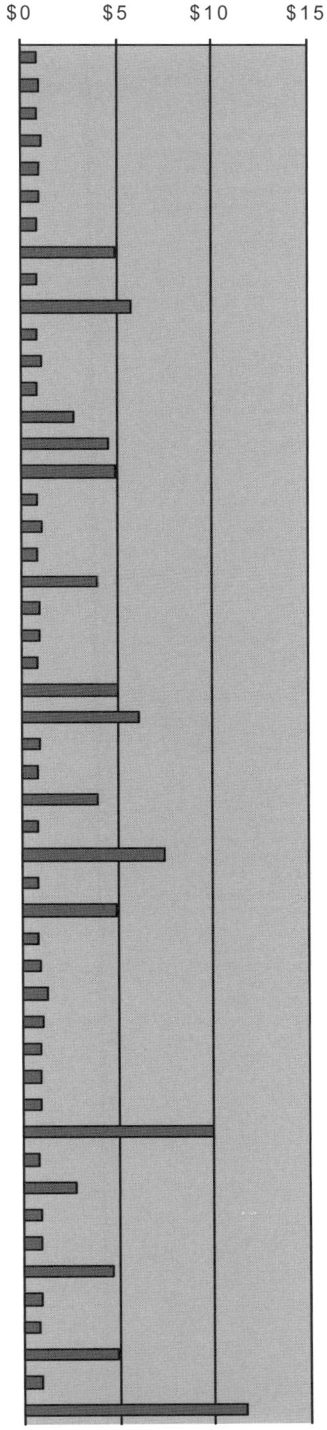

A value of "0.00" means a cost of more than $.000 but less than $.005 per gross square foot.

Post Office

Gross Sqft:	13,000
Height in Ft:	14
Exterior:	Brick Veneer
Floor coverings:	Finished Concrete/Vinyl Tile
HVAC:	Electric Cool, Gas Heating, Singlezone Air
Occupancy:	1300
Replacement Cost:	$1,123,638

Components

Component	Quantity	Unit
B20 Exterior Enclosure		
Clay Brick, Exterior, 1st Floor	5242	Sq Ft
Aluminum Fixed Window, 12 sf, 1st Floor	50	Each
Steel w/ Safety Glass, Painted, Exterior Door	4	Each
Aluminum & Glass Revolving Door	2	Each
B30 Roofing		
Concrete Decking	1000	Sq Ft
Built-up Roof	13000	Sq Ft
C10 Interior Construction		
Wood, Hollow Core, Painted, Interior Door	87	Each
C30 Interior Finishes		
Sheetrock, Stippled, Interior Wall Finish	24276	Sq Ft
Concrete, Finished Flooring	6500	Sq Ft
Vinyl Tile Flooring	6500	Sq Ft
Metal Ceiling, Painted	9750	Sq Ft
Acoustical Tile Ceiling	3250	Sq Ft
D20 Plumbing		
Tankless Water Closet	5	Each
Urinal, Vitreous China	2	Each
Lavatory, Vitreous China	9	Each
Drinking Fountain, Refrigerated	2	Each
Pipe & Fittings, 3/4" Copper, Cold Water	0.64	K Ln Ft
Pipe & Fittings, 4" Steel	0.968	K Ln Ft
Pipe & Fittings, 2" Copper, Cold Water	0.852	K Ln Ft
Pipe & Fittings, 3/4" Copper, Hot Water	0.27	K Ln Ft
Pipe Insulation, Cold Water	1.124	K Ln Ft
Pipe Insulation, Hot Water	0.526	K Ln Ft
Water Heater, Gas/Oil 75 Gph	1	Each
Pipe & Fittings, 6" Cast Iron	0.392	K Ln Ft
Pipe & Fittings, 10" Cast Iron	0.228	K Ln Ft
Pipe & Fittings, 4" DWV PVC	0.105	K Ln Ft
Roof Drain, 4-6"	6	Each
D30 HVAC		
Exhaust Fan, Ceiling, 200-500 Cfm	2	Each
Air Conditioner, Rooftop, 50 Ton	1	Each
D40 Fire Protection		
Fire Sprinkler System	1	Each
Fire Sprinkler Head	93	Each
D50 Electrical		
Safety Switch, Fused, 400 Amp., 3 Ph.	1	Each
Main Switchgear, <1200 Amp.	1	Each
Distribution Panel Board	3	Each
Grounded Fault Circuit Interruptor, 15 Amps	2	Each
Emergency Horn & Strobe	6	Each
Exit Lighting Fixture, w/ Battery	6	Each
Fluorescent Lighting Fixture, 160w	130	Each
Wiring Device, Switch	33	Each
Receptacle, 120V, 15 Amp	104	Each
Annunciation Panel	1	Each
Fire Alarm Bell, 6"	4	Each
Fire Alarm Control Panel	1	Each
Manual Pull Station	4	Each
Heat Detector	2	Each
Smoke Detector	9	Each

50-Year M&R Cost Summary

	50 Year Total	Annual Cost per Sqft	Annual cost as % replacement
PM & Minor Repair	$501,993	$0.77	0.89%
Unscheduled Maintenance	$577,340	$0.89	1.03%
Renewal & Replacement	$1,305,326	$2.01	2.32%
Total M&R Costs	$2,384,659	$3.67	4.24%

Distribution of Repair Costs

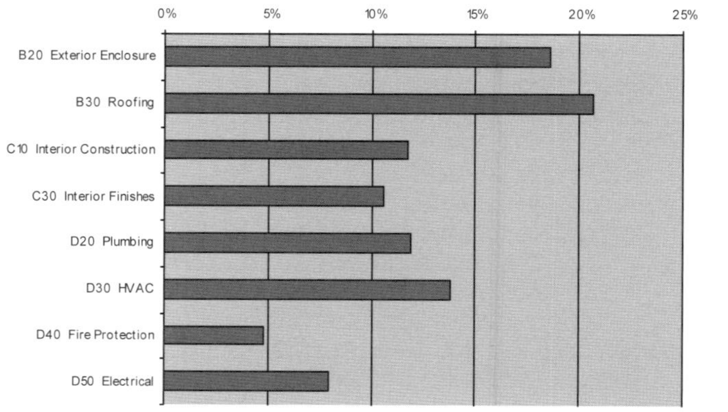

Thirty Most Costly Repair Tasks

Major Repair Task	Task Cost*	Pct.**
Maintain Built-up Roof	14.62	10.5%
Replace Aluminum & Glass Revolving Door	12.96	9.3%
Replace Air Conditioner, Rooftop, 50 Ton	11.86	8.5%
Refinish Sheetrock, Stippled, Interior Wall Finish	11.34	8.2%
Repair Air Conditioner, Rooftop, 50 Ton	7.26	5.2%
Fire Sprinkler System, Annual PM	6.65	4.8%
Clean & Reseal Clay Brick, Exterior, 1st Floor	5.99	4.3%
Maintain Wood, Hollow Core, Painted, Door Locks	5.58	4.0%
Replace Pipe & Fittings, 2" Copper, Cold Water	4.44	3.2%
Repoint (50% surface) Clay Brick, Exterior, 1st Floor	3.58	2.6%
Place New Membrane Over Existing, Built-up Roof	3.49	2.5%
Remove & Replace Membrane, Built-up Roof	3.45	2.5%
Replace Wood, Hollow Core, Painted, Door Locks	3.38	2.4%
Maintain Air Conditioner, Rooftop, 50 Ton	3.38	2.4%
Replace Pipe & Fittings, 3/4" Copper, Cold Water	2.93	2.1%
Replace Ballast & Lamps, Fluorescent Lighting Fixture, 160w	2.76	2.0%
Replace Vinyl Tile Flooring	2.10	1.5%
Replace Wood, Hollow Core, Painted, Interior Door	2.00	1.4%
Replace Fluorescent Lighting Fixture, 160w	1.96	1.4%
Refinish Metal Ceiling, Painted	1.33	1.0%
Replace Pipe & Fittings, 3/4" Copper, Hot Water	1.23	0.9%
Minor Repair, Sheetrock, Stippled, Interior Wall Finish	1.11	0.8%
Annual PM, Distribution Panel Board	1.09	0.8%
Refinish Wood, Hollow Core, Painted, Interior Door	.99	0.7%
Replace Water Heater, Gas/Oil 75 Gph	.91	0.7%
Minor Repair, Metal Ceiling, Painted	.91	0.7%
Maintain Roof Drain, 4-6"	.84	0.6%
Non-Destructive Moisture Inspection	.81	0.6%
Repair Clay Brick, Exterior, 1st Floor	.77	0.6%
Replace Faucet Washer & Clean Trap, Lavatory, Vitreous China	.76	0.5%

**Task cost ($2004) per gross square foot over 50 years.*

***Percent of total M&R costs.*

Cost per Sqft. by System

Building Age	Exterior Closure	Roofing	Interior Constrution	Stairways	Interior Finishes	Conveying Systems	Plumbing	HVAC Systems	Fire Protection	Electrical	Equipment	Total per Sqft.
1	.04	.59	.31				.10	.13	.14	.12		**1.42**
2	.04	.59	.31				.24	.13	.14	.12		**1.57**
3	.04	.64	.31				.10	.13	.14	.12		**1.47**
4	.05	.59	.39		.95		.24	.13	.14	.12		**2.60**
5	.04	.59	.31		.13		.15	.13	.20	.25		**1.80**
6	.04	.64	.31				.24	.13	.14	.12		**1.62**
7	.04	.68	.31				.12	.13	.14	.12		**1.53**
8	.05	.59	.39		1.17		.35	.13	.14	.12		**2.93**
9	.04	.64	.31		.11		.10	.13	.14	.12		**1.58**
10	2.90	.59	1.15		.13		.45	2.55	.43	1.28		**9.47**
11	.04	.59	.31				.10	.13	.14	.12		**1.42**
12	.05	.64	.39		.95		.24	.13	.14	.12		**2.65**
13	.04	.59	.31				.23	.13	.14	.12		**1.56**
14	.14	2.42	.31				.26	.13	.14	.12		**3.52**
15	4.33	.64	.31		.32		.12	3.96	.20	.57		**10.44**
16	.05	.59	.39		1.17		.35	.13	.14	.12		**2.93**
17	.04	.59	.31				.10	.13	.14	.12		**1.42**
18	.04	.64	.31		1.14		.24	.13	.14	.12		**2.76**
19	.04	.59	.31				.10	.13	.14	.12		**1.42**
20	2.94	1.02	1.23		1.88		.48	.19	.39	1.78		**9.92**
21	.04	.73	.31				.15	.13	.14	.12		**1.62**
22	.04	.59	.31				.24	.13	.14	.12		**1.57**
23	.04	.59	.31				.10	.13	.14	.12		**1.42**
24	.05	.64	.39		1.17		.35	.13	.14	.12		**2.98**
25	2.22	.59	.31		.47		5.30	2.55	.20	.56		**12.19**
26	.04	.59	.31				.32	.13	.14	.12		**1.65**
27	.04	.64	.31		.11		.11	.13	.14	.12		**1.59**
28	.15	3.54	.39		.95		.25	.13	.14	.12		**5.67**
29	.04	.59	.31				.14	.13	.14	.12		**1.47**
30	7.19	.59	2.07		.32		.44	3.96	.43	1.49		**16.48**
31	.04	.64	.31				.11	.13	.14	.12		**1.48**
32	.05	.59	.31		1.17		.34	.13	.14	.12		**2.84**
33	.04	.59	.31				.11	.13	.14	.12		**1.43**
34	.04	.64	.39				.23	.13	.14	.12		**1.69**
35	.04	.68	.31		.13		.67	.13	.20	.25		**2.40**
36	.05	.59	.31		2.08		.15	.13	.14	.12		**3.56**
37	.04	.64	.31				.23	.13	.14	.12		**1.60**
38	.04	.59	.39				.16	.13	.14	.12		**1.56**
39	.04	.59	.31				.28	.13	.14	.12		**1.61**
40	2.94	1.07	1.15		2.10		.41	2.62	.39	1.88		**12.56**
41	.04	.59	.31				.19	.13	.14	.12		**1.52**
42	.14	2.42	.39				.17	.13	.14	.12		**3.51**
43	.04	.64	.31				.19	.13	.14	.12		**1.57**
44	.05	.59	.31		.95		.15	.13	.14	.12		**2.42**
45	4.33	.59	.31		.42		.41	3.96	.20	.57		**10.79**
46	.04	.64	.39				.15	.13	.14	.12		**1.61**
47	.04	.59	.31				.19	.13	.14	.12		**1.51**
48	.05	.59	.31		1.17		.26	.13	.14	.12		**2.75**
49	.04	.73	.31				.21	.13	.14	.12		**1.68**
50	5.07	.59	1.23		.47		5.41	.13	.43	1.37		**14.70**
Total	34.10	37.94	21.52		19.44		21.74	25.38	8.72	14.61		183.44

50-Year Total M&R Cost Profile per Sqft.

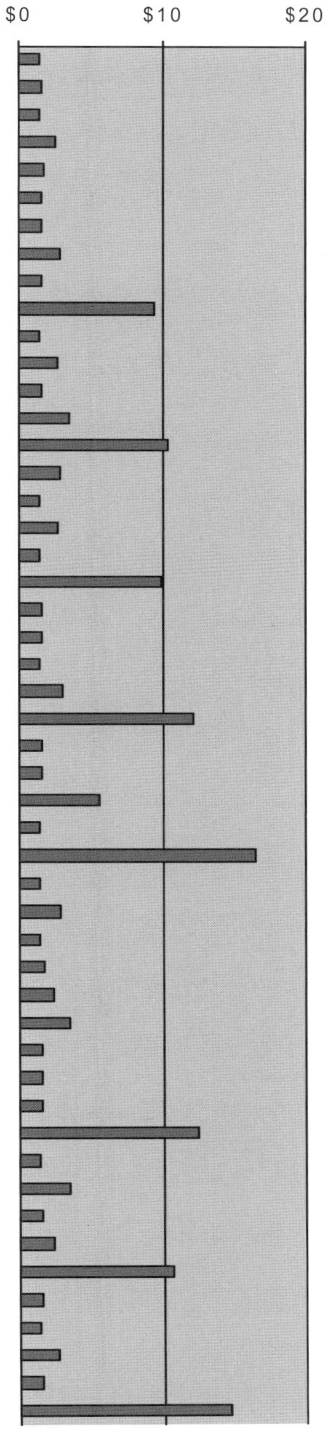

A value of "0.00" means a cost of more than $.000 but less than $.005 per gross square foot.

Public Library, 3 Story

Gross Sqft:	60,000
Height in Ft:	36
Exterior:	Concrete Block
Floor coverings:	Finished Concrete/Quarry Tile/Carpet
HVAC:	Chilled Water, Gas Boiler, Multizone Air
Occupancy:	900
Replacement Cost:	$6,603,063

Components

Component	Quantity	Unit
B20 Exterior Enclosure		
Concrete, Painted, Exterior, 1st Floor	5750	Sq Ft
Concrete, Painted, Exterior, 2nd Floor	5750	Sq Ft
Concrete, Painted, Exterior, 3rd+ Floor	5750	Sq Ft
Aluminum Fixed Window, 24 sf, 1st Floor	50	Each
Aluminum Fixed Window, 24 sf, 2nd Floor	40	Each
Aluminum Fixed Window, 24 sf, 3rd+ Floor	40	Each
Steel, Painted, Exterior Door	6	Each
Aluminum Frame, Fully Glazed, Exterior Door	4	Each
Steel Single 12'x12', Painted, Roll-up Door	1	Each
B30 Roofing		
Built-up Roof	20000	Sq Ft
C10 Interior Construction		
Wood, Solid Core, Painted, Interior Door	48	Each
Wood, Solid Core w/ Safety Glass, Painted, Interior Door	24	Each
C30 Interior Finishes		
Sheetrock, Unstippled, Interior Wall Finish	36590	Sq Ft
Concrete Block, Painted, Interior Wall Finish	1200	Sq Ft
Concrete, Finished Flooring	3000	Sq Ft
Quarry Tile Flooring	12000	Sq Ft
Carpet, Nylon 20 oz., High Traffic	45000	Sq Ft
Metal Ceiling, Painted	3000	Sq Ft
Acoustical Tile, Dropped, Ceiling	57000	Sq Ft
D10 Conveying		
Elevator, Hydraulic, 2500 lbs, 3-5 Floor, 200 fpm	2	Each
Escalator, 16 ft rise, 100 fpm	4	Each
D20 Plumbing		
Tankless Water Closet	21	Each
Urinal, Vitreous China	3	Each
Lavatory, Vitreous China	24	Each
Service Sink, Iron Enamel	3	Each
Drinking Fountain, Refrigerated	12	Each
Pipe & Fittings, 3/4" Copper, Cold Water	2.1	K Ln Ft
Pipe & Fittings, 2" Copper, Cold Water	2.6	K Ln Ft
Pipe & Fittings, 3/4" Copper, Hot Water	0.9	K Ln Ft
Pipe Insulation, Cold Water	2.4	K Ln Ft
Pipe Insulation, Hot Water	1.4	K Ln Ft
Circulator Pump, 1/2 HP, Hot Water	1	Each
Water Heater, Gas/Oil 75 Gph	1	Each
Pipe & Fittings, 6" Cast Iron	1	K Ln Ft
Pipe & Fittings, 10" Cast Iron	0.3	K Ln Ft
Pipe & Fittings, 4" DWV PVC	0.5	K Ln Ft
Roof Drain, 4-6"	6	Each
D30 HVAC		
Pipe & Fittings, 4" Steel, Gas	1.3	K Ln Ft
Gas Boiler, 2000 Mbh	1	Each
Chemical Feed System	1	Each
Circulation Pump 5 HP, Hot Water	2	Each
Expansion Tank, 100 Gal.	2	Each
Circulation Pump, 5 HP, Chiller & Condenser Water	2	Each
Chiller, Reciprocal Water-Cooled Hermetic, 100 Ton	2	Each
Air Handler, Multizone, 25,000 Cfm	3	Each
D40 Fire Protection		
Fire Sprinkler Head	462	Each
D50 Electrical		
Secondary Transformer, Dry, 150 KVA	4	Each
Main Switchgear, <1200 Amp.	1	Each
Distribution Switch, Fused, <600 V	8	Each
Motor Starter, 5-20 HP, <600 V	5	Each
Motor Starter, 21-50 HP, <600 V	1	Each
Motor Starter, 50-75 HP, <600 V	2	Each
Fluorescent Lighting Fixture, 160w	750	Each
Wiring Device, Switch	150	Each
Receptacle, 120V, 15 Amp	600	Each
Annunciation Panel	1	Each
Fire Alarm Bell, 6"	12	Each
Fire Alarm Control Panel	1	Each
Smoke Detector	120	Each

50-Year M&R Cost Summary

	50 Year Total	Annual Cost per Sqft	Annual cost as % replacement
PM & Minor Repair	$1,147,933	$0.38	0.35%
Unscheduled Maintenance	$1,032,347	$0.34	0.31%
Renewal & Replacement	$8,649,462	$2.88	2.62%
Total M&R Costs	$10,829,742	$3.61	3.28%

Distribution of Repair Costs

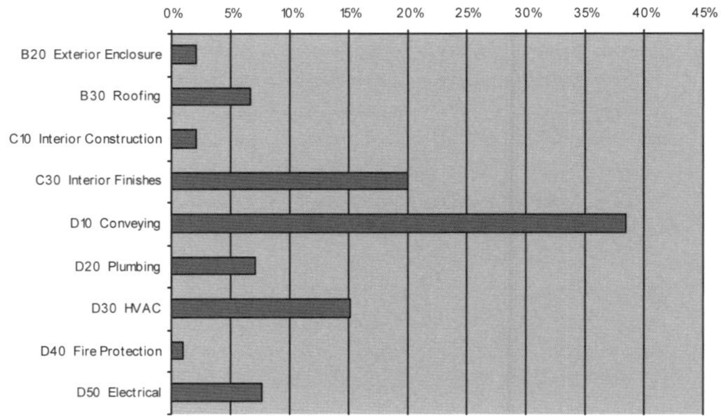

Thirty Most Costly Repair Tasks

Major Repair Task	Task Cost*	Pct.**
Maintain Escalator, 16 ft rise, 100 fpm	59.26	36.3%
Replace Carpet, Nylon 20 oz., High Traffic	21.76	13.3%
Maintain Elevator, Hydraulic, 2500 lbs, 3-5 Floor, 200 fpm	8.65	5.3%
Replace Quarry Tile Flooring	8.47	5.2%
Replace Chiller, Reciprocal Water-Cooled Hermetic, 100 Ton	4.90	3.0%
Maintain Built-up Roof	4.87	3.0%
Replace Air Handler, Multizone, 25,000 Cfm	3.98	2.4%
Refinish Sheetrock, Unstippled, Interior Wall Finish	3.70	2.3%
Replace Ballast & Lamps, Fluorescent Lighting Fixture, 160w	3.44	2.1%
Maintain Chiller, Reciprocal Water-Cooled Hermetic, 100 Ton	3.31	2.0%
Replace Pipe & Fittings, 2" Copper, Cold Water	2.93	1.8%
Repair Chiller, Reciprocal Water-Cooled Hermetic, 100 Ton	2.79	1.7%
Replace Fluorescent Lighting Fixture, 160w	2.45	1.5%
Inspect & Test Gas Boiler, 2000 Mbh	2.18	1.3%
Replace Pipe & Fittings, 3/4" Copper, Cold Water	2.09	1.3%
Maintain Air Handler, Multizone, 25,000 Cfm	1.94	1.2%
Renovate Elevator, Hydraulic, 2500 lbs, 3-5 Floor, 200 fpm	1.55	1.0%
Check Operation, Smoke Detector	1.33	0.8%
Place New Membrane Over Existing, Built-up Roof	1.16	0.7%
Remove & Replace Membrane, Built-up Roof	1.15	0.7%
Refinish Concrete, Painted, Exterior, 3rd+ Floor	1.12	0.7%
Refinish Concrete, Painted, Exterior, 2nd Floor	.94	0.6%
Replace Pipe & Fittings, 3/4" Copper, Hot Water	.89	0.5%
Refinish Concrete, Painted, Exterior, 1st Floor	.76	0.5%
Repair Fire Sprinkler Head	.74	0.5%
Replace Drinking Fountain, Refrigerated	.72	0.4%
Replace Smoke Detector	.70	0.4%
Maintain Wood, Solid Core, Painted, Door Locks	.67	0.4%
Replace Receptacle	.63	0.4%
Minor Repair, Acoustic Tile, Dropped, Ceiling	.59	0.4%

**Task cost ($2004) per gross square foot over 50 years.*

***Percent of total M&R costs.*

Cost per Sqft. by System

Building Age	Exterior Closure	Roofing	Interior Constrution	Stairways	Interior Finishes	Conveying Systems	Plumbing	HVAC Systems	Fire Protection	Electrical	Equipment	Total per Sqft.
1	.01	.20	.06			1.36	.03	.29	.01	.08		**2.03**
2	.01	.20	.06		.02	1.36	.15	.29	.01	.08		**2.18**
3	.01	.21	.06			1.36	.03	.29	.01	.09		**2.05**
4	.01	.20	.07		.34	1.36	.15	.29	.01	.08		**2.51**
5	.01	.20	.06		.01	1.36	.05	.29	.01	.15		**2.13**
6	.01	.21	.06		.02	1.36	.15	.29	.01	.09		**2.20**
7	.01	.20	.06			1.36	.04	.34	.01	.08		**2.10**
8	.01	.20	.07		3.96	1.36	.18	.29	.01	.08		**6.16**
9	.01	.21	.06			1.36	.03	.29	.01	.09		**2.05**
10	.58	.20	.18		.15	1.36	.37	1.29	.26	1.44		**5.82**
11	.01	.20	.06			1.36	.03	.29	.01	.08		**2.03**
12	.03	.21	.07		.34	1.36	.15	.30	.01	.09		**2.56**
13	.01	.20	.06			1.36	.06	.29	.01	.08		**2.06**
14	.04	.78	.06		.02	1.36	.16	.35	.01	.08		**2.86**
15	.01	.21	.06		.07	1.36	.04	1.69	.01	.45		**3.89**
16	.01	.20	.07		3.96	1.36	.18	.29	.01	.08		**6.16**
17	.01	.20	.06			1.36	.03	.29	.01	.08		**2.03**
18	.01	.21	.06		.02	1.36	.15	.29	.01	.22		**2.33**
19	.01	.20	.06			1.36	.03	.29	.01	.08		**2.03**
20	.62	.20	.19		.73	1.36	.39	2.62	.21	1.70		**8.03**
21	.01	.21	.06			1.36	.05	.34	.01	.09		**2.12**
22	.01	.20	.06		.02	1.36	.15	.29	.01	.08		**2.18**
23	.01	.20	.06			1.36	.03	.29	.01	.13		**2.07**
24	.03	.21	.07		3.97	1.36	.18	.30	.01	.09		**6.22**
25	.01	.20	.06		.04	2.14	3.28	.36	.01	.23		**6.32**
26	.01	.20	.06		.02	1.36	.16	.29	.01	.08		**2.19**
27	.01	.21	.06			1.36	.03	.29	.01	.09		**2.05**
28	.05	1.15	.07		.34	1.36	.16	.35	.01	.13		**3.61**
29	.01	.20	.06			1.36	.04	.29	.01	.08		**2.04**
30	.58	.20	.38		.21	1.36	.36	3.07	.26	2.19		**8.60**
31	.01	.21	.06			1.36	.03	.29	.01	.08		**2.05**
32	.01	.20	.06		3.96	1.36	.17	.29	.01	.08		**6.15**
33	.01	.20	.06			1.36	.03	.29	.01	.13		**2.08**
34	.01	.21	.07		.02	1.36	.15	.29	.01	.08		**2.20**
35	.05	.20	.06		.01	1.36	.43	.29	.01	.11		**2.51**
36	.03	.20	.06		.34	1.36	.10	.30	.01	.22		**2.61**
37	.01	.21	.06			1.36	.09	.34	.01	.08		**2.17**
38	.01	.20	.07		.02	1.36	.10	.29	.01	.08		**2.14**
39	.01	.20	.06			1.36	.10	.29	.01	.09		**2.10**
40	.62	.21	.31		4.35	1.36	.25	2.69	.21	1.82		**11.83**
41	.01	.20	.06			1.36	.09	.29	.01	.13		**2.13**
42	.04	.73	.07		.02	1.36	.10	.29	.01	.09		**2.76**
43	.01	.21	.06			1.36	.09	.29	.01	.08		**2.11**
44	.01	.20	.06		.34	1.36	.10	.35	.01	.08		**2.50**
45	.02	.20	.06		.07	1.36	.23	1.69	.01	.40		**4.02**
46	.01	.21	.07		.02	1.36	.10	.29	.01	.13		**2.19**
47	.01	.20	.06			1.36	.09	.29	.01	.08		**2.09**
48	.03	.20	.06		3.97	1.36	.12	.30	.01	.09		**6.13**
49	.01	.21	.06			1.36	.09	.29	.01	.08		**2.11**
50	.58	.20	.19		8.65	2.14	3.45	1.26	.26	1.27		**17.99**
Total	3.70	12.14	3.87		36.05	69.46	12.73	27.24	1.64	13.67		180.50

A value of "0.00" means a cost of more than $.000 but less than $.005 per gross square foot.

50-Year Total M&R Cost Profile per Sqft.

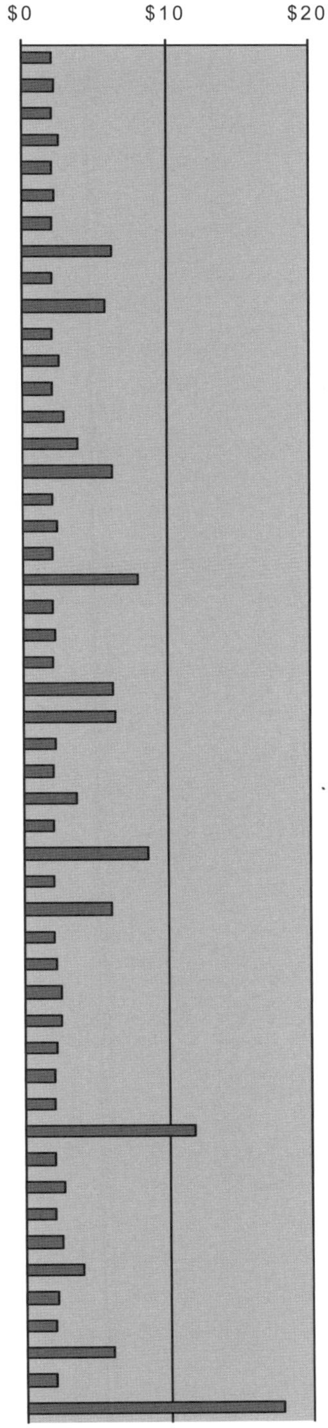

Religious Education

Gross Sqft:	10,000
Height in Ft:	12
Exterior:	Brick Veneer
Floor coverings:	Carpet/Vinyl Tile
HVAC:	Oil Heat Wall Fin Radiation, Split System
Occupancy:	500
Replacement Cost:	$1,110,196

50-Year M&R Cost Summary

	50 Year Total	Annual Cost per Sqft	Annual cost as % replacement
PM & Minor Repair	$580,873	$1.16	1.05%
Unscheduled Maintenance	$570,692	$1.14	1.03%
Renewal & Replacement	$1,342,347	$2.68	2.42%
Total M&R Costs	$2,493,912	$4.99	4.49%

Components

Component	Quantity	Unit
B20 Exterior Enclosure		
Clay Brick, Exterior, 1st Floor	4335	Sq Ft
Aluminum Operable Window, 12 sf, 1st Floor	26	Each
Steel w/ Safety Glass, Painted, Exterior Door	4	Each
B30 Roofing		
Concrete Decking	24	Sq Ft
Built-up Roof	10000	Sq Ft
C10 Interior Construction		
Steel, Painted, Interior Door	14	Each
C20 Stairs		
Concrete Interior Steps	24	Sq Ft
C30 Interior Finishes		
Concrete, Painted, Interior Wall Finish	30000	Sq Ft
Vinyl Tile Flooring	5000	Sq Ft
Carpet, Nylon 20 oz., High Traffic	5000	Sq Ft
Acoustical Tile Ceiling	10000	Sq Ft
D20 Plumbing		
Tankless Water Closet	12	Each
Urinal, Vitreous China	3	Each
Lavatory, Vitreous China	12	Each
Sink, Stainless Steel	4	Each
Drinking Fountain, Refrigerated	3	Each
Pipe & Fittings, 4" Steel	0.848	K Ln Ft
Pipe & Fittings, 3/4" Copper, Cold Water	1.15	K Ln Ft
Pipe & Fittings, 2" Copper, Cold Water	1.172	K Ln Ft
Pipe & Fittings, 3/4" Copper, Hot Water	0.5	K Ln Ft
Pipe Insulation, Cold Water	1.574	K Ln Ft
Pipe Insulation, Hot Water	0.724	K Ln Ft
Water Heater, Gas/Oil 75 Gph	2	Each
Pipe & Fittings, 6" Cast Iron	0.634	K Ln Ft
Pipe & Fittings, 10" Cast Iron	0.2	K Ln Ft
Pipe & Fittings, 4" DWV PVC	0.069	K Ln Ft
Roof Drain, 4-6"	5	Each
D30 HVAC		
Gas Boiler, 2000 Mbh	1	Each
Circulation Pump 5 HP, Hot Water	1	Each
Expansion Tank, 250 Gal.	1	Each
Condenser, Air-Cooled, 20 Ton	2	Each
Air Handler, Singlezone, 6500 Cfm	2	Each
Exhaust Fan, Ceiling, 200-500 Cfm	4	Each
D40 Fire Protection		
Fire Sprinkler Head	71	Each
D50 Electrical		
Safety Switch, Fused, 400 Amp., 3 Ph.	5	Each
Main Switchgear, <1200 Amp.	1	Each
Distribution Panel Board	2	Each
Grounded Fault Circuit Interruptor, 15 Amps	4	Each
Emergency Horn & Strobe	35	Each
Exit Lighting Fixture, w/ Battery	6	Each
Fluorescent Lighting Fixture, 160w	125	Each
Wiring Device, Switch	30	Each
Receptacle, 120V, 15 Amp	50	Each
Annunciation Panel	1	Each
Fire Alarm Bell, 6"	8	Each
Fire Alarm Control Panel	1	Each
Manual Pull Station	8	Each
Smoke Detector	7	Each

Distribution of Repair Costs

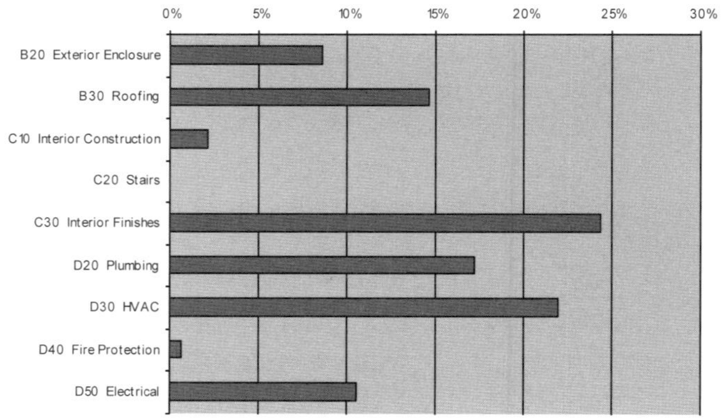

Thirty Most Costly Repair Tasks

Major Repair Task	Task Cost*	Pct.**
Refinish Concrete, Painted, Interior Wall Finish	23.02	12.0%
Minor Repair, Concrete, Painted, Interior Wall Finish	17.54	9.1%
Maintain Built-up Roof	14.62	7.6%
Replace Carpet, Nylon 20 oz., High Traffic	14.51	7.5%
Inspect & Test Gas Boiler, 2000 Mbh	13.09	6.8%
Replace Pipe & Fittings, 2" Copper, Cold Water	7.93	4.1%
Replace Pipe & Fittings, 3/4" Copper, Cold Water	6.85	3.6%
Clean & Reseal Clay Brick, Exterior, 1st Floor	6.44	3.3%
Maintain Air Handler, Singlezone, 6500 Cfm	5.45	2.8%
Replace Air Handler, Singlezone, 6500 Cfm	5.24	2.7%
Replace Condenser, Air-Cooled, 20 Ton	4.36	2.3%
Repoint (50% surface) Clay Brick, Exterior, 1st Floor	3.84	2.0%
Place New Membrane Over Existing, Built-up Roof	3.49	1.8%
Remove & Replace Membrane, Built-up Roof	3.45	1.8%
Replace Ballast & Lamps, Fluorescent Lighting Fixture, 160w	3.44	1.8%
Replace Pipe & Fittings, 3/4" Copper, Hot Water	2.96	1.5%
Replace Safety Switch, Fused, 400 Amp., 3 Ph.	2.69	1.4%
Replace Gas Boiler, 2000 Mbh	2.63	1.4%
Replace Fluorescent Lighting Fixture, 160w	2.45	1.3%
Annual PM, Emergency Horn & Strobe	2.38	1.2%
Replace Water Heater, Gas/Oil 75 Gph	2.38	1.2%
Replace Vinyl Tile Flooring	2.10	1.1%
Repair Gas Boiler, 2000 Mbh	2.04	1.1%
Clean Water Heater, Gas/Oil 75 Gph	1.92	1.0%
Maintain Condenser, Air-Cooled, 20 Ton	1.84	1.0%
Replace Steel, Painted, Door Locks	1.79	0.9%
Minor Repair, Acoustic Tile Ceiling	1.78	0.9%
Finish Repaired Concrete, Painted, Interior Wall Finish	1.54	0.8%
Replace Emergency Horn & Strobe	1.52	0.8%
Annual PM, Safety Switch, Fused, 400 Amp., 3 Ph.	1.38	0.7%

**Task cost ($2004) per gross square foot over 50 years.*

***Percent of total M&R costs.*

Cost per Sqft. by System

Building Age	Exterior Closure	Roofing	Interior Constrution	Stairways	Interior Finishes	Conveying Systems	Plumbing	HVAC Systems	Fire Protection	Electrical	Equipment	Total per Sqft.
1	.02	.59	.06				.17	.78	.01	.24		**1.87**
2	.02	.59	.06		.01		.46	.80	.01	.24		**2.19**
3	.02	.64	.06				.17	.78	.01	.24		**1.92**
4	.02	.59	.08		1.93		.46	.80	.01	.24		**4.13**
5	.02	.59	.06				.30	.79	.01	.42		**2.18**
6	.02	.64	.06		.01		.46	.80	.01	.24		**2.24**
7	.02	.59	.06				.21	1.12	.01	.24		**2.25**
8	.02	.59	.08		4.34		.76	.80	.01	.25		**6.85**
9	.02	.64	.06		.37		.17	.78	.01	.24		**2.29**
10	3.08	.59	.42		.01		.95	1.15	.24	1.70		**8.14**
11	.02	.59	.06				.17	.78	.01	.24		**1.87**
12	.02	.64	.08		1.93		.46	.80	.01	.24		**4.18**
13	.02	.59	.06				.41	.78	.01	.24		**2.10**
14	.15	2.33	.06		.01		.49	1.14	.01	.24		**4.44**
15	.02	.64	.06	.00	6.36		.20	3.93	.01	.83		**12.05**
16	.02	.59	.08		4.34		.76	.80	.01	.25		**6.85**
17	.02	.59	.06				.17	.78	.01	.24		**1.87**
18	.02	.64	.06		1.42		.46	.80	.01	.24		**3.64**
19	.02	.59	.06				.17	.78	.01	.24		**1.87**
20	3.12	.60	.44		1.93		1.02	.98	.20	2.78		**11.06**
21	.02	.64	.06				.30	1.12	.01	.24		**2.39**
22	.02	.59	.06		.01		.46	.80	.01	.24		**2.19**
23	.02	.59	.06				.17	.78	.01	.24		**1.87**
24	.02	.64	.08		4.34		.76	.80	.01	.25		**6.90**
25	2.35	.59	.06				10.45	1.13	.01	1.85		**16.44**
26	.02	.59	.06		.01		.57	.80	.01	.24		**2.30**
27	.02	.64	.06		.37		.19	.78	.01	.24		**2.31**
28	.15	3.46	.08		1.93		.48	1.14	.01	.24		**7.49**
29	.02	.59	.06				.28	.78	.01	.24		**1.98**
30	3.08	.59	.42	.00	6.37		.93	6.08	.24	1.98		**19.69**
31	.02	.64	.06				.19	.78	.01	.24		**1.93**
32	.02	.59	.08		4.34		.75	.80	.01	.24		**6.82**
33	.02	.59	.06				.19	.78	.01	.25		**1.90**
34	.02	.64	.06		.01		.44	.80	.01	.24		**2.22**
35	.02	.59	.06				1.47	.79	.01	.42		**3.35**
36	.02	.59	.08		3.34		.30	.80	.01	.24		**5.37**
37	.02	.64	.06				.43	1.12	.01	.24		**2.52**
38	.02	.59	.06		.01		.31	.80	.01	.24		**2.04**
39	.02	.59	.06				.46	.78	.01	.24		**2.15**
40	3.12	.65	.44		4.34		1.01	1.32	.20	2.91		**13.99**
41	.02	.59	.06				.33	.78	.01	.25		**2.04**
42	.15	2.33	.06		.01		.33	.80	.01	.24		**3.94**
43	.02	.64	.06				.33	.78	.01	.24		**2.08**
44	.02	.59	.08		1.93		.30	1.14	.01	.24		**4.31**
45	.02	.59	.06	.00	6.73		.83	3.93	.01	.83		**13.00**
46	.02	.64	.06		.01		.30	.80	.01	.24		**2.08**
47	.02	.59	.06				.33	.78	.01	.24		**2.03**
48	.02	.59	.08		4.34		.60	.80	.01	.24		**6.67**
49	.02	.64	.06				.37	.78	.01	.25		**2.13**
50	5.42	.59	.42		.01		10.73	1.01	.24	2.86		**21.27**
Total	21.39	36.47	5.23	.01	60.81		43.01	54.61	1.51	26.34		249.39

50-Year Total M&R Cost Profile per Sqft.

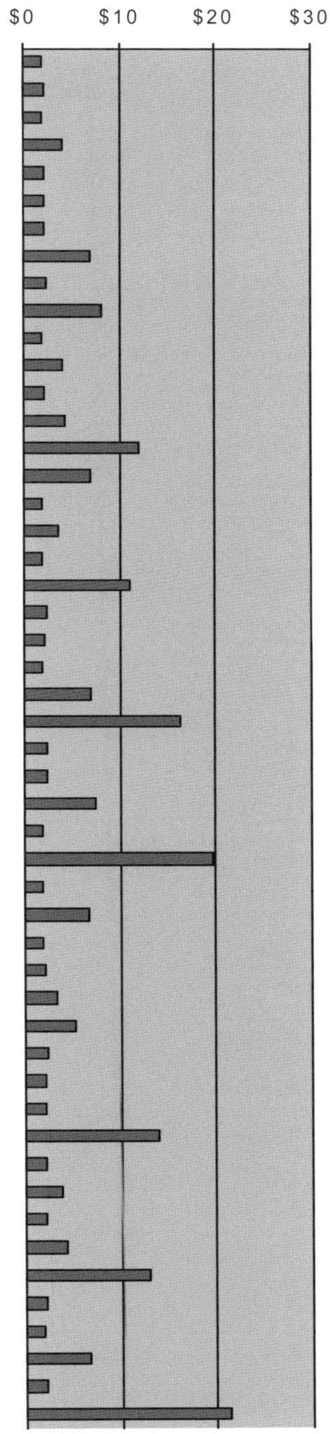

A value of "0.00" means a cost of more than $.000 but less than $.005 per gross square foot.

Restaurant, Fast Food

Gross Sqft:	4,000
Height in Ft:	10
Exterior:	Stucco
Floor coverings:	Finished Concrete/Quarry Tile
HVAC:	Single Zone Rooftop Units
Occupancy:	280
Replacement Cost:	$475,169

Components

Component	Quantity	Unit
B20 Exterior Enclosure		
Stucco, Exterior, 1st Floor	2150	Sq Ft
Aluminum Fixed Window, 24 sf, 1st Floor	16	Each
Steel, Painted, Exterior Door	2	Each
Aluminum Frame, Fully Glazed, Exterior Door	4	Each
B30 Roofing		
Clay Tile Roof	4870	Sq Ft
C10 Interior Construction		
Wood, Solid Core, Painted, Interior Door	15	Each
C30 Interior Finishes		
Sheetrock, Unstippled, Interior Wall Finish	1060	Sq Ft
Vinyl Interior Wall Finish	5260	Sq Ft
Ceramic Tile Wall Finish	278	Sq Ft
Concrete, Finished Flooring	500	Sq Ft
Quarry Tile Flooring	3500	Sq Ft
Acoustical Tile Ceiling	1200	Sq Ft
Wood Finished Ceiling	3670	Sq Ft
D20 Plumbing		
Tankless Water Closet	8	Each
Urinal, Vitreous China	2	Each
Lavatory, Vitreous China	9	Each
Sink, Stainless Steel	2	Each
Service Sink, Iron Enamel	1	Each
Drinking Fountain, Refrigerated	1	Each
Pipe & Fittings, 3/4" Copper, Hot Water	0.4	K Ln Ft
Pipe & Fittings, 3/4" Copper, Cold Water	0.8	K Ln Ft
Pipe & Fittings, 2" Copper, Cold Water	0.8	K Ln Ft
Pipe Insulation, Cold Water	0.8	K Ln Ft
Pipe Insulation, Hot Water	0.4	K Ln Ft
Circulator Pump, 1/12 HP, Hot Water	1	Each
Water Heater, Gas/Oil 275 Gph	1	Each
Pipe & Fittings, 6" Cast Iron	0.5	K Ln Ft
Pipe & Fittings, 10" Cast Iron	0.2	K Ln Ft
Aluminum Gutter, Downspouts, Fittings	0.3	K Ln Ft
D30 HVAC		
Pipe & Fittings, 4" Steel, Gas	0.2	K Ln Ft
Exhaust Fan, Roof Mounted, 10,000 Cfm	1	Each
Air Conditioner, Rooftop, 20 Ton	2	Each
D40 Fire Protection		
Fire Sprinkler Head	20	Each
D50 Electrical		
Distribution Switch, Fused, <600 V	4	Each
Main Switchgear, <1200 Amp.	1	Each
Motor Starter, 5-20 HP, <600 V	2	Each
Fluorescent Lighting Fixture, 160w	50	Each
Wiring Device, Switch	10	Each
Receptacle, 120V, 15 Amp	40	Each
Smoke Detector	8	Each

50-Year M&R Cost Summary

	50 Year Total	Annual Cost per Sqft	Annual cost as % replacement
PM & Minor Repair	$193,496	$0.97	0.81%
Unscheduled Maintenance	$190,823	$0.95	0.80%
Renewal & Replacement	$786,085	$3.93	3.31%
Total M&R Costs	$1,170,404	$5.85	4.93%

Distribution of Repair Costs

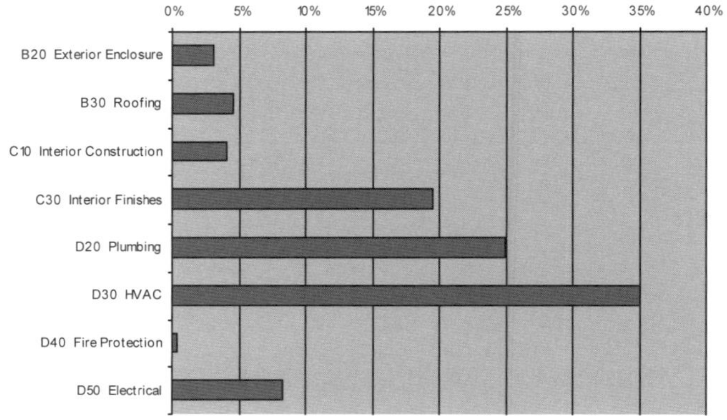

Thirty Most Costly Repair Tasks

Major Repair Task	Task Cost*	Pct.**
Replace Air Conditioner, Rooftop, 20 Ton	45.61	18.6%
Replace Quarry Tile Flooring	37.06	15.1%
Repair Air Conditioner, Rooftop, 20 Ton	18.28	7.5%
Maintain Air Conditioner, Rooftop, 20 Ton	17.00	6.9%
Replace Pipe & Fittings, 2" Copper, Cold Water	13.54	5.5%
Replace Pipe & Fittings, 3/4" Copper, Cold Water	11.92	4.9%
Replace Vinyl Interior Wall Finish	10.85	4.4%
Replace Water Heater, Gas/Oil 275 Gph	8.13	3.3%
Maintain Clay Tile Roof	6.24	2.5%
Replace Pipe & Fittings, 3/4" Copper, Hot Water	5.92	2.4%
Refinish Stucco, Exterior, 1st Floor	4.33	1.8%
Maintain Exhaust Fan, Roof Mounted, 10,000 Cfm	3.91	1.6%
Refinish Wood Finished Ceiling	3.72	1.5%
Replace Ballast & Lamps, Fluorescent Lighting Fixture, 160w	3.44	1.4%
Maintain Wood, Solid Core, Painted, Door Locks	3.12	1.3%
Replace Faucet Washer & Clean Trap, Lavatory, Vitreous China	2.48	1.0%
Replace Fluorescent Lighting Fixture, 160w	2.45	1.0%
Clean Water Heater, Gas/Oil 275 Gph	2.41	1.0%
Replace Distribution Switch, Fused, <600 V	2.30	0.9%
Inspect & Clean Motor Starter, 5-20 HP, <600 V	2.26	0.9%
Repair Main Switchgear, <1200 Amp.	1.97	0.8%
Replace Exhaust Fan, Roof Mounted, 10,000 Cfm	1.97	0.8%
Refinish Sheetrock, Unstippled, Interior Wall Finish	1.61	0.7%
Replace Wood, Solid Core, Painted, Interior Door	1.40	0.6%
Minor Repair, Vinyl Interior Wall Finish	1.39	0.6%
Replace Solid Core, Painted, Door Locks	1.39	0.6%
Replace Main Switchgear, <1200 Amp.	1.38	0.6%
Replace Pipe Insulation, Cold Water	1.33	0.5%
Check Operation, Smoke Detector	1.33	0.5%
Replace Aluminum Gutter, Downspouts, Fittings	1.32	0.5%

**Task cost ($2004) per gross square foot over 50 years.*

***Percent of total M&R costs.*

Cost per Sqft. by System

Building Age	Exterior Closure	Roofing	Interior Construction	Stairways	Interior Finishes	Conveying Systems	Plumbing	HVAC Systems	Fire Protection	Electrical	Equipment	Total per Sqft.
1	.06	.26	.17				.22	.77	.01	.17		**1.66**
2	.06	.26	.17				.64	.77	.01	.17		**2.08**
3	.06	.26	.17				.22	.77	.01	.17		**1.66**
4	.07	.26	.22		.13		.64	.77	.01	.17		**2.27**
5	.06	.26	.17		.37		.42	.77	.01	.68		**2.74**
6	.06	.26	.17		.46		.64	.77	.01	.17		**2.55**
7	.06	.26	.17				.28	.77	.01	.17		**1.72**
8	.07	.26	.22		.13		1.89	.77	.01	.17		**3.52**
9	.06	.26	.17		.11		.22	.77	.01	.17		**1.77**
10	.93	.26	.52		2.25		1.49	6.93	.17	1.96		**14.51**
11	.06	.26	.17				.22	.77	.01	.17		**1.66**
12	.27	.26	.22		.60		.64	.80	.01	.17		**2.97**
13	.06	.26	.17				.50	.77	.01	.17		**1.94**
14	.23	.26	.17				.70	.77	.01	.17		**2.31**
15	.06	.26	.17		.59		.30	15.35	.01	.94		**17.68**
16	.07	.26	.22		.13		1.89	.77	.01	.17		**3.52**
17	.06	.26	.17				.22	.77	.01	.17		**1.66**
18	.06	.26	.17		.57		.64	.77	.01	.34		**2.82**
19	.06	.26	.17				.22	.77	.01	.17		**1.66**
20	1.00	.57	.57		2.50		2.21	1.61	.14	2.38		**10.97**
21	.06	.26	.17				.40	.77	.01	.17		**1.84**
22	.06	.26	.17				.64	.77	.01	.17		**2.08**
23	.06	.26	.17				.22	.77	.01	.32		**1.81**
24	.27	.26	.22		.60		1.89	.80	.01	.17		**4.22**
25	.06	.26	.17		.46		17.75	6.86	.01	.65		**26.22**
26	.06	.26	.17				.76	.77	.01	.17		**2.20**
27	.06	.26	.17		.11		.24	.77	.01	.17		**1.79**
28	.23	.26	.22		.13		.68	.77	.01	.32		**2.63**
29	.06	.26	.17				.36	.77	.01	.17		**1.79**
30	.93	.26	1.45		2.93		1.47	15.63	.17	4.23		**27.07**
31	.06	.26	.17				.24	.77	.01	.17		**1.68**
32	.07	.26	.17		.13		1.87	.77	.01	.17		**3.45**
33	.06	.26	.17				.24	.77	.01	.33		**1.83**
34	.06	.26	.22				.62	.77	.01	.17		**2.11**
35	.06	.26	.17		.37		2.60	.77	.01	.53		**4.76**
36	.27	.26	.17		.71		.33	.80	.01	.34		**2.88**
37	.06	.26	.17				.65	.77	.01	.17		**2.08**
38	.06	.26	.22				.36	.77	.01	.17		**1.85**
39	.06	.26	.17				.67	.77	.01	.17		**2.11**
40	1.00	.57	.52		2.50		2.64	7.70	.14	2.51		**17.57**
41	.06	.26	.17				.53	.77	.01	.32		**2.12**
42	.23	.26	.22		.46		.39	.77	.01	.17		**2.51**
43	.06	.26	.17				.54	.77	.01	.17		**1.98**
44	.07	.26	.17		.13		.33	.77	.01	.17		**1.91**
45	.06	.26	.17		.70		1.41	15.35	.01	.79		**18.74**
46	.06	.26	.22				.33	.77	.01	.32		**1.97**
47	.06	.26	.17				.53	.77	.01	.17		**1.97**
48	.27	.26	.17		.60		1.57	.80	.01	.17		**3.85**
49	.06	.26	.17				.59	.77	.01	.17		**2.03**
50	.93	.26	.57		39.40		18.08	.84	.17	1.69		**61.92**
Total	9.06	13.53	11.85		57.08		73.11	102.63	1.07	24.27		292.60

50-Year Total M&R Cost Profile per Sqft.

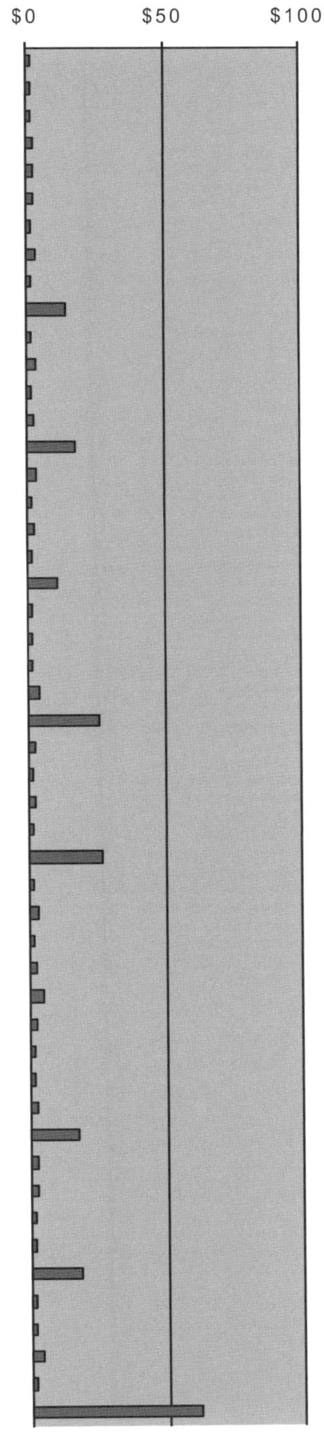

A value of "0.00" means a cost of more than $.000 but less than $.005 per gross square foot.

Restaurant, Large

Gross Sqft:	10,000
Height in Ft:	12
Exterior:	Clay Brick
Floor coverings:	Finished Concrete/Quarry Tile/Carpet
HVAC:	Single Zone Rooftop Units
Occupancy:	410
Replacement Cost:	$1,263,065

50-Year M&R Cost Summary

	50 Year Total	Annual Cost per Sqft	Annual cost as % replacement
PM & Minor Repair	$544,164	$1.09	0.86%
Unscheduled Maintenance	$598,028	$1.20	0.95%
Renewal & Replacement	$1,501,631	$3.00	2.38%
Total M&R Costs	$2,643,823	$5.29	4.19%

Components

Component	Quantity	Unit
B20 Exterior Enclosure		
Clay Brick, Exterior, 1st Floor	4100	Sq Ft
Steel Frame, Painted, Operable Window, 12 sf, 1st Floor	60	Each
Steel, Painted, Exterior Door	2	Each
Wood, Solid Core, Painted, Exterior Door	1	Each
B30 Roofing		
Built-up Roof	10000	Sq Ft
C10 Interior Construction		
Wood, Solid Core, Painted, Interior Door	40	Each
C30 Interior Finishes		
Plaster Interior Wall Finish	13660	Sq Ft
Concrete, Finished Flooring	500	Sq Ft
Quarry Tile Flooring	1500	Sq Ft
Carpet, Nylon 20 oz., High Traffic	8000	Sq Ft
Plaster Ceiling	10000	Sq Ft
D20 Plumbing		
Tankless Water Closet	14	Each
Urinal, Vitreous China	3	Each
Lavatory, Vitreous China	12	Each
Sink, Stainless Steel	5	Each
Service Sink, Iron Enamel	1	Each
Drinking Fountain, Refrigerated	1	Each
Pipe & Fittings, 3/4" Copper, Cold Water	1.2	K Ln Ft
Pipe & Fittings, 2" Copper, Cold Water	1.1	K Ln Ft
Pipe & Fittings, 3/4" Copper, Hot Water	0.6	K Ln Ft
Pipe Insulation, Cold Water	1.1	K Ln Ft
Pipe Insulation, Hot Water	0.6	K Ln Ft
Water Storage Tank, 250 Gal.	1	Each
Circulator Pump, 1/12 HP, Hot Water	2	Each
Water Heater, Gas/Oil 575 Gph	1	Each
Pipe & Fittings, 6" Cast Iron	0.7	K Ln Ft
Pipe & Fittings, 10" Cast Iron	0.2	K Ln Ft
Pipe & Fittings, 4" DWV PVC	1	K Ln Ft
Roof Drain, 4-6"	4	Each
D30 HVAC		
Pipe & Fittings, 4" Steel, Gas	0.2	K Ln Ft
Exhaust Fan, Roof Mounted, 10,000 Cfm	3	Each
Air Conditioner, Rooftop, 20 Ton	3	Each
D40 Fire Protection		
Fire Sprinkler Head	50	Each
D50 Electrical		
Main Switchgear, <1200 Amp.	1	Each
Distribution Switch, Fused, <600 V	5	Each
Motor Starter, 5-20 HP, <600 V	3	Each
Fluorescent Lighting Fixture, 160w	106	Each
Wiring Device, Switch	16	Each
Receptacle, 120V, 15 Amp	75	Each
Heat Detector	20	Each

Distribution of Repair Costs

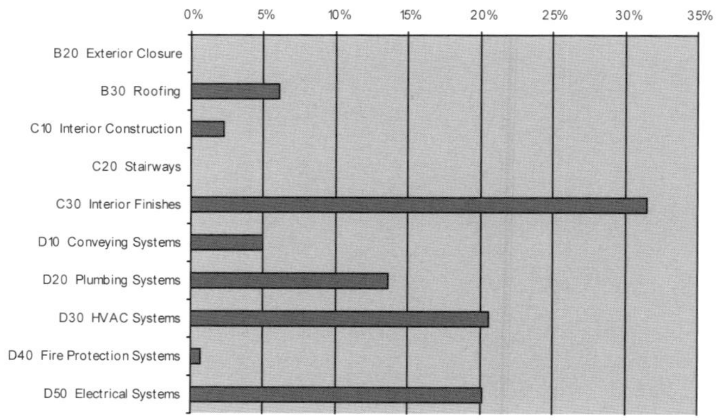

Thirty Most Costly Repair Tasks

Major Repair Task	Task Cost*	Pct.**
Replace Air Conditioner, Rooftop, 20 Ton	27.37	13.4%
Replace Carpet, Nylon 20 oz., High Traffic	23.22	11.3%
Maintain Built-up Roof	14.62	7.1%
Repair Air Conditioner, Rooftop, 20 Ton	10.97	5.4%
Maintain Air Conditioner, Rooftop, 20 Ton	10.20	5.0%
Refinish Plaster Interior Wall Finish	8.30	4.1%
Replace Pipe & Fittings, 2" Copper, Cold Water	7.45	3.6%
Replace Pipe & Fittings, 3/4" Copper, Cold Water	7.15	3.5%
Replace Quarry Tile Flooring	6.35	3.1%
Refinish Plaster Ceiling	6.10	3.0%
Clean & Reseal Clay Brick, Exterior, 1st Floor	6.09	3.0%
Replace Water Heater, Gas/Oil 575 Gph	5.24	2.6%
Minor Repair, Plaster Ceiling	4.95	2.4%
Maintain Exhaust Fan, Roof Mounted, 10,000 Cfm	4.70	2.3%
Clean & Service Water Heater, Gas/Oil 575 Gph	4.68	2.3%
Repoint (50% surface) Clay Brick, Exterior, 1st Floor	3.64	1.8%
Replace Pipe & Fittings, 3/4" Copper, Hot Water	3.55	1.7%
Place New Membrane Over Existing, Built-up Roof	3.49	1.7%
Remove & Replace Membrane, Built-up Roof	3.45	1.7%
Maintain Wood, Solid Core, Painted, Door Locks	3.34	1.6%
Replace Ballast & Lamps, Fluorescent Lighting Fixture, 160w	2.92	1.4%
Replace Exhaust Fan, Roof Mounted, 10,000 Cfm	2.36	1.2%
Replace Fluorescent Lighting Fixture, 160w	2.07	1.0%
Minor Repair, Plaster Interior Wall Finish	2.00	1.0%
Maintain Pipe & Fittings, 4" DWV PVC	1.83	0.9%
Replace Wood, Solid Core, Painted, Interior Door	1.49	0.7%
Replace Solid Core, Painted, Door Locks	1.48	0.7%
Inspect & Clean Motor Starter, 5-20 HP, <600 V	1.36	0.7%
Check Operation, Heat Detector	1.33	0.7%
Replace Faucet Washer & Clean Trap, Lavatory, Vitreous China	1.32	0.6%

**Task cost ($2004) per gross square foot over 50 years.*

***Percent of total M&R costs.*

Cost per Sqft. by System

Building Age	Exterior Closure	Roofing	Interior Construtiion	Stairways	Interior Finishes	Conveying Systems	Plumbing	HVAC Systems	Fire Protection	Electrical	Equipment	Total per Sqft.
1	.01	.59	.18				.17	.55	.01	.12		**1.63**
2	.01	.59	.18		.02		1.08	.55	.01	.12		**2.55**
3	.01	.64	.18				.17	.55	.01	.12		**1.68**
4	.02	.59	.23		1.22		1.08	.55	.01	.12		**3.81**
5	.07	.59	.18				.28	.55	.01	.35		**2.02**
6	.01	.64	.18		.02		1.08	.55	.01	.12		**2.61**
7	.03	.59	.18				.21	.55	.01	.12		**1.68**
8	.02	.59	.23		5.07		1.33	.55	.01	.12		**7.90**
9	.01	.64	.18				.17	.55	.01	.12		**1.68**
10	2.97	.59	.55		.02		1.62	4.29	.17	1.44		**11.64**
11	.01	.59	.18				.17	.55	.01	.12		**1.63**
12	.02	.64	.23		3.26		1.08	.56	.01	.12		**5.91**
13	.01	.59	.18				.31	.55	.01	.12		**1.76**
14	.09	2.33	.18		.02		1.11	.55	.01	.12		**4.41**
15	.07	.64	.18		.05		.23	9.30	.01	.46		**10.93**
16	.02	.59	.23		5.07		1.33	.55	.01	.12		**7.90**
17	.01	.59	.18				.17	.55	.01	.12		**1.63**
18	.01	.64	.18		.02		1.08	.55	.01	.22		**2.70**
19	.01	.59	.18				.17	.55	.01	.12		**1.63**
20	3.03	.59	.60		1.22		1.69	1.55	.14	1.66		**10.49**
21	.03	.64	.18				.25	.55	.01	.12		**1.77**
22	.01	.59	.18		.02		1.08	.55	.01	.12		**2.55**
23	.01	.59	.18				.17	.55	.01	.21		**1.72**
24	.02	.64	.23		7.11		1.33	.56	.01	.12		**10.01**
25	2.28	.59	.18		.03		10.32	4.20	.01	.37		**17.97**
26	.01	.59	.18		.02		1.14	.55	.01	.12		**2.61**
27	.01	.64	.18				.18	.55	.01	.12		**1.69**
28	.10	3.45	.23		1.22		1.10	.55	.01	.21		**6.87**
29	.01	.59	.18				.23	.55	.01	.12		**1.69**
30	2.97	.59	1.54		.07		1.60	9.47	.17	2.49		**18.90**
31	.01	.64	.18				.18	.55	.01	.12		**1.69**
32	.02	.59	.18		5.07		1.32	.55	.01	.12		**7.84**
33	.01	.59	.18				.18	.55	.01	.21		**1.73**
34	.01	.64	.23		.02		1.07	.55	.01	.12		**2.64**
35	.08	.59	.18				1.64	.55	.01	.26		**3.31**
36	.02	.59	.18		3.26		.91	.56	.01	.22		**5.75**
37	.01	.64	.18				.38	.55	.01	.12		**1.89**
38	.01	.59	.23		.02		.93	.55	.01	.12		**2.46**
39	.01	.59	.18				.41	.55	.01	.12		**1.86**
40	3.10	.64	.55		5.07		1.60	5.21	.14	1.77		**18.08**
41	.01	.59	.18				.34	.55	.01	.21		**1.88**
42	.08	2.33	.23		.02		.95	.55	.01	.12		**4.28**
43	.01	.64	.18				.34	.55	.01	.12		**1.85**
44	.02	.59	.18		1.22		.91	.55	.01	.12		**3.59**
45	.07	.59	.18		.05		.85	9.30	.01	.37		**11.41**
46	.01	.64	.23		.02		.91	.55	.01	.21		**2.58**
47	.03	.59	.18				.34	.55	.01	.12		**1.81**
48	.02	.59	.18		7.11		1.16	.56	.01	.12		**9.74**
49	.01	.64	.18				.37	.55	.01	.12		**1.88**
50	5.18	.59	.60		6.41		11.33	.63	.17	1.24		**26.14**
Total	20.64	36.43	12.63		52.77		58.06	66.97	1.07	15.81		264.38

50-Year Total M&R Cost Profile per Sqft.

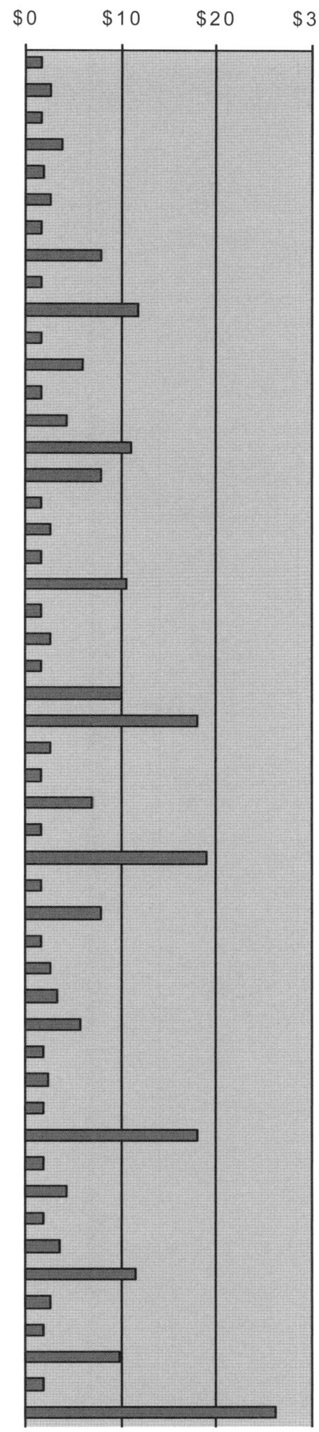

A value cf "0.00" means a cost of more than $.000 but less than $.005 per gross square foot.

Store, Convenience

Gross Sqft:	4,000
Height in Ft:	12
Exterior:	Wood Siding
Floor coverings:	Vinyl Tile
HVAC:	Singlezone Rooftop Unit
Occupancy:	120
Replacement Cost:	$299,732

Components

Component	Quantity	Unit
B20 Exterior Enclosure		
Concrete Block, Exterior, 1st Floor	2440	Sq Ft
Aluminum Operable Window, 24 sf, 1st Floor	10	Each
Wood, Solid Core, Painted, Exterior Door	2	Each
Aluminum Frame, Fully Glazed, Exterior Door	1	Each
B30 Roofing		
Asphalt Shingle Roof	4216	Sq Ft
C10 Interior Construction		
Wood, Solid Core w/ Safety Glass, Painted, Interior Door	3	Each
C30 Interior Finishes		
Sheetrock, Stippled, Interior Wall Finish	1600	Sq Ft
Concrete Block, Painted, Interior Wall Finish	2429	Sq Ft
Vinyl Tile Flooring	4000	Sq Ft
Acoustical Tile Ceiling	4000	Sq Ft
D20 Plumbing		
Tankless Water Closet	4	Each
Lavatory, Vitreous China	4	Each
Sink, Stainless Steel	1	Each
Pipe & Fittings, 3/4" Copper, Cold Water	0.355	K Ln Ft
Pipe & Fittings, 4" Steel	0.554	K Ln Ft
Pipe & Fittings, 2" Copper, Cold Water	0.368	K Ln Ft
Pipe & Fittings, 3/4" Copper, Hot Water	0.155	K Ln Ft
Pipe Insulation, Cold Water	0.631	K Ln Ft
Pipe Insulation, Hot Water	0.305	K Ln Ft
Water Heater, Gas/Oil 75 Gph	1	Each
Pipe & Fittings, 6" Cast Iron	0.207	K Ln Ft
Pipe & Fittings, 10" Cast Iron	0.126	K Ln Ft
Pipe & Fittings, 4" DWV PVC	0.024	K Ln Ft
Roof Drain, 2"	2	Each
Aluminum Gutter, Downspouts, Fittings	0.063	K Ln Ft
D30 HVAC		
Exhaust Fan, Ceiling, 200-500 Cfm	2	Each
Air Conditioner, Rooftop, 20 Ton	1	Each
D40 Fire Protection		
Fire Sprinkler System	1	Each
Fire Sprinkler Head	30	Each
D50 Electrical		
Safety Switch, Fused, 400 Amp., 3 Ph.	1	Each
Main Switchgear, <1200 Amp.	1	Each
Distribution Panel Board	2	Each
Emergency Horn & Strobe	3	Each
Exit Lighting Fixture, w/ Battery	2	Each
Fluorescent Lighting Fixture, 160w	40	Each
Wiring Device, Switch	4	Each
Receptacle, 120V, 15 Amp	10	Each
Annunciation Panel	1	Each
Fire Alarm Control Panel	1	Each
Heat Detector	2	Each
Smoke Detector	2	Each
Generator Switchgear	1	Each

50-Year M&R Cost Summary

	50 Year Total	Annual Cost per Sqft	Annual cost as % replacement
PM & Minor Repair	$199,465	$1.00	1.33%
Unscheduled Maintenance	$205,217	$1.03	1.37%
Renewal & Replacement	$497,324	$2.49	3.32%
Total M&R Costs	$902,006	$4.51	6.02%

Distribution of Repair Costs

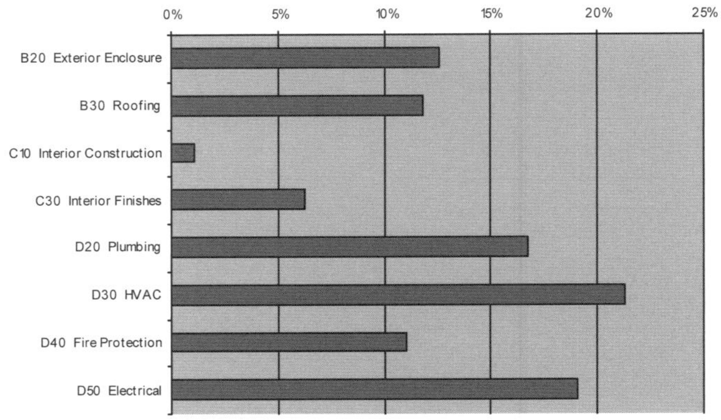

Thirty Most Costly Repair Tasks

Major Repair Task	Task Cost*	Pct.**
Replace Air Conditioner, Rooftop, 20 Ton	22.81	13.1%
Fire Sprinkler System, Annual PM	21.60	12.4%
Repair Air Conditioner, Rooftop, 20 Ton	9.14	5.2%
Clean & Seal Concrete Block, Exterior, 1st Floor	9.10	5.2%
Maintain Asphalt Shingle Roof	8.77	5.0%
Maintain Air Conditioner, Rooftop, 20 Ton	8.51	4.9%
Annual PM, Generator Switchgear	6.78	3.9%
Replace Pipe & Fittings, 2" Copper, Cold Water	6.23	3.6%
Minor Replacement, Asphalt Shingle Roof	5.57	3.2%
Replace Pipe & Fittings, 3/4" Copper, Cold Water	5.29	3.0%
Refinish Concrete Block, Painted, Interior Wall Finish	4.66	2.7%
Replace Vinyl Tile Flooring	4.20	2.4%
Replace Generator Switchgear	3.85	2.2%
Replace Water Heater, Gas/Oil 75 Gph	2.97	1.7%
Replace Ballast & Lamps, Fluorescent Lighting Fixture, 160w	2.76	1.6%
Repoint (50% surface) Concrete Block, Exterior, 1st Floor	2.62	1.5%
Refinish Sheetrock, Stippled, Interior Wall Finish	2.43	1.4%
Clean Water Heater, Gas/Oil 75 Gph	2.41	1.4%
Annual PM, Distribution Panel Board	2.35	1.3%
Replace Pipe & Fittings, 3/4" Copper, Hot Water	2.30	1.3%
Remove & Replace Asphalt Shingle Roof	2.24	1.3%
Repair Main Switchgear, <1200 Amp.	1.97	1.1%
Replace Fluorescent Lighting Fixture, 160w	1.96	1.1%
Minor Repair, Acoustic Tile Ceiling	1.78	1.0%
Fire Sprinkler System, Test Gages & Valves	1.76	1.0%
Place New Over Existing, Asphalt Shingle Roof	1.76	1.0%
Replace Fire Alarm Control Panel	1.43	0.8%
Replace Main Switchgear, <1200 Amp.	1.38	0.8%
Replace Safety Switch, Fused, 400 Amp., 3 Ph.	1.35	0.8%
Repair Concrete Block, Exterior, 1st Floor	1.28	0.7%

**Task cost ($2004) per gross square foot over 50 years.*

***Percent of total M&R costs.*

Cost per Sqft. by System

Building Age	Exterior Closure	Roofing	Interior Constrution	Stairways	Interior Finishes	Conveying Systems	Plumbing	HVAC Systems	Fire Protection	Electrical	Equipment	Total per Sqft.
1	.03	.35	.03				.21	.33	.44	.53		**1.93**
2	.03	.35	.03				.37	.33	.44	.53		**2.09**
3	.03	.35	.03				.21	.33	.44	.53		**1.93**
4	.04	.35	.04		.59		.37	.33	.44	.53		**2.69**
5	.03	.37	.03				.36	.33	.62	.93		**3.28**
6	.03	.35	.03				.37	.33	.44	.53		**2.09**
7	.11	.35	.03				.23	.33	.44	.53		**2.03**
8	.04	.35	.04		.59		.73	.33	.44	.54		**3.06**
9	.03	.35	.03		.39		.21	.33	.44	.53		**2.33**
10	4.36	.97	.13				.67	3.38	.86	1.90		**12.27**
11	.03	.35	.03				.21	.33	.44	.53		**1.93**
12	.09	.35	.04		.59		.37	.33	.44	.53		**2.75**
13	.03	.35	.03				.51	.33	.44	.53		**2.23**
14	.11	.35	.03				.39	.33	.44	.53		**2.18**
15	.03	.97	.03				.25	7.62	.62	1.52		**11.04**
16	.04	.35	.04		.59		.73	.33	.44	.54		**3.06**
17	.03	.35	.03				.21	.33	.44	.53		**1.93**
18	.03	.35	.03		2.46		.37	.33	.44	2.19		**6.20**
19	.03	.35	.03				.21	.33	.44	.53		**1.93**
20	4.41	2.73	.14		.76		.86	.54	.82	2.52		**12.78**
21	.11	.35	.03				.35	.33	.44	.53		**2.15**
22	.03	.35	.03				.37	.33	.44	.53		**2.09**
23	.03	.35	.03				.21	.33	.44	.53		**1.93**
24	.09	.35	.04		.88		.73	.33	.44	.54		**3.40**
25	1.98	.97	.03				8.76	3.38	.62	1.63		**17.38**
26	.03	.35	.03				.52	.33	.44	.53		**2.24**
27	.03	.35	.03		.39		.23	.33	.44	.53		**2.34**
28	.12	.35	.04		.59		.37	.33	.44	.53		**2.78**
29	.03	.35	.03				.35	.33	.44	.53		**2.07**
30	4.36	.97	.13				.66	7.62	.86	2.44		**17.03**
31	.03	.35	.03				.23	.33	.44	.53		**1.95**
32	.04	.35	.04		.59		.72	.33	.44	.53		**3.04**
33	.03	.35	.03				.23	.33	.44	.54		**1.96**
34	.03	.35	.03				.35	.33	.44	.53		**2.07**
35	.11	.97	.03				1.18	.33	.62	.93		**4.17**
36	.09	.35	.04		3.05		.23	.33	.44	2.19		**6.72**
37	.03	.35	.03				.47	.33	.44	.53		**2.19**
38	.03	.35	.03				.24	.33	.44	.53		**1.96**
39	.03	.35	.03				.52	.33	.44	.53		**2.24**
40	4.76	2.24	.37		.76		.93	3.59	.82	2.58		**16.05**
41	.03	.35	.03				.35	.33	.44	.54		**2.08**
42	.03	.35	.03				.25	.33	.44	.53		**1.97**
43	.03	.35	.03				.35	.33	.44	.53		**2.07**
44	.04	.35	.04		.59		.23	.33	.44	.53		**2.56**
45	.03	.97	.03		.39		.81	7.62	.62	1.52		**12.00**
46	.03	.35	.03				.23	.33	.44	.53		**1.95**
47	.11	.35	.03				.35	.33	.44	.53		**2.15**
48	.09	.35	.04		.88		.59	.33	.44	.53		**3.26**
49	.03	.35	.03				.37	.33	.44	.54		**2.10**
50	6.31	.97	.13				8.80	.33	.86	2.49		**19.89**
Total	28.36	26.61	2.52		14.10		37.84	48.06	24.97	43.05		225.50

A value of "0.00" means a cost of more than $.000 but less than $.005 per gross square foot.

50-Year Total M&R Cost Profile per Sqft.

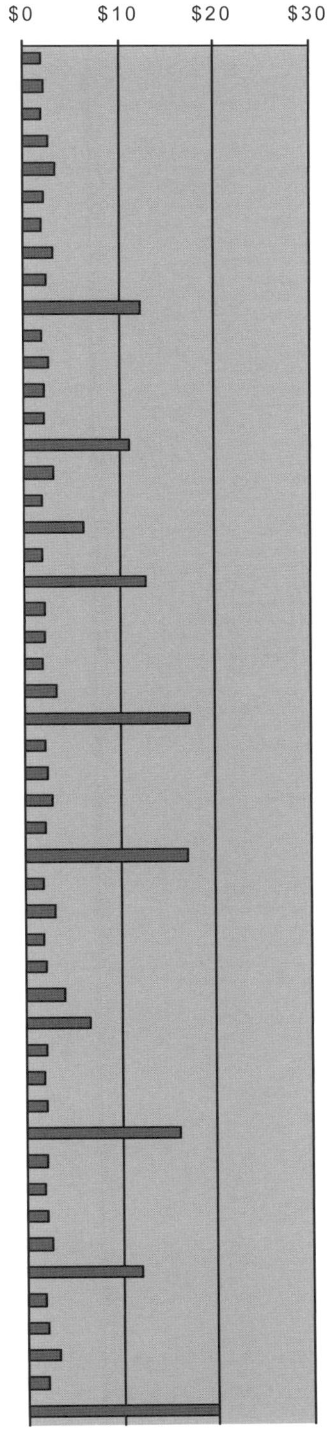

Store, Retail

Gross Sqft:	8,000
Height in Ft:	14
Exterior:	Decorative Concrete Block
Floor coverings:	Vinyl Tile
HVAC:	Singlezone Rooftop Unit
Occupancy:	250
Replacement Cost:	$623,999

Components

Component	Quantity	Unit
B20 Exterior Enclosure		
Concrete Block, Exterior, 1st Floor	4536	Sq Ft
Aluminum Operable Window, 24 sf, 1st Floor	17	Each
Steel w/ Safety Glass, Painted, Exterior Door	1	Each
Aluminum Frame, Fully Glazed, Exterior Door	1	Each
B30 Roofing		
Built-up Roof	8000	Sq Ft
Concrete Decking	800	Sq Ft
Concrete Steps	40	Sq Ft
C10 Interior Construction		
Steel, Painted, Interior Door	13	Each
C30 Interior Finishes		
Sheetrock, Stippled, Interior Wall Finish	3724	Sq Ft
Vinyl Tile Flooring	4000	Sq Ft
Carpet, Nylon 20 oz., High Traffic	4000	Sq Ft
Acoustical Tile Ceiling	8000	Sq Ft
D20 Plumbing		
Tankless Water Closet	5	Each
Urinal, Vitreous China	2	Each
Lavatory, Vitreous China	5	Each
Sink, Stainless Steel	2	Each
Drinking Fountain, Refrigerated	2	Each
Pipe & Fittings, 3/4" Copper, Cold Water	0.59	K Ln Ft
Pipe & Fittings, 3/4" Copper, Hot Water	0.22	K Ln Ft
Pipe & Fittings, 2" Copper, Cold Water	0.663	K Ln Ft
Pipe & Fittings, 4" Steel	0.772	K Ln Ft
Pipe Insulation, Cold Water	0.976	K Ln Ft
Pipe Insulation, Hot Water	0.427	K Ln Ft
Water Heater, Gas/Oil 75 Gph	1	Each
Pipe & Fittings, 6" Cast Iron	0.367	K Ln Ft
Pipe & Fittings, 10" Cast Iron	0.179	K Ln Ft
Roof Drain, 2"	8	Each
Pipe & Fittings, 4" DWV PVC	0.042	K Ln Ft
D30 HVAC		
Air Conditioner, Rooftop, 20 Ton	2	Each
D40 Fire Protection		
Fire Sprinkler System	1	Each
Fire Sprinkler Head	57	Each
D50 Electrical		
Safety Switch, Fused, 400 Amp., 3 Ph.	2	Each
Distribution Panel Board	2	Each
Main Switchgear, >1200 Amp.	1	Each
Exit Lighting Fixture, w/ Battery	4	Each
Emergency Horn & Strobe	6	Each
Fluorescent Lighting Fixture, 160w	100	Each
Wiring Device, Switch	16	Each
Grounded Fault Circuit Interruptor, 15 Amps	2	Each
Receptacle, 120V, 15 Amp	64	Each
Annunciation Panel	1	Each
Fire Alarm Bell, 6"	4	Each
Fire Alarm Control Panel	1	Each
Manual Pull Station	4	Each
Heat Detector	2	Each
Smoke Detector	5	Each

50-Year M&R Cost Summary

	50 Year Total	Annual Cost per Sqft	Annual cost as % replacement
PM & Minor Repair	$361,733	$0.90	1.16%
Unscheduled Maintenance	$382,688	$0.96	1.23%
Renewal & Replacement	$886,735	$2.22	2.84%
Total M&R Costs	$1,631,156	$4.08	5.23%

Distribution of Repair Costs

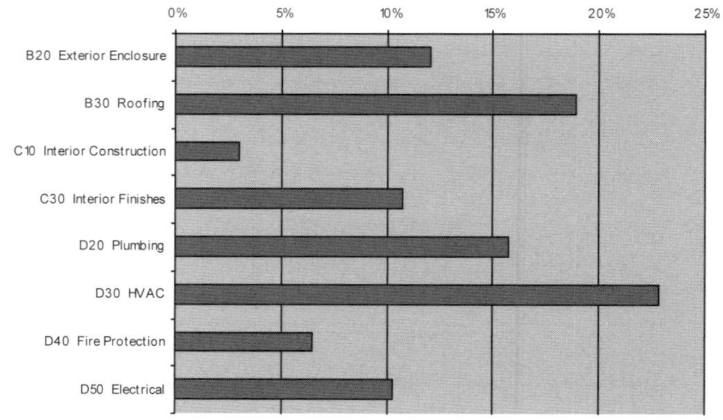

Thirty Most Costly Repair Tasks

Major Repair Task	Task Cost*	Pct.**
Replace Air Conditioner, Rooftop, 20 Ton	22.81	14.6%
Maintain Built-up Roof	14.61	9.4%
Replace Carpet, Nylon 20 oz., High Traffic	14.51	9.3%
Fire Sprinkler System, Annual PM	10.80	6.9%
Repair Air Conditioner, Rooftop, 20 Ton	9.14	5.9%
Maintain Air Conditioner, Rooftop, 20 Ton	8.50	5.4%
Clean & Seal Concrete Block, Exterior, 1st Floor	8.46	5.4%
Replace Pipe & Fittings, 2" Copper, Cold Water	5.61	3.6%
Replace Pipe & Fittings, 3/4" Copper, Cold Water	4.39	2.8%
Place New Membrane Over Existing, Built-up Roof	3.49	2.2%
Remove & Replace Membrane, Built-up Roof	3.45	2.2%
Replace Ballast & Lamps, Fluorescent Lighting Fixture, 160w	3.44	2.2%
Refinish Sheetrock, Stippled, Interior Wall Finish	2.83	1.8%
Replace Fluorescent Lighting Fixture, 160w	2.45	1.6%
Repoint (50% surface) Concrete Block, Exterior, 1st Floor	2.44	1.6%
Replace Vinyl Tile Flooring	2.10	1.3%
Replace Steel, Painted, Door Locks	2.08	1.3%
Maintain Roof Drain, 2"	1.83	1.2%
Minor Repair, Acoustic Tile Ceiling	1.78	1.1%
Replace Pipe & Fittings, 3/4" Copper, Hot Water	1.63	1.0%
Replace Water Heater, Gas/Oil 75 Gph	1.48	1.0%
Maintain Steel, Painted, Door Locks	1.38	0.9%
Replace Safety Switch, Fused, 400 Amp., 3 Ph.	1.35	0.9%
Clean Water Heater, Gas/Oil 75 Gph	1.20	0.8%
Repair Concrete Block, Exterior, 1st Floor	1.19	0.8%
Annual PM, Distribution Panel Board	1.18	0.8%
Replace Main Switchgear, >1200 Amp.	1.13	0.7%
Repair Main Switchgear, >1200 Amp.	.99	0.6%
Replace Drinking Fountain, Refrigerated	.90	0.6%
Fire Sprinkler System, Test Gages & Valves	.88	0.6%

**Task cost ($2004) per gross square foot over 50 years.*

***Percent of total M&R costs.*

Cost per Sqft. by System

Building Age	Exterior Closure	Roofing	Interior Constrution	Stairways	Interior Finishes	Conveying Systems	Plumbing	HVAC Systems	Fire Protection	Electrical	Equipment	Total per Sqft.
1	.01	.59	.07				.18	.31	.23	.16		**1.55**
2	.01	.59	.07		.01		.36	.31	.23	.16		**1.75**
3	.01	.64	.07				.18	.31	.23	.16		**1.60**
4	.01	.59	.10		.25		.36	.31	.23	.16		**2.01**
5	.01	.59	.07				.26	.31	.31	.37		**1.93**
6	.01	.64	.07		.01		.36	.31	.23	.16		**1.80**
7	.01	.71	.07				.20	.31	.23	.16		**1.70**
8	.01	.59	.10		2.65		.55	.31	.23	.17		**4.61**
9	.01	.64	.07		.37		.18	.31	.23	.16		**1.98**
10	4.03	.59	.49		.01		.67	3.36	.54	1.63		**11.31**
11	.01	.59	.07				.18	.31	.23	.16		**1.55**
12	.04	.64	.10		.25		.36	.31	.23	.16		**2.08**
13	.01	.59	.07				.37	.31	.23	.16		**1.74**
14	.05	2.46	.07		.01		.38	.31	.23	.16		**3.68**
15	.01	.64	.07				.20	7.60	.31	.78		**9.62**
16	.01	.59	.10		2.65		.55	.31	.23	.17		**4.61**
17	.01	.59	.07				.18	.31	.23	.16		**1.55**
18	.01	.64	.07		1.42		.36	.31	.23	.16		**3.21**
19	.01	.59	.07				.18	.31	.23	.16		**1.55**
20	4.07	1.18	.51		.45		.72	.31	.50	2.51		**10.25**
21	.01	.76	.07				.26	.31	.23	.16		**1.81**
22	.01	.59	.07		.01		.36	.31	.23	.16		**1.75**
23	.01	.59	.07				.18	.31	.23	.16		**1.55**
24	.04	.64	.10		2.65		.55	.31	.23	.17		**4.68**
25	1.83	.59	.07				7.31	3.36	.31	1.12		**14.59**
26	.01	.59	.07		.01		.47	.31	.23	.16		**1.86**
27	.01	.64	.07		.37		.19	.31	.23	.16		**1.99**
28	.05	3.58	.10		.25		.37	.31	.23	.16		**5.05**
29	.01	.59	.07				.25	.31	.23	.16		**1.62**
30	4.03	.59	.49		.01		.65	7.60	.54	1.93		**15.85**
31	.01	.64	.07				.19	.31	.23	.16		**1.61**
32	.01	.59	.10		2.65		.54	.31	.23	.16		**4.59**
33	.01	.59	.07				.19	.31	.23	.17		**1.57**
34	.01	.64	.07		.01		.35	.31	.23	.16		**1.79**
35	.01	.71	.07				.91	.31	.31	.37		**2.70**
36	.04	.59	.10		1.66		.28	.31	.23	.16		**3.35**
37	.01	.64	.07				.32	.31	.23	.16		**1.75**
38	.01	.59	.07		.01		.28	.31	.23	.16		**1.67**
39	.01	.59	.07				.39	.31	.23	.16		**1.76**
40	4.07	1.23	.51		2.85		.82	3.36	.50	2.61		**15.96**
41	.01	.59	.07				.26	.31	.23	.17		**1.64**
42	.05	2.46	.07		.01		.30	.31	.23	.16		**3.59**
43	.01	.64	.07				.26	.31	.23	.16		**1.69**
44	.01	.59	.10		.25		.28	.31	.23	.16		**1.92**
45	.01	.59	.07		.37		.55	7.60	.31	.78		**10.30**
46	.01	.64	.07		.01		.28	.31	.23	.16		**1.71**
47	.01	.59	.07				.26	.31	.23	.16		**1.64**
48	.04	.59	.10		2.65		.46	.31	.23	.16		**4.54**
49	.01	.76	.07				.29	.31	.23	.17		**1.84**
50	5.84	.59	.49		.01		7.53	.31	.54	2.17		**17.48**
Total	24.59	38.50	6.07		21.94		32.10	46.59	13.20	20.91		203.89

A value of "0.00" means a cost of more than $.000 but less than $.005 per gross square foot.

50-Year Total M&R Cost Profile per Sqft.

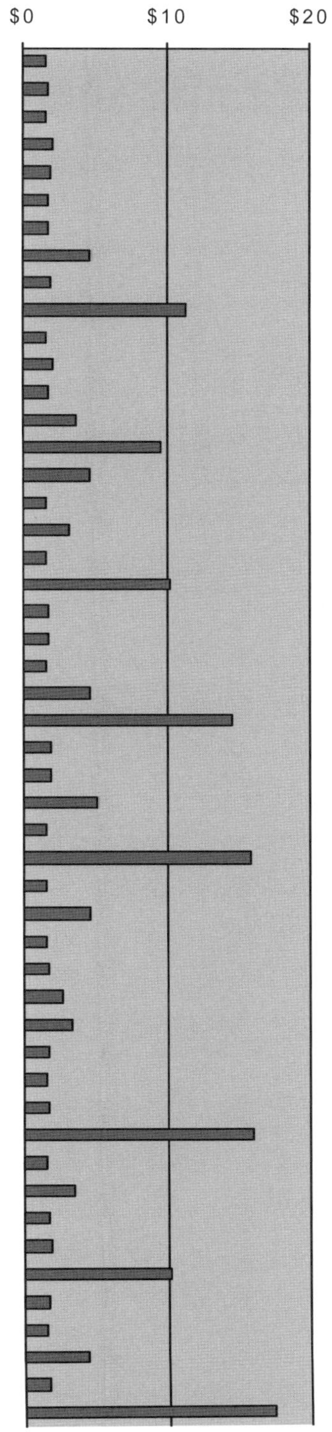

Supermarket

Gross Sqft:	96,000
Height in Ft:	20
Exterior:	Clay Brick
Floor coverings:	Finished Concrete/Vinyl Tile/Ceramic T
HVAC:	Single Zone Rooftop Units
Occupancy:	610
Replacement Cost:	$6,853,088

Components

Component	Quantity	Unit
B20 Exterior Enclosure		
Clay Brick, Exterior, 1st Floor	21074	Sq Ft
Aluminum Fixed Window, 24 sf, 1st Floor	75	Each
Aluminum Fixed Window, 24 sf, 2nd Floor	79	Each
Steel w/ Safety Glass, Painted, Exterior Door	4	Each
Steel, Painted, Exterior Door	12	Each
Aluminum Frame, Fully Glazed, Exterior Door	8	Each
Steel Single 12'x12', Painted, Roll-up Door	4	Each
B30 Roofing		
Built-up Roof	96000	Sq Ft
C10 Interior Construction		
Wood, Solid Core, Painted, Interior Door	40	Each
Aluminum, w/ Safety Glass, Plain or Anodized, Interior Door	40	Each
C30 Interior Finishes		
Sheetrock, Unstippled, Interior Wall Finish	76800	Sq Ft
Concrete Block, Painted, Interior Wall Finish	40274	Sq Ft
Concrete, Finished Flooring	19200	Sq Ft
Ceramic Tile Flooring	6000	Sq Ft
Vinyl Tile Flooring	77050	Sq Ft
Sheetrock, Unstippled, Ceiling	6250	Sq Ft
Acoustical Tile, Dropped, Ceiling	76800	Sq Ft
Metal Ceiling, Painted	19200	Sq Ft
D20 Plumbing		
Tankless Water Closet	11	Each
Urinal, Vitreous China	3	Each
Lavatory, Vitreous China	11	Each
Sink, Stainless Steel	6	Each
Service Sink, Iron Enamel	3	Each
Drinking Fountain, Vitreous China	4	Each
Pipe & Fittings, 3/4" Copper, Cold Water	1.3	K Ln Ft
Pipe & Fittings, 3/4" Copper, Hot Water	0.7	K Ln Ft
Pipe & Fittings, 2" Copper, Cold Water	1.8	K Ln Ft
Pipe Insulation, Cold Water	1.3	K Ln Ft
Pipe Insulation, Hot Water	0.7	K Ln Ft
Circulator Pump, 1/12 HP, Hot Water	1	Each
Water Heater, Gas/Oil 75 Gph	1	Each
Pipe & Fittings, 6" Cast Iron	0.9	K Ln Ft
Pipe & Fittings, 10" Cast Iron	0.7	K Ln Ft
Roof Drain, 4-6"	28	Each
Pipe & Fittings, 4" DWV PVC	1.2	K Ln Ft
D30 HVAC		
Pipe & Fittings, 4" Steel, Gas	0.7	K Ln Ft
Air Conditioner, Rooftop, 20 Ton	1	Each
Air Conditioner, Rooftop, 50 Ton	4	Each
D40 Fire Protection		
Fire Sprinkler Head	520	Each
D50 Electrical		
Distribution Switch, Fused, <600 V	16	Each
Main Switchgear, <1200 Amp.	1	Each
Incandescent Lighting Fixture, Basic, 100w	20	Each
Fluorescent Lighting Fixture, 160w	280	Each
Metal Halide Lighting Fixture, Low Bay, 250w	975	Each
Wiring Device, Switch	130	Each
Receptacle, 120V, 15 Amp	520	Each
Annunciation Panel	1	Each
Fire Alarm Control Panel	1	Each
Manual Pull Station	20	Each
Smoke Detector	200	Each

50-Year M&R Cost Summary

	50 Year Total	Annual Cost per Sqft	Annual cost as % replacement
PM & Minor Repair	$2,706,654	$0.56	0.79%
Unscheduled Maintenance	$2,456,198	$0.51	0.72%
Renewal & Replacement	$5,565,368	$1.16	1.62%
Total M&R Costs	$10,728,220	$2.24	3.13%

Distribution of Repair Costs

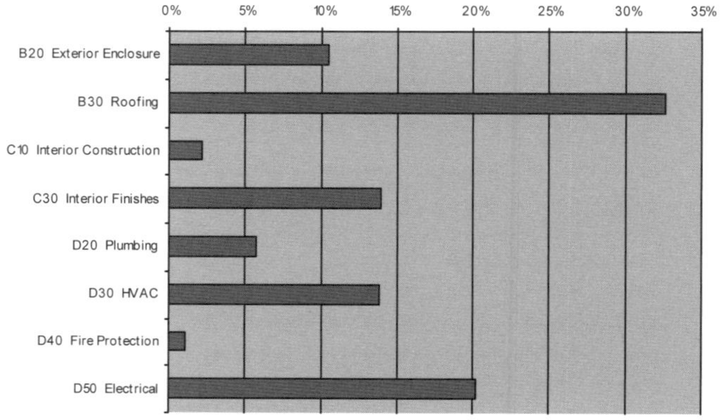

Thirty Most Costly Repair Tasks

Major Repair Task	Task Cost*	Pct.**
Maintain Built-up Roof	14.62	17.0%
Replace Metal Halide Lighting Fixture, Low Bay, 250w	7.18	8.3%
Replace Air Conditioner, Rooftop, 50 Ton	6.43	7.5%
Refinish Sheetrock, Unstippled, Interior Wall Finish	4.86	5.6%
Replace Ballast, Metal Halide Lighting Fixture, Low Bay, 250w	4.53	5.3%
Replace Lamp, Metal Halide Lighting Fixture, Low Bay, 250w	4.05	4.7%
Repair Air Conditioner, Rooftop, 50 Ton	3.93	4.6%
Place New Membrane Over Existing, Built-up Roof	3.49	4.1%
Remove & Replace Membrane, Built-up Roof	3.45	4.0%
Replace Vinyl Tile Flooring	3.37	3.9%
Clean & Reseal Clay Brick, Exterior, 1st Floor	3.26	3.8%
Refinish Concrete Block, Painted, Interior Wall Finish	3.22	3.7%
Repoint (50% surface) Clay Brick, Exterior, 1st Floor	1.95	2.3%
Maintain Air Conditioner, Rooftop, 50 Ton	1.83	2.1%
Check Operation, Smoke Detector	1.38	1.6%
Replace Pipe & Fittings, 2" Copper, Cold Water	1.27	1.5%
Replace Air Conditioner, Rooftop, 20 Ton	.95	1.1%
Non-Destructive Moisture Inspection	.81	0.9%
Replace Pipe & Fittings, 3/4" Copper, Cold Water	.81	0.9%
Replace Ballast & Lamps, Fluorescent Lighting Fixture, 160w	.80	0.9%
Replace Ceramic Tile Flooring	.80	0.9%
Replace Smoke Detector	.73	0.8%
Replace Fluorescent Lighting Fixture, 160w	.57	0.7%
Maintain Roof Drain, 4-6"	.53	0.6%
Repair Fire Sprinkler Head	.52	0.6%
Minor Repair, Acoustic Tile, Dropped, Ceiling	.50	0.6%
Minor Repair, Sheetrock, Unstippled, Interior Wall Finish	.48	0.6%
Replace Pipe & Fittings, 3/4" Copper, Hot Water	.43	0.5%
Repair Clay Brick, Exterior, 1st Floor	.42	0.5%
Refinish Sheetrock, Unstippled, Ceiling	.40	0.5%

**Task cost ($2004) per gross square foot over 50 years.*

***Percent of total M&R costs.*

Cost per Sqft. by System

Building Age	Exterior Closure	Roofing	Interior Constrution	Stairways	Interior Finishes	Conveying Systems	Plumbing	HVAC Systems	Fire Protection	Electrical	Equipment	Total per Sqft.
1	.02	.59	.04				.05	.08	.01	.06		**.84**
2	.02	.59	.04				.07	.08	.01	.06		**.86**
3	.02	.64	.04				.05	.08	.01	.06		**.89**
4	.02	.59	.04		.71		.07	.08	.01	.06		**1.58**
5	.02	.59	.04		.04		.06	.08	.01	.63		**1.46**
6	.02	.64	.04				.07	.08	.01	.06		**.91**
7	.02	.59	.04				.05	.08	.01	.06		**.84**
8	.02	.59	.04		.77		.09	.08	.01	.06		**1.66**
9	.02	.64	.04		.03		.05	.08	.01	.06		**.92**
10	1.59	.59	.09		.14		.12	1.52	.18	2.51		**6.73**
11	.02	.59	.04				.05	.08	.01	.06		**.84**
12	.04	.64	.04		.71		.07	.08	.01	.06		**1.65**
13	.02	.59	.04				.07	.08	.01	.06		**.86**
14	.08	2.33	.04				.08	.08	.01	.06		**2.67**
15	.02	.64	.04		.11		.05	2.46	.01	.89		**4.22**
16	.02	.59	.04		.77		.09	.08	.01	.06		**1.66**
17	.02	.59	.04				.05	.08	.01	.06		**.84**
18	.02	.64	.04		1.69		.07	.08	.01	.06		**2.60**
19	.02	.59	.04				.05	.08	.01	.06		**.84**
20	1.62	.59	.10		1.23		.13	.08	.15	4.16		**8.05**
21	.02	.64	.04				.06	.08	.01	.06		**.90**
22	.02	.59	.04				.07	.08	.01	.06		**.86**
23	.02	.59	.04				.05	.08	.01	.06		**.84**
24	.04	.64	.04		.96		.09	.08	.01	.06		**1.93**
25	1.21	.59	.04		.17		1.45	1.52	.01	.74		**5.71**
26	.02	.59	.04				.08	.08	.01	.06		**.88**
27	.02	.64	.04		.03		.05	.08	.01	.06		**.92**
28	.08	3.45	.04		.71		.07	.08	.01	.06		**4.50**
29	.02	.59	.04				.05	.08	.01	.06		**.85**
30	1.59	.59	.20		.21		.12	2.49	.18	3.05		**8.42**
31	.02	.64	.04				.05	.08	.01	.06		**.89**
32	.02	.59	.04		.77		.09	.08	.01	.06		**1.65**
33	.02	.59	.04				.05	.08	.01	.06		**.84**
34	.02	.64	.04				.07	.08	.01	.06		**.92**
35	.11	.59	.04		.04		.21	.08	.01	.63		**1.70**
36	.04	.59	.04		2.39		.05	.08	.01	.06		**3.26**
37	.02	.64	.04				.07	.08	.01	.06		**.92**
38	.02	.59	.04				.05	.08	.01	.06		**.85**
39	.02	.59	.04				.08	.08	.01	.06		**.87**
40	1.61	.64	.09		1.29		.15	1.52	.15	4.26		**9.72**
41	.02	.59	.04				.07	.08	.01	.06		**.86**
42	.08	2.33	.04				.06	.08	.01	.06		**2.65**
43	.02	.64	.04				.07	.08	.01	.06		**.91**
44	.02	.59	.04		.71		.05	.08	.01	.06		**1.55**
45	.03	.59	.04		.14		.12	2.46	.01	.89		**4.28**
46	.02	.64	.04				.05	.08	.01	.06		**.90**
47	.02	.59	.04				.07	.08	.01	.06		**.86**
48	.04	.59	.04		.96		.07	.08	.01	.06		**1.85**
49	.02	.64	.04				.07	.08	.01	.06		**.91**
50	2.76	.59	.10		1.07		1.46	.08	.18	2.41		**8.64**
Total	11.66	36.42	2.36		15.62		6.46	15.49	1.16	22.58		111.75

A value of "0.00" means a cost of more than $.000 but less than $.005 per gross square foot.

50-Year Total M&R Cost Profile per Sqft.

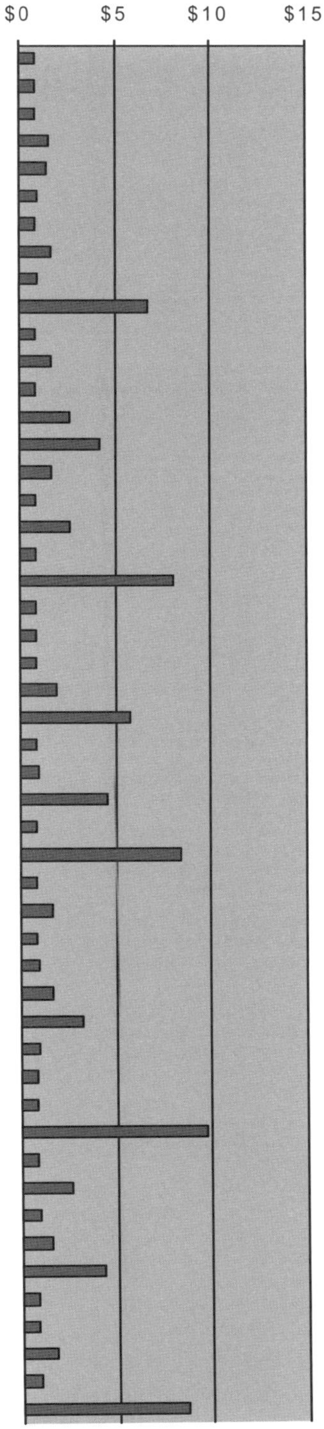

Telephone Exchange

Gross Sqft:	5,000
Height in Ft:	12
Exterior:	Brick Veneer
Floor coverings:	Carpet/Vinyl Tile
HVAC:	Singlezone Rooftop Unit
Occupancy:	50
Replacement Cost:	$549,219

Components

Component	Quantity	Unit
B20 Exterior Enclosure		
Clay Brick, Exterior, 1st Floor	2746	Sq Ft
Aluminum Fixed Window, 12 sf, 1st Floor	34	Each
Steel w/ Safety Glass, Painted, Exterior Door	1	Each
Aluminum Frame, Fully Glazed, Exterior Door	1	Each
B30 Roofing		
Built-up Roof	5000	Sq Ft
C10 Interior Construction		
Steel, Painted, Interior Door	14	Each
C30 Interior Finishes		
Concrete, Painted, Interior Wall Finish	2745	Sq Ft
Sheetrock, Stippled, Interior Wall Finish	6000	Sq Ft
Vinyl Tile Flooring	500	Sq Ft
Carpet, Nylon 20 oz., Low Traffic	4500	Sq Ft
Acoustical Tile Ceiling	5000	Sq Ft
D20 Plumbing		
Tankless Water Closet	3	Each
Urinal, Vitreous China	2	Each
Lavatory, Vitreous China	3	Each
Sink, Stainless Steel	2	Each
Drinking Fountain, Refrigerated	2	Each
Pipe & Fittings, 3/4" Copper, Cold Water	0.47	K Ln Ft
Pipe & Fittings, 4" Steel	0.614	K Ln Ft
Pipe & Fittings, 2" Copper, Cold Water	0.523	K Ln Ft
Pipe & Fittings, 3/4" Copper, Hot Water	0.16	K Ln Ft
Pipe Insulation, Cold Water	0.776	K Ln Ft
Pipe Insulation, Hot Water	0.225	K Ln Ft
Water Heater, Gas/Oil 75 Gph	1	Each
Pipe & Fittings, 6" Cast Iron	0.285	K Ln Ft
Pipe & Fittings, 10" Cast Iron	0.141	K Ln Ft
Roof Drain, 2"	5	Each
Pipe & Fittings, 4" DWV PVC	0.035	K Ln Ft
D30 HVAC		
Exhaust Fan, Ceiling, 200-500 Cfm	2	Each
Air Conditioner, Rooftop, 20 Ton	1	Each
D40 Fire Protection		
Fire Sprinkler System	1	Each
Fire Sprinkler Head	38	Each
D50 Electrical		
Safety Switch, Fused, 400 Amp., 3 Ph.	1	Each
Secondary Transformer, Dry, 150 KVA	1	Each
Main Switchgear, <1200 Amp.	1	Each
Distribution Panel Board	2	Each
Grounded Fault Circuit Interruptor, 15 Amps	2	Each
Exit Lighting Fixture, w/ Battery	2	Each
Fluorescent Lighting Fixture, 160w	50	Each
Receptacle, 120V, 15 Amp	25	Each
Annunciation Panel	1	Each
Fire Alarm Control Panel	1	Each
Heat Detector	2	Each
Smoke Detector	6	Each
Uninterruptable Power Supply	1	Each

50-Year M&R Cost Summary

	50 Year Total	Annual Cost per Sqft	Annual cost as % replacement
PM & Minor Repair	$247,766	$0.99	0.90%
Unscheduled Maintenance	$267,068	$1.07	0.97%
Renewal & Replacement	$832,088	$3.33	3.03%
Total M&R Costs	$1,346,922	$5.39	4.90%

Distribution of Repair Costs

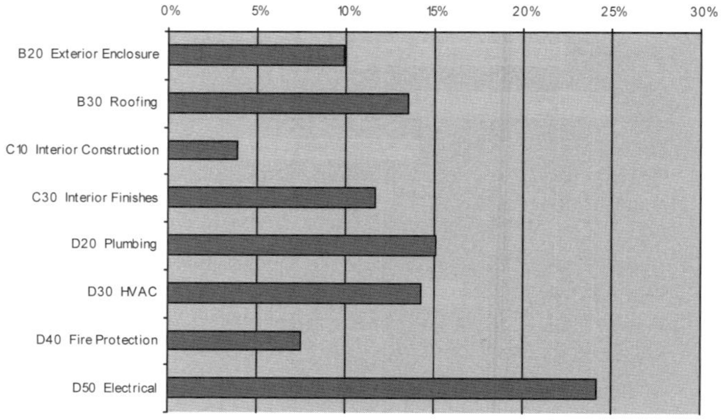

Thirty Most Costly Repair Tasks

Major Repair Task	Task Cost*	Pct.**
Replace Uninterruptable Power Supply Motor Generator	36.03	16.7%
Replace Air Conditioner, Rooftop, 20 Ton	18.25	8.4%
Fire Sprinkler System, Annual PM	17.28	8.0%
Maintain Built-up Roof	14.61	6.8%
Replace Carpet, Nylon 20 oz., Low Traffic	13.06	6.0%
Clean & Reseal Clay Brick, Exterior, 1st Floor	8.16	3.8%
Repair Air Conditioner, Rooftop, 20 Ton	7.31	3.4%
Refinish Sheetrock, Stippled, Interior Wall Finish	7.29	3.4%
Replace Pipe & Fittings, 2" Copper, Cold Water	7.08	3.3%
Maintain Air Conditioner, Rooftop, 20 Ton	6.81	3.2%
Replace Pipe & Fittings, 3/4" Copper, Cold Water	5.60	2.6%
Repoint (50% surface) Clay Brick, Exterior, 1st Floor	4.87	2.3%
Refinish Concrete, Painted, Interior Wall Finish	4.21	2.0%
Replace Steel, Painted, Door Locks	3.58	1.7%
Place New Membrane Over Existing, Built-up Roof	3.49	1.6%
Remove & Replace Membrane, Built-up Roof	3.45	1.6%
Minor Repair, Concrete, Painted, Interior Wall Finish	3.21	1.5%
Replace Ballast & Lamps, Fluorescent Lighting Fixture, 160w	2.76	1.3%
Annual PM, Uninterruptable Power Supply	2.66	1.2%
Maintain Steel, Painted, Door Locks	2.38	1.1%
Replace Water Heater, Gas/Oil 75 Gph	2.37	1.1%
Replace Fluorescent Lighting Fixture, 160w	1.96	0.9%
Clean Water Heater, Gas/Oil 75 Gph	1.93	0.9%
Replace Pipe & Fittings, 3/4" Copper, Hot Water	1.90	0.9%
Annual PM, Distribution Panel Board	1.88	0.9%
Maintain Roof Drain, 2"	1.83	0.8%
Minor Repair, Acoustic Tile Ceiling	1.78	0.8%
Repair Main Switchgear, <1200 Amp.	1.58	0.7%
Replace Drinking Fountain, Refrigerated	1.44	0.7%
Fire Sprinkler System, Test Gages & Valves	1.41	0.7%

**Task cost ($2004) per gross square foot over 50 years.*

***Percent of total M&R costs.*

Cost per Sqft. by System

Building Age	Exterior Closure	Roofing	Interior Constrution	Stairways	Interior Finishes	Conveying Systems	Plumbing	HVAC Systems	Fire Protection	Electrical	Equipment	Total per Sqft.
1	.02	.59	.13				.22	.27	.36	.32		**1.90**
2	.02	.59	.13				.46	.27	.36	.32		**2.14**
3	.02	.64	.13				.22	.27	.36	.32		**1.95**
4	.02	.59	.17		.96		.46	.27	.36	.32		**3.14**
5	.02	.59	.13		.03		.35	.27	.50	.64		**2.51**
6	.02	.64	.13				.46	.27	.36	.32		**2.19**
7	.02	.59	.13				.25	.27	.36	.32		**1.93**
8	.02	.59	.17		.96		.77	.27	.36	.33		**3.44**
9	.02	.64	.13		.36		.22	.27	.36	.32		**2.31**
10	3.90	.59	.84		.03		.84	2.70	.74	1.73		**11.37**
11	.02	.59	.13				.22	.27	.36	.32		**1.90**
12	.06	.64	.17		.96		.46	.27	.36	.32		**3.23**
13	.02	.59	.13				.48	.27	.36	.32		**2.16**
14	.08	2.33	.13				.49	.27	.36	.32		**3.98**
15	.02	.64	.13		5.52		.25	6.10	.50	13.07		**26.21**
16	.02	.59	.17		.96		.77	.27	.36	.33		**3.44**
17	.02	.59	.13				.22	.27	.36	.32		**1.90**
18	.02	.64	.13		.57		.46	.27	.36	.32		**2.76**
19	.02	.59	.13				.22	.27	.36	.32		**1.90**
20	3.97	.59	.88		1.50		.90	.43	.70	2.16		**11.13**
21	.02	.64	.13				.34	.27	.36	.32		**2.07**
22	.02	.59	.13				.46	.27	.36	.32		**2.14**
23	.02	.59	.13				.22	.27	.36	.32		**1.90**
24	.06	.64	.17		.96		.77	.27	.36	.33		**3.54**
25	2.98	.59	.13		.03		8.98	2.70	.50	1.29		**17.18**
26	.02	.59	.13				.61	.27	.36	.32		**2.28**
27	.02	.64	.13		.36		.23	.27	.36	.32		**2.32**
28	.09	3.45	.17		.96		.48	.27	.36	.32		**6.09**
29	.02	.59	.13				.33	.27	.36	.32		**2.00**
30	3.90	.59	.84		5.52		.83	6.10	.74	15.00		**33.51**
31	.02	.64	.13				.23	.27	.36	.32		**1.96**
32	.02	.59	.17		.96		.76	.27	.36	.32		**3.43**
33	.02	.59	.13				.23	.27	.36	.33		**1.91**
34	.02	.64	.13				.45	.27	.36	.32		**2.18**
35	.02	.59	.13		.03		1.01	.27	.50	.64		**3.17**
36	.06	.59	.17		1.52		.38	.27	.36	.32		**3.66**
37	.02	.64	.13				.40	.27	.36	.32		**2.13**
38	.02	.59	.13				.39	.27	.36	.32		**2.07**
39	.02	.59	.13				.46	.27	.36	.32		**2.14**
40	3.97	.64	.88		1.50		1.17	2.87	.70	2.29		**14.02**
41	.02	.59	.13				.31	.27	.36	.33		**1.99**
42	.08	2.33	.13				.41	.27	.36	.32		**3.89**
43	.02	.64	.13				.31	.27	.36	.32		**2.03**
44	.02	.59	.17		.96		.38	.27	.36	.32		**3.05**
45	.02	.59	.13		5.88		.64	6.10	.50	13.07		**26.91**
46	.02	.64	.13				.38	.27	.36	.32		**2.11**
47	.02	.59	.13				.31	.27	.36	.32		**1.98**
48	.06	.59	.17		.96		.68	.27	.36	.32		**3.40**
49	.02	.64	.13				.33	.27	.36	.33		**2.06**
50	6.86	.59	.84		.03		9.33	.27	.74	2.12		**20.78**
Total	26.73	36.42	10.45		31.47		40.54	38.45	20.31	65.01		269.38

50-Year Total M&R Cost Profile per Sqft.

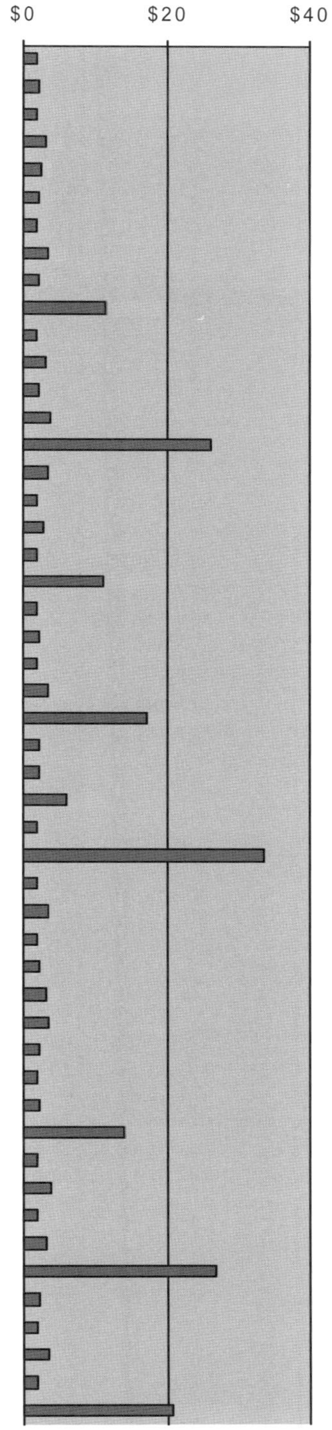

A value of "0.00" means a cost of more than $.000 but less than $.005 per gross square foot.

Town Hall, 1 Story

Gross Sqft:	11,000
Height in Ft:	12
Exterior:	Brick Veneer
Floor coverings:	Carpet/Vinyl Tile/Terrazzo Tile
HVAC:	Electric Cooling, Gas Heating
Occupancy:	770
Replacement Cost:	$998,001

Components

Component	Quantity	Unit
B20 Exterior Enclosure		
Clay Brick, Exterior, 1st Floor	3654	Sq Ft
Aluminum Fixed Window, 12 sf, 1st Floor	60	Each
Steel w/ Safety Glass, Painted, Exterior Door	1	Each
Aluminum Frame, Fully Glazed, Exterior Door	3	Each
B30 Roofing		
Built-up Roof	11000	Sq Ft
C10 Interior Construction		
Wood, Hollow Core, Painted, Interior Door	55	Each
C30 Interior Finishes		
Sheetrock, Stippled, Interior Wall Finish	11880	Sq Ft
Ceramic Tile Wall Finish	1320	Sq Ft
Terrazzo Flooring	1650	Sq Ft
Vinyl Tile Flooring	1650	Sq Ft
Carpet, Nylon 20 oz., Low Traffic	7700	Sq Ft
Metal Ceiling, Painted	3600	Sq Ft
Acoustical Tile Ceiling	7400	Sq Ft
D20 Plumbing		
Tankless Water Closet	11	Each
Urinal, Vitreous China	5	Each
Lavatory, Vitreous China	13	Each
Sink, Enameled Steel	8	Each
Sink, Stainless Steel	2	Each
Drinking Fountain, Refrigerated	6	Each
Pipe & Fittings, 3/4" Copper, Cold Water	1.47	K Ln Ft
Pipe & Fittings, 4" Steel	0.887	K Ln Ft
Pipe & Fittings, 2" Copper, Cold Water	1.592	K Ln Ft
Pipe & Fittings, 3/4" Copper, Hot Water	0.67	K Ln Ft
Pipe Insulation, Cold Water	1.914	K Ln Ft
Pipe Insulation, Hot Water	0.904	K Ln Ft
Water Heater, Gas/Oil 75 Gph	1	Each
Pipe & Fittings, 6" Cast Iron	0.709	K Ln Ft
Pipe & Fittings, 10" Cast Iron	0.21	K Ln Ft
Roof Drain, 2"	5	Each
Pipe & Fittings, 4" DWV PVC	0.076	K Ln Ft
D30 HVAC		
Exhaust Fan, Ceiling, 200-500 Cfm	2	Each
Air Conditioner, Rooftop, 50 Ton	1	Each
D40 Fire Protection		
Fire Sprinkler System	1	Each
Fire Sprinkler Head	79	Each
D50 Electrical		
Safety Switch, Fused, 400 Amp., 3 Ph.	1	Each
Main Switchgear, <1200 Amp.	1	Each
Distribution Panel Board	3	Each
Grounded Fault Circuit Interrupter, 15 Amps	2	Each
Emergency Horn & Strobe	6	Each
Exit Lighting Fixture, w/ Battery	6	Each
Fluorescent Lighting Fixture, 160w	110	Each
Wiring Device, Switch	55	Each
Receptacle, 120V, 15 Amp	178	Each
Annunciation Panel	1	Each
Fire Alarm Bell, 6"	4	Each
Fire Alarm Control Panel	1	Each
Manual Pull Station	4	Each
Smoke Detector	11	Each

50-Year M&R Cost Summary

	50 Year Total	Annual Cost per Sqft	Annual cost as % replacement
PM & Minor Repair	$425,740	$0.77	0.85%
Unscheduled Maintenance	$479,971	$0.87	0.96%
Renewal & Replacement	$1,225,646	$2.23	2.46%
Total M&R Costs	$2,131,357	$3.88	4.27%

Distribution of Repair Costs

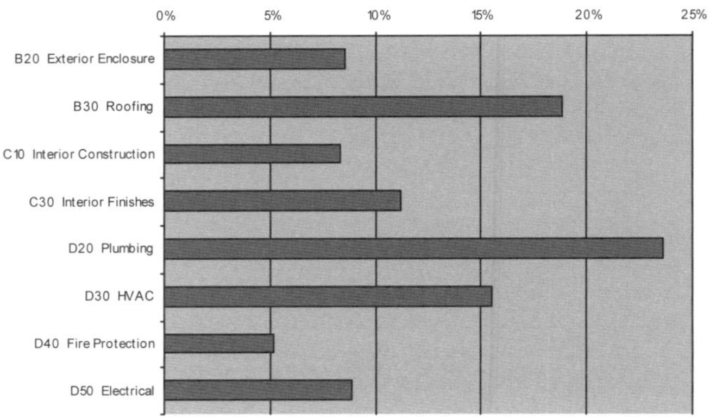

Thirty Most Costly Repair Tasks

Major Repair Task	Task Cost*	Pct.**
Maintain Built-up Roof	14.62	9.7%
Replace Air Conditioner, Rooftop, 50 Ton	14.02	9.3%
Replace Carpet, Nylon 20 oz., Low Traffic	10.16	6.8%
Replace Pipe & Fittings, 2" Copper, Cold Water	9.80	6.5%
Repair Air Conditioner, Rooftop, 50 Ton	8.58	5.7%
Replace Pipe & Fittings, 3/4" Copper, Cold Water	7.96	5.3%
Fire Sprinkler System, Annual PM	7.85	5.2%
Refinish Sheetrock, Stippled, Interior Wall Finish	6.56	4.4%
Clean & Reseal Clay Brick, Exterior, 1st Floor	4.93	3.3%
Maintain Wood, Hollow Core, Painted, Door Locks	4.17	2.8%
Maintain Air Conditioner, Rooftop, 50 Ton	3.99	2.7%
Replace Pipe & Fittings, 3/4" Copper, Hot Water	3.61	2.4%
Place New Membrane Over Existing, Built-up Roof	3.49	2.3%
Remove & Replace Membrane, Built-up Roof	3.45	2.3%
Repoint (50% surface) Clay Brick, Exterior, 1st Floor	2.95	2.0%
Replace Ballast & Lamps, Fluorescent Lighting Fixture, 160w	2.76	1.8%
Replace Wood, Hollow Core, Painted, Door Locks	2.52	1.7%
Replace Drinking Fountain, Refrigerated	1.96	1.3%
Replace Fluorescent Lighting Fixture, 160w	1.96	1.3%
Replace Wood, Hollow Core, Painted, Interior Door	1.50	1.0%
Replace Coolant & Adjust Drinking Fountain, Refrigerated	1.38	0.9%
Replace Faucet Washer & Clean Trap, Lavatory, Vitreous China	1.30	0.9%
Annual PM, Distribution Panel Board	1.29	0.9%
Minor Repair, Acoustic Tile Ceiling	1.20	0.8%
Replace Pipe Insulation, Cold Water	1.16	0.8%
Replace Water Heater, Gas/Oil 75 Gph	1.08	0.7%
Replace Receptacle	1.04	0.7%
Clean Water Heater, Gas/Oil 75 Gph	.88	0.6%
Maintain Roof Drain, 2"	.83	0.6%
Non-Destructive Moisture Inspection	.81	0.5%

**Task cost ($2004) per gross square foot over 50 years.*

***Percent of total M&R costs.*

Cost per Sqft. by System

Building Age	Exterior Closure	Roofing	Interior Construtiion	Stairways	Interior Finishes	Conveying Systems	Plumbing	HVAC Systems	Fire Protection	Electrical	Equipment	Total per Sqft.
1	.02	.59	.23				.11	.15	.17	.14		**1.39**
2	.02	.59	.23				.50	.15	.17	.14		**1.79**
3	.02	.64	.23				.11	.15	.17	.14		**1.44**
4	.02	.59	.29		.55		.50	.15	.17	.14		**2.40**
5	.02	.59	.23		.08		.18	.15	.23	.30		**1.77**
6	.02	.64	.23				.50	.15	.17	.14		**1.84**
7	.02	.59	.23				.14	.15	.17	.14		**1.43**
8	.02	.59	.29		.64		.68	.15	.17	.14		**2.68**
9	.02	.64	.23		.25		.11	.15	.17	.14		**1.69**
10	2.37	.59	.86		.22		1.11	3.01	.46	1.39		**10.00**
11	.02	.59	.23				.11	.15	.17	.14		**1.39**
12	.07	.64	.29		.55		.50	.15	.17	.14		**2.50**
13	.02	.59	.23				.29	.15	.17	.14		**1.57**
14	.05	2.33	.23				.54	.15	.17	.14		**3.60**
15	.02	.64	.23		3.52		.14	4.68	.23	.75		**10.21**
16	.02	.59	.29		.64		.68	.15	.17	.14		**2.68**
17	.02	.59	.23				.11	.15	.17	.14		**1.39**
18	.02	.64	.23		.55		.50	.15	.17	.14		**2.39**
19	.02	.59	.23				.11	.15	.17	.14		**1.39**
20	2.42	.59	.92		1.23		1.20	.23	.42	2.13		**9.14**
21	.02	.64	.23				.19	.15	.17	.14		**1.52**
22	.02	.59	.23				.50	.15	.17	.14		**1.79**
23	.02	.59	.23				.11	.15	.17	.14		**1.39**
24	.07	.64	.29		.64		.68	.15	.17	.14		**2.78**
25	1.81	.59	.23		.08		12.26	3.01	.23	.71		**18.91**
26	.02	.59	.23				.61	.15	.17	.14		**1.89**
27	.02	.64	.23		.25		.12	.15	.17	.14		**1.70**
28	.05	3.45	.29		.55		.53	.15	.17	.14		**5.32**
29	.02	.59	.23				.16	.15	.17	.14		**1.45**
30	2.37	.59	1.55		3.66		1.09	4.68	.46	1.67		**16.06**
31	.02	.64	.23				.12	.15	.17	.14		**1.46**
32	.02	.59	.23		.64		.67	.15	.17	.14		**2.60**
33	.02	.59	.23				.12	.15	.17	.14		**1.41**
34	.02	.64	.29				.49	.15	.17	.14		**1.88**
35	.02	.59	.23		.08		1.61	.15	.23	.30		**3.19**
36	.07	.59	.23		1.10		.30	.15	.17	.14		**2.74**
37	.02	.64	.23				.35	.15	.17	.14		**1.69**
38	.02	.59	.29				.32	.15	.17	.14		**1.66**
39	.02	.59	.23				.43	.15	.17	.14		**1.72**
40	2.42	.64	.86		1.32		.82	3.09	.42	2.31		**11.88**
41	.02	.59	.23				.31	.15	.17	.14		**1.60**
42	.05	2.33	.29				.34	.15	.17	.14		**3.46**
43	.02	.64	.23				.35	.15	.17	.14		**1.69**
44	.02	.59	.23		.55		.30	.15	.17	.14		**2.13**
45	.02	.59	.23		3.77		.82	4.68	.23	.75		**11.08**
46	.02	.64	.29				.30	.15	.17	.14		**1.70**
47	.02	.59	.23				.31	.15	.17	.14		**1.60**
48	.07	.59	.23		.64		.43	.15	.17	.14		**2.42**
49	.02	.64	.23				.35	.15	.17	.14		**1.69**
50	4.16	.59	.92		.22		12.68	.15	.46	1.45		**20.62**
Total	16.54	36.43	16.08		21.72		45.76	29.99	10.03	17.21		193.76

50-Year Total M&R Cost Profile per Sqft.

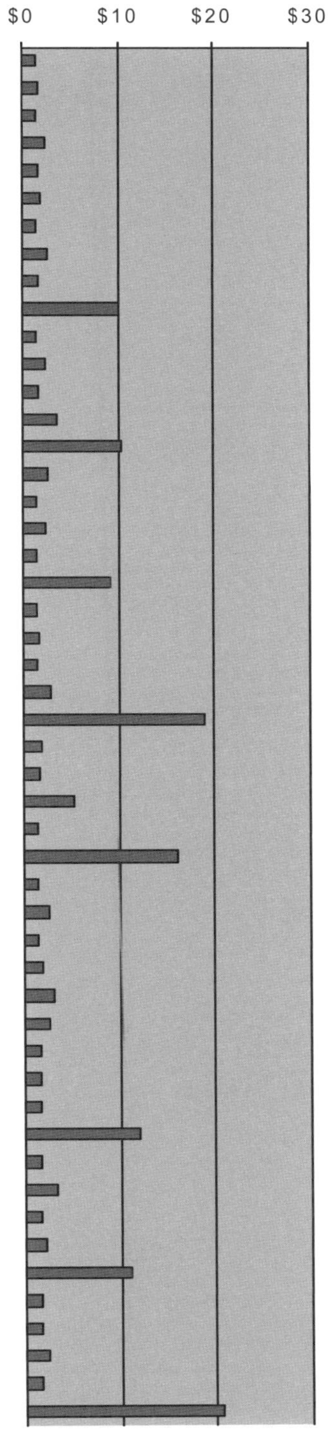

A value of "0.00" means a cost of more than $.000 but less than $.005 per gross square foot.

Warehouse, Large

Gross Sqft:	80,000
Height in Ft:	24
Exterior:	Concrete Block
Floor coverings:	Vinyl Tile/Carpet/Finished Concrete
HVAC:	Pkgd DX AC, Gas Boiler, Unit Heaters
Occupancy:	400
Replacement Cost:	$3,782,837

Components

Component	Quantity	Unit
B20 Exterior Enclosure		
Concrete Block, Painted, Exterior, 1st Floor	13580	Sq Ft
Concrete Block, Painted, Exterior, 2nd Floor	12580	Sq Ft
Aluminum Fixed Window, 12 sf, 1st Floor	8	Each
Steel, Painted, Exterior Door	4	Each
Steel Single 12'x12', Painted, Roll-up Door	10	Each
B30 Roofing		
Built-up Roof	80000	Sq Ft
C10 Interior Construction		
Wood, Solid Core, Painted, Interior Door	20	Each
Steel, Painted, Interior Door	10	Each
C30 Interior Finishes		
Sheetrock, Unstippled, Interior Wall Finish	6400	Sq Ft
Concrete Block, Painted, Interior Wall Finish	31360	Sq Ft
Concrete, Finished Flooring	70600	Sq Ft
Vinyl Tile Flooring	3000	Sq Ft
Carpet, Nylon 20 oz., Low Traffic	6400	Sq Ft
Sheetrock, Unstippled, Ceiling	3000	Sq Ft
Acoustical Tile, Dropped, Ceiling	6400	Sq Ft
D20 Plumbing		
Tankless Water Closet	18	Each
Urinal, Vitreous China	7	Each
Lavatory, Vitreous China	30	Each
Service Sink, Iron Enamel	3	Each
Drinking Fountain, Refrigerated	6	Each
Pipe & Fittings, 3/4" Copper, Cold Water	2.1	K Ln Ft
Pipe & Fittings, 2" Copper, Cold Water	2.7	K Ln Ft
Pipe & Fittings, 3/4" Copper, Hot Water	1.1	K Ln Ft
Pipe Insulation, Cold Water	2.7	K Ln Ft
Pipe Insulation, Hot Water	2.1	K Ln Ft
Circulator Pump, 1/12 HP, Hot Water	2	Each
Water Heater, Gas/Oil 175 Gph	1	Each
Pipe & Fittings, 6" Cast Iron	1.3	K Ln Ft
Pipe & Fittings, 10" Cast Iron	0.6	K Ln Ft
Pipe & Fittings, 4" DWV PVC	1.2	K Ln Ft
Roof Drain, 4-6"	23	Each
D30 HVAC		
Pipe & Fittings, 4" Steel, Gas	2.4	K Ln Ft
Gas Boiler, 2000 Mbh	3	Each
Chemical Feed System	1	Each
Circulation Pump 5 HP, Hot Water	2	Each
Expansion Tank, 100 Gal.	1	Each
Air Conditioner, Indoor DX Packaged, Air Cooled, 20 Ton	1	Each
Suspended Heater, Hot Water, 4000 Cfm, 180 Mbh	30	Each
D40 Fire Protection		
Fire Sprinkler Head	840	Each
D50 Electrical		
Main Switchgear, <1200 Amp.	1	Each
Distribution Switch, Fused, <600 V	5	Each
Motor Starter, 5-20 HP, <600 V	4	Each
Fluorescent Lighting Fixture, 160w	600	Each
Wiring Device, Switch	63	Each
Receptacle, 120V, 15 Amp	250	Each
Fire Alarm Bell, 6"	40	Each
Fire Alarm Control Panel	1	Each
Manual Pull Station	40	Each
Smoke Detector	160	Each

50-Year M&R Cost Summary

	50 Year Total	Annual Cost per Sqft	Annual cost as % replacement
PM & Minor Repair	$2,232,187	$0.56	1.18%
Unscheduled Maintenance	$2,084,949	$0.52	1.10%
Renewal & Replacement	$3,287,833	$0.82	1.74%
Total M&R Costs	$7,604,969	$1.90	4.02%

Distribution of Repair Costs

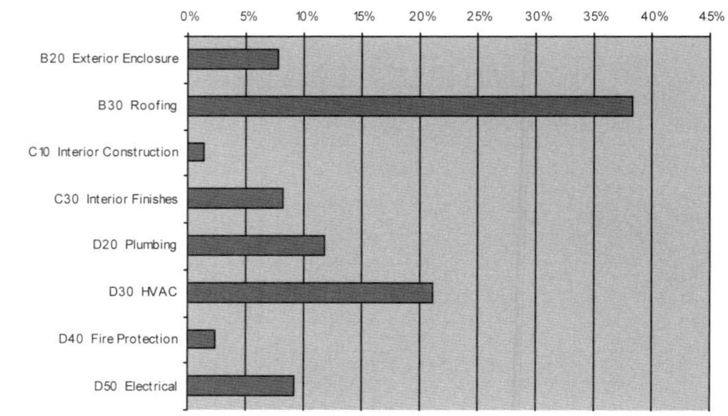

Thirty Most Costly Repair Tasks

Major Repair Task	Task Cost*	Pct.**
Maintain Built-up Roof	14.62	21.2%
Inspect & Test Gas Boiler, 2000 Mbh	4.91	7.1%
Place New Membrane Over Existing, Built-up Roof	3.49	5.1%
Remove & Replace Membrane, Built-up Roof	3.45	5.0%
Refinish Concrete Block, Painted, Interior Wall Finish	3.01	4.4%
Maintain Suspended Heater, Hot Water, 4000 Cfm, 180 Mbh	2.72	3.9%
Replace Suspended Heater, Hot Water, 4000 Cfm, 180 Mbh	2.35	3.4%
Replace Pipe & Fittings, 2" Copper, Cold Water	2.28	3.3%
Replace Ballast & Lamps, Fluorescent Lighting Fixture, 160w	2.07	3.0%
Refinish Concrete Block, Painted, Exterior, 2nd Floor	1.68	2.4%
Replace Pipe & Fittings, 3/4" Copper, Cold Water	1.56	2.3%
Refinish Concrete Block, Painted, Exterior, 1st Floor	1.49	2.2%
Replace Fluorescent Lighting Fixture, 160w	1.47	2.1%
Check Operation, Smoke Detector	1.33	1.9%
Refinish Concrete, Finished Flooring	1.21	1.7%
Replace Carpet, Nylon 20 oz., Low Traffic	1.16	1.7%
Repair Fire Sprinkler Head	1.01	1.5%
Replace Gas Boiler, 2000 Mbh	.99	1.4%
Minor Repair, Concrete, Finished Flooring	.97	1.4%
Repoint (50% surface) Concrete Block, Painted, Exterior, 2nd Floor	.91	1.3%
Repoint (50% surface) Concrete Block, Painted, Exterior, 1st Floor	.91	1.3%
Replace Pipe & Fittings, 3/4" Copper, Hot Water	.81	1.2%
Non-Destructive Moisture Inspection	.81	1.2%
Repair Gas Boiler, 2000 Mbh	.76	1.1%
Replace Smoke Detector	.70	1.0%
Maintain Steel Single 12'x12', Painted, Roll-up Door	.60	0.9%
Replace Fire Sprinkler Head	.59	0.8%
Replace Air Conditioner, Indoor DX Packaged, Air Cooled, 20 Ton	.53	0.8%
Maintain Roof Drain, 4-6"	.53	0.8%
Inspect & Test, Sprinkler Head	.49	0.7%

**Task cost ($2004) per gross square foot over 50 years.*

***Percent of total M&R costs.*

Cost per Sqft. by System

Building Age	Exterior Closure	Roofing	Interior Constrution	Stairways	Interior Finishes	Conveying Systems	Plumbing	HVAC Systems	Fire Protection	Electrical	Equipment	Total per Sqft.
1	.03	.59	.02				.05	.29	.01	.07		**1.06**
2	.03	.59	.02				.13	.30	.01	.07		**1.15**
3	.03	.64	.02				.05	.29	.01	.07		**1.11**
4	.03	.59	.02		.31		.13	.30	.01	.07		**1.46**
5	.04	.59	.02		.00		.07	.30	.01	.10		**1.13**
6	.03	.64	.02				.13	.30	.01	.07		**1.20**
7	.03	.59	.02				.06	.42	.01	.07		**1.20**
8	.03	.59	.02		.31		.19	.30	.01	.07		**1.52**
9	.03	.64	.02		.00		.05	.29	.01	.07		**1.11**
10	.71	.59	.07		.01		.25	.50	.35	.89		**3.38**
11	.03	.59	.02				.05	.29	.01	.07		**1.06**
12	.03	.64	.02		.31		.13	.32	.01	.07		**1.53**
13	.03	.59	.02				.08	.29	.01	.07		**1.09**
14	.05	2.33	.02				.14	.43	.01	.07		**3.05**
15	.04	.64	.02		.71		.06	1.05	.01	.36		**2.89**
16	.03	.59	.02		.31		.19	.30	.01	.07		**1.52**
17	.03	.59	.02				.05	.29	.01	.07		**1.06**
18	.03	.64	.02		.08		.13	.30	.01	.08		**1.29**
19	.03	.59	.02				.05	.29	.01	.07		**1.06**
20	.71	.59	.08		.39		.27	.56	.29	.97		**3.85**
21	.03	.64	.02				.07	.42	.01	.07		**1.26**
22	.03	.59	.02				.13	.30	.01	.07		**1.15**
23	.03	.59	.02				.05	.29	.01	.08		**1.08**
24	.03	.64	.02		.50		.19	.32	.01	.07		**1.77**
25	1.04	.59	.02		.60		2.69	.40	.01	.19		**5.55**
26	.03	.59	.02				.14	.30	.01	.07		**1.16**
27	.03	.64	.02		.00		.06	.29	.01	.07		**1.12**
28	.05	3.45	.02		.31		.14	.43	.01	.08		**4.49**
29	.03	.59	.02				.06	.29	.01	.07		**1.07**
30	.71	.59	.13		.72		.25	2.08	.35	1.17		**5.99**
31	.03	.64	.02				.06	.29	.01	.07		**1.12**
32	.03	.59	.02		.31		.18	.30	.01	.07		**1.51**
33	.03	.59	.02				.06	.29	.01	.08		**1.08**
34	.03	.64	.02				.12	.30	.01	.07		**1.19**
35	.29	.59	.02		.00		.39	.30	.01	.08		**1.68**
36	.03	.59	.02		.39		.08	.32	.01	.08		**1.52**
37	.03	.64	.02				.11	.42	.01	.07		**1.30**
38	.03	.59	.02				.08	.30	.01	.07		**1.10**
39	.03	.59	.02				.12	.29	.01	.07		**1.13**
40	.68	.64	.07		.39		.22	.66	.29	1.08		**4.03**
41	.03	.59	.02				.11	.29	.01	.08		**1.13**
42	.05	2.33	.02				.09	.30	.01	.07		**2.87**
43	.03	.64	.02				.11	.29	.01	.07		**1.17**
44	.03	.59	.02		.31		.08	.43	.01	.07		**1.53**
45	.07	.59	.02		.71		.22	1.05	.01	.34		**3.01**
46	.03	.64	.02				.08	.30	.01	.08		**1.16**
47	.03	.59	.02				.11	.29	.01	.07		**1.11**
48	.03	.59	.02		.50		.13	.32	.01	.07		**1.66**
49	.03	.64	.02				.12	.29	.01	.07		**1.18**
50	1.67	.59	.08		.61		2.75	.41	.35	.77		**7.24**
Total	7.34	36.42	1.26		7.78		11.24	20.07	2.24	8.70		95.06

50-Year Total M&R Cost Profile per Sqft.

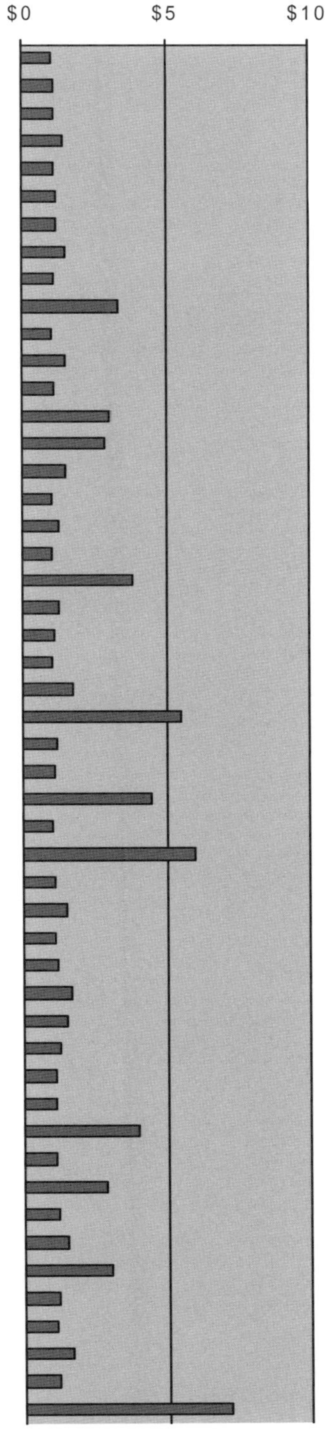

A value of "0.00" means a cost of more than $.000 but less than $.005 per gross square foot.

Warehouse, Self-storage

Gross Sqft:	24,000
Height in Ft:	12
Exterior:	Stucco
Floor coverings:	Unfinished Concrete/Linoleum
HVAC:	Window AC, Gas Boiler, Single Zone Air
Occupancy:	8
Replacement Cost:	$1,569,209

Components

Component	Quantity	Unit
B20 Exterior Enclosure		
Stucco, Exterior, 1st Floor	10300	Sq Ft
Aluminum Operable Window, 12 sf, 1st Floor	4	Each
Steel, Painted, Exterior Door	4	Each
Steel Single 12'x12', Painted, Roll-up Door	160	Each
B30 Roofing		
Built-up Roof	24000	Sq Ft
C30 Interior Finishes		
Concrete Block, Painted, Interior Wall Finish	50304	Sq Ft
Concrete, Unfinished Flooring	23760	Sq Ft
Linoleum Flooring	240	Sq Ft
Metal Unpainted Interior Ceiling	23760	Sq Ft
Acoustical Tile, Dropped, Ceiling	240	Sq Ft
D20 Plumbing		
Flush Tank Water Closet	1	Each
Lavatory, Vitreous China	1	Each
Service Sink, Iron Enamel	4	Each
Pipe & Fittings, 3/4" Copper, Cold Water	0.3	K Ln Ft
Water Heater, Gas/Oil 65 Gph	1	Each
Pipe & Fittings, 2" Copper, Cold Water	0.9	K Ln Ft
Pipe & Fittings, 3/4" Copper, Hot Water	0.2	K Ln Ft
Pipe Insulation, Cold Water	0.3	K Ln Ft
Pipe Insulation, Hot Water	0.2	K Ln Ft
Pipe & Fittings, 6" Cast Iron	0.4	K Ln Ft
Pipe & Fittings, 10" Cast Iron	0.7	K Ln Ft
Pipe & Fittings, 4" DWV PVC	0.3	K Ln Ft
Roof Drain, 4-6"	8	Each
D30 HVAC		
Pipe & Fittings, 4" Steel, Gas	1.3	K Ln Ft
Gas Boiler, 250 Mbh	4	Each
Chemical Feed System	4	Each
Circulation Pump 5 HP, Hot Water	4	Each
Expansion Tank, 30 Gal.	4	Each
Air Handler, Singlezone, 6500 Cfm	4	Each
Air Conditioner, Window, 2 Ton	1	Each
D40 Fire Protection		
Fire Sprinkler Head	1	Each
D50 Electrical		
Main Switchgear, <1200 Amp.	1	Each
Distribution Switch, Fused, <600 V	4	Each
Motor Starter, 5-20 HP, <600 V	12	Each
Fluorescent Lighting Fixture, 160w	162	Each
Wiring Device, Switch	162	Each
Receptacle, 120V, 15 Amp	170	Each
Fire Alarm Control Panel	1	Each
Smoke Detector	170	Each

50-Year M&R Cost Summary

	50 Year Total	Annual Cost per Sqft	Annual cost as % replacement
PM & Minor Repair	$1,584,617	$1.32	2.02%
Unscheduled Maintenance	$1,902,092	$1.59	2.42%
Renewal & Replacement	$1,983,507	$1.65	2.53%
Total M&R Costs	$5,470,216	$4.56	6.97%

Distribution of Repair Costs

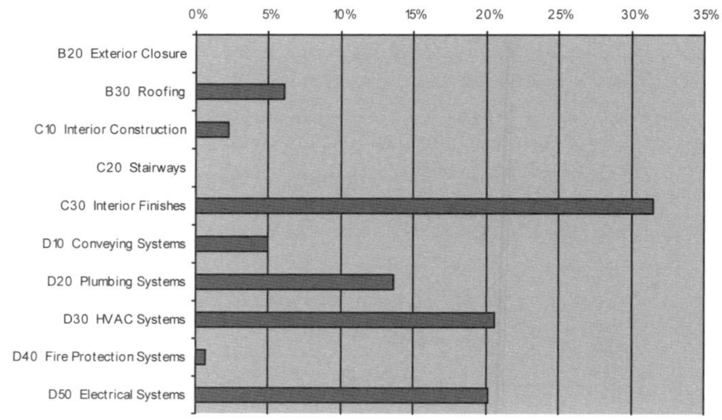

Thirty Most Costly Repair Tasks

Major Repair Task	Task Cost*	Pct.**
Maintain Steel Single 12'x12', Painted, Roll-up Door	31.93	21.5%
Refinish Concrete Block, Painted, Interior Wall Finish	16.09	10.8%
Replace Steel Single 12'x12', Painted, Roll-up Door	14.86	10.0%
Maintain Built-up Roof	14.62	9.8%
Repair Steel Single 12'x12', Painted, Roll-up Door	6.37	4.3%
Refinish Steel Single 12'x12', Painted, Roll-up Door	5.47	3.7%
Check Operation, Smoke Detector	4.71	3.2%
Maintain Air Handler, Singlezone, 6500 Cfm	4.55	3.1%
Replace Air Handler, Singlezone, 6500 Cfm	4.36	2.9%
Place New Membrane Over Existing, Built-up Roof	3.49	2.3%
Refinish Stucco, Exterior, 1st Floor	3.45	2.3%
Remove & Replace Membrane, Built-up Roof	3.45	2.3%
Inspect & Test Gas Boiler, 250 Mbh	2.73	1.8%
Replace Pipe & Fittings, 2" Copper, Cold Water	2.54	1.7%
Replace Smoke Detector	2.48	1.7%
Inspect & Clean Motor Starter, 5-20 HP, <600 V	2.26	1.5%
Replace Ballast & Lamps, Fluorescent Lighting Fixture, 160w	1.86	1.3%
Replace Fluorescent Lighting Fixture, 160w	1.32	0.9%
Minor Repair, Concrete Block, Painted, Interior Wall Finish	1.27	0.9%
Replace Gas Boiler, 250 Mbh	1.26	0.8%
Minor Repair, Concrete, Unfinished Flooring	1.09	0.7%
Replace Circulation Pump 5 HP, Hot Water	1.06	0.7%
Repair Metal Unpainted Interior Ceiling	1.01	0.7%
Annual PM, Expansion Tank, 30 Gal.	1.00	0.7%
Non-Destructive Moisture Inspection	.81	0.5%
Repair Gas Boiler, 250 Mbh	.79	0.5%
Replace Pipe & Fittings, 3/4" Copper, Cold Water	.74	0.5%
Replace Coil, Motor Starter, 5-20 HP, <600 V	.72	0.5%
Finish Repaired Concrete Block, Painted, Interior Wall Finish	.72	0.5%
Replace Wiring Device, Switch	.65	0.4%

**Task cost ($2004) per gross square foot over 50 years.*

***Percent of total M&R costs.*

Cost per Sqft. by System

Building Age	Exterior Closure	Roofing	Interior Construction	Stairways	Interior Finishes	Conveying Systems	Plumbing	HVAC Systems	Fire Protection	Electrical	Equipment	Total per Sqft.
1	1.65	.59					.07	.33	.00	.29		**2.92**
2	1.65	.59					.09	.36	.00	.29		**2.97**
3	1.65	.64					.07	.33	.00	.29		**2.97**
4	1.65	.59			1.34		.09	.36	.00	.29		**4.31**
5	2.26	.59					.09	.34	.00	.51		**3.79**
6	1.65	.64					.09	.36	.00	.29		**3.02**
7	1.65	.59					.07	.46	.00	.29		**3.05**
8	1.65	.59			1.34		.15	.36	.00	.29		**4.38**
9	1.65	.64			.00		.07	.33	.00	.29		**2.97**
10	4.54	.59			.00		.11	.54	.00	1.51		**7.30**
11	1.65	.59					.07	.33	.00	.29		**2.92**
12	1.65	.64			1.34		.09	.39	.00	.29		**4.40**
13	1.65	.59					.13	.33	.00	.29		**2.98**
14	1.70	2.33					.09	.49	.00	.29		**4.90**
15	2.26	.64			.36		.08	2.11	.00	1.43		**6.88**
16	1.65	.59			1.34		.15	.36	.00	.29		**4.38**
17	1.65	.59					.07	.33	.00	.29		**2.92**
18	1.65	.64			.04		.09	.36	.00	.46		**3.22**
19	1.65	.59					.07	.33	.00	.29		**2.92**
20	4.54	.59			1.34		.13	.44	.00	1.29		**8.33**
21	1.65	.64					.09	.46	.00	.29		**3.12**
22	1.65	.59					.09	.36	.00	.29		**2.97**
23	1.65	.59					.07	.33	.00	.44		**3.07**
24	1.65	.64			2.84		.15	.39	.00	.29		**5.96**
25	2.26	.59					2.21	.45	.00	.74		**6.23**
26	1.65	.59					.13	.36	.00	.29		**3.01**
27	1.65	.64			.00		.07	.33	.00	.29		**2.97**
28	1.71	3.45			1.34		.09	.49	.00	.44		**7.52**
29	1.65	.59					.09	.33	.00	.29		**2.94**
30	4.54	.59			.36		.11	3.57	.00	2.28		**11.45**
31	1.65	.64					.07	.33	.00	.29		**2.97**
32	1.65	.59			1.34		.15	.36	.00	.29		**4.38**
33	1.65	.59					.07	.33	.00	.44		**3.08**
34	1.65	.64					.09	.36	.00	.29		**3.02**
35	15.46	.59					.19	.34	.00	.35		**16.94**
36	1.65	.59			1.38		.07	.39	.00	.46		**4.53**
37	1.65	.64					.11	.46	.00	.29		**3.14**
38	1.65	.59					.07	.36	.00	.29		**2.95**
39	1.65	.59					.13	.33	.00	.29		**2.98**
40	2.95	.64			1.34		.19	.54	.00	1.67		**7.33**
41	1.65	.59					.09	.33	.00	.44		**3.09**
42	1.70	2.33					.07	.36	.00	.29		**4.75**
43	1.65	.64					.10	.33	.00	.29		**3.00**
44	1.65	.59			1.34		.07	.49	.00	.29		**4.43**
45	3.85	.59			.36		.15	2.11	.00	1.28		**8.34**
46	1.65	.64					.07	.36	.00	.44		**3.16**
47	1.65	.59					.09	.33	.00	.29		**2.94**
48	1.65	.59			2.84		.13	.39	.00	.29		**5.88**
49	1.65	.64					.09	.33	.00	.29		**2.99**
50	2.95	.59			.00		2.21	.51	.00	.97		**7.23**
Total	111.77	36.43			20.25		9.16	25.57	.01	24.75		227.93

50-Year Total M&R Cost Profile per Sqft.

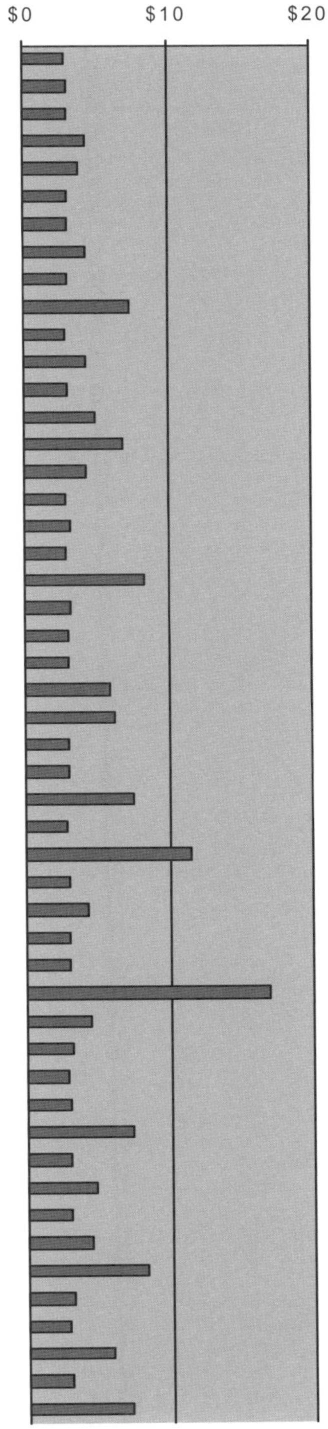

A value of "0.00" means a cost of more than $.000 but less than $.005 per gross square foot.

Chapter 3
Local M&R Costs

Local Maintenance Cost Indexes

In-House Shop Rates

Contract Labor

3. Local M&R Costs

The statistics in this chapter focus on local maintenance costs for 210 major U.S. and Canadian metropolitan areas. Three types of measures are presented:

- **Local maintenance cost indexes** measure relative maintenance and repair (M&R) costs across metro areas;
- **In-house shop rates** for trades and supervisory positions common to the in-house M&R staff;
- **Contract labor rates** for trades common in M&R construction.

The local maintenance cost index is based on the M&R costs of a two-story office building (shown in Chapter 2) standardized to the Washington DC area. The range of the index is considerable, as indicated in Table 3.1. Costs in New York, NY are an estimated 59% higher than those in Washington DC for the same building. In the other direction, M&R costs in Beaufort, SC are an estimated 45% lower than the Washington DC value. This index can be used for simple comparisons among metro areas, and also used to adjust the cost profiles in Chapter 2 for metro areas other than Washington DC.

Table 3-1
Comparison of M&R Costs by Metro Area

Metro Area	Local Maintenance Cost Index*	Metro Area	Local Maintenance Cost Index*	Metro Area	Local Maintenance Cost Index*	Metro Area	Local Maintenance Cost Index*
New York, NY	158.9	Milwaukee, WI	103.6	Spokane, WA	90.9	Virginia Beach, VA	76.5
Yonkers, NY	143.2	Fresno, CA	102.8	Cumberland, MD	90.2	Burlington, VT	75.7
San Francisco, CA	138.9	Toledo, OH	102.7	Utica, NY	90.0	Rutland, VT	75.6
San Jose, CA	133.3	Bakersfield, CA	102.6	Boulder, CO	89.7	Knoxville, TN	75.6
Philadelphia, PA	131.3	Buffalo, NY	102.3	Pueblo, CO	88.7	Fort Smith, AR	75.4
Trenton, NJ	130.3	Olympia, WA	101.9	Colorado Springs, CO	88.4	Mobile, AL	75.4
Oakland, CA	128.4	Tacoma, WA	101.9	Omaha, NE	88.0	Biloxi, MS	75.4
Stamford, CT	127.6	Charleston, WV	101.6	Watertown, NY	88.0	Lexington, KY	75.4
Hilo, HI	127.4	Springfield, IL	101.3	Springfield, MO	87.4	Bismarck, ND	74.8
Honolulu, HI	127.4	Scranton, PA	100.9	Memphis, TN	87.3	Tampa, FL	74.2
Jersey City, NJ	126.7	Medford, OR	100.7	Kalamazoo, MI	87.2	Newport News, VA	74.1
Newark, NJ	126.7	Duluth, MN	100.6	Bowling Green, KY	86.8	Tuscaloosa, AL	73.8
Atlantic City, NJ	124.5	Springfield, MA	100.1	Concord, NH	86.5	Wichita Falls, TX	73.7
Camden, NJ	124.5	Reading, PA	100.0	Manchester, NH	86.5	Orlando, FL	73.6
Boston, MA	122.4	Washington DC	100.0	New Orleans, LA	86.4	Sioux Falls, SD	73.6
New Brunswick, NJ	121.0	Indianapolis, IN	99.9	Owensboro, KY	86.1	Macon, GA	73.3
Lowell, MA	119.8	Columbus, OH	99.7	Boise, ID	86.1	Amarillo, TX	73.0
Chicago, IL	119.1	Ann Arbor, MI	99.3	Marquette, MI	85.8	Corpus Christi, TX	72.1
Norwalk, CT	115.5	Harrisburg, PA	99.2	Fort Lauderdale, FL	85.5	Grand Rapids, MI	70.8
Danbury, CT	114.3	Akron, OH	99.2	Houston, TX	85.1	Hampton, VA	70.7
Kansas City, MO	113.5	Baltimore, MD	98.9	Albuquerque, NM	85.1	El Paso, TX	70.7
Detroit, MI	113.4	Rochester, MN	98.7	Miami, FL	85.0	Chattanooga, TN	70.1
Minneapolis, MN	112.9	Youngstown, OH	98.0	Daytona Beach, FL	84.8	Tulsa, OK	70.1
Pittsburgh, PA	112.1	Peoria, IL	97.2	Phoenix, AZ	84.6	Savannah, GA	69.2
Wilmington, DE	111.6	Springfield, OH	96.6	Portland, ME	84.0	Cheyenne, WY	67.5
Sacramento, CA	111.3	Flint, MI	96.6	Wichita, KS	83.6	Lubbock, TX	66.1
Fairbanks, AK	111.1	Richland, WA	96.5	Sioux City, IA	83.4	Rapid City, SD	64.1
Juneau, AK	110.9	Terre Haute, IN	96.1	Nashville, TN	83.3	Columbus, GA	63.8
Stockton, CA	110.8	Lansing, MI	96.0	Salt Lake City, UT	82.3	Tallahassee, FL	63.1
Los Angeles, CA	110.8	Rochester, NY	96.0	Pocatello, ID	81.9	Roanoke, VA	63.1
St. Louis, MO	110.5	Davenport, IA	96.0	Huntsville, AL	81.6	Raleigh-Durham, NC	62.3
New Haven, CT	110.4	Cincinnati, OH	96.0	Billings, MT	81.5	Winston-Salem, NC	62.2
Anaheim, CA	110.3	Moline, IL	95.5	Birmingham, AL	81.3	Charlotte, NC	61.2
Norwich, CT	110.2	Erie, PA	95.4	San Antonio, TX	81.2	Greensboro, NC	61.2
Santa Barbara, CA	109.8	Kokomo, IN	95.3	Lewiston, ME	81.2	Jackson, MS	61.0
Anchorage, AK	109.7	Madison, WI	95.2	Little Rock, AR	81.1	Columbia, SC	60.3
Waterbury, CT	109.3	Dayton, OH	94.6	Altus, OK	81.0	Charleston, SC	60.0
Oxnard, CA	109.2	Evansville, IN	94.4	Lawton, OK	81.0	Beaufort, SC	54.8
Seattle, WA	108.9	Carson City, NV	94.0	Austin, TX	80.9		
Hartford, CT	108.0	Reno, NV	94.0	Ogden, UT	80.9	**Canadian Cities+**	
Riverside, CA	108.0	Eau Claire, WI	93.7	Beaumont, TX	80.8		
Rockford, IL	107.5	Albany, NY	93.5	Richmond, VA	80.6	Toronto, Ontario	78.0
Portland, OR	107.0	Louisville, KY	92.7	Great Falls, MT	80.5	Hamilton, Ontario	75.4
Cleveland, OH	107.0	Battle Creek, MI	92.6	Tucson, AZ	80.0	London, Ontario	74.2
Providence, RI	106.9	Denver, CO	92.3	Oklahoma City, OK	80.0	Calgary, Alberta	73.9
Worcester, MA	106.9	South Bend, IN	92.1	Alamogordo, NM	79.2	Edmonton, Alberta	73.1
Salem, OR	106.6	Green Bay, WI	91.9	Las Cruces, NM	79.2	Ottawa, Ontario	72.7
Brockton, MA	106.2	Muncie, IN	91.9	Jacksonville, FL	78.7	Vancouver, B.C.	69.5
Eugene, OR	106.2	Saginaw, MI	91.9	Dallas, TX	78.6	Quebec, Quebec	68.8
Las Vegas, NV	106.1	Syracuse, NY	91.3	Fargo, ND	78.3	Montreal, Quebec	68.0
Fall River, MA	105.5	Atlanta, GA	91.2	Shreveport, LA	77.0	Winnipeg, Manitoba	58.4
San Diego, CA	105.3	Cedar Rapids, IA	91.1	Norfolk, VA	77.0		
Parkersburg, WV	104.6	Topeka, KS	91.1	Fort Worth, TX	77.0		
Gary, IN	103.9	Des Moines, IA	91.0	Waco, TX	76.6		

*Total average cost, Washington DC=100

+Adjusted using 6/30/04 exchange rate from the Federal Reserve Bank of New York

3.1 Local Maintenance Cost Indexes, Selected Metro Areas

Area	Cost per Sqft.	Local Index	200 Area Ranking
Akron, OH			
PM & Minor Repair	$ .37	95.4	78
Unscheduled Maintenance	.37	94.6	78
Renewal & Replacement	1.37	101.7	69
Total Average Cost	2.11	99.2	74
Alamogordo, NM			
PM & Minor Repair	.27	69.6	169
Unscheduled Maintenance	.25	65.3	169
Renewal & Replacement	1.16	86.1	138
Total Average Cost	1.69	79.2	154
Albany, NY			
PM & Minor Repair	.38	96.9	67
Unscheduled Maintenance	.37	96.1	67
Renewal & Replacement	1.24	91.7	109
Total Average Cost	1.99	93.5	96
Albuquerque, NM			
PM & Minor Repair	.33	84.7	123
Unscheduled Maintenance	.32	82.5	123
Renewal & Replacement	1.16	86.0	139
Total Average Cost	1.81	85.1	129
Altus, OK			
PM & Minor Repair	.30	75.9	147
Unscheduled Maintenance	.28	72.6	146
Renewal & Replacement	1.15	84.9	145
Total Average Cost	1.73	81.0	145
Amarillo, TX			
PM & Minor Repair	.27	69.0	173
Unscheduled Maintenance	.25	64.7	173
Renewal & Replacement	1.03	76.5	177
Total Average Cost	1.55	73.0	179
Anaheim, CA			
PM & Minor Repair	.43	110.4	35
Unscheduled Maintenance	.43	111.6	35
Renewal & Replacement	1.48	109.9	27
Total Average Cost	2.35	110.3	33
Anchorage, AK			
PM & Minor Repair	.46	116.9	22
Unscheduled Maintenance	.46	118.5	22
Renewal & Replacement	1.42	105.1	56
Total Average Cost	2.34	109.7	36
Ann Arbor, MI			
PM & Minor Repair	.37	95.4	80
Unscheduled Maintenance	.37	94.3	80
Renewal & Replacement	1.38	101.9	68
Total Average Cost	2.12	99.3	72
Atlanta, GA			
PM & Minor Repair	$ .34	87.0	112
Unscheduled Maintenance	.33	85.1	111
Renewal & Replacement	1.27	94.1	100
Total Average Cost	1.94	91.2	105
Atlantic City, NJ			
PM & Minor Repair	.49	125.9	13
Unscheduled Maintenance	.50	129.2	13
Renewal & Replacement	1.66	122.7	10
Total Average Cost	2.65	124.5	13
Austin, TX			
PM & Minor Repair	.30	76.0	146
Unscheduled Maintenance	.28	72.8	144
Renewal & Replacement	1.14	84.6	148
Total Average Cost	1.72	80.9	147
Bakersfield, CA			
PM & Minor Repair	.38	97.7	65
Unscheduled Maintenance	.38	97.1	65
Renewal & Replacement	1.43	105.6	53
Total Average Cost	2.19	102.6	58
Baltimore, MD			
PM & Minor Repair	.38	96.5	70
Unscheduled Maintenance	.37	96.0	68
Renewal & Replacement	1.36	100.4	72
Total Average Cost	2.11	98.9	75
Battle Creek, MI			
PM & Minor Repair	.36	91.2	95
Unscheduled Maintenance	.35	89.7	96
Renewal & Replacement	1.27	93.9	101
Total Average Cost	1.97	92.6	98
Beaufort, SC			
PM & Minor Repair	.14	36.6	200
Unscheduled Maintenance	.11	28.1	200
Renewal & Replacement	.92	67.8	199
Total Average Cost	1.17	54.8	200
Beaumont, TX			
PM & Minor Repair	.28	72.4	155
Unscheduled Maintenance	.27	68.7	155
Renewal & Replacement	1.17	86.8	134
Total Average Cost	1.72	80.8	149
Billings, MT			
PM & Minor Repair	.33	83.7	127
Unscheduled Maintenance	.32	81.2	129
Renewal & Replacement	1.09	81.0	162
Total Average Cost	1.74	81.5	140

Note: Costs per Sqft. are the annual average costs, over a 50 year service life, of maintaining the two-story office building shown in Chapter 2. Local Indexes are standardized (equal 100) for the Washington DC area.

3.1 Local Maintenance Cost Indexes, Selected Metro Areas

Area	Cost per Sqft.	Local Index	200 Area Ranking
Biloxi, MS			
PM & Minor Repair...............	$ .27	69.1	172
Unscheduled Maintenance...	.25	65.0	172
Renewal & Replacement......	1.08	80.2	165
Total Average Cost.............	1.61	75.4	169
Birmingham, AL			
PM & Minor Repair...............	.27	69.8	167
Unscheduled Maintenance...	.26	65.5	168
Renewal & Replacement......	1.20	89.1	124
Total Average Cost.............	1.73	81.3	141
Bismarck, ND			
PM & Minor Repair...............	.31	79.4	137
Unscheduled Maintenance...	.30	76.3	137
Renewal & Replacement......	.99	73.0	188
Total Average Cost.............	1.59	74.8	171
Boise, ID			
PM & Minor Repair...............	.34	87.4	110
Unscheduled Maintenance...	.33	85.4	110
Renewal & Replacement......	1.16	85.9	141
Total Average Cost.............	1.83	86.1	125
Boston, MA			
PM & Minor Repair...............	.48	124.0	17
Unscheduled Maintenance...	.49	126.9	17
Renewal & Replacement......	1.63	120.7	14
Total Average Cost.............	2.61	122.4	15
Boulder, CO			
PM & Minor Repair...............	.38	97.7	62
Unscheduled Maintenance...	.38	97.1	64
Renewal & Replacement......	1.15	85.3	143
Total Average Cost.............	1.91	89.7	112
Bowling Green, KY			
PM & Minor Repair...............	.34	86.4	117
Unscheduled Maintenance...	.33	84.4	118
Renewal & Replacement......	1.18	87.7	128
Total Average Cost.............	1.85	86.8	120
Brockton, MA			
PM & Minor Repair...............	.41	104.3	54
Unscheduled Maintenance...	.41	104.6	54
Renewal & Replacement......	1.45	107.2	43
Total Average Cost.............	2.26	106.2	48
Buffalo, NY			
PM & Minor Repair...............	.40	101.8	59
Unscheduled Maintenance...	.40	101.5	59
Renewal & Replacement......	1.39	102.7	62
Total Average Cost.............	2.18	102.3	59
Burlington, VT			
PM & Minor Repair..............	$ .30	76.2	144
Unscheduled Maintenance...	.28	72.6	145
Renewal & Replacement......	1.03	76.5	178
Total Average Cost.............	1.61	75.7	164
Camden, NJ			
PM & Minor Repair..............	.49	125.9	13
Unscheduled Maintenance...	.50	129.2	13
Renewal & Replacement......	1.66	122.7	10
Total Average Cost.............	2.65	124.5	13
Carson City, NV			
PM & Minor Repair..............	.36	92.7	89
Unscheduled Maintenance...	.36	91.5	89
Renewal & Replacement......	1.28	95.1	96
Total Average Cost.............	2.00	94.0	93
Cedar Rapids, IA			
PM & Minor Repair..............	.36	91.0	97
Unscheduled Maintenance...	.35	89.6	97
Renewal & Replacement......	1.24	91.6	111
Total Average Cost.............	1.94	91.1	106
Charleston, SC			
PM & Minor Repair..............	.17	43.0	199
Unscheduled Maintenance...	.14	35.2	199
Renewal & Replacement......	.97	72.0	194
Total Average Cost.............	1.28	60.0	199
Charleston, WV			
PM & Minor Repair..............	.36	91.4	94
Unscheduled Maintenance...	.35	90.1	94
Renewal & Replacement......	1.46	107.8	40
Total Average Cost.............	2.16	101.6	62
Charlotte, NC			
PM & Minor Repair..............	.18	45.5	196
Unscheduled Maintenance...	.15	38.0	196
Renewal & Replacement......	.98	72.5	192
Total Average Cost.............	1.30	61.2	195
Chattanooga, TN			
PM & Minor Repair..............	.25	63.2	179
Unscheduled Maintenance...	.23	58.1	179
Renewal & Replacement......	1.02	75.6	181
Total Average Cost.............	1.49	70.1	184
Cheyenne, WY			
PM & Minor Repair..............	.24	61.6	180
Unscheduled Maintenance...	.22	56.1	181
Renewal & Replacement......	.98	72.5	193
Total Average Cost.............	1.44	67.5	187

Note: Costs per Sqft. are the annual average costs, over a 50 year service life, of maintaining the two-story office building shown in Chapter 2. Local Indexes are standardized (equal 100) for the Washington DC area.

3.1 Local Maintenance Cost Indexes, Selected Metro Areas

Area	Cost per Sqft.	Local Index	200 Area Ranking
Chicago, IL			
PM & Minor Repair	$ .48	122.2	18
Unscheduled Maintenance	.49	124.9	18
Renewal & Replacement	1.57	116.5	18
Total Average Cost	2.54	119.1	18
Cincinnati, OH			
PM & Minor Repair	.34	86.4	118
Unscheduled Maintenance	.33	84.5	117
Renewal & Replacement	1.38	102.1	67
Total Average Cost	2.05	96.0	86
Cleveland, OH			
PM & Minor Repair	.43	109.0	40
Unscheduled Maintenance	.43	110.0	40
Renewal & Replacement	1.42	105.5	54
Total Average Cost	2.28	107.0	44
Colorado Springs, CO			
PM & Minor Repair	.37	95.4	79
Unscheduled Maintenance	.37	94.6	77
Renewal & Replacement	1.14	84.5	149
Total Average Cost	1.88	88.4	114
Columbia, SC			
PM & Minor Repair	.17	44.7	198
Unscheduled Maintenance	.14	37.2	198
Renewal & Replacement	.97	71.5	195
Total Average Cost	1.29	60.3	198
Columbus, GA			
PM & Minor Repair	.19	47.8	191
Unscheduled Maintenance	.16	40.6	191
Renewal & Replacement	1.01	75.1	184
Total Average Cost	1.36	63.8	190
Columbus, OH			
PM & Minor Repair	.36	91.9	93
Unscheduled Maintenance	.35	90.7	92
Renewal & Replacement	1.41	104.6	57
Total Average Cost	2.12	99.7	71
Concord, NH			
PM & Minor Repair	.32	82.2	131
Unscheduled Maintenance	.31	79.5	131
Renewal & Replacement	1.21	89.7	119
Total Average Cost	1.84	86.5	121
Corpus Christi, TX			
PM & Minor Repair	.24	61.4	181
Unscheduled Maintenance	.22	56.2	180
Renewal & Replacement	1.08	79.9	169
Total Average Cost	1.54	72.1	180
Cumberland, MD			
PM & Minor Repair	$ .34	86.8	115
Unscheduled Maintenance	.33	85.0	112
Renewal & Replacement	1.25	92.7	105
Total Average Cost	1.92	90.2	110
Dallas, TX			
PM & Minor Repair	.28	71.9	157
Unscheduled Maintenance	.27	68.0	157
Renewal & Replacement	1.13	83.6	151
Total Average Cost	1.67	78.6	157
Danbury, CT			
PM & Minor Repair	.43	109.7	38
Unscheduled Maintenance	.43	110.9	38
Renewal & Replacement	1.57	116.6	17
Total Average Cost	2.44	114.3	20
Davenport, IA			
PM & Minor Repair	.38	97.7	63
Unscheduled Maintenance	.38	97.2	63
Renewal & Replacement	1.29	95.2	95
Total Average Cost	2.05	96.0	85
Dayton, OH			
PM & Minor Repair	.34	86.0	121
Unscheduled Maintenance	.33	84.1	121
Renewal & Replacement	1.35	100.2	74
Total Average Cost	2.02	94.6	91
Daytona Beach, FL			
PM & Minor Repair	.27	69.8	166
Unscheduled Maintenance	.26	65.7	166
Renewal & Replacement	1.28	94.7	98
Total Average Cost	1.81	84.8	131
Denver, CO			
PM & Minor Repair	.38	96.3	72
Unscheduled Maintenance	.37	95.6	72
Renewal & Replacement	1.22	90.2	116
Total Average Cost	1.97	92.3	99
Des Moines, IA			
PM & Minor Repair	.36	92.0	91
Unscheduled Maintenance	.35	90.8	91
Renewal & Replacement	1.22	90.7	115
Total Average Cost	1.94	91.0	108
Detroit, MI			
PM & Minor Repair	.47	119.6	19
Unscheduled Maintenance	.47	121.8	19
Renewal & Replacement	1.47	109.1	32
Total Average Cost	2.42	113.4	22

Note: Costs per Sqft. are the annual average costs, over a 50 year service life, of maintaining the two-story office building shown in Chapter 2. Local Indexes are standardized (equal 100) for the Washington DC area.

Area	Cost per Sqft.	Local Index	200 Area Ranking	Area	Cost per Sqft.	Local Index	200 Area Ranking
Duluth, MN				**Flint, MI**			
PM & Minor Repair	$ .40	102.0	58	PM & Minor Repair	$ .38	96.3	73
Unscheduled Maintenance	.40	101.7	58	Unscheduled Maintenance	.37	95.3	76
Renewal & Replacement	1.35	99.8	77	Renewal & Replacement	1.31	97.0	86
Total Average Cost	2.14	100.6	66	Total Average Cost	2.06	96.6	80
Eau Claire, WI				**Fort Lauderdale, FL**			
PM & Minor Repair	.37	94.3	81	PM & Minor Repair	.31	80.1	136
Unscheduled Maintenance	.36	93.1	82	Unscheduled Maintenance	.30	77.4	135
Renewal & Replacement	1.26	93.6	103	Renewal & Replacement	1.21	89.3	123
Total Average Cost	2.00	93.7	95	Total Average Cost	1.82	85.5	127
El Paso, TX				**Fort Smith, AR**			
PM & Minor Repair	.25	64.7	178	PM & Minor Repair	.28	71.9	156
Unscheduled Maintenance	.23	59.9	178	Unscheduled Maintenance	.27	68.1	156
Renewal & Replacement	1.02	75.5	182	Renewal & Replacement	1.06	78.5	174
Total Average Cost	1.51	70.7	183	Total Average Cost	1.61	75.4	167
Erie, PA				**Fort Worth, TX**			
PM & Minor Repair	.35	89.9	103	PM & Minor Repair	.27	69.7	168
Unscheduled Maintenance	.34	88.2	103	Unscheduled Maintenance	.26	65.6	167
Renewal & Replacement	1.34	99.1	79	Renewal & Replacement	1.11	82.4	154
Total Average Cost	2.03	95.4	88	Total Average Cost	1.64	77.0	161
Eugene, OR				**Fresno, CA**			
PM & Minor Repair	.41	105.6	50	PM & Minor Repair	.38	96.3	74
Unscheduled Maintenance	.41	106.3	50	Unscheduled Maintenance	.37	95.5	74
Renewal & Replacement	1.44	106.4	50	Renewal & Replacement	1.44	106.8	48
Total Average Cost	2.26	106.2	49	Total Average Cost	2.19	102.8	56
Evansville, IN				**Gary, IN**			
PM & Minor Repair	.37	93.5	84	PM & Minor Repair	.44	112.0	33
Unscheduled Maintenance	.36	92.5	84	Unscheduled Maintenance	.44	113.4	33
Renewal & Replacement	1.29	95.3	94	Renewal & Replacement	1.33	98.8	80
Total Average Cost	2.01	94.4	92	Total Average Cost	2.21	103.9	54
Fairbanks, AK				**Grand Rapids, MI**			
PM & Minor Repair	.46	117.0	21	PM & Minor Repair	.27	70.1	160
Unscheduled Maintenance	.46	118.6	21	Unscheduled Maintenance	.26	65.7	165
Renewal & Replacement	1.45	107.2	44	Renewal & Replacement	.98	72.5	191
Total Average Cost	2.37	111.1	27	Total Average Cost	1.51	70.8	181
Fall River, MA				**Great Falls, MT**			
PM & Minor Repair	.41	105.0	53	PM & Minor Repair	.33	84.1	125
Unscheduled Maintenance	.41	105.0	53	Unscheduled Maintenance	.32	81.7	126
Renewal & Replacement	1.43	105.8	52	Renewal & Replacement	1.07	79.1	170
Total Average Cost	2.25	105.5	51	Total Average Cost	1.72	80.5	151
Fargo, ND				**Green Bay, WI**			
PM & Minor Repair	.31	79.1	138	PM & Minor Repair	.37	94.0	83
Unscheduled Maintenance	.30	76.0	138	Unscheduled Maintenance	.36	92.7	83
Renewal & Replacement	1.06	78.7	172	Renewal & Replacement	1.23	91.1	113
Total Average Cost	1.67	78.3	158	Total Average Cost	1.96	91.9	101

Note: Costs per Sqft. are the annual average costs, over a 50 year service life, of maintaining the two-story office building shown in Chapter 2. Local Indexes are standardized (equal 100) for the Washington DC area.

3.1 Local Maintenance Cost Indexes, Selected Metro Areas

Area	Cost per Sqft.	Local Index	200 Area Ranking	Area	Cost per Sqft.	Local Index	200 Area Ranking
Greensboro, NC				**Jackson, MS**			
PM & Minor Repair	$ .19	47.7	192	PM & Minor Repair	$ .19	47.7	193
Unscheduled Maintenance	.16	40.5	193	Unscheduled Maintenance	.16	40.5	192
Renewal & Replacement	.96	71.1	196	Renewal & Replacement	.96	70.8	197
Total Average Cost	1.30	61.2	196	Total Average Cost	1.30	61.0	197
Hampton, VA				**Jacksonville, FL**			
PM & Minor Repair	.27	68.0	175	PM & Minor Repair	.27	70.2	159
Unscheduled Maintenance	.25	63.6	175	Unscheduled Maintenance	.26	66.1	159
Renewal & Replacement	.99	73.6	186	Renewal & Replacement	1.15	84.8	147
Total Average Cost	1.51	70.7	182	Total Average Cost	1.68	73.7	156
Harrisburg, PA				**Jersey City, NJ**			
PM & Minor Repair	.35	89.4	105	PM & Minor Repair	.51	131.6	8
Unscheduled Maintenance	.34	87.9	104	Unscheduled Maintenance	.53	135.8	8
Renewal & Replacement	1.42	105.4	55	Renewal & Replacement	1.66	122.7	12
Total Average Cost	2.11	99.2	73	Total Average Cost	2.70	126.7	11
Hartford, CT				**Juneau, AK**			
PM & Minor Repair	.42	107.6	44	PM & Minor Repair	.46	116.7	23
Unscheduled Maintenance	.42	108.4	44	Unscheduled Maintenance	.46	118.4	23
Renewal & Replacement	1.46	108.0	39	Renewal & Replacement	1.44	107.0	46
Total Average Cost	2.30	108.0	40	Total Average Cost	2.36	110.9	28
Hilo, HI				**Kalamazoo, MI**			
PM & Minor Repair	.49	125.1	15	PM & Minor Repair	.31	80.1	135
Unscheduled Maintenance	.50	128.2	15	Unscheduled Maintenance	.30	77.1	136
Renewal & Replacement	1.73	127.8	5	Renewal & Replacement	1.24	92.1	107
Total Average Cost	2.71	127.4	9	Total Average Cost	1.86	87.2	119
Honolulu, HI				**Kansas City, MO**			
PM & Minor Repair	.49	125.1	15	PM & Minor Repair	.44	113.6	27
Unscheduled Maintenance	.50	128.2	15	Unscheduled Maintenance	.45	115.4	26
Renewal & Replacement	1.73	127.8	5	Renewal & Replacement	1.52	112.9	24
Total Average Cost	2.71	127.4	9	Total Average Cost	2.42	113.5	21
Houston, TX				**Knoxville, TN**			
PM & Minor Repair	.29	75.3	150	PM & Minor Repair	.27	69.9	164
Unscheduled Maintenance	.28	72.0	150	Unscheduled Maintenance	.26	65.7	164
Renewal & Replacement	1.24	91.8	108	Renewal & Replacement	1.08	80.1	166
Total Average Cost	1.81	85.1	128	Total Average Cost	1.61	75.6	166
Huntsville, AL				**Kokomo, IN**			
PM & Minor Repair	.29	73.6	153	PM & Minor Repair	.36	91.1	96
Unscheduled Maintenance	.27	69.9	153	Unscheduled Maintenance	.35	89.8	95
Renewal & Replacement	1.18	87.3	130	Renewal & Replacement	1.32	98.0	83
Total Average Cost	1.74	81.6	139	Total Average Cost	2.03	95.3	89
Indianapolis, IN				**Lansing, MI**			
PM & Minor Repair	.38	96.2	75	PM & Minor Repair	.37	95.5	77
Unscheduled Maintenance	.37	95.6	71	Unscheduled Maintenance	.37	94.5	79
Renewal & Replacement	1.38	102.3	64	Renewal & Replacement	1.30	96.6	88
Total Average Cost	2.13	99.9	70	Total Average Cost	2.05	96.0	83

Note: Costs per Sqft. are the annual average costs, over a 50 year service life, of maintaining the two-story office building shown in Chapter 2. Local Indexes are standardized (equal 100) for the Washington DC area.

3.1 Local Maintenance Cost Indexes, Selected Metro Areas

Area	Cost per Sqft.	Local Index	200 Area Ranking
Las Cruces, NM			
PM & Minor Repair	$ .27	69.6	169
Unscheduled Maintenance	.25	65.3	169
Renewal & Replacement	1.16	86.0	140
Total Average Cost	1.69	79.2	155
Las Vegas, NV			
PM & Minor Repair	.42	108.5	41
Unscheduled Maintenance	.43	109.4	41
Renewal & Replacement	1.41	104.5	58
Total Average Cost	2.26	106.1	50
Lawton, OK			
PM & Minor Repair	.30	75.9	147
Unscheduled Maintenance	.28	72.6	146
Renewal & Replacement	1.15	84.9	145
Total Average Cost	1.73	81.0	145
Lewiston, ME			
PM & Minor Repair	.35	89.1	106
Unscheduled Maintenance	.34	87.2	107
Renewal & Replacement	1.04	77.2	176
Total Average Cost	1.73	81.2	143
Lexington, KY			
PM & Minor Repair	.23	58.6	185
Unscheduled Maintenance	.21	52.9	185
Renewal & Replacement	1.17	86.7	137
Total Average Cost	1.61	75.4	170
Little Rock, AR			
PM & Minor Repair	.30	77.4	141
Unscheduled Maintenance	.29	74.2	141
Renewal & Replacement	1.14	84.2	150
Total Average Cost	1.73	81.1	144
Los Angeles, CA			
PM & Minor Repair	.43	110.9	34
Unscheduled Maintenance	.44	112.1	34
Renewal & Replacement	1.49	110.4	26
Total Average Cost	2.36	110.8	30
Louisville, KY			
PM & Minor Repair	.34	86.1	120
Unscheduled Maintenance	.33	84.2	120
Renewal & Replacement	1.31	97.0	87
Total Average Cost	1.97	92.7	97
Lowell, MA			
PM & Minor Repair	.50	126.8	12
Unscheduled Maintenance	.51	130.1	12
Renewal & Replacement	1.55	114.8	21
Total Average Cost	2.55	119.8	17
Lubbock, TX			
PM & Minor Repair	$ .22	55.5	187
Unscheduled Maintenance	.19	49.4	187
Renewal & Replacement	1.00	74.0	185
Total Average Cost	1.41	66.1	188
Macon, GA			
PM & Minor Repair	.26	66.4	176
Unscheduled Maintenance	.24	61.7	176
Renewal & Replacement	1.06	78.6	173
Total Average Cost	1.56	73.3	178
Madison, WI			
PM & Minor Repair	.37	94.3	82
Unscheduled Maintenance	.36	93.1	81
Renewal & Replacement	1.30	96.1	90
Total Average Cost	2.03	95.2	90
Manchester, NH			
PM & Minor Repair	.32	82.2	131
Unscheduled Maintenance	.31	79.5	131
Renewal & Replacement	1.21	89.7	119
Total Average Cost	1.84	86.5	121
Marquette, MI			
PM & Minor Repair	.34	86.9	113
Unscheduled Maintenance	.33	84.7	115
Renewal & Replacement	1.16	85.8	142
Total Average Cost	1.83	85.8	126
Medford, OR			
PM & Minor Repair	.38	96.1	76
Unscheduled Maintenance	.37	95.4	75
Renewal & Replacement	1.40	103.6	60
Total Average Cost	2.15	100.7	65
Memphis, TN			
PM & Minor Repair	.33	84.4	124
Unscheduled Maintenance	.32	82.2	124
Renewal & Replacement	1.21	89.5	122
Total Average Cost	1.86	87.3	118
Miami, FL			
PM & Minor Repair	.31	78.4	140
Unscheduled Maintenance	.29	75.5	140
Renewal & Replacement	1.21	89.7	121
Total Average Cost	1.81	85.0	130
Milwaukee, WI			
PM & Minor Repair	.42	106.8	47
Unscheduled Maintenance	.42	107.3	48
Renewal & Replacement	1.37	101.6	70
Total Average Cost	2.21	103.6	55

Note: Costs per Sqft. are the annual average costs, over a 50 year service life, of maintaining the two-story office building shown in Chapter 2. Local Indexes are standardized (equal 100) for the Washington DC area.

3.1 Local Maintenance Cost Indexes, Selected Metro Areas

Area	Cost per Sqft.	Local Index	200 Area Ranking
Minneapolis, MN			
PM & Minor Repair	$ .45	114.0	25
Unscheduled Maintenance	.45	115.2	27
Renewal & Replacement	1.51	111.9	25
Total Average Cost	2.41	112.9	23
Mobile, AL			
PM & Minor Repair	.23	58.7	184
Unscheduled Maintenance	.21	53.1	184
Renewal & Replacement	1.17	86.7	136
Total Average Cost	1.61	75.4	168
Moline, IL			
PM & Minor Repair	.36	91.9	92
Unscheduled Maintenance	.35	90.6	93
Renewal & Replacement	1.32	97.9	84
Total Average Cost	2.03	95.5	87
Muncie, IN			
PM & Minor Repair	.34	86.2	119
Unscheduled Maintenance	.33	84.3	119
Renewal & Replacement	1.29	95.8	93
Total Average Cost	1.96	91.9	102
Nashville, TN			
PM & Minor Repair	.30	76.7	142
Unscheduled Maintenance	.29	73.3	143
Renewal & Replacement	1.19	88.2	127
Total Average Cost	1.78	83.3	136
New Brunswick, NJ			
PM & Minor Repair	.50	127.5	11
Unscheduled Maintenance	.51	131.2	11
Renewal & Replacement	1.57	116.2	19
Total Average Cost	2.58	121.0	16
New Haven, CT			
PM & Minor Repair	.44	112.5	29
Unscheduled Maintenance	.44	114.0	29
Renewal & Replacement	1.47	108.7	34
Total Average Cost	2.35	110.4	32
New Orleans, LA			
PM & Minor Repair	.29	75.2	151
Unscheduled Maintenance	.28	71.8	151
Renewal & Replacement	1.27	93.8	102
Total Average Cost	1.84	86.4	123
New York, NY			
PM & Minor Repair	.68	173.0	1
Unscheduled Maintenance	.71	182.7	1
Renewal & Replacement	2.00	148.0	1
Total Average Cost	3.39	158.9	1
Newark, NJ			
PM & Minor Repair	$ .51	131.6	8
Unscheduled Maintenance	.53	135.8	8
Renewal & Replacement	1.66	122.7	12
Total Average Cost	2.70	126.7	11
Newport News, VA			
PM & Minor Repair	.27	68.2	174
Unscheduled Maintenance	.25	63.7	174
Renewal & Replacement	1.06	78.9	171
Total Average Cost	1.58	74.1	173
Norfolk, VA			
PM & Minor Repair	.27	70.1	161
Unscheduled Maintenance	.26	65.9	160
Renewal & Replacement	1.11	82.2	156
Total Average Cost	1.64	77.0	160
Norwalk, CT			
PM & Minor Repair	.44	112.1	32
Unscheduled Maintenance	.44	113.7	32
Renewal & Replacement	1.58	116.9	16
Total Average Cost	2.46	115.5	19
Norwich, CT			
PM & Minor Repair	.44	112.5	29
Unscheduled Maintenance	.44	114.0	29
Renewal & Replacement	1.46	108.4	37
Total Average Cost	2.35	110.2	34
Oakland, CA			
PM & Minor Repair	.53	135.1	7
Unscheduled Maintenance	.54	139.5	7
Renewal & Replacement	1.67	123.3	9
Total Average Cost	2.74	128.4	7
Ogden, UT			
PM & Minor Repair	.33	83.5	128
Unscheduled Maintenance	.32	81.2	127
Renewal & Replacement	1.08	80.0	168
Total Average Cost	1.72	80.9	148
Oklahoma City, OK			
PM & Minor Repair	.30	75.9	147
Unscheduled Maintenance	.28	72.6	148
Renewal & Replacement	1.12	83.2	152
Total Average Cost	1.70	80.0	153
Olympia, WA			
PM & Minor Repair	.42	107.8	42
Unscheduled Maintenance	.42	108.4	42
Renewal & Replacement	1.33	98.3	81
Total Average Cost	2.17	101.9	60

Note: Costs per Sqft. are the annual average costs, over a 50 year service life, of maintaining the two-story office building shown in Chapter 2. Local Indexes are standardized (equal 100) for the Washington DC area.

Area	Cost per Sqft.	Local Index	200 Area Ranking	Area	Cost per Sqft.	Local Index	200 Area Ranking
Omaha, NE				**Pocatello, ID**			
PM & Minor Repair	$ .35	90.1	102	PM & Minor Repair	$ .33	83.4	130
Unscheduled Maintenance	.35	88.7	102	Unscheduled Maintenance	.31	80.7	130
Renewal & Replacement	1.18	87.2	131	Renewal & Replacement	1.11	81.9	158
Total Average Cost	1.88	88.0	115	Total Average Cost	1.75	81.9	138
Orlando, FL				**Portland, ME**			
PM & Minor Repair	.21	52.5	189	PM & Minor Repair	.35	88.8	108
Unscheduled Maintenance	.18	46.0	189	Unscheduled Maintenance	.34	87.1	108
Renewal & Replacement	1.18	87.6	129	Renewal & Replacement	1.10	81.8	159
Total Average Cost	1.57	73.6	176	Total Average Cost	1.79	84.0	133
Owensboro, KY				**Portland, OR**			
PM & Minor Repair	.33	85.2	122	PM & Minor Repair	.44	113.7	26
Unscheduled Maintenance	.32	83.1	122	Unscheduled Maintenance	.45	115.4	25
Renewal & Replacement	1.18	87.2	132	Renewal & Replacement	1.39	102.7	61
Total Average Cost	1.83	86.1	124	Total Average Cost	2.28	107.0	43
Oxnard, CA				**Providence, RI**			
PM & Minor Repair	.43	109.0	39	PM & Minor Repair	.40	102.8	56
Unscheduled Maintenance	.43	110.0	39	Unscheduled Maintenance	.40	102.8	56
Renewal & Replacement	1.47	109.1	33	Renewal & Replacement	1.47	109.2	31
Total Average Cost	2.33	109.2	38	Total Average Cost	2.28	106.9	45
Parkersburg, WV				**Pueblo, CO**			
PM & Minor Repair	.38	97.7	64	PM & Minor Repair	.34	87.6	109
Unscheduled Maintenance	.38	97.3	62	Unscheduled Maintenance	.33	85.9	109
Renewal & Replacement	1.47	108.7	35	Renewal & Replacement	1.21	89.8	118
Total Average Cost	2.23	104.6	53	Total Average Cost	1.89	88.7	113
Peoria, IL				**Raleigh-Durham, NC**			
PM & Minor Repair	.37	93.5	85	PM & Minor Repair	.19	47.6	194
Unscheduled Maintenance	.36	92.4	86	Unscheduled Maintenance	.16	40.5	194
Renewal & Replacement	1.35	99.7	78	Renewal & Replacement	.98	72.9	189
Total Average Cost	2.07	97.2	78	Total Average Cost	1.33	62.3	193
Philadelphia, PA				**Rapid City, SD**			
PM & Minor Repair	.51	130.0	10	PM & Minor Repair	.24	61.1	182
Unscheduled Maintenance	.52	133.9	10	Unscheduled Maintenance	.22	55.6	182
Renewal & Replacement	1.77	130.9	4	Renewal & Replacement	.91	67.5	200
Total Average Cost	2.80	131.3	5	Total Average Cost	1.37	64.1	189
Phoenix, AZ				**Reading, PA**			
PM & Minor Repair	.29	74.5	152	PM & Minor Repair	.34	86.5	116
Unscheduled Maintenance	.28	71.1	152	Unscheduled Maintenance	.33	84.6	116
Renewal & Replacement	1.23	91.5	112	Renewal & Replacement	1.46	108.4	38
Total Average Cost	1.80	84.6	132	Total Average Cost	2.13	100.0	68
Pittsburgh, PA				**Reno, NV**			
PM & Minor Repair	.42	106.6	48	PM & Minor Repair	.36	92.7	89
Unscheduled Maintenance	.42	107.5	47	Unscheduled Maintenance	.36	91.5	89
Renewal & Replacement	1.55	115.1	20	Renewal & Replacement	1.28	95.1	96
Total Average Cost	2.39	112.1	24	Total Average Cost	2.00	94.0	93

Note: Costs per Sqft. are the annual average costs, over a 50 year service life, of maintaining the two-story office building shown in Chapter 2. Local Indexes are standardized (equal 100) for the Washington DC area.

3.1 Local Maintenance Cost Indexes, Selected Metro Areas

Area	Cost per Sqft.	Local Index	200 Area Ranking
Richland, WA			
PM & Minor Repair	$ .38	97.1	66
Unscheduled Maintenance	.38	96.6	66
Renewal & Replacement	1.30	96.3	89
Total Average Cost	2.06	96.5	81
Richmond, VA			
PM & Minor Repair	.31	78.8	139
Unscheduled Maintenance	.30	75.8	139
Renewal & Replacement	1.11	82.5	153
Total Average Cost	1.72	80.6	150
Riverside, CA			
PM & Minor Repair	.41	105.5	52
Unscheduled Maintenance	.41	106.0	52
Renewal & Replacement	1.48	109.3	30
Total Average Cost	2.30	108.0	41
Roanoke, VA			
PM & Minor Repair	.18	45.3	197
Unscheduled Maintenance	.15	38.0	197
Renewal & Replacement	1.02	75.5	183
Total Average Cost	1.34	63.1	192
Rochester, MN			
PM & Minor Repair	.38	96.5	71
Unscheduled Maintenance	.37	95.5	73
Renewal & Replacement	1.35	100.3	73
Total Average Cost	2.10	98.7	76
Rochester, NY			
PM & Minor Repair	.38	96.6	68
Unscheduled Maintenance	.37	95.8	69
Renewal & Replacement	1.30	95.9	91
Total Average Cost	2.05	96.0	84
Rockford, IL			
PM & Minor Repair	.44	113.3	28
Unscheduled Maintenance	.45	114.7	28
Renewal & Replacement	1.40	103.7	59
Total Average Cost	2.29	107.5	42
Rutland, VT			
PM & Minor Repair	.30	76.1	145
Unscheduled Maintenance	.28	72.6	149
Renewal & Replacement	1.03	76.4	179
Total Average Cost	1.61	75.6	165
Sacramento, CA			
PM & Minor Repair	.41	106.1	49
Unscheduled Maintenance	.42	106.6	49
Renewal & Replacement	1.54	114.2	22
Total Average Cost	2.37	111.3	26
Saginaw, MI			
PM & Minor Repair	$ .36	90.9	98
Unscheduled Maintenance	.35	89.2	98
Renewal & Replacement	1.25	92.9	104
Total Average Cost	1.96	91.9	103
Salem, OR			
PM & Minor Repair	.41	105.6	50
Unscheduled Maintenance	.41	106.3	50
Renewal & Replacement	1.44	106.9	47
Total Average Cost	2.27	106.6	47
Salt Lake City, UT			
PM & Minor Repair	.33	83.5	128
Unscheduled Maintenance	.32	81.2	127
Renewal & Replacement	1.11	82.3	155
Total Average Cost	1.75	82.3	137
San Antonio, TX			
PM & Minor Repair	.27	69.9	165
Unscheduled Maintenance	.26	65.8	163
Renewal & Replacement	1.20	89.0	126
Total Average Cost	1.73	81.2	142
San Diego, CA			
PM & Minor Repair	.40	102.3	57
Unscheduled Maintenance	.40	102.4	57
Renewal & Replacement	1.45	107.1	45
Total Average Cost	2.24	105.3	52
San Francisco, CA			
PM & Minor Repair	.58	148.1	2
Unscheduled Maintenance	.60	154.3	2
Renewal & Replacement	1.78	131.8	3
Total Average Cost	2.96	138.9	3
San Jose, CA			
PM & Minor Repair	.55	141.2	4
Unscheduled Maintenance	.57	146.4	4
Renewal & Replacement	1.72	127.3	7
Total Average Cost	2.84	133.3	4
Santa Barbara, CA			
PM & Minor Repair	.43	110.1	37
Unscheduled Maintenance	.43	111.2	37
Renewal & Replacement	1.48	109.3	29
Total Average Cost	2.34	109.8	35
Savannah, GA			
PM & Minor Repair	.23	59.9	183
Unscheduled Maintenance	.21	54.4	183
Renewal & Replacement	1.03	76.1	180
Total Average Cost	1.47	69.2	186

Note: Costs per Sqft. are the annual average costs, over a 50 year service life, of maintaining the two-story office building shown in Chapter 2. Local Indexes are standardized (equal 100) for the Washington DC area.

3.1 Local Maintenance Cost Indexes, Selected Metro Areas

Area	Cost per Sqft.	Local Index	200 Area Ranking	Area	Cost per Sqft.	Local Index	200 Area Ranking
Scranton, PA				Springfield, MO			
PM & Minor Repair	$ .35	90.4	100	PM & Minor Repair	$ .30	76.6	143
Unscheduled Maintenance	.35	89.0	99	Unscheduled Maintenance	.29	73.4	142
Renewal & Replacement	1.45	107.4	42	Renewal & Replacement	1.28	94.6	99
Total Average Cost	2.15	100.9	64	Total Average Cost	1.86	87.4	117
Seattle, WA				Springfield, OH			
PM & Minor Repair	.47	119.5	20	PM & Minor Repair	.35	89.1	107
Unscheduled Maintenance	.47	121.7	20	Unscheduled Maintenance	.34	87.5	106
Renewal & Replacement	1.38	102.1	65	Renewal & Replacement	1.37	101.4	71
Total Average Cost	2.32	108.9	39	Total Average Cost	2.06	96.6	79
Shreveport, LA				St. Louis, MO			
PM & Minor Repair	.28	72.8	154	PM & Minor Repair	.45	114.7	24
Unscheduled Maintenance	.27	69.1	154	Unscheduled Maintenance	.45	116.4	24
Renewal & Replacement	1.09	80.6	164	Renewal & Replacement	1.45	107.6	41
Total Average Cost	1.64	77.0	159	Total Average Cost	2.35	110.5	31
Sioux City, IA				Stamford, CT			
PM & Minor Repair	.34	86.8	114	PM & Minor Repair	.54	139.0	5
Unscheduled Maintenance	.33	84.7	114	Unscheduled Maintenance	.56	144.3	5
Renewal & Replacement	1.11	82.0	157	Renewal & Replacement	1.61	119.5	15
Total Average Cost	1.78	83.4	135	Total Average Cost	2.72	127.6	8
Sioux Falls, SD				Stockton, CA			
PM & Minor Repair	.32	81.7	134	PM & Minor Repair	.44	112.3	31
Unscheduled Maintenance	.31	79.0	134	Unscheduled Maintenance	.44	113.7	31
Renewal & Replacement	.94	69.6	198	Renewal & Replacement	1.48	109.6	28
Total Average Cost	1.57	73.6	177	Total Average Cost	2.36	110.8	29
South Bend, IN				Syracuse, NY			
PM & Minor Repair	.36	93.3	88	PM & Minor Repair	.35	90.4	99
Unscheduled Maintenance	.36	92.2	87	Unscheduled Maintenance	.35	88.7	101
Renewal & Replacement	1.24	91.6	110	Renewal & Replacement	1.25	92.3	106
Total Average Cost	1.96	92.1	100	Total Average Cost	1.95	91.3	104
Spokane, WA				Tacoma, WA			
PM & Minor Repair	.36	93.4	87	PM & Minor Repair	.42	107.8	42
Unscheduled Maintenance	.36	92.1	88	Unscheduled Maintenance	.42	108.4	42
Renewal & Replacement	1.21	89.8	117	Renewal & Replacement	1.33	98.3	81
Total Average Cost	1.94	90.9	109	Total Average Cost	2.17	101.9	60
Springfield, IL				Tallahassee, FL			
PM & Minor Repair	.37	93.5	86	PM & Minor Repair	.19	49.3	190
Unscheduled Maintenance	.36	92.5	85	Unscheduled Maintenance	.17	42.4	190
Renewal & Replacement	1.43	106.1	51	Renewal & Replacement	.99	73.2	187
Total Average Cost	2.16	101.3	63	Total Average Cost	1.35	63.1	191
Springfield, MA				Tampa, FL			
PM & Minor Repair	.38	96.5	69	PM & Minor Repair	.22	55.7	186
Unscheduled Maintenance	.37	95.7	70	Unscheduled Maintenance	.19	49.6	186
Renewal & Replacement	1.38	102.4	63	Renewal & Replacement	1.17	86.7	135
Total Average Cost	2.13	100.1	67	Total Average Cost	1.58	74.2	172

Note: Costs per Sqft. are the annual average costs, over a 50 year service life, of maintaining the two-story office building shown in Chapter 2. Local Indexes are standardized (equal 100) for the Washington DC area.

3.1 Local Maintenance Cost Indexes, Selected Metro Areas

Area	Cost per Sqft.	Local Index	200 Area Ranking
Terre Haute, IN			
PM & Minor Repair	$ .35	90.2	101
Unscheduled Maintenance	.35	88.8	100
Renewal & Replacement	1.35	99.9	76
Total Average Cost	2.05	96.1	82
Toledo, OH			
PM & Minor Repair	.40	103.5	55
Unscheduled Maintenance	.40	103.8	55
Renewal & Replacement	1.38	102.1	66
Total Average Cost	2.19	102.7	57
Topeka, KS			
PM & Minor Repair	.33	84.0	126
Unscheduled Maintenance	.32	81.9	125
Renewal & Replacement	1.29	95.8	92
Total Average Cost	1.94	91.1	107
Trenton, NJ			
PM & Minor Repair	.54	137.6	6
Unscheduled Maintenance	.56	142.6	6
Renewal & Replacement	1.68	124.6	8
Total Average Cost	2.78	130.3	6
Tucson, AZ			
PM & Minor Repair	.27	70.0	163
Unscheduled Maintenance	.26	65.9	161
Renewal & Replacement	1.17	86.9	133
Total Average Cost	1.70	80.0	152
Tulsa, OK			
PM & Minor Repair	.21	54.7	188
Unscheduled Maintenance	.19	48.5	188
Renewal & Replacement	1.09	80.7	163
Total Average Cost	1.49	70.1	185
Tuscaloosa, AL			
PM & Minor Repair	.27	69.4	171
Unscheduled Maintenance	.25	65.3	171
Renewal & Replacement	1.05	77.4	175
Total Average Cost	1.57	73.8	174
Utica, NY			
PM & Minor Repair	.35	89.6	104
Unscheduled Maintenance	.34	87.8	105
Renewal & Replacement	1.23	90.8	114
Total Average Cost	1.92	90.0	111
Virginia Beach, VA			
PM & Minor Repair	.27	70.0	162
Unscheduled Maintenance	.26	65.9	162
Renewal & Replacement	1.10	81.5	160
Total Average Cost	1.63	76.5	163
Waco, TX			
PM & Minor Repair	$ .27	70.2	158
Unscheduled Maintenance	.26	66.2	158
Renewal & Replacement	1.10	81.5	161
Total Average Cost	1.63	76.6	162
Washington DC			
PM & Minor Repair	.39	100.0	60
Unscheduled Maintenance	.39	100.0	60
Renewal & Replacement	1.35	100.0	75
Total Average Cost	2.13	100.0	69
Waterbury, CT			
PM & Minor Repair	.43	110.1	36
Unscheduled Maintenance	.43	111.2	36
Renewal & Replacement	1.46	108.5	36
Total Average Cost	2.33	109.3	37
Watertown, NY			
PM & Minor Repair	.34	87.1	111
Unscheduled Maintenance	.33	84.9	113
Renewal & Replacement	1.20	89.1	125
Total Average Cost	1.87	88.0	116
Wichita Falls, TX			
PM & Minor Repair	.25	65.1	177
Unscheduled Maintenance	.24	60.4	177
Renewal & Replacement	1.08	80.0	167
Total Average Cost	1.57	73.7	175
Wichita, KS			
PM & Minor Repair	.32	82.0	133
Unscheduled Maintenance	.31	79.4	133
Renewal & Replacement	1.15	85.2	144
Total Average Cost	1.78	83.6	134
Wilmington, DE			
PM & Minor Repair	.42	106.9	46
Unscheduled Maintenance	.42	107.8	45
Renewal & Replacement	1.54	114.1	23
Total Average Cost	2.38	111.6	25
Winston-Salem, NC			
PM & Minor Repair	.19	47.6	194
Unscheduled Maintenance	.16	40.5	194
Renewal & Replacement	.98	72.7	190
Total Average Cost	1.33	62.2	194
Worcester, MA			
PM & Minor Repair	.42	107.0	45
Unscheduled Maintenance	.42	107.5	46
Renewal & Replacement	1.44	106.6	49
Total Average Cost	2.28	106.9	46

Note: Costs per Sqft. are the annual average costs, over a 50 year service life, of maintaining the two-story office building shown in Chapter 2. Local Indexes are standardized (equal 100) for the Washington DC area.

Area	Cost per Sqft.	Local Index	200 Area Ranking
Yonkers, NY			
PM & Minor Repair	$ .58	147.4	3
Unscheduled Maintenance	.60	153.6	3
Renewal & Replacement	1.88	139.0	2
Total Average Cost	3.05	143.2	2
Youngstown, OH			
PM & Minor Repair	.39	99.9	61
Unscheduled Maintenance	.39	99.6	61
Renewal & Replacement	1.31	97.0	85
Total Average Cost	2.09	98.0	77

Canadian Cities

Area	Cost per Sqft.	Local Index	200 Area Ranking
Calgary, Alberta			
PM & Minor Repair	.42	80.3	
Unscheduled Maintenance	.42	80.8	
Renewal & Replacement	1.27	70.1	
Total Average Cost	2.11	73.9	
Edmonton, Alberta			
PM & Minor Repair	.42	80.5	
Unscheduled Maintenance	.42	80.9	
Renewal & Replacement	1.24	68.6	
Total Average Cost	2.09	73.1	
Hamilton, Ontario			
PM & Minor Repair	.43	81.6	
Unscheduled Maintenance	.43	82.2	
Renewal & Replacement	1.30	71.7	
Total Average Cost	2.15	75.4	
London, Ontario			
PM & Minor Repair	.42	79.4	
Unscheduled Maintenance	.42	79.8	
Renewal & Replacement	1.29	71.0	
Total Average Cost	2.12	74.2	
Montreal, Quebec			
PM & Minor Repair	.37	70.6	
Unscheduled Maintenance	.36	69.6	
Renewal & Replacement	1.21	66.7	
Total Average Cost	1.94	68.0	
Ottawa, Ontario			
PM & Minor Repair	.42	80.4	
Unscheduled Maintenance	.42	80.7	
Renewal & Replacement	1.23	68.2	
Total Average Cost	2.08	72.7	
Quebec, Quebec			
PM & Minor Repair	$ .37	70.4	
Unscheduled Maintenance	.36	69.6	
Renewal & Replacement	1.23	68.1	
Total Average Cost	1.97	68.8	
Toronto, Ontario			
PM & Minor Repair	.43	82.0	
Unscheduled Maintenance	.43	82.8	
Renewal & Replacement	1.37	75.4	
Total Average Cost	2.23	78.0	
Vancouver, B.C.			
PM & Minor Repair	.36	69.2	
Unscheduled Maintenance	.36	68.3	
Renewal & Replacement	1.26	69.9	
Total Average Cost	1.98	69.5	
Winnipeg, Manitoba			
PM & Minor Repair	.33	63.6	
Unscheduled Maintenance	.32	61.7	
Renewal & Replacement	1.01	55.9	
Total Average Cost	1.67	58.4	

Note: Costs per Sqft. are the annual average costs, over a 50 year service life, of maintaining the two-story office building shown in Chapter 2. Local Indexes are standardized (equal 100) for the Washington DC area. Canadian cities are in Canadian dollars.

8.2 In-House Shop Rates, Selected Metro Areas

Area	HVAC Technician	Maintenance Electrician	Maintenance Plumber	Maintenance Carpenter	Maintenance Painter	General Maintenance Worker	Trade Shop Supervisor
Akron, OH							
Direct Wage	29.64	29.59	29.72	28.90	26.50	24.75	35.42
Markup	59.28	59.19	59.43	57.80	53.01	49.51	70.84
Total Shop Rate	88.92	88.78	89.15	86.70	79.51	74.26	106.26
Alamogordo, NM							
Direct Wage	20.47	20.44	20.52	19.96	18.31	17.10	24.46
Markup	40.94	40.88	41.05	39.92	36.61	34.19	48.93
Total Shop Rate	61.41	61.32	61.57	59.88	54.92	51.29	73.39
Albany, NY							
Direct Wage	30.09	30.04	30.17	29.34	26.91	25.13	35.96
Markup	60.17	60.08	60.33	58.67	53.81	50.25	71.91
Total Shop Rate	90.26	90.12	90.50	88.01	80.72	75.38	107.87
Albuquerque, NM							
Direct Wage	25.88	25.85	25.95	25.24	23.15	21.62	30.94
Markup	51.77	51.69	51.91	50.48	46.30	43.24	61.87
Total Shop Rate	77.65	77.54	77.86	75.72	69.44	64.85	92.81
Altus, OK							
Direct Wage	22.81	22.77	22.87	22.24	20.40	19.05	27.26
Markup	45.61	45.54	45.73	44.48	40.79	38.10	54.51
Total Shop Rate	68.42	68.32	68.60	66.71	61.19	57.14	81.77
Amarillo, TX							
Direct Wage	20.30	20.27	20.35	19.79	18.15	16.95	24.26
Markup	40.60	40.54	40.71	39.59	36.31	33.91	48.52
Total Shop Rate	60.90	60.80	61.06	59.38	54.46	50.86	72.78
Anaheim, CA							
Direct Wage	34.98	34.93	35.07	34.11	31.28	29.21	41.81
Markup	69.96	69.85	70.15	68.21	62.56	58.43	83.61
Total Shop Rate	104.94	104.78	105.22	102.32	93.84	87.64	125.42
Anchorage, AK							
Direct Wage	37.08	37.02	37.18	36.16	33.16	30.97	44.32
Markup	74.16	74.05	74.36	72.31	66.32	61.94	88.63
Total Shop Rate	111.24	111.07	111.54	108.47	99.48	92.91	132.95
Ann Arbor, MI							
Direct Wage	29.48	29.43	29.55	28.74	26.36	24.62	35.23
Markup	58.95	58.86	59.11	57.48	52.72	49.23	70.45
Total Shop Rate	88.43	88.29	88.66	86.22	79.08	73.85	105.68
Atlanta, GA							
Direct Wage	26.72	26.67	26.79	26.05	23.89	22.31	31.93
Markup	53.43	53.35	53.57	52.10	47.78	44.62	63.86
Total Shop Rate	80.15	80.02	80.36	78.15	71.67	66.94	95.78
Atlantic City, NJ							
Direct Wage	40.61	40.55	40.72	39.60	36.32	33.92	48.54
Markup	81.23	81.11	81.44	79.20	72.64	67.84	97.08
Total Shop Rate	121.84	121.66	122.17	118.80	108.96	101.76	145.62

Note: Total shop rates are expressed as costs per hour. Shop markup includes all benefits, non-productive time, support, and supervision.

3.2 In-House Shop Rates, Selected Metro Areas

Area	HVAC Technician	Maintenance Electrician	Maintenance Plumber	Maintenance Carpenter	Maintenance Painter	General Maintenance Worker	Trade Shop Supervisor
Austin, TX							
Direct Wage	22.91	22.88	22.97	22.34	20.49	19.14	27.38
Markup	45.83	45.76	45.95	44.68	40.98	38.27	54.77
Total Shop Rate	68.74	68.63	68.92	67.02	61.47	57.41	82.15
Bakersfield, CA							
Direct Wage	30.40	30.36	30.48	29.65	27.19	25.39	36.34
Markup	60.81	60.72	60.97	59.29	54.38	50.78	72.67
Total Shop Rate	91.21	91.07	91.45	88.94	81.57	76.18	109.01
Baltimore, MD							
Direct Wage	30.21	30.16	30.29	29.45	27.01	25.23	36.10
Markup	60.42	60.33	60.58	58.91	54.03	50.46	72.21
Total Shop Rate	90.63	90.49	90.87	88.36	81.04	75.69	108.31
Battle Creek, MI							
Direct Wage	28.07	28.02	28.14	27.37	25.10	23.44	33.54
Markup	56.13	56.05	56.28	54.73	50.20	46.88	67.09
Total Shop Rate	84.20	84.07	84.42	82.10	75.30	70.32	100.63
Beaufort, SC							
Direct Wage	8.80	8.79	8.83	8.58	7.87	7.35	10.52
Markup	17.60	17.58	17.65	17.16	15.74	14.70	21.04
Total Shop Rate	26.41	26.37	26.48	25.75	23.61	22.05	31.56
Beaumont, TX							
Direct Wage	21.57	21.54	21.63	21.03	19.29	18.01	25.78
Markup	43.14	43.07	43.25	42.06	38.58	36.03	51.56
Total Shop Rate	64.71	64.61	64.88	63.09	57.87	54.04	77.33
Billings, MT							
Direct Wage	25.45	25.41	25.51	24.81	22.75	21.25	30.41
Markup	50.89	50.81	51.03	49.62	45.51	42.50	60.82
Total Shop Rate	76.34	76.22	76.54	74.43	68.26	63.75	91.23
Biloxi, MS							
Direct Wage	20.45	20.42	20.51	19.94	18.29	17.08	24.44
Markup	40.91	40.85	41.02	39.89	36.58	34.16	48.89
Total Shop Rate	61.36	61.27	61.52	59.83	54.87	51.25	73.33
Birmingham, AL							
Direct Wage	20.53	20.50	20.58	20.02	18.36	17.14	24.53
Markup	41.05	40.99	41.16	40.03	36.71	34.29	49.06
Total Shop Rate	61.58	61.49	61.74	60.05	55.07	51.43	73.60
Bismarck, ND							
Direct Wage	23.91	23.88	23.98	23.32	21.39	19.97	28.58
Markup	47.83	47.76	47.96	46.64	42.77	39.94	57.16
Total Shop Rate	71.74	71.63	71.93	69.95	64.16	59.92	85.74
Boise, ID							
Direct Wage	26.77	26.73	26.84	26.10	23.94	22.36	32.00
Markup	53.54	53.46	53.69	52.21	47.88	44.72	63.99
Total Shop Rate	80.32	80.20	80.53	78.31	71.82	67.08	95.99

Note: Total shop rates are expressed as costs per hour. Shop markup includes all benefits, non-productive time, support, and supervision.

3.2 In-House Shop Rates, Selected Metro Areas

Area	HVAC Technician	Maintenance Electrician	Maintenance Plumber	Maintenance Carpenter	Maintenance Painter	General Maintenance Worker	Trade Shop Supervisor
Boston, MA							
Direct Wage	39.78	39.72	39.88	38.78	35.57	33.22	47.54
Markup	79.55	79.43	79.76	77.57	71.14	66.44	95.07
Total Shop Rate	119.33	119.15	119.64	116.35	106.71	99.66	142.61
Boulder, CO							
Direct Wage	30.44	30.40	30.53	29.69	27.23	25.43	36.39
Markup	60.89	60.80	61.05	59.37	54.45	50.85	72.77
Total Shop Rate	91.33	91.20	91.58	89.06	81.68	76.28	109.16
Bowling Green, KY							
Direct Wage	26.45	26.41	26.52	25.79	23.66	22.09	31.62
Markup	52.91	52.83	53.05	51.59	47.32	44.19	63.23
Total Shop Rate	79.36	79.24	79.57	77.38	70.97	66.28	94.85
Brockton, MA							
Direct Wage	32.78	32.73	32.87	31.96	29.32	27.38	39.18
Markup	65.56	65.46	65.74	63.93	58.63	54.76	78.36
Total Shop Rate	98.34	98.20	98.61	95.89	87.95	82.13	117.53
Buffalo, NY							
Direct Wage	31.75	31.70	31.83	30.96	28.39	26.51	37.94
Markup	63.49	63.40	63.66	61.91	56.78	53.03	75.88
Total Shop Rate	95.24	95.10	95.49	92.87	85.17	79.54	113.83
Burlington, VT							
Direct Wage	22.77	22.73	22.83	22.20	20.36	19.01	27.21
Markup	45.53	45.46	45.65	44.40	40.72	38.03	54.42
Total Shop Rate	68.30	68.20	68.48	66.59	61.08	57.04	81.63
Camden, NJ							
Direct Wage	40.61	40.55	40.72	39.60	36.32	33.92	48.54
Markup	81.23	81.11	81.44	79.20	72.64	67.84	97.08
Total Shop Rate	121.84	121.66	122.17	118.80	108.96	101.76	145.62
Carson City, NV							
Direct Wage	28.65	28.61	28.73	27.94	25.62	23.93	34.24
Markup	57.31	57.22	57.46	55.88	51.25	47.86	68.49
Total Shop Rate	85.96	85.83	86.19	83.82	76.87	71.79	102.73
Cedar Rapids, IA							
Direct Wage	28.09	28.05	28.17	27.39	25.12	23.46	33.57
Markup	56.18	56.10	56.33	54.78	50.24	46.92	67.15
Total Shop Rate	84.27	84.15	84.50	82.17	75.36	70.38	100.72
Charleston, SC							
Direct Wage	11.04	11.02	11.07	10.77	9.87	9.22	13.20
Markup	22.08	22.05	22.14	21.53	19.75	18.44	26.39
Total Shop Rate	33.12	33.07	33.21	32.30	29.62	27.66	39.59
Charleston, WV							
Direct Wage	28.29	28.25	28.37	27.59	25.30	23.63	33.82
Markup	56.59	56.50	56.74	55.18	50.61	47.26	67.63
Total Shop Rate	84.88	84.76	85.11	82.77	75.91	70.89	101.45

Note: Total shop rates are expressed as costs per hour. Shop markup includes all benefits, non-productive time, support, and supervision.

Area	HVAC Technician	Maintenance Electrician	Maintenance Plumber	Maintenance Carpenter	Maintenance Painter	General Maintenance Worker	Trade Shop Supervisor
Charlotte, NC							
Direct Wage	11.90	11.89	11.94	11.61	10.65	9.94	14.23
Markup	23.81	23.77	23.87	23.21	21.29	19.88	28.45
Total Shop Rate	35.71	35.66	35.81	34.82	31.94	29.83	42.68
Chattanooga, TN							
Direct Wage	18.22	18.20	18.27	17.77	16.30	15.22	21.78
Markup	36.45	36.39	36.54	35.54	32.59	30.44	43.56
Total Shop Rate	54.67	54.59	54.81	53.30	48.89	45.66	65.34
Cheyenne, WY							
Direct Wage	17.59	17.56	17.63	17.15	15.73	14.69	21.02
Markup	35.18	35.12	35.27	34.30	31.46	29.38	42.04
Total Shop Rate	52.76	52.68	52.90	51.45	47.18	44.07	63.06
Chicago, IL							
Direct Wage	39.14	39.08	39.24	38.16	35.00	32.69	46.78
Markup	78.28	78.16	78.49	76.33	70.00	65.38	93.56
Total Shop Rate	117.42	117.24	117.73	114.49	105.01	98.07	140.33
Cincinnati, OH							
Direct Wage	26.57	26.53	26.64	25.91	23.76	22.19	31.75
Markup	53.14	53.06	53.28	51.81	47.52	44.38	63.51
Total Shop Rate	79.71	79.59	79.92	77.72	71.28	66.57	95.26
Cleveland, OH							
Direct Wage	34.49	34.44	34.58	33.63	30.84	28.81	41.22
Markup	68.98	68.88	69.17	67.26	61.69	57.61	82.44
Total Shop Rate	103.47	103.32	103.75	100.89	92.53	86.42	123.66
Colorado Springs, CO							
Direct Wage	29.70	29.66	29.78	28.96	26.56	24.81	35.50
Markup	59.41	59.32	59.56	57.93	53.13	49.61	71.00
Total Shop Rate	89.11	88.98	89.35	86.89	79.69	74.42	106.50
Columbia, SC							
Direct Wage	11.64	11.63	11.67	11.35	10.41	9.72	13.92
Markup	23.29	23.25	23.35	22.71	20.83	19.45	27.83
Total Shop Rate	34.93	34.88	35.02	34.06	31.24	29.17	41.75
Columbus, GA							
Direct Wage	12.69	12.67	12.73	12.38	11.35	10.60	15.17
Markup	25.39	25.35	25.46	24.75	22.70	21.20	30.34
Total Shop Rate	38.08	38.02	38.18	37.13	34.06	31.80	45.51
Columbus, OH							
Direct Wage	28.53	28.49	28.61	27.82	25.51	23.83	34.10
Markup	57.06	56.98	57.21	55.64	51.03	47.66	68.20
Total Shop Rate	85.59	85.46	85.82	83.46	76.54	71.48	102.29
Concord, NH							
Direct Wage	24.91	24.87	24.97	24.29	22.27	20.80	29.77
Markup	49.82	49.74	49.95	48.57	44.55	41.60	59.54
Total Shop Rate	74.72	74.61	74.92	72.86	66.82	62.41	89.30

Note: Total shop rates are expressed as costs per hour. Shop markup includes all benefits, non-productive time, support, and supervision.

Area	HVAC Technician	Maintenance Electrician	Maintenance Plumber	Maintenance Carpenter	Maintenance Painter	General Maintenance Worker	Trade Shop Supervisor
Corpus Christi, TX							
Direct Wage	17.66	17.63	17.71	17.22	15.79	14.75	21.11
Markup	35.32	35.27	35.42	34.44	31.59	29.50	42.21
Total Shop Rate	52.98	52.90	53.12	51.66	47.38	44.25	63.32
Cumberland, MD							
Direct Wage	26.73	26.69	26.80	26.06	23.91	22.33	31.95
Markup	53.46	53.38	53.61	52.13	47.81	44.65	63.90
Total Shop Rate	80.19	80.07	80.41	78.19	71.72	66.98	95.84
Dallas, TX							
Direct Wage	21.34	21.31	21.40	20.81	19.08	17.82	25.51
Markup	42.68	42.62	42.80	41.62	38.17	35.65	51.01
Total Shop Rate	64.02	63.93	64.19	62.43	57.25	53.47	76.52
Danbury, CT							
Direct Wage	34.86	34.80	34.95	33.99	31.17	29.11	41.66
Markup	69.72	69.61	69.90	67.98	62.34	58.22	83.32
Total Shop Rate	104.57	104.41	104.85	101.96	93.52	87.34	124.98
Davenport, IA							
Direct Wage	30.49	30.45	30.57	29.73	27.27	25.47	36.44
Markup	60.99	60.89	61.15	59.47	54.54	50.93	72.89
Total Shop Rate	91.48	91.34	91.72	89.20	81.81	76.40	109.33
Dayton, OH							
Direct Wage	26.43	26.39	26.50	25.77	23.64	22.07	31.59
Markup	52.86	52.78	53.00	51.54	47.27	44.15	63.18
Total Shop Rate	79.29	79.17	79.50	77.31	70.91	66.22	94.76
Daytona Beach, FL							
Direct Wage	20.59	20.56	20.65	20.08	18.41	17.20	24.61
Markup	41.18	41.12	41.29	40.16	36.83	34.40	49.22
Total Shop Rate	61.78	61.68	61.94	60.24	55.24	51.59	73.83
Denver, CO							
Direct Wage	30.00	29.95	30.08	29.25	26.83	25.05	35.85
Markup	59.99	59.90	60.15	58.50	53.65	50.10	71.70
Total Shop Rate	89.99	89.85	90.23	87.75	80.48	75.16	107.55
Des Moines, IA							
Direct Wage	28.47	28.43	28.55	27.76	25.46	23.78	34.03
Markup	56.95	56.86	57.10	55.53	50.93	47.56	68.06
Total Shop Rate	85.42	85.29	85.65	83.29	76.39	71.34	102.09
Detroit, MI							
Direct Wage	38.11	38.06	38.22	37.16	34.08	31.83	45.55
Markup	76.23	76.11	76.43	74.33	68.17	63.66	91.10
Total Shop Rate	114.34	114.17	114.65	111.49	102.25	95.50	136.66
Duluth, MN							
Direct Wage	31.82	31.77	31.91	31.03	28.46	26.58	38.03
Markup	63.64	63.54	63.81	62.05	56.91	53.15	76.06
Total Shop Rate	95.46	95.32	95.72	93.08	85.37	79.73	114.09

Note: Total shop rates are expressed as costs per hour. Shop markup includes all benefits, non-productive time, support, and supervision.

Area	HVAC Technician	Maintenance Electrician	Maintenance Plumber	Maintenance Carpenter	Maintenance Painter	General Maintenance Worker	Trade Shop Supervisor
Eau Claire, WI							
Direct Wage	29.11	29.07	29.19	28.38	26.03	24.31	34.79
Markup	58.22	58.13	58.37	56.77	52.06	48.62	69.58
Total Shop Rate	87.33	87.20	87.56	85.15	78.09	72.93	104.37
El Paso, TX							
Direct Wage	18.78	18.76	18.83	18.32	16.80	15.69	22.45
Markup	37.57	37.51	37.67	36.63	33.60	31.38	44.90
Total Shop Rate	56.35	56.27	56.50	54.95	50.40	47.06	67.35
Erie, PA							
Direct Wage	27.64	27.60	27.72	26.95	24.72	23.09	33.04
Markup	55.29	55.20	55.43	53.91	49.44	46.17	66.07
Total Shop Rate	82.93	82.80	83.15	80.86	74.16	69.26	99.11
Eugene, OR							
Direct Wage	33.40	33.35	33.49	32.57	29.87	27.89	39.92
Markup	66.80	66.70	66.98	65.13	59.74	55.79	79.83
Total Shop Rate	100.20	100.05	100.47	97.70	89.61	83.68	119.75
Evansville, IN							
Direct Wage	29.03	28.98	29.10	28.30	25.96	24.24	34.69
Markup	58.06	57.97	58.21	56.61	51.92	48.49	69.38
Total Shop Rate	87.08	86.95	87.31	84.91	77.88	72.73	104.08
Fairbanks, AK							
Direct Wage	37.08	37.02	37.18	36.16	33.16	30.97	44.32
Markup	74.16	74.05	74.36	72.31	66.32	61.94	88.63
Total Shop Rate	111.24	111.07	111.54	108.47	99.48	92.91	132.95
Fall River, MA							
Direct Wage	32.78	32.73	32.87	31.96	29.32	27.38	39.18
Markup	65.56	65.46	65.74	63.93	58.63	54.76	78.36
Total Shop Rate	98.34	98.20	98.61	95.89	87.95	82.13	117.53
Fargo, ND							
Direct Wage	23.82	23.78	23.88	23.22	21.30	19.89	28.46
Markup	47.63	47.56	47.76	46.44	42.60	39.78	56.93
Total Shop Rate	71.45	71.34	71.64	69.67	63.90	59.67	85.39
Flint, MI							
Direct Wage	29.83	29.78	29.90	29.08	26.67	24.91	35.65
Markup	59.65	59.56	59.81	58.16	53.34	49.82	71.29
Total Shop Rate	89.48	89.34	89.71	87.25	80.02	74.73	106.94
Fort Lauderdale, FL							
Direct Wage	24.29	24.25	24.35	23.68	21.72	20.29	29.03
Markup	48.58	48.50	48.71	47.37	43.44	40.57	58.06
Total Shop Rate	72.87	72.76	73.06	71.05	65.16	60.86	87.08
Fort Smith, AR							
Direct Wage	21.41	21.37	21.46	20.87	19.14	17.88	25.58
Markup	42.81	42.75	42.93	41.74	38.29	35.76	51.17
Total Shop Rate	64.22	64.12	64.39	62.62	57.43	53.63	76.75

Note: Total shop rates are expressed as costs per hour. Shop markup includes all benefits, non-productive time, support, and supervision.

Area	HVAC Technician	Maintenance Electrician	Maintenance Plumber	Maintenance Carpenter	Maintenance Painter	General Maintenance Worker	Trade Shop Supervisor
Fort Worth, TX							
Direct Wage	20.58	20.54	20.63	20.06	18.40	17.18	24.59
Markup	41.15	41.09	41.26	40.13	36.80	34.37	49.18
Total Shop Rate	61.73	61.63	61.89	60.19	55.20	51.55	73.77
Fresno, CA							
Direct Wage	29.90	29.85	29.98	29.15	26.74	24.97	35.73
Markup	59.80	59.71	59.96	58.31	53.48	49.94	71.47
Total Shop Rate	89.70	89.56	89.94	87.46	80.21	74.91	107.20
Gary, IN							
Direct Wage	35.60	35.54	35.69	34.71	31.83	29.73	42.54
Markup	71.20	71.09	71.39	69.42	63.67	59.46	85.09
Total Shop Rate	106.80	106.63	107.08	104.13	95.50	89.19	127.63
Grand Rapids, MI							
Direct Wage	20.56	20.53	20.61	20.05	18.39	17.17	24.57
Markup	41.12	41.06	41.23	40.09	36.77	34.34	49.14
Total Shop Rate	61.68	61.59	61.84	60.14	55.16	51.51	73.71
Great Falls, MT							
Direct Wage	25.59	25.55	25.66	24.95	22.89	21.37	30.59
Markup	51.18	51.11	51.32	49.91	45.77	42.75	61.17
Total Shop Rate	76.77	76.66	76.98	74.86	68.66	64.12	91.76
Green Bay, WI							
Direct Wage	29.00	28.95	29.07	28.27	25.93	24.22	34.65
Markup	57.99	57.90	58.14	56.54	51.86	48.43	69.31
Total Shop Rate	86.99	86.85	87.22	84.82	77.79	72.65	103.96
Greensboro, NC							
Direct Wage	12.69	12.67	12.72	12.37	11.34	10.59	15.16
Markup	25.37	25.33	25.44	24.74	22.69	21.19	30.32
Total Shop Rate	38.06	38.00	38.16	37.11	34.03	31.78	45.48
Hampton, VA							
Direct Wage	19.97	19.94	20.03	19.48	17.86	16.68	23.87
Markup	39.95	39.89	40.05	38.95	35.72	33.36	47.74
Total Shop Rate	59.92	59.83	60.08	58.43	53.58	50.04	71.61
Harrisburg, PA							
Direct Wage	27.64	27.60	27.72	26.95	24.72	23.09	33.04
Markup	55.29	55.20	55.43	53.91	49.44	46.17	66.07
Total Shop Rate	82.93	82.80	83.15	80.86	74.16	69.26	99.11
Hartford, CT							
Direct Wage	33.97	33.92	34.06	33.12	30.38	28.37	40.60
Markup	67.94	67.84	68.12	66.25	60.76	56.74	81.20
Total Shop Rate	101.91	101.76	102.18	99.37	91.14	85.11	121.80
Hilo, HI							
Direct Wage	40.22	40.15	40.32	39.21	35.96	33.59	48.06
Markup	80.43	80.31	80.64	78.42	71.93	67.17	96.12
Total Shop Rate	120.65	120.46	120.97	117.64	107.89	100.76	144.19

Note: Total shop rates are expressed as costs per hour. Shop markup includes all benefits, non-productive time, support, and supervision.

Area	HVAC Technician	Maintenance Electrician	Maintenance Plumber	Maintenance Carpenter	Maintenance Painter	General Maintenance Worker	Trade Shop Supervisor
Honolulu, HI							
Direct Wage	40.22	40.15	40.32	39.21	35.96	33.59	48.06
Markup	80.43	80.31	80.64	78.42	71.93	67.17	96.12
Total Shop Rate	120.65	120.46	120.97	117.64	107.89	100.76	144.19
Houston, TX							
Direct Wage	22.64	22.60	22.70	22.07	20.24	18.90	27.05
Markup	45.27	45.20	45.39	44.14	40.49	37.81	54.11
Total Shop Rate	67.91	67.80	68.09	66.21	60.73	56.71	81.16
Huntsville, AL							
Direct Wage	21.92	21.89	21.98	21.37	19.60	18.31	26.20
Markup	43.84	43.77	43.96	42.75	39.20	36.61	52.39
Total Shop Rate	65.76	65.66	65.93	64.12	58.81	54.92	78.59
Indianapolis, IN							
Direct Wage	30.07	30.02	30.15	29.32	26.89	25.11	35.94
Markup	60.14	60.05	60.30	58.64	53.78	50.23	71.87
Total Shop Rate	90.21	90.07	90.45	87.96	80.67	75.34	107.81
Jackson, MS							
Direct Wage	12.70	12.68	12.74	12.39	11.36	10.61	15.18
Markup	25.40	25.37	25.47	24.77	22.72	21.22	30.36
Total Shop Rate	38.11	38.05	38.21	37.16	34.08	31.83	45.54
Jacksonville, FL							
Direct Wage	20.80	20.77	20.86	20.28	18.60	17.37	24.86
Markup	41.61	41.54	41.72	40.57	37.21	34.75	49.73
Total Shop Rate	62.41	62.32	62.58	60.85	55.81	52.12	74.59
Jersey City, NJ							
Direct Wage	42.67	42.60	42.78	41.60	38.16	35.63	50.99
Markup	85.33	85.20	85.56	83.20	76.31	71.27	101.98
Total Shop Rate	128.00	127.80	128.34	124.81	114.47	106.90	152.97
Juneau, AK							
Direct Wage	37.08	37.02	37.18	36.16	33.16	30.97	44.32
Markup	74.16	74.05	74.36	72.31	66.32	61.94	88.63
Total Shop Rate	111.24	111.07	111.54	108.47	99.48	92.91	132.95
Kalamazoo, MI							
Direct Wage	24.09	24.06	24.16	23.49	21.55	20.12	28.79
Markup	48.19	48.11	48.31	46.98	43.09	40.24	57.59
Total Shop Rate	72.28	72.17	72.47	70.48	64.64	60.37	86.38
Kansas City, MO							
Direct Wage	36.27	36.22	36.37	35.37	32.44	30.30	43.35
Markup	72.55	72.44	72.74	70.74	64.88	60.59	86.71
Total Shop Rate	108.82	108.66	109.11	106.11	97.32	90.89	130.06
Knoxville, TN							
Direct Wage	20.62	20.59	20.68	20.11	18.44	17.23	24.65
Markup	41.25	41.19	41.36	40.22	36.89	34.45	49.30
Total Shop Rate	61.87	61.78	62.04	60.33	55.33	51.68	73.95

Note: Total shop rates are expressed as costs per hour. Shop markup includes all benefits, non-productive time, support, and supervision.

.2 In-House Shop Rates, Selected Metro Areas

Area	HVAC Technician	Maintenance Electrician	Maintenance Plumber	Maintenance Carpenter	Maintenance Painter	General Maintenance Worker	Trade Shop Supervisor
Kokomo, IN							
Direct Wage	28.26	28.22	28.34	27.56	25.27	23.60	33.78
Markup	56.52	56.44	56.67	55.11	50.55	47.21	67.55
Total Shop Rate	84.79	84.66	85.01	82.67	75.82	70.81	101.33
Lansing, MI							
Direct Wage	29.58	29.54	29.66	28.84	26.45	24.71	35.35
Markup	59.16	59.07	59.32	57.69	52.91	49.41	70.71
Total Shop Rate	88.74	88.61	88.98	86.53	79.36	74.12	106.06
Las Cruces, NM							
Direct Wage	20.47	20.44	20.52	19.96	18.31	17.10	24.46
Markup	40.94	40.88	41.05	39.92	36.61	34.19	48.93
Total Shop Rate	61.41	61.32	61.57	59.88	54.92	51.29	73.39
Las Vegas, NV							
Direct Wage	34.35	34.30	34.44	33.50	30.72	28.69	41.06
Markup	68.71	68.60	68.89	66.99	61.44	57.38	82.11
Total Shop Rate	103.06	102.90	103.33	100.49	92.16	86.07	123.17
Lawton, OK							
Direct Wage	22.81	22.77	22.87	22.24	20.40	19.05	27.26
Markup	45.61	45.54	45.73	44.48	40.79	38.10	54.51
Total Shop Rate	68.42	68.32	68.60	66.71	61.19	57.14	81.77
Lewiston, ME							
Direct Wage	27.30	27.26	27.37	26.62	24.41	22.80	32.63
Markup	54.60	54.52	54.75	53.24	48.83	45.60	65.26
Total Shop Rate	81.90	81.78	82.12	79.86	73.24	68.40	97.89
Lexington, KY							
Direct Wage	16.61	16.59	16.65	16.20	14.85	13.87	19.85
Markup	33.22	33.17	33.31	32.39	29.71	27.75	39.70
Total Shop Rate	49.83	49.76	49.96	48.59	44.56	41.62	59.56
Little Rock, AR							
Direct Wage	23.28	23.24	23.34	22.70	20.82	19.44	27.82
Markup	46.56	46.49	46.68	45.40	41.64	38.88	55.64
Total Shop Rate	69.84	69.73	70.02	68.10	62.45	58.33	83.46
Los Angeles, CA							
Direct Wage	35.14	35.09	35.24	34.27	31.43	29.35	42.00
Markup	70.29	70.18	70.47	68.53	62.85	58.70	84.00
Total Shop Rate	105.43	105.27	105.71	102.80	94.28	88.05	126.00
Louisville, KY							
Direct Wage	26.45	26.41	26.52	25.79	23.66	22.09	31.62
Markup	52.91	52.83	53.05	51.59	47.32	44.19	63.23
Total Shop Rate	79.36	79.24	79.57	77.38	70.97	66.28	94.85
Lowell, MA							
Direct Wage	40.79	40.73	40.90	39.78	36.48	34.07	48.75
Markup	81.59	81.46	81.80	79.55	72.96	68.14	97.51
Total Shop Rate	122.38	122.20	122.71	119.33	109.44	102.21	146.26

Note: Total shop rates are expressed as costs per hour. Shop markup includes all benefits, non-productive time, support, and supervision.

3.2 In-House Shop Rates, Selected Metro Areas

Area	HVAC Technician	Maintenance Electrician	Maintenance Plumber	Maintenance Carpenter	Maintenance Painter	General Maintenance Worker	Trade Shop Supervisor
Lubbock, TX							
Direct Wage	15.49	15.46	15.53	15.10	13.85	12.93	18.51
Markup	30.97	30.93	31.06	30.20	27.70	25.87	37.02
Total Shop Rate	46.46	46.39	46.58	45.30	41.55	38.80	55.53
Macon, GA							
Direct Wage	19.37	19.34	19.42	18.89	17.32	16.18	23.15
Markup	38.74	38.68	38.84	37.78	34.65	32.36	46.30
Total Shop Rate	58.11	58.02	58.27	56.66	51.97	48.53	69.45
Madison, WI							
Direct Wage	29.13	29.09	29.21	28.41	26.05	24.33	34.82
Markup	58.27	58.18	58.42	56.81	52.11	48.66	69.64
Total Shop Rate	87.40	87.27	87.63	85.22	78.16	72.99	104.45
Manchester, NH							
Direct Wage	24.91	24.87	24.97	24.29	22.27	20.80	29.77
Markup	49.82	49.74	49.95	48.57	44.55	41.60	59.54
Total Shop Rate	74.72	74.61	74.92	72.86	66.82	62.41	89.30
Marquette, MI							
Direct Wage	26.48	26.44	26.55	25.82	23.68	22.11	31.65
Markup	52.96	52.88	53.10	51.64	47.36	44.23	63.29
Total Shop Rate	79.44	79.32	79.65	77.46	71.04	66.34	94.94
Medford, OR							
Direct Wage	29.97	29.93	30.05	29.22	26.80	25.03	35.82
Markup	59.94	59.85	60.10	58.45	53.61	50.06	71.64
Total Shop Rate	89.92	89.78	90.16	87.67	80.41	75.10	107.46
Memphis, TN							
Direct Wage	25.80	25.76	25.87	25.16	23.08	21.55	30.84
Markup	51.61	51.53	51.74	50.32	46.15	43.10	61.68
Total Shop Rate	77.41	77.29	77.62	75.48	69.23	64.65	92.51
Miami, FL							
Direct Wage	23.69	23.66	23.76	23.10	21.19	19.79	28.32
Markup	47.39	47.32	47.51	46.21	42.38	39.58	56.64
Total Shop Rate	71.08	70.98	71.27	69.31	63.57	59.37	84.95
Milwaukee, WI							
Direct Wage	33.57	33.52	33.66	32.73	30.02	28.04	40.12
Markup	67.14	67.04	67.32	65.47	60.04	56.08	80.24
Total Shop Rate	100.71	100.56	100.98	98.20	90.07	84.11	120.37
Minneapolis, MN							
Direct Wage	36.03	35.98	36.13	35.13	32.22	30.09	43.06
Markup	72.06	71.95	72.25	70.26	64.44	60.18	86.12
Total Shop Rate	108.09	107.93	108.38	105.39	96.66	90.27	129.18
Mobile, AL							
Direct Wage	16.68	16.65	16.72	16.26	14.91	13.93	19.93
Markup	33.35	33.30	33.44	32.52	29.83	27.85	39.86
Total Shop Rate	50.03	49.95	50.16	48.78	44.74	41.78	59.79

Note: Total shop rates are expressed as costs per hour. Shop markup includes all benefits, non-productive time, support, and supervision.

3.2 In-House Shop Rates, Selected Metro Areas

Area	HVAC Technician	Maintenance Electrician	Maintenance Plumber	Maintenance Carpenter	Maintenance Painter	General Maintenance Worker	Trade Shop Supervisor
Moline, IL							
Direct Wage	28.38	28.34	28.46	27.68	25.38	23.71	33.92
Markup	56.77	56.68	56.92	55.35	50.77	47.41	67.85
Total Shop Rate	85.15	85.02	85.38	83.03	76.15	71.12	101.77
Muncie, IN							
Direct Wage	26.50	26.46	26.57	25.84	23.70	22.13	31.68
Markup	53.01	52.93	53.15	51.68	47.40	44.27	63.35
Total Shop Rate	79.51	79.39	79.72	77.53	71.10	66.40	95.03
Nashville, TN							
Direct Wage	22.99	22.95	23.05	22.41	20.56	19.20	27.47
Markup	45.97	45.90	46.09	44.83	41.11	38.39	54.94
Total Shop Rate	68.96	68.85	69.14	67.24	61.67	57.59	82.41
New Brunswick, NJ							
Direct Wage	41.27	41.20	41.38	40.24	36.90	34.46	49.32
Markup	82.53	82.41	82.75	80.47	73.81	68.93	98.64
Total Shop Rate	123.80	123.61	124.13	120.71	110.71	103.39	147.95
New Haven, CT							
Direct Wage	35.73	35.67	35.82	34.84	31.95	29.84	42.70
Markup	71.46	71.35	71.65	69.68	63.90	59.68	85.40
Total Shop Rate	107.19	107.02	107.47	104.51	95.85	89.52	128.10
New Orleans, LA							
Direct Wage	22.51	22.48	22.57	21.95	20.13	18.80	26.91
Markup	45.03	44.96	45.15	43.90	40.27	37.61	53.81
Total Shop Rate	67.54	67.44	67.72	65.86	60.40	56.41	80.72
New York, NY							
Direct Wage	57.38	57.29	57.53	55.95	51.31	47.92	68.58
Markup	114.76	114.59	115.06	111.90	102.63	95.84	137.15
Total Shop Rate	172.14	171.88	172.60	167.85	153.94	143.76	205.73
Newark, NJ							
Direct Wage	42.67	42.60	42.78	41.60	38.16	35.63	50.99
Markup	85.33	85.20	85.56	83.20	76.31	71.27	101.98
Total Shop Rate	128.00	127.80	128.34	124.81	114.47	106.90	152.97
Newport News, VA							
Direct Wage	19.97	19.94	20.03	19.48	17.86	16.68	23.87
Markup	39.95	39.89	40.05	38.95	35.72	33.36	47.74
Total Shop Rate	59.92	59.83	60.08	58.43	53.58	50.04	71.61
Norfolk, VA							
Direct Wage	20.66	20.63	20.71	20.14	18.47	17.25	24.69
Markup	41.31	41.25	41.42	40.28	36.95	34.50	49.38
Total Shop Rate	61.97	61.88	62.14	60.43	55.42	51.76	74.06
Norwalk, CT							
Direct Wage	35.73	35.67	35.82	34.84	31.95	29.84	42.70
Markup	71.46	71.35	71.65	69.68	63.90	59.68	85.40
Total Shop Rate	107.19	107.02	107.47	104.51	95.85	89.52	128.10

Note: Total shop rates are expressed as costs per hour. Shop markup includes all benefits, non-productive time, support, and supervision.

Area	HVAC Technician	Maintenance Electrician	Maintenance Plumber	Maintenance Carpenter	Maintenance Painter	General Maintenance Worker	Trade Shop Supervisor
Norwich, CT							
Direct Wage	35.73	35.67	35.82	34.84	31.95	29.84	42.70
Markup	71.46	71.35	71.65	69.68	63.90	59.68	85.40
Total Shop Rate	107.19	107.02	107.47	104.51	95.85	89.52	128.10
Oakland, CA							
Direct Wage	43.71	43.64	43.82	42.62	39.09	36.50	52.24
Markup	87.42	87.28	87.65	85.24	78.17	73.01	104.47
Total Shop Rate	131.13	130.93	131.47	127.85	117.26	109.51	156.71
Ogden, UT							
Direct Wage	25.49	25.46	25.56	24.86	22.80	21.29	30.47
Markup	50.99	50.91	51.12	49.72	45.60	42.58	60.94
Total Shop Rate	76.48	76.37	76.68	74.57	68.40	63.88	91.41
Oklahoma City, OK							
Direct Wage	22.81	22.77	22.87	22.24	20.40	19.05	27.26
Markup	45.61	45.54	45.73	44.48	40.79	38.10	54.51
Total Shop Rate	68.42	68.32	68.60	66.71	61.19	57.14	81.77
Olympia, WA							
Direct Wage	33.94	33.89	34.03	33.09	30.35	28.34	40.56
Markup	67.88	67.77	68.06	66.18	60.70	56.69	81.12
Total Shop Rate	101.81	101.66	102.08	99.27	91.05	85.03	121.68
Omaha, NE							
Direct Wage	27.82	27.78	27.90	27.13	24.88	23.24	33.25
Markup	55.65	55.56	55.79	54.26	49.76	46.47	66.50
Total Shop Rate	83.47	83.34	83.69	81.39	74.64	69.71	99.75
Orlando, FL							
Direct Wage	14.36	14.34	14.40	14.00	12.84	12.00	17.17
Markup	28.73	28.68	28.80	28.01	25.69	23.99	34.33
Total Shop Rate	43.09	43.02	43.20	42.01	38.53	35.99	51.50
Owensboro, KY							
Direct Wage	26.09	26.05	26.16	25.44	23.33	21.79	31.18
Markup	52.18	52.10	52.32	50.88	46.66	43.58	62.36
Total Shop Rate	78.27	78.15	78.47	76.31	69.99	65.36	93.54
Oxnard, CA							
Direct Wage	34.48	34.43	34.57	33.62	30.84	28.80	41.21
Markup	68.97	68.86	69.15	67.25	61.67	57.60	82.42
Total Shop Rate	103.45	103.29	103.72	100.87	92.51	86.40	123.64
Parkersburg, WV							
Direct Wage	30.58	30.54	30.66	29.82	27.35	25.54	36.55
Markup	61.17	61.07	61.33	59.64	54.70	51.08	73.10
Total Shop Rate	91.75	91.61	91.99	89.46	82.05	76.63	109.65
Peoria, IL							
Direct Wage	28.93	28.89	29.01	28.21	25.87	24.16	34.58
Markup	57.86	57.77	58.01	56.42	51.74	48.32	69.15
Total Shop Rate	86.79	86.66	87.02	84.63	77.61	72.48	103.73

Note: Total shop rates are expressed as costs per hour. Shop markup includes all benefits, non-productive time, support, and supervision.

Area	HVAC Technician	Maintenance Electrician	Maintenance Plumber	Maintenance Carpenter	Maintenance Painter	General Maintenance Worker	Trade Shop Supervisor
Philadelphia, PA							
Direct Wage	42.09	42.02	42.20	41.04	37.64	35.15	50.30
Markup	84.18	84.05	84.40	82.08	75.28	70.30	100.60
Total Shop Rate	126.26	126.07	126.60	123.11	112.91	105.45	150.90
Phoenix, AZ							
Direct Wage	22.33	22.29	22.39	21.77	19.97	18.65	26.68
Markup	44.65	44.59	44.77	43.54	39.93	37.29	53.37
Total Shop Rate	66.98	66.88	67.16	65.31	59.90	55.94	80.05
Pittsburgh, PA							
Direct Wage	33.79	33.74	33.88	32.95	30.22	28.22	40.38
Markup	67.58	67.48	67.76	65.90	60.44	56.44	80.77
Total Shop Rate	101.37	101.22	101.64	98.84	90.65	84.66	121.15
Pocatello, ID							
Direct Wage	25.26	25.22	25.32	24.63	22.59	21.09	30.19
Markup	50.52	50.44	50.65	49.26	45.17	42.19	60.37
Total Shop Rate	75.77	75.66	75.97	73.88	67.76	63.28	90.56
Portland, ME							
Direct Wage	27.30	27.26	27.37	26.62	24.41	22.80	32.63
Markup	54.60	54.52	54.75	53.24	48.83	45.60	65.26
Total Shop Rate	81.90	81.78	82.12	79.86	73.24	68.40	97.89
Portland, OR							
Direct Wage	36.18	36.12	36.27	35.27	32.35	30.21	43.24
Markup	72.35	72.24	72.55	70.55	64.70	60.43	86.47
Total Shop Rate	108.53	108.37	108.82	105.82	97.06	90.64	129.71
Providence, RI							
Direct Wage	32.21	32.16	32.30	31.41	28.81	26.90	38.50
Markup	64.42	64.33	64.59	62.82	57.61	53.80	76.99
Total Shop Rate	96.63	96.49	96.89	94.22	86.42	80.71	115.49
Pueblo, CO							
Direct Wage	27.02	26.98	27.09	26.34	24.16	22.56	32.29
Markup	54.03	53.95	54.18	52.69	48.32	45.13	64.58
Total Shop Rate	81.05	80.93	81.26	79.03	72.48	67.69	96.86
Raleigh-Durham, NC							
Direct Wage	12.69	12.67	12.72	12.37	11.34	10.59	15.16
Markup	25.37	25.33	25.44	24.74	22.69	21.19	30.32
Total Shop Rate	38.06	38.00	38.16	37.11	34.03	31.78	45.48
Rapid City, SD							
Direct Wage	17.45	17.42	17.50	17.01	15.60	14.57	20.85
Markup	34.90	34.85	34.99	34.03	31.21	29.15	41.71
Total Shop Rate	52.35	52.27	52.49	51.04	46.81	43.72	62.56
Reading, PA							
Direct Wage	26.59	26.55	26.66	25.93	23.78	22.21	31.78
Markup	53.19	53.11	53.33	51.86	47.56	44.42	63.56
Total Shop Rate	79.78	79.66	79.99	77.79	71.34	66.63	95.35

Note: Total shop rates are expressed as costs per hour. Shop markup includes all benefits, non-productive time, support, and supervision.

Area	HVAC Technician	Maintenance Electrician	Maintenance Plumber	Maintenance Carpenter	Maintenance Painter	General Maintenance Worker	Trade Shop Supervisor
Reno, NV							
Direct Wage	28.65	28.61	28.73	27.94	25.62	23.93	34.24
Markup	57.31	57.22	57.46	55.88	51.25	47.86	68.49
Total Shop Rate	85.96	85.83	86.19	83.82	76.87	71.79	102.73
Richland, WA							
Direct Wage	30.30	30.25	30.38	29.54	27.09	25.30	36.21
Markup	60.60	60.50	60.76	59.08	54.19	50.61	72.42
Total Shop Rate	90.89	90.76	91.14	88.63	81.28	75.91	108.63
Richmond, VA							
Direct Wage	23.78	23.74	23.84	23.18	21.26	19.86	28.42
Markup	47.55	47.48	47.68	46.37	42.52	39.71	56.83
Total Shop Rate	71.33	71.22	71.52	69.55	63.79	59.57	85.25
Riverside, CA							
Direct Wage	33.21	33.16	33.30	32.38	29.70	27.74	39.69
Markup	66.43	66.33	66.60	64.77	59.40	55.48	79.39
Total Shop Rate	99.64	99.49	99.90	97.15	89.10	83.21	119.08
Roanoke, VA							
Direct Wage	11.93	11.91	11.96	11.63	10.67	9.96	14.26
Markup	23.86	23.82	23.92	23.26	21.33	19.92	28.51
Total Shop Rate	35.79	35.73	35.88	34.89	32.00	29.89	42.77
Rochester, MN							
Direct Wage	29.88	29.84	29.96	29.14	26.72	24.96	35.71
Markup	59.77	59.67	59.92	58.27	53.45	49.91	71.43
Total Shop Rate	89.65	89.51	89.89	87.41	80.17	74.87	107.14
Rochester, NY							
Direct Wage	29.97	29.93	30.05	29.22	26.80	25.03	35.82
Markup	59.94	59.85	60.10	58.45	53.61	50.06	71.64
Total Shop Rate	89.92	89.78	90.16	87.67	80.41	75.10	107.46
Rockford, IL							
Direct Wage	35.89	35.84	35.99	35.00	32.10	29.98	42.90
Markup	71.78	71.67	71.97	69.99	64.19	59.95	85.79
Total Shop Rate	107.68	107.51	107.96	104.99	96.29	89.93	128.69
Rutland, VT							
Direct Wage	22.77	22.73	22.83	22.20	20.36	19.01	27.21
Markup	45.53	45.46	45.65	44.40	40.72	38.03	54.42
Total Shop Rate	68.30	68.20	68.48	66.59	61.08	57.04	81.63
Sacramento, CA							
Direct Wage	33.41	33.36	33.50	32.58	29.88	27.90	39.93
Markup	66.82	66.72	66.99	65.15	59.75	55.80	79.85
Total Shop Rate	100.23	100.07	100.49	97.73	89.63	83.70	119.78
Saginaw, MI							
Direct Wage	27.91	27.87	27.99	27.22	24.96	23.31	33.36
Markup	55.82	55.74	55.97	54.43	49.92	46.62	66.72
Total Shop Rate	83.74	83.61	83.96	81.65	74.88	69.93	100.08

Note: Total shop rates are expressed as costs per hour. Shop markup includes all benefits, non-productive time, support, and supervision.

3.2 In-House Shop Rates, Selected Metro Areas

Area	HVAC Technician	Maintenance Electrician	Maintenance Plumber	Maintenance Carpenter	Maintenance Painter	General Maintenance Worker	Trade Shop Supervisor
Salem, OR							
Direct Wage	33.40	33.35	33.49	32.57	29.87	27.89	39.92
Markup	66.80	66.70	66.98	65.13	59.74	55.79	79.83
Total Shop Rate	100.20	100.05	100.47	97.70	89.61	83.68	119.75
Salt Lake City, UT							
Direct Wage	25.49	25.46	25.56	24.86	22.80	21.29	30.47
Markup	50.99	50.91	51.12	49.72	45.60	42.58	60.94
Total Shop Rate	76.48	76.37	76.68	74.57	68.40	63.88	91.41
San Antonio, TX							
Direct Wage	20.69	20.66	20.74	20.17	18.50	17.28	24.73
Markup	41.38	41.32	41.49	40.35	37.00	34.56	49.45
Total Shop Rate	62.07	61.98	62.23	60.52	55.51	51.84	74.18
San Diego, CA							
Direct Wage	32.10	32.05	32.18	31.30	28.70	26.81	38.36
Markup	64.19	64.10	64.37	62.59	57.41	53.61	76.72
Total Shop Rate	96.29	96.15	96.55	93.89	86.11	80.42	115.08
San Francisco, CA							
Direct Wage	48.40	48.33	48.53	47.19	43.28	40.42	57.84
Markup	96.80	96.65	97.05	94.38	86.56	80.84	115.68
Total Shop Rate	145.20	144.98	145.58	141.57	129.84	121.26	173.53
San Jose, CA							
Direct Wage	45.90	45.83	46.02	44.75	41.05	38.33	54.85
Markup	91.80	91.66	92.04	89.51	82.09	76.67	109.71
Total Shop Rate	137.70	137.49	138.06	134.26	123.14	115.00	164.56
Santa Barbara, CA							
Direct Wage	34.86	34.80	34.95	33.99	31.17	29.11	41.66
Markup	69.72	69.61	69.90	67.98	62.34	58.22	83.32
Total Shop Rate	104.57	104.41	104.85	101.96	93.52	87.34	124.98
Savannah, GA							
Direct Wage	17.07	17.05	17.12	16.65	15.27	14.26	20.41
Markup	34.15	34.10	34.24	33.30	30.54	28.52	40.81
Total Shop Rate	51.22	51.15	51.36	49.95	45.81	42.78	61.22
Scranton, PA							
Direct Wage	27.99	27.95	28.07	27.30	25.03	23.38	33.46
Markup	55.99	55.90	56.14	54.59	50.07	46.76	66.91
Total Shop Rate	83.98	83.85	84.20	81.89	75.10	70.14	100.37
Seattle, WA							
Direct Wage	38.13	38.07	38.23	37.18	34.10	31.85	45.57
Markup	76.26	76.15	76.46	74.36	68.20	63.69	91.14
Total Shop Rate	114.39	114.22	114.70	111.54	102.30	95.54	136.71
Shreveport, LA							
Direct Wage	21.68	21.64	21.73	21.13	19.38	18.10	25.90
Markup	43.35	43.28	43.47	42.27	38.77	36.20	51.81
Total Shop Rate	65.03	64.93	65.20	63.40	58.15	54.31	77.71

Note: Total shop rates are expressed as costs per hour. Shop markup includes all benefits, non-productive time, support, and supervision.

3.2 In-House Shop Rates, Selected Metro Areas

Area	HVAC Technician	Maintenance Electrician	Maintenance Plumber	Maintenance Carpenter	Maintenance Painter	General Maintenance Worker	Trade Shop Supervisor
Sioux City, IA							
Direct Wage	26.57	26.53	26.64	25.91	23.76	22.19	31.75
Markup	53.14	53.06	53.28	51.81	47.52	44.38	63.51
Total Shop Rate	79.71	79.59	79.92	77.72	71.28	66.57	95.26
Sioux Falls, SD							
Direct Wage	24.80	24.76	24.87	24.18	22.18	20.71	29.64
Markup	49.60	49.53	49.74	48.37	44.36	41.43	59.28
Total Shop Rate	74.41	74.29	74.60	72.55	66.54	62.14	88.92
South Bend, IN							
Direct Wage	28.92	28.88	29.00	28.20	25.86	24.15	34.57
Markup	57.84	57.76	58.00	56.40	51.73	48.31	69.13
Total Shop Rate	86.77	86.63	87.00	84.60	77.59	72.46	103.70
Spokane, WA							
Direct Wage	28.84	28.80	28.92	28.12	25.79	24.09	34.47
Markup	57.68	57.59	57.83	56.24	51.58	48.17	68.94
Total Shop Rate	86.52	86.39	86.75	84.36	77.37	72.26	103.40
Springfield, IL							
Direct Wage	29.04	29.00	29.12	28.32	25.97	24.26	34.71
Markup	58.09	58.00	58.24	56.64	51.95	48.51	69.42
Total Shop Rate	87.13	87.00	87.36	84.96	77.92	72.77	104.13
Springfield, MA							
Direct Wage	29.96	29.92	30.04	29.22	26.80	25.03	35.81
Markup	59.93	59.84	60.09	58.43	53.59	50.05	71.62
Total Shop Rate	89.89	89.76	90.13	87.65	80.39	75.08	107.43
Springfield, MO							
Direct Wage	23.05	23.02	23.11	22.48	20.61	19.25	27.55
Markup	46.10	46.03	46.22	44.95	41.23	38.50	55.10
Total Shop Rate	69.15	69.05	69.34	67.43	61.84	57.75	82.65
Springfield, OH							
Direct Wage	27.51	27.47	27.59	26.83	24.60	22.98	32.88
Markup	55.03	54.94	55.17	53.65	49.21	45.96	65.76
Total Shop Rate	82.54	82.41	82.76	80.48	73.81	68.93	98.65
St. Louis, MO							
Direct Wage	36.51	36.46	36.61	35.60	32.65	30.49	43.63
Markup	73.02	72.91	73.22	71.20	65.30	60.98	87.27
Total Shop Rate	109.53	109.37	109.82	106.80	97.95	91.48	130.90
Stamford, CT							
Direct Wage	45.40	45.33	45.52	44.27	40.60	37.92	54.26
Markup	90.80	90.67	91.05	88.54	81.20	75.84	108.52
Total Shop Rate	136.21	136.00	136.57	132.81	121.81	113.76	162.78
Stockton, CA							
Direct Wage	35.64	35.59	35.73	34.75	31.87	29.76	42.59
Markup	71.28	71.17	71.47	69.50	63.74	59.53	85.19
Total Shop Rate	106.92	106.76	107.20	104.25	95.61	89.29	127.78

Note: Total shop rates are expressed as costs per hour. Shop markup includes all benefits, non-productive time, support, and supervision.

Area	HVAC Technician	Maintenance Electrician	Maintenance Plumber	Maintenance Carpenter	Maintenance Painter	General Maintenance Worker	Trade Shop Supervisor
Syracuse, NY							
Direct Wage	27.77	27.72	27.84	27.07	24.83	23.19	33.18
Markup	55.53	55.45	55.68	54.15	49.66	46.38	66.37
Total Shop Rate	83.30	83.17	83.52	81.22	74.49	69.57	99.55
Tacoma, WA							
Direct Wage	33.94	33.89	34.03	33.09	30.35	28.34	40.56
Markup	67.88	67.77	68.06	66.18	60.70	56.69	81.12
Total Shop Rate	101.81	101.66	102.08	99.27	91.05	85.03	121.68
Tallahassee, FL							
Direct Wage	13.30	13.28	13.33	12.96	11.89	11.10	15.89
Markup	26.59	26.55	26.66	25.93	23.78	22.21	31.78
Total Shop Rate	39.89	39.83	40.00	38.89	35.67	33.31	47.67
Tampa, FL							
Direct Wage	15.53	15.50	15.57	15.14	13.89	12.97	18.56
Markup	31.06	31.01	31.14	30.28	27.77	25.94	37.11
Total Shop Rate	46.58	46.51	46.71	45.42	41.66	38.90	55.67
Terre Haute, IN							
Direct Wage	27.94	27.90	28.02	27.25	24.99	23.34	33.40
Markup	55.89	55.80	56.04	54.50	49.98	46.68	66.80
Total Shop Rate	83.83	83.71	84.06	81.74	74.97	70.02	100.19
Toledo, OH							
Direct Wage	32.55	32.50	32.63	31.73	29.10	27.18	38.90
Markup	65.09	64.99	65.26	63.47	58.21	54.36	77.79
Total Shop Rate	97.64	97.49	97.89	95.20	87.31	81.54	116.69
Topeka, KS							
Direct Wage	25.80	25.76	25.86	25.15	23.07	21.54	30.83
Markup	51.59	51.51	51.73	50.30	46.14	43.09	61.66
Total Shop Rate	77.39	77.27	77.59	75.46	69.20	64.63	92.49
Trenton, NJ							
Direct Wage	44.83	44.76	44.95	43.71	40.09	37.44	53.58
Markup	89.66	89.53	89.90	87.43	80.18	74.88	107.16
Total Shop Rate	134.50	134.29	134.85	131.14	120.28	112.33	160.74
Tucson, AZ							
Direct Wage	20.69	20.66	20.74	20.17	18.50	17.28	24.73
Markup	41.38	41.32	41.49	40.35	37.00	34.56	49.45
Total Shop Rate	62.07	61.98	62.23	60.52	55.51	51.84	74.18
Tulsa, OK							
Direct Wage	15.20	15.18	15.24	14.82	13.59	12.70	18.17
Markup	30.40	30.36	30.48	29.65	27.19	25.39	36.34
Total Shop Rate	45.61	45.54	45.73	44.47	40.78	38.09	54.50
Tuscaloosa, AL							
Direct Wage	20.53	20.50	20.58	20.02	18.36	17.14	24.53
Markup	41.05	40.99	41.16	40.03	36.71	34.29	49.06
Total Shop Rate	61.58	61.49	61.74	60.05	55.07	51.43	73.60

Note: Total shop rates are expressed as costs per hour. Shop markup includes all benefits, non-productive time, support, and supervision.

Area	HVAC Technician	Maintenance Electrician	Maintenance Plumber	Maintenance Carpenter	Maintenance Painter	General Maintenance Worker	Trade Shop Supervisor
Utica, NY							
Direct Wage	27.48	27.44	27.55	26.80	24.58	22.95	32.84
Markup	54.96	54.88	55.11	53.59	49.15	45.90	65.69
Total Shop Rate	82.44	82.32	82.66	80.39	73.73	68.85	98.53
Virginia Beach, VA							
Direct Wage	20.66	20.63	20.71	20.14	18.47	17.25	24.69
Markup	41.31	41.25	41.42	40.28	36.95	34.50	49.38
Total Shop Rate	61.97	61.88	62.14	60.43	55.42	51.76	74.06
Waco, TX							
Direct Wage	20.86	20.83	20.92	20.34	18.66	17.42	24.93
Markup	41.72	41.66	41.83	40.68	37.31	34.84	49.86
Total Shop Rate	62.58	62.49	62.75	61.02	55.97	52.27	74.79
Washington, DC							
Direct Wage	31.46	31.41	31.55	30.68	28.14	26.28	37.60
Markup	62.92	62.83	63.09	61.36	56.27	52.55	75.20
Total Shop Rate	94.39	94.24	94.64	92.03	84.41	78.83	112.80
Waterbury, CT							
Direct Wage	34.86	34.80	34.95	33.99	31.17	29.11	41.66
Markup	69.72	69.61	69.90	67.98	62.34	58.22	83.32
Total Shop Rate	104.57	104.41	104.85	101.96	93.52	87.34	124.98
Watertown, NY							
Direct Wage	26.57	26.53	26.64	25.91	23.76	22.19	31.75
Markup	53.14	53.06	53.28	51.81	47.52	44.38	63.51
Total Shop Rate	79.71	79.59	79.92	77.72	71.28	66.57	95.26
Wichita Falls, TX							
Direct Wage	18.96	18.93	19.01	18.48	16.95	15.83	22.65
Markup	37.91	37.85	38.01	36.97	33.90	31.66	45.31
Total Shop Rate	56.87	56.78	57.02	55.45	50.85	47.49	67.96
Wichita, KS							
Direct Wage	24.93	24.89	25.00	24.31	22.30	20.82	29.80
Markup	49.86	49.79	50.00	48.62	44.59	41.64	59.59
Total Shop Rate	74.80	74.68	74.99	72.93	66.89	62.47	89.39
Wilmington, DE							
Direct Wage	33.87	33.82	33.96	33.03	30.29	28.29	40.48
Markup	67.75	67.64	67.92	66.06	60.58	56.58	80.96
Total Shop Rate	101.62	101.46	101.89	99.08	90.87	84.87	121.45
Winston-Salem, NC							
Direct Wage	12.69	12.67	12.72	12.37	11.34	10.59	15.16
Markup	25.37	25.33	25.44	24.74	22.69	21.19	30.32
Total Shop Rate	38.06	38.00	38.16	37.11	34.03	31.78	45.48
Worcester, MA							
Direct Wage	33.60	33.55	33.69	32.77	30.05	28.06	40.16
Markup	67.21	67.11	67.39	65.53	60.10	56.13	80.32
Total Shop Rate	100.81	100.66	101.08	98.30	90.15	84.19	120.48

Note: Total shop rates are expressed as costs per hour. Shop markup includes all benefits, non-productive time, support, and supervision.

Area	HVAC Technician	Maintenance Electrician	Maintenance Plumber	Maintenance Carpenter	Maintenance Painter	General Maintenance Worker	Trade Shop Supervisor
Yonkers, NY							
Direct Wage	48.24	48.16	48.36	47.03	43.14	40.28	57.65
Markup	96.47	96.33	96.73	94.07	86.27	80.57	115.30
Total Shop Rate	144.71	144.49	145.09	141.10	129.41	120.85	172.94
Youngstown, OH							
Direct Wage	31.26	31.21	31.34	30.48	27.95	26.11	37.36
Markup	62.52	62.42	62.68	60.96	55.91	52.21	74.72
Total Shop Rate	93.78	93.63	94.03	91.44	83.86	78.32	112.07
Canadian Cities							
Calgary, Alberta							
Direct Wage	33.94	33.89	34.03	33.09	30.35	28.34	40.56
Markup	67.88	67.77	68.06	66.18	60.70	56.69	81.12
Total Shop Rate	101.81	101.66	102.08	99.27	91.05	85.03	121.68
Edmonton, Alberta							
Direct Wage	33.94	33.89	34.03	33.09	30.35	28.34	40.56
Markup	67.88	67.77	68.06	66.18	60.70	56.69	81.12
Total Shop Rate	101.81	101.66	102.08	99.27	91.05	85.03	121.68
Hamilton, Ontario							
Direct Wage	34.52	34.47	34.62	33.66	30.87	28.83	41.26
Markup	69.05	68.94	69.23	67.33	61.75	57.67	82.52
Total Shop Rate	103.57	103.41	103.85	100.99	92.62	86.50	123.78
London, Ontario							
Direct Wage	33.52	33.47	33.61	32.69	29.98	28.00	40.06
Markup	67.04	66.94	67.22	65.37	59.96	55.99	80.13
Total Shop Rate	100.57	100.41	100.83	98.06	89.93	83.99	120.19
Montreal, Quebec							
Direct Wage	29.20	29.15	29.28	28.47	26.11	24.39	34.90
Markup	58.40	58.31	58.55	56.94	52.22	48.77	69.79
Total Shop Rate	87.60	87.46	87.83	85.41	78.33	73.16	104.69
Ottawa, Ontario							
Direct Wage	33.87	33.82	33.96	33.03	30.29	28.29	40.48
Markup	67.75	67.64	67.92	66.06	60.58	56.58	80.96
Total Shop Rate	101.62	101.46	101.89	99.08	90.87	84.87	121.45
Quebec, Quebec							
Direct Wage	29.20	29.15	29.28	28.47	26.11	24.39	34.90
Markup	58.40	58.31	58.55	56.94	52.22	48.77	69.79
Total Shop Rate	87.60	87.46	87.83	85.41	78.33	73.16	104.69
Toronto, Ontario							
Direct Wage	34.82	34.76	34.91	33.95	31.14	29.08	41.61
Markup	69.63	69.53	69.82	67.90	62.27	58.16	83.22
Total Shop Rate	104.45	104.29	104.73	101.85	93.41	87.23	124.83

Note: Total shop rates are expressed as costs per hour. Shop markup includes all benefits, non-productive time, support, and supervision.

Area	HVAC Technician	Maintenance Electrician	Maintenance Plumber	Maintenance Carpenter	Maintenance Painter	General Maintenance Worker	Trade Shop Supervisor
Vancouver, B.C.							
Direct Wage	28.73	28.68	28.80	28.01	25.69	23.99	34.33
Markup	57.45	57.37	57.61	56.02	51.38	47.98	68.66
Total Shop Rate	86.18	86.05	86.41	84.03	77.07	71.97	103.00
Winnipeg, Manitoba							
Direct Wage	25.91	25.87	25.98	25.26	23.17	21.64	30.96
Markup	51.82	51.74	51.96	50.53	46.34	43.28	61.93
Total Shop Rate	77.73	77.61	77.93	75.79	69.51	64.92	92.89

Note: Total shop rates are expressed as costs per hour. Shop markup includes all benefits, non-productive time, support, and supervision.

.3 Contract Labor Rates, Selected Metro Areas

Area	Carpenter	Carpet Layer	Electrician	Plumber	Painter	Elevator Mechanic	Brick Layer	Roofer	Crane Operator	Fork Lift Driver
Akron, OH										
Basic Wage & Fringe.....	33.13	33.13	36.40	41.50	34.08	37.96	34.80	26.96	35.20	35.10
Comp. Insurance..........	3.21	3.21	2.02	2.66	5.13	1.68	3.98	5.84	2.97	2.96
Other Req'd Benefits.....	5.47	5.47	6.01	6.85	5.62	6.26	5.74	4.45	5.81	5.79
Total Contract Rate.......	41.80	41.80	44.43	51.00	44.83	45.90	44.53	37.25	43.98	43.85
Alamogordo, NM										
Basic Wage & Fringe.....	24.86	22.06	25.14	30.93	19.43	34.41	23.09	15.29	25.20	22.66
Comp. Insurance..........	3.64	3.23	1.61	2.38	1.99	2.84	2.98	4.40	2.08	1.87
Other Req'd Benefits.....	4.10	3.64	4.15	5.10	3.21	5.68	3.81	2.52	4.16	3.74
Total Contract Rate.......	32.61	28.93	30.90	38.42	24.62	42.93	29.88	22.21	31.44	28.27
Albany, NY										
Basic Wage & Fringe.....	31.04	31.04	36.95	35.94	24.29	35.57	33.19	26.77	37.79	35.79
Comp. Insurance..........	4.62	4.62	2.17	2.68	3.01	1.73	5.87	7.75	3.12	2.95
Other Req'd Benefits.....	5.12	5.12	6.10	5.93	4.01	5.87	5.48	4.42	6.24	5.91
Total Contract Rate.......	40.78	40.78	45.21	44.55	31.30	43.17	44.54	38.93	47.14	44.65
Albuquerque, NM										
Basic Wage & Fringe.....	24.86	22.06	31.79	30.93	19.43	34.41	25.04	15.29	25.20	22.66
Comp. Insurance..........	3.64	3.23	2.04	2.38	1.99	2.84	3.24	4.40	2.08	1.87
Other Req'd Benefits.....	4.10	3.64	5.25	5.10	3.21	5.68	4.13	2.52	4.16	3.74
Total Contract Rate.......	32.61	28.93	39.07	38.42	24.62	42.93	32.41	22.21	31.44	28.27
Altus, OK										
Basic Wage & Fringe.....	19.70	14.92	28.01	29.35	17.15	31.62	26.96	19.50	24.10	23.60
Comp. Insurance..........	3.81	2.88	2.08	2.56	2.93	3.88	4.53	8.24	4.45	4.35
Other Req'd Benefits.....	3.25	2.46	4.62	4.84	2.83	5.22	4.45	3.22	3.98	3.89
Total Contract Rate.......	26.76	20.27	34.71	36.75	22.91	40.72	35.94	30.96	32.52	31.85
Amarillo, TX										
Basic Wage & Fringe.....	19.79	19.79	24.93	23.69	12.47	32.17	22.49	8.95	25.30	24.30
Comp. Insurance..........	2.62	2.62	1.83	1.91	1.23	2.25	3.08	2.04	2.65	2.55
Other Req'd Benefits.....	3.27	3.27	4.11	3.91	2.06	5.31	3.71	1.48	4.17	4.01
Total Contract Rate.......	25.68	25.68	30.88	29.51	15.76	39.73	29.28	12.47	32.13	30.86
Anaheim, CA										
Basic Wage & Fringe.....	37.45	32.95	42.96	41.21	30.91	44.90	35.86	31.22	44.04	43.93
Comp. Insurance..........	10.59	9.32	4.27	4.78	6.01	2.59	5.22	12.42	3.47	3.46
Other Req'd Benefits.....	6.18	5.44	7.09	6.80	5.10	7.41	5.92	5.15	7.27	7.25
Total Contract Rate.......	54.22	47.71	54.31	52.79	42.02	54.90	46.99	48.79	54.78	54.64
Anchorage, AK										
Basic Wage & Fringe.....	40.85	34.27	45.54	43.30	35.18	48.36	42.45	39.12	43.26	42.30
Comp. Insurance..........	4.89	4.10	4.39	3.79	4.90	5.30	7.11	14.46	8.02	7.84
Other Req'd Benefits.....	6.74	5.65	7.51	7.14	5.80	7.98	7.00	6.45	7.14	6.98
Total Contract Rate.......	52.48	44.03	57.45	54.23	45.88	61.64	56.56	60.04	58.41	57.12
Ann Arbor, MI										
Basic Wage & Fringe.....	37.90	33.67	36.20	41.75	32.14	47.33	35.76	31.63	41.70	39.99
Comp. Insurance..........	7.50	6.67	2.07	3.44	4.80	3.76	6.91	13.07	4.87	4.67
Other Req'd Benefits.....	6.25	5.56	5.97	6.89	5.30	7.81	5.90	5.22	6.88	6.60
Total Contract Rate.......	51.66	45.89	44.24	52.08	42.24	58.90	48.57	49.92	53.46	51.26

Note: Contract labor rates are defined as prevailing hourly rates.

3.3 Contract Labor Rates, Selected Metro Areas

Area	Carpenter	Carpet Layer	Electrician	Plumber	Painter	Elevator Mechanic	Brick Layer	Roofer	Crane Operator	Fork Lift Driver
Atlanta, GA										
Basic Wage & Fringe.....	25.00	21.14	32.81	33.41	24.50	38.68	24.62	19.60	25.96	24.52
Comp. Insurance..........	6.17	5.21	3.00	3.77	4.53	3.32	5.24	7.89	3.83	3.62
Other Req'd Benefits.....	4.13	3.49	5.41	5.51	4.04	6.38	4.06	3.23	4.28	4.05
Total Contract Rate.......	35.29	29.84	41.22	42.69	33.07	48.38	33.92	30.72	34.08	32.18
Atlantic City, NJ										
Basic Wage & Fringe.....	45.65	45.65	49.88	44.81	39.05	48.72	42.92	39.05	50.02	48.43
Comp. Insurance..........	5.11	5.11	1.92	2.58	3.50	2.28	5.16	11.07	3.49	3.38
Other Req'd Benefits.....	7.53	7.53	8.23	7.39	6.44	8.04	7.08	6.44	8.25	7.99
Total Contract Rate.......	58.29	58.29	60.03	54.78	49.00	59.03	55.16	56.57	61.76	59.80
Austin, TX										
Basic Wage & Fringe.....	21.31	18.49	28.14	25.73	14.00	31.93	24.50	11.13	18.00	17.36
Comp. Insurance..........	2.82	2.45	2.07	2.07	1.38	2.23	3.35	2.54	1.89	1.82
Other Req'd Benefits.....	3.52	3.05	4.64	4.25	2.31	5.27	4.04	1.84	2.97	2.86
Total Contract Rate.......	27.65	23.99	34.85	32.05	17.69	39.43	31.89	15.50	22.86	22.05
Bakersfield, CA										
Basic Wage & Fringe.....	37.45	32.95	37.34	36.54	27.21	52.40	35.85	29.15	44.04	43.93
Comp. Insurance..........	10.59	9.32	3.71	4.23	5.29	3.02	5.22	11.60	3.47	3.46
Other Req'd Benefits.....	6.18	5.44	6.16	6.03	4.49	8.65	5.92	4.81	7.27	7.25
Total Contract Rate.......	54.22	47.71	47.21	46.80	36.99	64.07	46.98	45.56	54.78	54.64
Baltimore, MD										
Basic Wage & Fringe.....	26.96	26.96	37.10	37.29	23.41	43.20	29.95	27.81	30.66	28.60
Comp. Insurance..........	2.84	2.84	1.91	2.07	1.58	1.53	3.40	6.34	2.84	2.65
Other Req'd Benefits.....	4.45	4.45	6.12	6.15	3.86	7.13	4.94	4.59	5.06	4.72
Total Contract Rate.......	34.25	34.25	45.13	45.51	28.85	51.86	38.29	38.74	38.55	35.96
Battle Creek, MI										
Basic Wage & Fringe.....	28.98	28.98	34.47	38.36	26.95	41.20	29.97	25.74	36.91	35.91
Comp. Insurance..........	5.74	5.74	1.97	3.16	4.02	3.28	5.79	10.64	4.31	4.20
Other Req'd Benefits.....	4.78	4.78	5.69	6.33	4.45	6.80	4.95	4.25	6.09	5.93
Total Contract Rate.......	39.50	39.50	42.13	47.85	35.42	51.27	40.71	40.63	47.31	46.03
Beaufort, SC										
Basic Wage & Fringe.....	9.57	12.30	10.81	11.02	8.52	17.28	9.82	8.07	10.05	13.87
Comp. Insurance..........	1.84	2.37	.76	.76	.95	1.08	.94	2.34	.84	1.15
Other Req'd Benefits.....	1.58	2.03	1.78	1.82	1.41	2.85	1.62	1.33	1.66	2.29
Total Contract Rate.......	12.99	16.70	13.35	13.60	10.88	21.21	12.38	11.74	12.54	17.31
Beaumont, TX										
Basic Wage & Fringe.....	19.59	18.00	26.49	29.07	16.82	39.67	24.47	14.08	20.10	19.63
Comp. Insurance..........	2.60	2.39	1.95	2.34	1.66	2.77	3.35	3.21	2.11	2.06
Other Req'd Benefits.....	3.23	2.97	4.37	4.80	2.78	6.55	4.04	2.32	3.32	3.24
Total Contract Rate.......	25.42	23.36	32.81	36.21	21.26	48.99	31.86	19.61	25.52	24.93
Billings, MT										
Basic Wage & Fringe.....	22.05	14.88	31.25	34.25	19.02	32.43	24.95	16.20	22.11	21.81
Comp. Insurance..........	4.41	2.98	1.73	3.06	2.29	2.13	3.64	9.59	3.80	3.75
Other Req'd Benefits.....	3.64	2.46	5.16	5.65	3.14	5.35	4.12	2.67	3.65	3.60
Total Contract Rate.......	30.10	20.31	38.14	42.96	24.45	39.91	32.70	28.46	29.56	29.16

Note: Contract labor rates are defined as prevailing hourly rates.

Area	Carpenter	Carpet Layer	Electrician	Plumber	Painter	Elevator Mechanic	Brick Layer	Roofer	Crane Operator	Fork Lift Driver
Biloxi, MS										
Basic Wage & Fringe.....	14.35	14.35	25.12	25.80	11.32	25.45	13.85	12.33	14.18	13.30
Comp. Insurance...........	2.83	2.83	1.81	2.04	1.44	2.72	1.89	3.91	1.56	1.46
Other Req'd Benefits.....	2.37	2.37	4.14	4.26	1.87	4.20	2.29	2.03	2.34	2.19
Total Contract Rate..	19.55	19.55	31.07	32.10	14.63	32.37	18.03	18.27	18.08	16.95
Birmingham, AL										
Basic Wage & Fringe.....	22.76	22.76	25.21	26.57	18.80	38.05	23.67	17.96	20.52	19.80
Comp. Insurance......	7.62	7.62	2.31	3.16	5.42	3.47	7.33	12.35	3.11	3.00
Other Req'd Benefits	3.76	3.76	4.16	4.38	3.10	6.28	3.91	2.96	3.39	3.27
Total Contract Rate.......	34.14	34.14	31.68	34.11	27.32	47.79	34.91	33.28	27.01	26.06
Bismarck, ND										
Basic Wage & Fringe	16.54	13.28	29.37	26.85	12.72	35.57	26.20	12.25	25.65	25.40
Comp. Insurance...........	3.15	2.53	1.17	1.44	.80	2.95	2.22	2.64	1.38	1.37
Other Req'd Benefits.....	2.73	2.19	4.85	4.43	2.10	5.87	4.32	2.02	4.23	4.19
Total Contract Rate.......	22.42	18.00	35.39	32.72	15.62	44.39	32.74	16.91	31.26	30.96
Boise, ID										
Basic Wage & Fringe.....	29.17	29.17	32.88	31.71	17.06	37.69	24.25	22.40	21.25	21.07
Comp. Insurance...........	4.08	4.08	1.63	1.24	1.21	3.49	1.94	6.13	1.14	1.13
Other Req'd Benefits.....	4.81	4.81	5.43	5.23	2.81	6.22	4.00	3.70	3.51	3.48
Total Contract Rate... ...	38.06	38.06	39.94	38.19	21.09	47.40	30.19	32.22	25.90	25.68
Boston, MA										
Basic Wage & Fringe.....	46.10	44.70	48.85	50.70	46.07	52.12	50.22	37.86	43.44	43.23
Comp. Insurance....... ...	7.65	7.42	1.80	2.58	3.81	4.45	7.25	12.60	2.78	2.77
Other Req'd Benefits. ...	7.61	7.38	8.06	8.37	7.60	8.60	8.29	6.25	7.17	7.13
Total Contract Rate.......	61.36	59.50	58.71	61.65	57.48	65.17	65.76	56.71	53.39	53.13
Boulder, CO										
Basic Wage & Fringe. ...	28.30	15.98	37.39	34.13	22.43	41.70	28.22	18.68	18.31	18.41
Comp. Insurance...........	3.44	1.94	2.13	2.84	2.52	3.34	3.98	4.67	2.18	2.19
Other Req'd Benefits.....	4.67	2.64	6.17	5.63	3.70	6.88	4.66	3.08	3.02	3.04
Total Contract Rate.......	36.41	20.56	45.69	42.60	28.65	51.92	36.86	26.44	23.51	23.64
Bowling Green, KY										
Basic Wage & Fringe.....	26.07	26.07	32.49	31.32	21.93	21.20	21.76	20.37	28.65	28.65
Comp. Insurance...........	5.01	5.01	2.23	1.72	2.39	1.59	2.35	3.53	5.15	5.15
Other Req'd Benefits.....	4.30	4.30	5.36	5.17	3.62	3.50	3.59	3.36	4.73	4.73
Total Contract Rate.... ..	35.38	35.38	40.08	38.20	27.94	26.29	27.70	27.26	38.53	38.53
Brockton, MA										
Basic Wage & Fringe.....	40.41	28.45	40.26	42.37	37.43	52.12	44.68	34.61	43.44	43.23
Comp. Insurance...........	6.71	4.72	1.49	2.16	3.09	4.45	6.45	11.52	2.78	2.77
Other Req'd Benefits.. ..	6.67	4.69	6.64	6.99	6.18	8.60	7.37	5.71	7.17	7.13
Total Contract Rate.......	53.79	37.87	48.39	51.52	46.70	65.17	58.50	51.84	53.39	53.13
Buffalo, NY										
Basic Wage & Fringe.....	39.05	39.05	38.99	37.39	34.49	44.84	39.24	32.41	40.07	40.07
Comp. Insurance........ ..	5.81	5.81	2.28	2.79	4.27	2.18	6.95	9.38	3.31	3.31
Other Req'd Benefits.....	6.44	6.44	6.43	6.17	5.69	7.40	6.47	5.35	6.61	6.61
Total Contract Rate.......	51.30	51.30	47.71	46.35	44.45	54.42	52.66	47.14	49.99	49.99

Note: Contract labor rates are defined as prevailing hourly rates.

3.3 Contract Labor Rates, Selected Metro Areas

Area	Carpenter	Carpet Layer	Electrician	Plumber	Painter	Elevator Mechanic	Brick Layer	Roofer	Crane Operator	Fork Lift Driver
Burlington, VT										
Basic Wage & Fringe.....	16.41	16.41	27.96	30.05	13.15	41.83	24.05	17.25	20.33	20.13
Comp. Insurance...........	3.19	3.19	1.71	3.04	1.37	1.65	4.70	4.96	2.07	2.05
Other Req'd Benefits.....	2.71	2.71	4.61	4.96	2.17	6.90	3.97	2.85	3.35	3.32
Total Contract Rate.......	22.31	22.31	34.28	38.04	16.69	50.38	32.72	25.05	25.75	25.50
Camden, NJ										
Basic Wage & Fringe.....	45.65	45.65	49.88	44.81	39.05	48.72	42.92	39.05	50.02	48.43
Comp. Insurance...........	5.11	5.11	1.92	2.58	3.50	2.28	5.16	11.07	3.49	3.38
Other Req'd Benefits.....	7.53	7.53	8.23	7.39	6.44	8.04	7.08	6.44	8.25	7.99
Total Contract Rate.......	58.29	58.29	60.03	54.78	49.00	59.03	55.16	56.57	61.76	59.80
Carson City, NV										
Basic Wage & Fringe.....	38.02	25.37	35.19	36.90	25.59	53.67	27.02	24.60	42.31	41.99
Comp. Insurance...........	5.28	3.52	2.70	3.75	2.88	3.65	3.29	5.83	4.85	4.82
Other Req'd Benefits.....	6.27	4.19	5.81	6.09	4.22	8.86	4.46	4.06	6.98	6.93
Total Contract Rate.......	49.57	33.08	43.70	46.74	32.69	66.18	34.76	34.48	54.14	53.73
Cedar Rapids, IA										
Basic Wage & Fringe.....	27.62	27.62	34.50	35.52	24.90	35.08	29.15	29.80	28.70	27.32
Comp. Insurance...........	3.18	3.18	1.37	1.72	2.06	1.28	2.28	5.33	1.81	1.72
Other Req'd Benefits.....	4.56	4.56	5.69	5.86	4.11	5.79	4.81	4.92	4.74	4.51
Total Contract Rate.......	35.35	35.35	41.56	43.10	31.07	42.15	36.24	40.04	35.24	33.55
Charleston, SC										
Basic Wage & Fringe.....	12.30	12.30	13.56	19.72	12.50	31.67	14.00	11.90	16.40	14.68
Comp. Insurance...........	2.37	2.37	.95	1.36	1.40	1.97	1.34	3.45	1.36	1.22
Other Req'd Benefits.....	2.03	2.03	2.24	3.25	2.06	5.23	2.31	1.96	2.71	2.42
Total Contract Rate.......	16.70	16.70	16.75	24.33	15.96	38.87	17.65	17.31	20.47	18.32
Charleston, WV										
Basic Wage & Fringe.....	34.09	34.09	34.75	33.95	30.26	36.94	34.22	29.23	35.29	34.09
Comp. Insurance...........	4.09	4.09	17.83	1.62	3.69	1.95	3.96	3.93	2.73	2.64
Other Req'd Benefits.....	5.62	5.62	5.73	5.60	4.99	6.10	5.65	4.82	5.82	5.62
Total Contract Rate.......	43.81	43.81	58.31	41.17	38.94	44.99	43.83	37.98	43.84	42.35
Charlotte, NC										
Basic Wage & Fringe.....	14.91	14.91	14.62	19.12	12.50	29.87	14.84	10.00	18.75	18.50
Comp. Insurance...........	2.69	2.69	1.13	1.54	1.20	1.85	1.40	2.57	1.58	1.56
Other Req'd Benefits.....	2.46	2.46	2.41	3.15	2.06	4.93	2.45	1.65	3.09	3.05
Total Contract Rate.......	20.06	20.06	18.16	23.82	15.76	36.65	18.69	14.22	23.43	23.11
Chattanooga, TN										
Basic Wage & Fringe.....	16.97	16.97	22.38	18.50	14.83	25.80	17.45	16.75	15.56	14.56
Comp. Insurance...........	3.46	3.46	1.64	1.76	1.89	2.07	2.50	5.60	1.91	1.79
Other Req'd Benefits.....	2.80	2.80	3.69	3.05	2.45	4.26	2.88	2.76	2.57	2.40
Total Contract Rate.......	23.23	23.23	27.71	23.31	19.16	32.13	22.82	25.12	20.04	18.75
Cheyenne, WY										
Basic Wage & Fringe.....	15.37	16.86	21.60	18.90	18.74	38.45	16.00	17.55	18.37	16.98
Comp. Insurance...........	1.12	1.23	1.57	1.38	1.37	2.80	1.17	1.28	1.34	1.24
Other Req'd Benefits.....	2.54	2.78	3.56	3.12	3.09	6.34	2.64	2.90	3.03	2.80
Total Contract Rate.......	19.03	20.87	26.74	23.40	23.20	47.60	19.81	21.73	22.74	21.02

Note: Contract labor rates are defined as prevailing hourly rates.

.3 Contract Labor Rates, Selected Metro Areas

Area	Carpenter	Carpet Layer	Electrician	Plumber	Painter	Elevator Mechanic	Brick Layer	Roofer	Crane Operator	Fork Lift Driver
Chicago, IL										
Basic Wage & Fringe	42.97	42.97	48.07	45.72	40.74	51.12	42.00	39.20	48.25	46.95
Comp. Insurance	8.06	8.06	3.96	4.51	4.11	1.56	7.07	11.52	4.55	4.43
Other Req'd Benefits	7.09	7.09	7.93	7.54	6.72	8.43	6.93	6.47	7.96	7.75
Total Contract Rate	58.12	58.12	59.96	57.78	51.57	61.12	56.00	57.18	60.76	59.12
Cincinnati, OH										
Basic Wage & Fringe	28.67	22.53	32.63	37.15	27.15	42.24	31.95	31.24	35.20	35.10
Comp. Insurance	2.78	2.18	1.81	2.38	4.08	1.87	3.66	6.77	2.97	2.96
Other Req'd Benefits	4.73	3.72	5.38	6.13	4.48	6.97	5.27	5.15	5.81	5.79
Total Contract Rate	36.18	28.43	39.82	45.66	35.71	51.08	40.88	43.16	43.98	43.85
Cleveland, OH										
Basic Wage & Fringe	34.70	34.70	42.36	41.91	34.08	46.51	35.08	35.53	36.80	36.65
Comp. Insurance	3.36	3.36	2.35	2.68	5.13	2.06	4.02	7.70	3.11	3.09
Other Req'd Benefits	5.73	5.73	6.99	6.92	5.62	7.67	5.79	5.86	6.07	6.05
Total Contract Rate	43.78	43.78	51.70	51.51	44.83	56.24	44.88	49.09	45.98	45.79
Colorado Springs, CO										
Basic Wage & Fringe	28.30	17.73	36.48	32.85	15.46	41.70	27.87	20.93	26.54	26.39
Comp. Insurance	3.44	2.15	2.08	2.74	1.74	3.34	3.93	5.24	3.16	3.14
Other Req'd Benefits	4.67	2.93	6.02	5.42	2.55	6.88	4.60	3.45	4.38	4.35
Total Contract Rate	36.41	22.81	44.58	41.01	19.75	51.92	36.40	29.62	34.08	33.88
Columbia, SC										
Basic Wage & Fringe	13.46	13.46	14.30	15.90	12.50	31.67	14.58	11.90	16.40	14.68
Comp. Insurance	2.59	2.59	1.00	1.09	1.40	1.97	1.39	3.45	1.36	1.22
Other Req'd Benefits	2.22	2.22	2.36	2.62	2.06	5.23	2.41	1.96	2.71	2.42
Total Contract Rate	18.27	18.27	17.66	19.62	15.96	38.87	18.38	17.31	20.47	18.32
Columbus, GA										
Basic Wage & Fringe	17.95	17.95	15.59	14.79	11.50	38.68	12.15	19.60	25.96	24.52
Comp. Insurance	4.43	4.43	1.42	1.67	2.12	3.32	2.59	7.89	3.83	3.62
Other Req'd Benefits	2.96	2.96	2.57	2.44	1.90	6.38	2.00	3.23	4.28	4.05
Total Contract Rate	25.34	25.34	19.59	18.90	15.52	48.38	16.74	30.72	34.08	32.18
Columbus, OH										
Basic Wage & Fringe	28.79	28.79	35.04	37.89	28.54	42.21	32.65	27.96	35.20	35.10
Comp. Insurance	2.79	2.79	1.94	2.42	4.29	1.87	3.74	6.06	2.97	2.96
Other Req'd Benefits	4.75	4.75	5.78	6.25	4.71	6.96	5.39	4.61	5.81	5.79
Total Contract Rate	36.33	36.33	42.77	46.57	37.54	51.04	41.78	38.63	43.98	43.85
Concord, NH										
Basic Wage & Fringe	18.81	18.81	30.59	34.25	27.20	34.74	32.93	28.21	21.66	21.66
Comp. Insurance	4.04	4.04	2.08	3.89	4.11	1.00	7.20	17.40	3.80	3.80
Other Req'd Benefits	3.10	3.10	5.05	5.65	4.49	5.73	5.43	4.65	3.57	3.57
Total Contract Rate	25.96	25.96	37.72	43.80	35.80	41.47	45.56	50.27	29.04	29.04
Corpus Christi, TX										
Basic Wage & Fringe	13.70	13.70	21.69	21.47	13.70	39.67	21.55	11.25	16.45	15.95
Comp. Insurance	1.82	1.82	1.60	1.73	1.35	2.77	2.95	2.57	1.73	1.67
Other Req'd Benefits	2.26	2.26	3.58	3.54	2.26	6.55	3.56	1.86	2.71	2.63
Total Contract Rate	17.78	17.78	26.87	26.74	17.32	48.99	28.05	15.67	20.89	20.25

Note: Contract labor rates are defined as prevailing hourly rates.

Area	Carpenter	Carpet Layer	Electrician	Plumber	Painter	Elevator Mechanic	Brick Layer	Roofer	Crane Operator	Fork Lift Driver
Cumberland, MD										
Basic Wage & Fringe.....	25.24	25.24	32.83	29.41	17.81	43.20	25.40	21.80	26.20	25.64
Comp. Insurance...........	2.66	2.66	1.69	1.63	1.20	1.53	2.88	4.97	2.42	2.37
Other Req'd Benefits.....	4.16	4.16	5.42	4.85	2.94	7.13	4.19	3.60	4.32	4.23
Total Contract Rate.......	32.07	32.07	39.94	35.89	21.95	51.86	32.47	30.37	32.95	32.24
Dallas, TX										
Basic Wage & Fringe.....	21.03	17.06	26.21	28.92	17.83	38.59	23.73	14.10	25.30	24.30
Comp. Insurance...........	2.79	2.26	1.93	2.33	1.76	2.70	3.25	3.21	2.65	2.55
Other Req'd Benefits.....	3.47	2.81	4.32	4.77	2.94	6.37	3.92	2.33	4.17	4.01
Total Contract Rate.......	27.29	22.14	32.46	36.02	22.54	47.65	30.89	19.64	32.13	30.86
Danbury, CT										
Basic Wage & Fringe.....	33.18	33.18	42.81	40.03	32.62	45.23	38.03	34.85	38.12	37.25
Comp. Insurance...........	9.92	9.92	4.09	4.59	6.14	4.88	10.69	18.31	4.91	4.80
Other Req'd Benefits.....	5.47	5.47	7.06	6.60	5.38	7.46	6.27	5.75	6.29	6.15
Total Contract Rate.......	48.58	48.58	53.96	51.23	44.14	57.57	55.00	58.91	49.32	48.19
Davenport, IA										
Basic Wage & Fringe.....	32.05	20.44	37.45	39.99	30.53	35.08	30.43	29.80	33.45	33.45
Comp. Insurance...........	3.69	2.35	1.48	1.94	2.52	1.28	2.38	5.33	2.11	2.11
Other Req'd Benefits.....	5.29	3.37	6.18	6.60	5.04	5.79	5.02	4.92	5.52	5.52
Total Contract Rate.......	41.02	26.16	45.11	48.53	38.09	42.15	37.83	40.04	41.08	41.08
Dayton, OH										
Basic Wage & Fringe.....	27.32	24.99	32.46	34.42	26.42	42.24	29.34	24.77	35.20	35.10
Comp. Insurance...........	2.64	2.42	1.80	2.20	3.97	1.87	3.36	5.37	2.97	2.96
Other Req'd Benefits.....	4.51	4.12	5.36	5.68	4.36	6.97	4.84	4.09	5.81	5.79
Total Contract Rate.......	34.47	31.53	39.62	42.30	34.75	51.08	37.54	34.22	43.98	43.85
Daytona Beach, FL										
Basic Wage & Fringe.....	21.98	21.98	25.29	31.62	20.50	39.58	22.35	19.35	23.43	22.59
Comp. Insurance...........	7.95	7.95	3.16	3.94	5.10	3.99	6.03	10.39	3.76	3.62
Other Req'd Benefits.....	3.63	3.63	4.17	5.22	3.38	6.53	3.69	3.19	3.87	3.73
Total Contract Rate.......	33.55	33.55	32.62	40.77	28.98	50.10	32.07	32.93	31.05	29.94
Denver, CO										
Basic Wage & Fringe.....	28.30	25.51	36.84	34.63	22.43	41.70	29.12	17.38	26.54	26.39
Comp. Insurance...........	3.44	3.10	2.10	2.88	2.52	3.34	4.11	4.35	3.16	3.14
Other Req'd Benefits.....	4.67	4.21	6.08	5.71	3.70	6.88	4.80	2.87	4.38	4.35
Total Contract Rate.......	36.41	32.82	45.02	43.23	28.65	51.92	38.03	24.60	34.08	33.88
Des Moines, IA										
Basic Wage & Fringe.....	27.55	27.55	34.97	33.96	25.25	35.08	29.46	24.26	28.10	27.82
Comp. Insurance...........	3.17	3.17	1.38	1.65	2.09	1.28	2.30	4.34	1.77	1.75
Other Req'd Benefits.....	4.55	4.55	5.77	5.60	4.17	5.79	4.86	4.00	4.64	4.59
Total Contract Rate.......	35.26	35.26	42.12	41.21	31.50	42.15	36.62	32.60	34.51	34.16
Detroit, MI										
Basic Wage & Fringe.....	41.31	34.90	46.81	47.36	34.16	47.33	40.13	40.41	41.70	39.99
Comp. Insurance...........	8.18	6.91	2.67	3.90	5.10	3.76	7.76	16.70	4.87	4.67
Other Req'd Benefits.....	6.82	5.76	7.72	7.81	5.64	7.81	6.62	6.67	6.88	6.60
Total Contract Rate.......	56.31	47.57	57.21	59.08	44.90	58.90	54.51	63.78	53.46	51.26

Note: Contract labor rates are defined as prevailing hourly rates.

.3 Contract Labor Rates, Selected Metro Areas

Area	Carpenter	Carpet Layer	Electrician	Plumber	Painter	Elevator Mechanic	Brick Layer	Roofer	Crane Operator	Fork Lift Driver
Duluth, MN										
Basic Wage & Fringe	35.57	27.35	39.08	35.50	30.46	39.28	36.34	32.35	40.29	39.78
Comp. Insurance	13.07	10.05	2.65	3.75	5.75	3.56	9.13	21.20	6.33	6.25
Other Req'd Benefits	5.87	4.51	6.45	5.86	5.03	6.48	6.00	5.34	6.65	6.56
Total Contract Rate	54.51	41.91	48.18	45.11	41.24	49.32	51.47	58.89	53.27	52.60
Eau Claire, WI										
Basic Wage & Fringe	32.07	32.07	35.75	34.87	26.91	40.99	33.48	23.14	37.87	36.84
Comp. Insurance	5.92	5.92	1.71	2.23	3.08	6.14	5.84	9.42	2.71	2.64
Other Req'd Benefits	5.29	5.29	5.90	5.75	4.44	6.76	5.52	3.82	6.25	6.08
Total Contract Rate	43.28	43.28	43.36	42.86	34.43	53.89	44.84	36.38	46.83	45.56
El Paso, TX										
Basic Wage & Fringe	18.47	18.47	23.07	15.34	11.98	32.17	22.40	15.29	16.03	14.36
Comp. Insurance	2.45	2.45	1.70	1.24	1.18	2.25	3.06	3.49	1.68	1.51
Other Req'd Benefits	3.05	3.05	3.81	2.53	1.98	5.31	3.70	2.52	2.64	2.37
Total Contract Rate	23.96	23.96	28.57	19.11	15.14	39.73	29.16	21.30	20.36	18.24
Erie, PA										
Basic Wage & Fringe	32.46	32.46	33.95	35.98	25.51	44.68	31.34	28.33	32.67	32.67
Comp. Insurance	3.98	3.98	21.93	2.58	3.33	6.08	3.65	7.33	2.58	2.58
Other Req'd Benefits	5.36	5.36	5.60	5.94	4.21	7.37	5.17	4.67	5.39	5.39
Total Contract Rate	41.80	41.80	61.48	44.50	33.05	58.13	40.16	40.33	40.64	40.64
Eugene, OR										
Basic Wage & Fringe	37.27	31.27	41.02	43.57	23.13	48.77	35.15	22.77	37.95	36.63
Comp. Insurance	6.33	5.31	2.05	2.64	3.27	2.35	5.32	5.23	4.00	3.86
Other Req'd Benefits	6.15	5.16	6.77	7.19	3.82	8.05	5.80	3.76	6.26	6.04
Total Contract Rate	49.75	41.74	49.84	53.40	30.22	59.16	46.27	31.76	48.22	46.54
Evansville, IN										
Basic Wage & Fringe	27.08	27.08	35.65	36.78	29.23	36.50	30.45	29.93	31.83	31.83
Comp. Insurance	1.88	1.88	.95	1.09	1.41	2.94	1.56	3.50	1.11	1.11
Other Req'd Benefits	4.47	4.47	5.88	6.07	4.82	6.02	5.02	4.94	5.25	5.25
Total Contract Rate	33.43	33.43	42.48	43.94	35.47	45.46	37.03	38.37	38.20	38.20
Fairbanks, AK										
Basic Wage & Fringe	43.65	34.27	45.54	47.41	35.18	48.36	42.45	41.12	43.26	42.30
Comp. Insurance	5.22	4.10	4.39	4.15	5.49	5.30	7.11	15.20	8.02	7.84
Other Req'd Benefits	7.20	5.65	7.51	7.82	6.51	7.98	7.00	6.78	7.14	6.98
Total Contract Rate	56.08	44.03	57.45	59.38	51.48	61.64	56.56	63.11	58.41	57.12
Fall River, MA										
Basic Wage & Fringe	40.41	44.70	40.26	42.37	37.43	52.12	46.36	34.61	43.44	43.23
Comp. Insurance	6.71	7.42	1.49	2.16	3.09	4.45	6.69	11.52	2.78	2.77
Other Req'd Benefits	6.67	7.38	6.64	6.99	6.18	8.60	7.65	5.71	7.17	7.13
Total Contract Rate	53.79	59.50	48.39	51.52	46.70	65.17	60.70	51.84	53.39	53.13
Fargo, ND										
Basic Wage & Fringe	16.80	13.89	29.25	29.86	24.08	35.57	15.65	12.25	25.65	25.40
Comp. Insurance	3.20	2.65	1.17	1.60	1.52	2.95	1.33	2.64	1.38	1.37
Other Req'd Benefits	2.77	2.29	4.83	4.93	3.97	5.87	2.58	2.02	4.23	4.19
Total Contract Rate	22.77	18.83	35.25	36.38	29.57	44.39	19.56	16.91	31.26	30.96

Note: Contract labor rates are defined as prevailing hourly rates.

3.3 Contract Labor Rates, Selected Metro Areas

Area	Carpenter	Carpet Layer	Electrician	Plumber	Painter	Elevator Mechanic	Brick Layer	Roofer	Crane Operator	Fork Lift Driver
Flint, MI										
Basic Wage & Fringe....	32.33	33.67	36.63	39.61	27.42	41.20	34.06	27.66	36.91	35.91
Comp. Insurance..........	6.40	6.67	2.09	3.26	4.09	3.28	6.58	11.43	4.31	4.20
Other Req'd Benefits.....	5.33	5.56	6.04	6.54	4.52	6.80	5.62	4.56	6.09	5.93
Total Contract Rate.......	44.07	45.89	44.77	49.41	36.04	51.27	46.26	43.66	47.31	46.03
Fort Lauderdale, FL										
Basic Wage & Fringe....	20.86	20.86	29.83	26.95	14.19	41.98	19.67	14.63	22.22	21.89
Comp. Insurance..........	7.54	7.54	3.73	3.36	3.53	4.23	5.31	7.85	3.56	3.51
Other Req'd Benefits.....	3.44	3.44	4.92	4.45	2.34	6.93	3.25	2.41	3.67	3.61
Total Contract Rate.......	31.84	31.84	38.48	34.75	20.06	53.14	28.23	24.90	29.45	29.01
Fort Smith, TX										
Basic Wage & Fringe....	14.04	14.04	26.29	18.41	18.41	32.44	23.71	12.36	16.35	14.98
Comp. Insurance..........	2.06	2.06	1.64	1.17	1.81	.99	2.55	2.74	1.48	1.36
Other Req'd Benefits.....	2.32	2.32	4.34	3.04	3.04	5.35	3.91	2.04	2.70	2.47
Total Contract Rate.......	18.41	18.41	32.27	22.62	23.26	38.78	30.17	17.13	20.53	18.81
Fort Worth, TX										
Basic Wage & Fringe....	21.03	16.00	25.27	26.45	17.78	38.59	23.73	13.90	25.30	24.30
Comp. Insurance..........	2.79	2.12	1.86	2.13	1.76	2.70	3.25	3.17	2.65	2.55
Other Req'd Benefits.....	3.47	2.64	4.17	4.36	2.93	6.37	3.92	2.29	4.17	4.01
Total Contract Rate.......	27.29	20.76	31.30	32.95	22.47	47.65	30.89	19.36	32.13	30.86
Fresno, CA										
Basic Wage & Fringe....	41.73	26.52	36.72	43.68	28.01	53.74	35.00	29.15	45.51	42.86
Comp. Insurance..........	11.80	7.50	3.65	5.06	5.45	3.10	5.09	11.60	3.59	3.38
Other Req'd Benefits.....	6.89	4.38	6.06	7.21	4.62	8.87	5.78	4.81	7.51	7.07
Total Contract Rate.......	60.42	38.40	46.43	55.95	38.08	65.71	45.87	45.56	56.61	53.31
Gary, IN										
Basic Wage & Fringe....	38.92	18.10	43.72	38.61	33.77	44.70	34.81	36.61	41.35	40.85
Comp. Insurance..........	2.71	1.26	1.17	1.15	1.63	3.60	1.78	4.28	1.45	1.43
Other Req'd Benefits.....	6.42	2.99	7.21	6.37	5.57	7.38	5.74	6.04	6.82	6.74
Total Contract Rate.......	48.05	22.35	52.10	46.13	40.98	55.68	42.34	46.93	49.62	49.02
Grand Rapids, MI										
Basic Wage & Fringe....	25.90	14.58	25.25	20.72	13.28	41.20	15.59	13.37	36.91	35.91
Comp. Insurance..........	5.13	2.89	1.44	1.71	1.98	3.28	3.01	5.53	4.31	4.20
Other Req'd Benefits.....	4.27	2.41	4.17	3.42	2.19	6.80	2.57	2.21	6.09	5.93
Total Contract Rate.......	35.30	19.87	30.86	25.85	17.45	51.27	21.18	21.10	47.31	46.03
Great Falls, MT										
Basic Wage & Fringe....	22.12	17.53	31.43	34.25	14.58	32.43	25.95	14.75	22.11	21.81
Comp. Insurance..........	4.42	3.51	1.74	3.06	1.75	2.13	3.78	8.73	3.80	3.75
Other Req'd Benefits.....	3.65	2.89	5.19	5.65	2.41	5.35	4.28	2.43	3.65	3.60
Total Contract Rate.......	30.19	23.93	38.36	42.96	18.74	39.91	34.02	25.92	29.56	29.16
Green Bay, WI										
Basic Wage & Fringe....	32.07	32.07	35.61	33.51	24.05	40.99	33.48	19.30	37.87	36.84
Comp. Insurance..........	5.92	5.92	1.70	2.14	2.75	6.14	5.84	7.86	2.71	2.64
Other Req'd Benefits.....	5.29	5.29	5.88	5.53	3.97	6.76	5.52	3.18	6.25	6.08
Total Contract Rate.......	43.28	43.28	43.19	41.18	30.77	53.89	44.84	30.35	46.83	45.56

Note: Contract labor rates are defined as prevailing hourly rates.

.3 Contract Labor Rates, Selected Metro Areas

Area	Carpenter	Carpet Layer	Electrician	Plumber	Painter	Elevator Mechanic	Brick Layer	Roofer	Crane Operator	Fork Lift Driver
Greensboro, NC										
Basic Wage & Fringe. ...	14.91	14.91	15.58	19.12	12.50	29.87	14.84	10.00	18.75	18.50
Comp. Insurance..........	2.69	2.69	1.20	1.54	1.20	1.85	1.40	2.57	1.58	1.56
Other Req'd Benefits.....	2.46	2.46	2.57	3.15	2.06	4.93	2.45	1.65	3.09	3.05
Total Contract Rate.... ..	20.06	20.06	19.35	23.82	15.76	36.65	18.69	14.22	23.43	23.11
Hampton, VA										
Basic Wage & Fringe.. ..	15.05	10.94	24.53	26.62	10.94	32.28	16.61	11.94	25.81	21.03
Comp. Insurance..........	1.71	1.24	1.02	1.61	1.11	2.03	1.33	2.67	1.92	1.56
Other Req'd Benefits.....	2.48	1.81	4.05	4.39	1.81	5.33	2.74	1.97	4.26	3.47
Total Contract Rate..... .	19.24	13.99	29.60	32.62	13.86	39.64	20.68	16.58	31.98	26.06
Harrisburg, PA										
Basic Wage & Fringe... .	27.51	27.51	33.95	35.12	24.27	36.73	29.45	39.05	33.94	30.16
Comp. Insurance..........	3.38	3.38	21.93	2.52	3.16	5.00	3.43	10.10	2.68	2.39
Other Req'd Benefits.....	4.54	4.54	5.60	5.79	4.00	6.06	4.86	6.44	5.60	4.98
Total Contract Rate......	35.42	35.42	61.48	43.44	31.44	47.79	37.74	55.60	42.22	37.52
Hartford, CT										
Basic Wage & Fringe....	33.18	33.18	41.72	40.03	32.62	45.23	38.03	32.99	38.12	37.25
Comp. Insurance..........	9.92	9.92	3.98	4.59	6.14	4.88	10.69	17.33	4.91	4.80
Other Req'd Benefits.....	5.47	5.47	6.88	6.60	5.38	7.46	6.27	5.44	6.29	6.15
Total Contract Rate.......	48.58	48.58	52.59	51.23	44.14	57.57	55.00	55.76	49.32	48.19
Hilo, HI										
Basic Wage & Fringe.....	47.97	39.50	49.39	47.10	46.40	46.85	44.32	42.13	48.16	47.73
Comp. Insurance..........	13.53	11.14	3.51	2.85	5.26	3.37	7.92	14.00	3.72	3.69
Other Req'd Benefits.....	7.92	6.52	8.15	7.77	7.66	7.73	7.31	6.95	7.95	7.88
Total Contract Rate.......	69.41	57.16	61.05	57.73	59.31	57.95	59.55	63.09	59.83	59.29
Honolulu, HI										
Basic Wage & Fringe.....	47.97	39.50	49.39	47.10	46.40	46.85	44.32	42.13	48.16	47.73
Comp. Insurance..........	13.53	11.14	3.51	2.85	5.26	3.37	7.92	14.00	3.72	3.69
Other Req'd Benefits.....	7.92	6.52	8.15	7.77	7.66	7.73	7.31	6.95	7.95	7.88
Total Contract Rate.......	69.41	57.16	61.05	57.73	59.31	57.95	59.55	63.09	59.83	59.29
Houston, TX										
Basic Wage & Fringe.....	23.69	21.16	27.80	30.98	20.35	39.67	23.34	18.08	23.53	23.53
Comp. Insurance..........	3.14	2.80	2.05	2.50	2.01	2.77	3.19	4.12	2.47	2.47
Other Req'd Benefits.....	3.91	3.49	4.59	5.11	3.36	6.55	3.85	2.98	3.88	3.88
Total Contract Rate.......	30.74	27.46	34.43	38.59	25.72	48.99	30.38	25.19	29.88	29.88
Huntsville, AL										
Basic Wage & Fringe.....	21.13	21.13	26.92	24.62	17.74	38.05	19.60	17.96	22.52	20.88
Comp. Insurance..........	7.08	7.08	2.47	2.92	5.11	3.47	6.07	12.35	3.41	3.16
Other Req'd Benefits.....	3.49	3.49	4.44	4.06	2.93	6.28	3.23	2.96	3.72	3.45
Total Contract Rate.......	31.69	31.69	33.83	31.61	25.78	47.79	28.90	33.28	29.65	27.49
Indianapolis, IN										
Basic Wage & Fringe.....	32.15	28.36	36.93	36.15	29.65	37.71	32.30	26.27	32.38	31.43
Comp. Insurance..........	2.24	1.97	.99	1.07	1.44	3.04	1.65	3.07	1.13	1.10
Other Req'd Benefits.....	5.30	4.68	6.09	5.96	4.89	6.22	5.33	4.33	5.34	5.19
Total Contract Rate.......	39.69	35.01	44.01	43.19	35.98	46.97	39.28	33.68	38.86	37.72

Note: Contract labor rates are defined as prevailing hourly rates.

Area	Carpenter	Carpet Layer	Electrician	Plumber	Painter	Elevator Mechanic	Brick Layer	Roofer	Crane Operator	Fork Lift Driver
Jackson, MS										
Basic Wage & Fringe.....	13.80	13.80	15.60	14.85	11.14	25.45	15.54	12.33	14.18	13.30
Comp. Insurance..........	2.72	2.72	1.12	1.18	1.42	2.72	2.12	3.91	1.56	1.46
Other Req'd Benefits.....	2.28	2.28	2.57	2.45	1.84	4.20	2.56	2.03	2.34	2.19
Total Contract Rate.......	18.80	18.80	19.30	18.48	14.40	32.37	20.23	18.27	18.08	16.95
Jacksonville, FL										
Basic Wage & Fringe.....	16.36	16.36	25.55	20.27	13.68	41.98	16.15	14.28	23.43	22.59
Comp. Insurance..........	5.92	5.92	3.19	2.52	3.40	3.81	4.36	7.67	3.76	3.62
Other Req'd Benefits.....	2.70	2.70	4.22	3.34	2.26	6.23	2.66	2.36	3.87	3.73
Total Contract Rate.......	24.98	24.98	32.96	26.14	19.34	47.81	23.18	24.30	31.05	29.94
Jersey City, NJ										
Basic Wage & Fringe.....	45.65	45.65	52.40	50.23	39.06	30.62	42.92	35.97	50.02	48.43
Comp. Insurance..........	5.11	5.11	2.02	2.89	3.50	1.43	5.16	10.20	3.49	3.38
Other Req'd Benefits.....	7.53	7.53	8.65	8.29	6.44	5.05	7.08	5.94	8.25	7.99
Total Contract Rate.......	58.29	58.29	63.06	61.41	49.01	37.10	55.16	52.11	61.76	59.80
Juneau, AK										
Basic Wage & Fringe.....	40.85	34.27	45.54	43.30	35.18	48.36	42.45	39.12	43.26	42.30
Comp. Insurance..........	4.89	4.10	4.39	3.79	4.90	5.30	7.11	14.46	8.02	7.84
Other Req'd Benefits.....	6.74	5.65	7.51	7.14	5.80	7.98	7.00	6.45	7.14	6.98
Total Contract Rate.......	52.48	44.03	57.45	54.23	45.88	61.64	56.56	60.04	58.41	57.12
Kalamazoo, MI										
Basic Wage & Fringe.....	28.98	28.98	29.59	35.84	26.95	41.20	29.97	25.74	36.91	35.91
Comp. Insurance..........	5.74	5.74	1.69	2.95	4.02	3.28	5.79	10.64	4.31	4.20
Other Req'd Benefits.....	4.78	4.78	4.88	5.91	4.45	6.80	4.95	4.25	6.09	5.93
Total Contract Rate.......	39.50	39.50	36.16	44.71	35.42	51.27	40.71	40.63	47.31	46.03
Kansas City, MO										
Basic Wage & Fringe.....	36.03	31.65	44.55	41.68	31.48	38.05	34.25	34.44	34.46	32.31
Comp. Insurance..........	6.20	5.45	3.51	4.03	4.74	3.62	6.30	12.69	3.97	3.73
Other Req'd Benefits.....	5.94	5.22	7.35	6.88	5.19	6.28	5.65	5.68	5.69	5.33
Total Contract Rate.......	48.18	42.32	55.41	52.58	41.41	47.95	46.20	52.81	44.12	41.37
Knoxville, TN										
Basic Wage & Fringe.....	17.38	17.38	25.33	26.25	17.40	31.62	17.94	13.87	15.56	14.56
Comp. Insurance..........	3.54	3.54	1.86	2.50	2.21	2.54	2.57	4.64	1.91	1.79
Other Req'd Benefits.....	2.87	2.87	4.18	4.33	2.87	5.22	2.96	2.29	2.57	2.40
Total Contract Rate.......	23.79	23.79	31.37	33.08	22.48	39.37	23.47	20.80	20.04	18.75
Kokomo, IN										
Basic Wage & Fringe.....	28.38	21.85	34.71	36.15	27.07	37.67	31.72	20.25	31.65	31.65
Comp. Insurance..........	1.98	1.52	.93	1.07	1.31	3.04	1.62	2.37	1.11	1.11
Other Req'd Benefits.....	4.68	3.61	5.73	5.96	4.47	6.22	5.23	3.34	5.22	5.22
Total Contract Rate.......	35.04	26.98	41.36	43.19	32.85	46.92	38.58	25.96	37.98	37.98
Lansing, MI										
Basic Wage & Fringe.....	30.70	30.70	36.33	38.36	29.38	41.20	32.74	27.35	36.91	35.91
Comp. Insurance..........	6.08	6.08	2.07	3.16	4.39	3.28	6.33	11.30	4.31	4.20
Other Req'd Benefits.....	5.07	5.07	5.99	6.33	4.85	6.80	5.40	4.51	6.09	5.93
Total Contract Rate.......	41.84	41.84	44.40	47.85	38.61	51.27	44.47	43.17	47.31	46.03

Note: Contract labor rates are defined as prevailing hourly rates.

.3 Contract Labor Rates, Selected Metro Areas

Area	Carpenter	Carpet Layer	Electrician	Plumber	Painter	Elevator Mechanic	Brick Layer	Roofer	Crane Operator	Fork Lift Driver
Las Cruces, NM										
Basic Wage & Fringe.....	24.86	22.06	25.14	30.93	19.43	34.41	22.34	15.29	25.20	22.66
Comp. Insurance..........	3.64	3.23	1.61	2.38	1.99	2.84	2.89	4.40	2.08	1.87
Other Req'd Benefits. ...	4.10	3.64	4.15	5.10	3.21	5.68	3.69	2.52	4.16	3.74
Total Contract Rate.......	32.61	28.93	30.90	38.42	24.62	42.93	28.91	22.21	31.44	28.27
Las Vegas, NV										
Basic Wage & Fringe. ...	38.02	22.06	42.19	43.62	32.96	47.67	26.29	24.60	43.95	43.62
Comp. Insurance..........	5.28	3.06	3.24	4.44	3.70	3.24	3.20	5.83	5.04	5.00
Other Req'd Benefits.....	6.27	3.64	6.96	7.20	5.44	7.87	4.34	4.06	7.25	7.20
Total Contract Rate.... ..	49.57	28.76	52.39	55.25	42.10	58.78	33.82	34.48	56.24	55.82
Lawton, OK										
Basic Wage & Fringe.....	19.70	14.92	28.01	29.35	17.15	31.62	26.96	19.50	24.10	23.60
Comp. Insurance........ ..	3.81	2.88	2.08	2.56	2.93	3.88	4.53	8.24	4.45	4.35
Other Req'd Benefits.....	3.25	2.46	4.62	4.84	2.83	5.22	4.45	3.22	3.98	3.89
Total Contract Rate.......	26.76	20.27	34.71	36.75	22.91	40.72	35.94	30.96	32.52	31.85
Lewiston, ME										
Basic Wage & Fringe.....	23.20	23.20	33.53	28.76	10.16	34.48	17.78	13.38	28.46	28.34
Comp. Insurance..........	10.11	10.11	1.69	2.11	1.63	1.18	3.07	4.29	3.47	3.45
Other Req'd Benefits... .	3.83	3.83	5.53	4.75	1.68	5.69	2.93	2.21	4.70	4.68
Total Contract Rate.......	37.14	37.14	40.75	35.61	13.47	41.35	23.78	19.88	36.63	36.47
Lexington, KY										
Basic Wage & Fringe... .	22.71	22.71	20.40	18.98	17.52	35.32	15.85	31.24	28.35	28.35
Comp. Insurance..........	4.36	4.36	1.40	1.04	1.91	2.66	1.71	5.41	5.10	5.10
Other Req'd Benefits.....	3.75	3.75	3.37	3.13	2.89	5.83	2.62	5.15	4.68	4.68
Total Contract Rate......	30.82	30.82	25.16	23.15	22.32	43.80	20.18	41.81	38.13	38.13
Little Rock, AR										
Basic Wage & Fringe....	22.29	22.29	28.59	26.79	18.89	32.44	23.71	12.36	16.35	14.98
Comp. Insurance..........	3.27	3.27	1.78	1.70	1.86	.99	2.55	2.74	1.48	1.36
Other Req'd Benefits.....	3.68	3.68	4.72	4.42	3.12	5.35	3.91	2.04	2.70	2.47
Total Contract Rate.......	29.24	29.24	35.09	32.91	23.87	38.78	30.17	17.13	20.53	18.81
Los Angeles, CA										
Basic Wage & Fringe.....	37.45	32.95	43.16	41.21	31.33	44.90	38.21	31.22	44.04	43.93
Comp. Insurance..........	10.59	9.32	4.29	4.78	6.09	2.59	5.56	12.42	3.47	3.46
Other Req'd Benefits.....	6.18	5.44	7.12	6.80	5.17	7.41	6.30	5.15	7.27	7.25
Total Contract Rate.......	54.22	47.71	54.57	52.79	42.59	54.90	50.07	48.79	54.78	54.64
Louisville, KY										
Basic Wage & Fringe.....	26.07	26.07	32.49	31.32	23.84	35.32	25.51	23.63	28.65	28.65
Comp. Insurance..........	5.01	5.01	2.23	1.72	2.60	2.66	2.76	4.10	5.15	5.15
Other Req'd Benefits.....	4.30	4.30	5.36	5.17	3.93	5.83	4.21	3.90	4.73	4.73
Total Contract Rate.......	35.38	35.38	40.08	38.20	30.37	43.80	32.48	31.62	38.53	38.53
Lowell, MA										
Basic Wage & Fringe.....	39.66	44.70	50.10	48.95	37.43	52.12	44.58	34.61	43.44	43.23
Comp. Insurance..........	6.58	7.42	1.85	2.49	3.09	4.45	6.44	11.52	2.78	2.77
Other Req'd Benefits.....	6.54	7.38	8.27	8.08	6.18	8.60	7.36	5.71	7.17	7.13
Total Contract Rate.......	52.79	59.50	60.22	59.52	46.70	65.17	58.37	51.84	53.39	53.13

Note: Contract labor rates are defined as prevailing hourly rates.

3.3 Contract Labor Rates, Selected Metro Areas

Area	Carpenter	Carpet Layer	Electrician	Plumber	Painter	Elevator Mechanic	Brick Layer	Roofer	Crane Operator	Fork Lift Driver
Lubbock, TX										
Basic Wage & Fringe.....	15.23	14.31	19.02	22.65	11.19	38.59	22.49	13.90	25.30	24.30
Comp. Insurance...........	2.02	1.90	1.40	1.83	1.11	2.70	3.08	3.17	2.65	2.55
Other Req'd Benefits.....	2.51	2.36	3.14	3.74	1.85	6.37	3.71	2.29	4.17	4.01
Total Contract Rate.......	19.76	18.57	23.56	28.21	14.14	47.65	29.28	19.36	32.13	30.86
Macon, GA										
Basic Wage & Fringe.....	17.60	17.60	23.79	16.41	14.05	38.68	12.25	19.60	25.96	24.52
Comp. Insurance...........	4.34	4.34	2.17	1.85	2.60	3.32	2.61	7.89	3.83	3.62
Other Req'd Benefits.....	2.90	2.90	3.93	2.71	2.32	6.38	2.02	3.23	4.28	4.05
Total Contract Rate.......	24.84	24.84	29.89	20.97	18.96	48.38	16.88	30.72	34.08	32.18
Madison, WI										
Basic Wage & Fringe.....	32.07	32.07	35.78	36.79	29.10	40.99	35.17	27.95	37.87	36.84
Comp. Insurance...........	5.92	5.92	1.71	2.35	3.33	6.14	6.13	11.38	2.71	2.64
Other Req'd Benefits.....	5.29	5.29	5.90	6.07	4.80	6.76	5.80	4.61	6.25	6.08
Total Contract Rate.......	43.28	43.28	43.39	45.21	37.23	53.89	47.11	43.95	46.83	45.56
Manchester, NH										
Basic Wage & Fringe.....	18.81	18.81	30.59	34.25	27.20	34.74	32.93	28.21	21.66	21.66
Comp. Insurance...........	4.04	4.04	2.08	3.89	4.11	1.00	7.20	17.40	3.80	3.80
Other Req'd Benefits.....	3.10	3.10	5.05	5.65	4.49	5.73	5.43	4.65	3.57	3.57
Total Contract Rate.......	25.96	25.96	37.72	43.80	35.80	41.47	45.56	50.27	29.04	29.04
Marquette, MI										
Basic Wage & Fringe.....	30.17	30.17	32.52	36.48	14.50	41.20	29.64	19.75	35.09	31.84
Comp. Insurance...........	5.97	5.97	1.86	3.01	2.16	3.28	5.73	8.16	4.10	3.72
Other Req'd Benefits.....	4.98	4.98	5.37	6.02	2.39	6.80	4.89	3.26	5.79	5.25
Total Contract Rate.......	41.12	41.12	39.74	45.51	19.06	51.27	40.26	31.17	44.98	40.82
Medford, OR										
Basic Wage & Fringe.....	37.27	31.27	36.81	43.57	19.46	48.77	35.15	22.77	37.95	36.63
Comp. Insurance...........	6.33	5.31	1.84	2.64	2.75	2.35	5.32	5.23	4.00	3.86
Other Req'd Benefits.....	6.15	5.16	6.07	7.19	3.21	8.05	5.80	3.76	6.26	6.04
Total Contract Rate.......	49.75	41.74	44.72	53.40	25.42	59.16	46.27	31.76	48.22	46.54
Memphis, TN										
Basic Wage & Fringe.....	23.35	23.35	31.69	30.31	18.69	33.47	32.50	21.27	24.60	23.21
Comp. Insurance...........	4.76	4.76	2.32	2.89	2.38	2.68	4.65	7.12	3.02	2.85
Other Req'd Benefits.....	3.85	3.85	5.23	5.00	3.08	5.52	5.36	3.51	4.06	3.83
Total Contract Rate.......	31.96	31.96	39.24	38.20	24.15	41.68	42.51	31.90	31.67	29.89
Miami, FL										
Basic Wage & Fringe.....	20.86	20.86	29.10	27.95	14.19	41.98	19.90	15.87	21.60	18.75
Comp. Insurance...........	7.54	7.54	3.63	3.48	3.53	4.23	5.37	8.52	3.46	3.01
Other Req'd Benefits.....	3.44	3.44	4.80	4.61	2.34	6.93	3.28	2.62	3.56	3.09
Total Contract Rate.......	31.84	31.84	37.54	36.04	20.06	53.14	28.56	27.01	28.63	24.85
Milwaukee, WI										
Basic Wage & Fringe.....	36.99	36.99	41.23	38.24	32.56	38.88	38.78	30.73	42.14	41.85
Comp. Insurance...........	6.83	6.83	1.97	2.45	3.72	5.82	6.76	12.52	3.02	3.00
Other Req'd Benefits.....	6.10	6.10	6.80	6.31	5.37	6.42	6.40	5.07	6.95	6.91
Total Contract Rate.......	49.92	49.92	50.00	47.00	41.66	51.12	51.94	48.32	52.11	51.75

Note: Contract labor rates are defined as prevailing hourly rates.

.3 Contract Labor Rates, Selected Metro Areas

Area	Carpenter	Carpet Layer	Electrician	Plumber	Painter	Elevator Mechanic	Brick Layer	Roofer	Crane Operator	Fork Lift Driver
Minneapolis, MN										
Basic Wage & Fringe.....	38.34	35.17	44.25	43.24	37.59	41.46	41.59	37.94	40.29	39.78
Comp. Insurance...........	14.09	12.92	3.00	4.57	7.10	3.76	10.45	24.87	6.33	6.25
Other Req'd Benefits. ...	6.33	5.80	7.30	7.13	6.20	6.84	6.86	6.26	6.65	6.56
Total Contract Rate.......	58.75	53.89	54.56	54.94	50.89	52.06	58.90	69.07	53.27	52.60
Mobile, AL										
Basic Wage & Fringe.....	16.13	16.13	20.48	26.11	15.73	31.48	17.16	17.96	21.58	16.30
Comp. Insurance...........	5.40	5.40	1.88	3.10	4.53	2.87	5.31	12.35	3.27	2.47
Other Req'd Benefits.. ..	2.66	2.66	3.38	4.31	2.60	5.19	2.83	2.96	3.56	2.69
Total Contract Rate.......	24.19	24.19	25.74	33.52	22.86	39.54	25.31	33.28	28.41	21.46
Moline, IL										
Basic Wage & Fringe.....	32.01	32.01	34.86	38.30	28.67	34.57	31.67	29.80	33.45	33.45
Comp. Insurance...........	6.01	6.01	2.87	3.78	2.89	1.06	5.33	8.76	3.15	3.15
Other Req'd Benefits... .	5.28	5.28	5.75	6.32	4.73	5.70	5.23	4.92	5.52	5.52
Total Contract Rate.......	43.30	43.30	43.48	48.40	36.29	41.33	42.23	43.47	42.12	42.12
Muncie, IN										
Basic Wage & Fringe.....	28.36	25.30	32.55	35.58	22.44	37.71	30.09	20.25	32.38	31.43
Comp. Insurance...........	1.97	1.76	.87	1.06	1.09	3.04	1.54	2.37	1.13	1.10
Other Req'd Benefits....	4.68	4.17	5.37	5.87	3.70	6.22	4.96	3.34	5.34	5.19
Total Contract Rate.......	35.01	31.24	38.79	42.51	27.23	46.97	36.60	25.96	38.86	37.72
Nashville, TN										
Basic Wage & Fringe.....	23.61	23.61	28.23	29.96	21.93	32.74	26.23	21.27	22.78	20.51
Comp. Insurance...........	4.81	4.81	2.07	2.85	2.79	2.63	3.75	7.12	2.79	2.51
Other Req'd Benefits.....	3.90	3.90	4.66	4.94	3.62	5.40	4.33	3.51	3.76	3.38
Total Contract Rate.......	32.32	32.32	34.96	37.76	28.34	40.77	34.31	31.90	29.33	26.41
New Brunswick, NJ										
Basic Wage & Fringe.....	45.65	20.85	50.68	48.38	39.06	30.62	42.92	35.97	50.02	48.43
Comp. Insurance...........	5.11	2.33	1.95	2.78	3.50	1.43	5.16	10.20	3.49	3.38
Other Req'd Benefits.....	7.53	3.44	8.36	7.98	6.44	5.05	7.08	5.94	8.25	7.99
Total Contract Rate.......	58.29	26.62	60.99	59.14	49.01	37.10	55.16	52.11	61.76	59.80
New Haven, CT										
Basic Wage & Fringe.....	33.18	33.18	43.88	40.03	32.62	45.23	38.03	34.85	38.12	37.25
Comp. Insurance...........	9.92	9.92	4.19	4.59	6.14	4.88	10.69	18.31	4.91	4.80
Other Req'd Benefits.....	5.47	5.47	7.24	6.60	5.38	7.46	6.27	5.75	6.29	6.15
Total Contract Rate.......	48.58	48.58	55.31	51.23	44.14	57.57	55.00	58.91	49.32	48.19
New Orleans, LA										
Basic Wage & Fringe.....	23.62	23.62	27.65	26.43	17.59	38.05	23.85	13.96	24.85	22.43
Comp. Insurance...........	12.56	12.56	2.73	2.28	5.20	4.48	6.76	10.77	4.35	3.92
Other Req'd Benefits.....	3.90	3.90	4.56	4.36	2.90	6.28	3.94	2.30	4.10	3.70
Total Contract Rate.......	40.08	40.08	34.94	33.07	25.70	48.81	34.54	27.03	33.30	30.05
New York, NY										
Basic Wage & Fringe.....	65.18	65.18	70.47	65.81	46.26	58.47	55.03	48.86	63.93	57.66
Comp. Insurance...........	9.69	9.69	4.13	4.91	5.73	2.85	9.74	14.14	5.27	4.76
Other Req'd Benefits.....	10.75	10.75	11.63	10.86	7.63	9.65	9.08	8.06	10.55	9.51
Total Contract Rate.......	85.63	85.63	86.23	81.58	59.62	70.97	73.85	71.06	79.75	71.93

Note: Contract labor rates are defined as prevailing hourly rates.

Area	Carpenter	Carpet Layer	Electrician	Plumber	Painter	Elevator Mechanic	Brick Layer	Roofer	Crane Operator	Fork Lift Driver
Newark, NJ										
Basic Wage & Fringe.....	45.65	45.65	52.40	50.23	39.06	30.62	42.92	35.97	50.02	48.43
Comp. Insurance...........	5.11	5.11	2.02	2.89	3.50	1.43	5.16	10.20	3.49	3.38
Other Req'd Benefits.....	7.53	7.53	8.65	8.29	6.44	5.05	7.08	5.94	8.25	7.99
Total Contract Rate.......	58.29	58.29	63.06	61.41	49.01	37.10	55.16	52.11	61.76	59.80
Newport News, VA										
Basic Wage & Fringe.....	21.78	21.78	24.53	26.62	13.90	32.28	23.74	13.24	23.96	20.08
Comp. Insurance...........	2.47	2.47	1.02	1.61	1.41	2.03	1.90	2.97	1.78	1.49
Other Req'd Benefits.....	3.59	3.59	4.05	4.39	2.29	5.33	3.92	2.18	3.95	3.31
Total Contract Rate.......	27.85	27.85	29.60	32.62	17.60	39.64	29.56	18.39	29.69	24.88
Norfolk, VA										
Basic Wage & Fringe.....	21.78	21.78	25.37	28.74	18.88	32.28	23.74	13.24	23.96	20.08
Comp. Insurance...........	2.47	2.47	1.05	1.74	1.92	2.03	1.90	2.97	1.78	1.49
Other Req'd Benefits.....	3.59	3.59	4.19	4.74	3.12	5.33	3.92	2.18	3.95	3.31
Total Contract Rate.......	27.85	27.85	30.61	35.22	23.91	39.64	29.56	18.39	29.69	24.88
Norwalk, CT										
Basic Wage & Fringe.....	33.18	33.18	43.68	40.03	32.62	45.23	38.93	34.85	38.12	37.25
Comp. Insurance...........	9.92	9.92	4.19	4.59	6.14	4.88	10.95	18.31	4.91	4.80
Other Req'd Benefits.....	5.47	5.47	7.24	6.60	5.38	7.46	6.42	5.75	6.29	6.15
Total Contract Rate.......	48.58	48.58	55.31	51.23	44.14	57.57	56.30	58.91	49.32	48.19
Norwich, CT										
Basic Wage & Fringe.....	33.18	33.18	43.88	40.03	32.62	45.23	38.03	32.99	38.12	37.25
Comp. Insurance...........	9.92	9.92	4.19	4.59	6.14	4.88	10.69	17.33	4.91	4.80
Other Req'd Benefits.....	5.47	5.47	7.24	6.60	5.38	7.46	6.27	5.44	6.29	6.15
Total Contract Rate.......	48.58	48.58	55.31	51.23	44.14	57.57	55.00	55.76	49.32	48.19
Oakland, CA										
Basic Wage & Fringe.....	42.52	44.28	53.68	50.56	40.12	53.74	42.00	32.65	45.51	42.86
Comp. Insurance...........	12.02	12.52	5.33	5.86	7.80	3.10	6.11	12.99	3.59	3.38
Other Req'd Benefits.....	7.02	7.31	8.86	8.34	6.62	8.87	6.93	5.39	7.51	7.07
Total Contract Rate.......	61.56	64.11	67.87	64.76	54.54	65.71	55.04	51.03	56.61	53.31
Ogden, UT										
Basic Wage & Fringe.....	21.55	13.57	31.31	30.91	15.13	34.53	22.30	19.65	24.73	24.61
Comp. Insurance...........	2.27	1.43	2.21	1.98	2.31	2.41	3.07	5.16	1.58	1.57
Other Req'd Benefits.....	3.56	2.24	5.17	5.10	2.50	5.70	3.68	3.24	4.08	4.06
Total Contract Rate.......	27.37	17.24	38.68	37.99	19.94	42.64	29.05	28.05	30.39	30.24
Oklahoma City, OK										
Basic Wage & Fringe.....	15.65	16.12	28.01	29.35	17.15	31.62	26.96	19.50	24.10	23.60
Comp. Insurance...........	3.03	3.12	2.08	2.56	2.93	3.88	4.53	8.24	4.45	4.35
Other Req'd Benefits.....	2.58	2.66	4.62	4.84	2.83	5.22	4.45	3.22	3.98	3.89
Total Contract Rate.......	21.26	21.90	34.71	36.75	22.91	40.72	35.94	30.96	32.52	31.85
Olympia, WA										
Basic Wage & Fringe.....	37.90	32.83	41.68	39.52	30.18	46.13	36.87	34.53	39.19	38.74
Comp. Insurance...........	3.28	2.84	1.19	1.85	2.74	1.15	4.73	6.78	3.26	3.22
Other Req'd Benefits.....	6.25	5.42	6.88	6.52	4.98	7.61	6.08	5.70	6.47	6.39
Total Contract Rate.......	47.44	41.09	49.75	47.89	37.90	54.89	47.68	47.01	48.91	48.35

Note: Contract labor rates are defined as prevailing hourly rates.

.3 Contract Labor Rates, Selected Metro Areas

Area	Carpenter	Carpet Layer	Electrician	Plumber	Painter	Elevator Mechanic	Brick Layer	Roofer	Crane Operator	Fork Lift Driver
Omaha, NE										
Basic Wage & Fringe.....	25.86	16.70	34.17	33.67	23.18	33.57	27.80	13.95	23.04	21.09
Comp. Insurance...........	5.42	3.50	2.84	3.76	2.86	2.03	5.14	5.12	2.81	2.57
Other Req'd Benefits.....	4.27	2.76	5.64	5.56	3.82	5.54	4.59	2.30	3.80	3.48
Total Contract Rate.......	35.54	22.95	42.65	42.99	29.86	41.14	37.53	21.37	29.65	27.14
Orlando, FL										
Basic Wage & Fringe.....	23.91	23.91	17.64	19.91	15.70	41.98	22.35	19.35	23.43	22.59
Comp. Insurance...........	8.65	8.65	2.20	2.48	3.90	3.99	6.03	10.39	3.76	3.62
Other Req'd Benefits.....	3.95	3.95	2.91	3.29	2.59	6.53	3.69	3.19	3.87	3.73
Total Contract Rate.......	36.50	36.50	22.75	25.67	22.19	50.10	32.07	32.93	31.05	29.94
Owensboro, KY										
Basic Wage & Fringe.....	20.41	20.41	32.04	20.30	29.23	36.50	15.87	19.25	25.75	25.75
Comp. Insurance...........	3.92	3.92	2.19	1.11	3.19	2.74	1.72	3.34	4.63	4.63
Other Req'd Benefits.....	3.37	3.37	5.29	3.35	4.82	6.02	2.62	3.18	4.25	4.25
Total Contract Rate.......	27.70	27.70	39.52	24.76	37.24	45.27	20.20	25.76	34.63	34.63
Oxnard, CA										
Basic Wage & Fringe.....	37.45	32.95	42.35	41.21	30.23	44.90	32.69	31.22	44.04	43.93
Comp. Insurance...........	10.59	9.32	4.21	4.78	5.88	2.59	4.76	12.42	3.47	3.46
Other Req'd Benefits.....	6.18	5.44	6.99	6.80	4.99	7.41	5.39	5.15	7.27	7.25
Total Contract Rate.......	54.22	47.71	53.54	52.79	41.10	54.90	42.84	48.79	54.78	54.64
Parkersburg, WV										
Basic Wage & Fringe.....	31.58	31.58	37.56	36.90	31.38	36.94	31.86	26.14	35.29	34.09
Comp. Insurance...........	3.79	3.79	19.27	1.76	3.83	1.95	3.69	3.51	2.73	2.64
Other Req'd Benefits.....	5.21	5.21	6.20	6.09	5.18	6.10	5.26	4.31	5.82	5.62
Total Contract Rate.......	40.58	40.58	63.03	44.75	40.39	44.99	40.80	33.97	43.84	42.35
Peoria, IL										
Basic Wage & Fringe.....	33.68	33.68	35.53	37.68	30.30	35.37	32.12	32.20	35.26	35.26
Comp. Insurance...........	6.32	6.32	2.92	3.72	3.06	1.08	5.41	9.46	3.33	3.33
Other Req'd Benefits.....	5.56	5.56	5.86	6.22	5.00	5.84	5.30	5.31	5.82	5.82
Total Contract Rate.......	45.56	45.56	44.32	47.62	38.36	42.29	42.83	46.97	44.40	44.40
Philadelphia, PA										
Basic Wage & Fringe.....	43.91	43.58	51.69	50.74	39.40	48.72	43.06	40.45	43.39	38.10
Comp. Insurance...........	5.39	5.35	33.39	3.64	5.14	6.63	5.02	10.46	3.43	3.01
Other Req'd Benefits.....	7.25	7.19	8.53	8.37	6.50	8.04	7.10	6.67	7.16	6.29
Total Contract Rate.......	56.54	56.12	93.61	62.76	51.04	63.38	55.18	57.59	53.98	47.40
Phoenix, AZ										
Basic Wage & Fringe.....	24.42	24.42	27.42	33.10	20.07	33.80	26.57	18.85	27.10	26.07
Comp. Insurance...........	3.21	3.21	.98	1.20	.85	1.82	1.64	2.11	1.40	1.35
Other Req'd Benefits.....	4.03	4.03	4.52	5.46	3.31	5.58	4.38	3.11	4.47	4.30
Total Contract Rate.......	31.66	31.66	32.92	39.77	24.23	41.20	32.60	24.07	32.97	31.72
Pittsburgh, PA										
Basic Wage & Fringe.....	32.44	32.46	41.50	39.91	32.78	44.68	34.16	30.39	35.90	35.90
Comp. Insurance...........	3.98	3.98	26.81	2.87	4.27	6.08	3.98	7.86	2.84	2.84
Other Req'd Benefits.....	5.35	5.36	6.85	6.59	5.41	7.37	5.64	5.01	5.92	5.92
Total Contract Rate.......	41.77	41.80	75.16	49.36	42.46	58.13	43.78	43.27	44.66	44.66

Note: Contract labor rates are defined as prevailing hourly rates.

3.3 Contract Labor Rates, Selected Metro Areas

Area	Carpenter	Carpet Layer	Electrician	Plumber	Painter	Elevator Mechanic	Brick Layer	Roofer	Crane Operator	Fork Lift Driver
Pocatello, ID										
Basic Wage & Fringe.....	29.17	29.17	31.02	31.71	17.76	37.69	19.22	17.17	21.25	21.07
Comp. Insurance...........	4.08	4.08	1.54	1.24	1.26	3.49	1.54	4.70	1.14	1.13
Other Req'd Benefits.....	4.81	4.81	5.12	5.23	2.93	6.22	3.17	2.83	3.51	3.48
Total Contract Rate.......	38.06	38.06	37.68	38.19	21.95	47.40	23.93	24.70	25.90	25.68
Portland, ME										
Basic Wage & Fringe.....	23.20	23.20	33.53	28.76	10.16	34.48	17.78	13.38	28.46	28.34
Comp. Insurance...........	10.11	10.11	1.69	2.11	1.63	1.18	3.07	4.29	3.47	3.45
Other Req'd Benefits.....	3.83	3.83	5.53	4.75	1.68	5.69	2.93	2.21	4.70	4.68
Total Contract Rate.......	37.14	37.14	40.75	35.61	13.47	41.35	23.78	19.88	36.63	36.47
Portland, OR										
Basic Wage & Fringe.....	37.27	31.27	44.43	45.81	23.13	48.77	37.82	30.58	37.95	36.63
Comp. Insurance...........	6.33	5.31	2.22	2.78	3.27	2.35	5.72	7.03	4.00	3.86
Other Req'd Benefits.....	6.15	5.16	7.33	7.56	3.82	8.05	6.24	5.05	6.26	6.04
Total Contract Rate.......	49.75	41.74	53.98	56.15	30.22	59.16	49.78	42.66	48.22	46.54
Providence, RI										
Basic Wage & Fringe.....	37.42	37.42	39.56	39.57	32.13	40.62	40.25	33.05	37.92	37.70
Comp. Insurance...........	6.76	6.76	17.53	3.37	7.75	2.94	10.11	11.21	3.94	3.91
Other Req'd Benefits.....	6.17	6.17	6.53	6.53	5.30	6.70	6.64	5.45	6.26	6.22
Total Contract Rate.......	50.36	50.36	63.61	49.47	45.18	50.27	57.00	49.71	48.11	47.83
Pueblo, CO										
Basic Wage & Fringe.....	28.30	17.73	33.18	27.67	14.32	41.70	28.22	20.93	26.54	26.39
Comp. Insurance...........	3.44	2.15	1.89	2.30	1.61	3.34	3.98	5.24	3.16	3.14
Other Req'd Benefits.....	4.67	2.93	5.47	4.57	2.36	6.88	4.66	3.45	4.38	4.35
Total Contract Rate.......	36.41	22.81	40.55	34.54	18.29	51.92	36.86	29.62	34.08	33.88
Raleigh-Durham, NC										
Basic Wage & Fringe.....	14.91	14.91	15.58	19.12	12.50	31.66	14.84	10.00	18.75	18.50
Comp. Insurance...........	2.69	2.69	1.20	1.54	1.20	1.96	1.40	2.57	1.58	1.56
Other Req'd Benefits.....	2.46	2.46	2.57	3.15	2.06	5.22	2.45	1.65	3.09	3.05
Total Contract Rate.......	20.06	20.06	19.35	23.82	15.76	38.85	18.69	14.22	23.43	23.11
Rapid City, SD										
Basic Wage & Fringe.....	12.20	12.20	21.43	15.20	12.50	35.08	25.10	12.00	17.12	15.31
Comp. Insurance...........	1.83	1.83	1.21	1.30	1.64	1.55	2.08	2.48	1.96	1.75
Other Req'd Benefits.....	2.01	2.01	3.54	2.51	2.06	5.79	4.14	1.98	2.82	2.53
Total Contract Rate.......	16.05	16.05	26.17	19.01	16.20	42.42	31.32	16.46	21.90	19.58
Reading, PA										
Basic Wage & Fringe.....	27.60	27.60	32.66	39.39	27.45	38.99	29.39	40.45	33.94	30.16
Comp. Insurance...........	3.39	3.39	21.10	2.83	3.58	5.30	3.42	10.46	2.68	2.39
Other Req'd Benefits.....	4.55	4.55	5.39	6.50	4.53	6.43	4.85	6.67	5.60	4.98
Total Contract Rate.......	35.54	35.54	59.15	48.72	35.56	50.73	37.66	57.59	42.22	37.52
Reno, NV										
Basic Wage & Fringe.....	38.02	25.37	35.19	36.90	25.59	53.67	27.02	24.60	42.31	41.99
Comp. Insurance...........	5.28	3.52	2.70	3.75	2.88	3.65	3.29	5.83	4.85	4.82
Other Req'd Benefits.....	6.27	4.19	5.81	6.09	4.22	8.86	4.46	4.06	6.98	6.93
Total Contract Rate.......	49.57	33.08	43.70	46.74	32.69	66.18	34.76	34.48	54.14	53.73

Note: Contract labor rates are defined as prevailing hourly rates.

.3 Contract Labor Rates, Selected Metro Areas

Area	Carpenter	Carpet Layer	Electrician	Plumber	Painter	Elevator Mechanic	Brick Layer	Roofer	Crane Operator	Fork Lift Driver
Richland, WA										
Basic Wage & Fringe.....	29.83	29.83	37.21	42.44	24.29	41.46	31.97	24.27	28.76	28.44
Comp. Insurance......	2.58	2.58	1.06	1.99	2.20	1.03	4.10	4.76	2.39	2.36
Other Req'd Benefits.....	4.92	4.92	6.14	7.00	4.01	6.84	5.28	4.00	4.75	4.69
Total Contract Rate.......	37.34	37.34	44.41	51.43	30.50	49.33	41.34	33.04	35.90	35.50
Richmond, VA										
Basic Wage & Fringe.....	21.03	21.03	29.20	28.60	18.17	35.65	21.70	13.24	23.96	20.08
Comp. Insurance....... ...	2.39	2.39	1.21	1.73	1.84	2.24	1.74	2.97	1.78	1.49
Other Req'd Benefits.....	3.47	3.47	4.82	4.72	3.00	5.88	3.58	2.18	3.95	3.31
Total Contract Rate.......	26.89	26.89	35.23	35.05	23.01	43.77	27.02	18.39	29.69	24.88
Riverside, CA										
Basic Wage & Fringe.....	37.45	32.95	40.79	41.21	30.91	44.90	35.62	30.17	44.04	43.93
Comp. Insurance........ ..	10.59	9.32	4.05	4.78	6.01	2.59	5.18	12.00	3.47	3.46
Other Req'd Benefits.....	6.18	5.44	6.73	6.80	5.10	7.41	5.88	4.98	7.27	7.25
Total Contract Rate.......	54.22	47.71	51.57	52.79	42.02	54.90	46.68	47.15	54.78	54.64
Roanoke, VA										
Basic Wage & Fringe.....	12.63	12.63	14.65	17.98	10.85	31.85	17.01	13.24	14.35	13.95
Comp. Insurance......... .	1.43	1.43	.61	1.09	1.10	2.00	1.36	2.97	1.06	1.04
Other Req'd Benefits.....	2.08	2.08	2.42	2.97	1.79	5.26	2.81	2.18	2.37	2.30
Total Contract Rate.......	16.15	16.15	17.68	22.03	13.74	39.11	21.18	18.39	17.78	17.29
Rochester, MN										
Basic Wage & Fringe.....	30.31	30.31	36.70	39.06	30.62	41.46	36.91	25.16	40.29	39.78
Comp. Insurance..........	11.14	11.14	2.49	4.12	5.78	3.76	9.28	16.49	6.33	6.25
Other Req'd Benefits.....	5.00	5.00	6.06	6.44	5.05	6.84	6.09	4.15	6.65	6.56
Total Contract Rate.......	46.45	46.45	45.25	49.63	41.46	52.06	52.28	45.80	53.27	52.60
Rochester, NY										
Basic Wage & Fringe.....	33.51	30.06	36.81	35.92	30.64	38.59	33.63	30.77	40.97	39.84
Comp. Insurance...........	4.98	4.47	2.16	2.68	3.79	1.88	5.95	8.90	3.38	3.29
Other Req'd Benefits.....	5.53	4.96	6.07	5.93	5.06	6.37	5.55	5.08	6.76	6.57
Total Contract Rate.......	44.02	39.49	45.04	44.53	39.49	46.84	45.13	44.75	51.11	49.70
Rockford, IL										
Basic Wage & Fringe.....	36.22	36.22	44.08	39.96	34.20	40.99	34.61	37.55	39.95	39.25
Comp. Insurance...........	6.79	6.79	3.63	3.94	3.45	1.25	5.82	11.03	3.77	3.70
Other Req'd Benefits.....	5.98	5.98	7.27	6.59	5.64	6.76	5.71	6.20	6.59	6.48
Total Contract Rate.......	48.99	48.99	54.98	50.50	43.29	49.01	46.15	54.78	50.31	49.43
Rutland, VT										
Basic Wage & Fringe.....	16.41	16.41	27.96	30.05	13.15	41.83	24.05	17.25	20.33	20.13
Comp. Insurance...........	3.19	3.19	1.71	3.04	1.37	1.65	4.70	4.96	2.07	2.05
Other Req'd Benefits.....	2.71	2.71	4.61	4.96	2.17	6.90	3.97	2.85	3.35	3.32
Total Contract Rate.......	22.31	22.31	34.28	38.04	16.69	50.38	32.72	25.05	25.75	25.50
Sacramento, CA										
Basic Wage & Fringe.....	41.73	33.87	41.03	43.37	34.19	53.74	36.00	30.64	45.51	42.86
Comp. Insurance...........	11.80	9.58	4.07	5.03	7.78	3.10	5.24	12.19	3.59	3.38
Other Req'd Benefits.....	6.89	5.59	6.77	7.16	6.60	8.87	5.94	5.06	7.51	7.07
Total Contract Rate.......	60.42	49.04	51.87	55.55	54.41	65.71	47.18	47.89	56.61	53.31

Note: Contract labor rates are defined as prevailing hourly rates.

Area	Carpenter	Carpet Layer	Electrician	Plumber	Painter	Elevator Mechanic	Brick Layer	Roofer	Crane Operator	Fork Lift Driver
Saginaw, MI										
Basic Wage & Fringe.....	31.73	27.68	34.28	37.92	25.23	41.20	29.38	27.66	36.91	35.91
Comp. Insurance...........	6.28	5.48	1.96	3.12	3.77	3.28	5.68	11.43	4.31	4.20
Other Req'd Benefits.....	5.24	4.57	5.66	6.26	4.16	6.80	4.85	4.56	6.09	5.93
Total Contract Rate.......	43.25	37.73	41.89	47.30	33.16	51.27	39.91	43.66	47.31	46.03
Salem, OR										
Basic Wage & Fringe.....	37.27	31.27	41.02	43.57	23.13	48.77	37.82	24.87	37.95	36.63
Comp. Insurance...........	6.33	5.31	2.05	2.64	3.27	2.35	5.72	5.72	4.00	3.86
Other Req'd Benefits.....	6.15	5.16	6.77	7.19	3.82	8.05	6.24	4.10	6.26	6.04
Total Contract Rate.......	49.75	41.74	49.84	53.40	30.22	59.16	49.78	34.69	48.22	46.54
Salt Lake City, UT										
Basic Wage & Fringe.....	21.55	13.57	31.31	30.91	19.30	34.53	22.30	19.65	24.73	24.61
Comp. Insurance...........	2.27	1.43	2.21	1.98	2.95	2.41	3.07	5.16	1.58	1.57
Other Req'd Benefits.....	3.56	2.24	5.17	5.10	3.18	5.70	3.68	3.24	4.08	4.06
Total Contract Rate.......	27.37	17.24	38.68	37.99	25.44	42.64	29.05	28.05	30.39	30.24
San Antonio, TX										
Basic Wage & Fringe.....	19.69	19.43	25.41	30.30	18.30	33.01	25.08	18.61	27.05	23.06
Comp. Insurance...........	2.61	2.57	1.87	2.44	1.81	2.31	3.43	4.24	2.84	2.42
Other Req'd Benefits.....	3.25	3.21	4.19	5.00	3.02	5.45	4.14	3.07	4.46	3.80
Total Contract Rate.......	25.55	25.21	31.47	37.74	23.13	40.76	32.65	25.92	34.35	29.28
San Diego, CA										
Basic Wage & Fringe.....	33.18	26.97	39.42	43.46	30.91	44.90	34.75	25.58	44.04	43.93
Comp. Insurance...........	10.47	7.63	3.91	5.04	6.01	2.59	5.06	10.18	3.47	3.46
Other Req'd Benefits.....	6.11	4.45	6.50	7.17	5.10	7.41	5.73	4.22	7.27	7.25
Total Contract Rate.......	53.58	39.05	49.84	55.67	42.02	54.90	45.54	39.98	54.78	54.64
San Francisco, CA										
Basic Wage & Fringe.....	42.52	44.28	59.44	63.51	44.64	53.74	45.80	35.89	45.51	42.86
Comp. Insurance...........	12.02	12.52	5.90	7.36	8.68	3.10	6.66	14.28	3.59	3.38
Other Req'd Benefits.....	7.02	7.31	9.81	10.48	7.37	8.87	7.56	5.92	7.51	7.07
Total Contract Rate.......	61.56	64.11	75.15	81.35	60.69	65.71	60.02	56.09	56.61	53.31
San Jose, CA										
Basic Wage & Fringe.....	42.52	44.28	56.37	58.13	41.02	53.74	41.70	36.77	45.51	42.86
Comp. Insurance...........	12.02	12.52	5.60	6.74	7.98	3.10	6.07	14.63	3.59	3.38
Other Req'd Benefits.....	7.02	7.31	9.30	9.59	6.77	8.87	6.88	6.07	7.51	7.07
Total Contract Rate.......	61.56	64.11	71.27	74.46	55.77	65.71	54.65	57.47	56.61	53.31
Santa Barbara, CA										
Basic Wage & Fringe.....	37.45	32.95	42.81	41.21	30.23	44.90	35.75	30.17	44.04	43.93
Comp. Insurance...........	10.59	9.32	4.25	4.78	5.88	2.59	5.20	12.00	3.47	3.46
Other Req'd Benefits.....	6.18	5.44	7.06	6.80	4.99	7.41	5.90	4.98	7.27	7.25
Total Contract Rate.......	54.22	47.71	54.12	52.79	41.10	54.90	46.85	47.15	54.78	54.64
Savannah, GA										
Basic Wage & Fringe.....	15.28	15.28	20.97	17.25	14.25	32.30	15.65	11.75	21.10	19.27
Comp. Insurance...........	3.77	3.77	1.91	1.94	2.63	2.77	3.33	4.73	3.11	2.84
Other Req'd Benefits.....	2.52	2.52	3.46	2.85	2.35	5.33	2.58	1.94	3.48	3.18
Total Contract Rate.......	21.57	21.57	26.34	22.04	19.23	40.40	21.56	18.42	27.70	25.29

Note: Contract labor rates are defined as prevailing hourly rates.

.3 Contract Labor Rates, Selected Metro Areas

Area	Carpenter	Carpet Layer	Electrician	Plumber	Painter	Elevator Mechanic	Brick Layer	Roofer	Crane Operator	Fork Lift Driver
Scranton, PA										
Basic Wage & Fringe.....	27.79	27.79	34.38	34.60	27.45	38.99	32.68	37.05	33.84	30.72
Comp. Insurance....... ...	3.41	3.41	22.21	2.48	3.58	5.30	3.81	9.58	2.68	2.43
Other Req'd Benefits.....	4.59	4.59	5.67	5.71	4.53	6.43	5.39	6.11	5.58	5.07
Total Contract Rate.......	35.79	35.79	62.26	42.79	35.56	50.73	41.88	52.75	42.10	38.22
Seattle, WA										
Basic Wage & Fringe.....	37.90	32.83	46.83	49.16	30.18	46.13	38.37	34.53	39.19	38.74
Comp. Insurance...........	3.28	2.84	1.33	2.30	2.74	1.15	4.92	6.78	3.26	3.22
Other Req'd Benefits.. ..	6.25	5.42	7.73	8.11	4.98	7.61	6.33	5.70	6.47	6.39
Total Contract Rate.......	47.44	41.09	55.89	59.57	37.90	54.89	49.62	47.01	48.91	48.35
Shreveport, LA										
Basic Wage & Fringe.. ..	15.45	15.45	26.62	23.45	11.60	32.47	20.75	12.85	14.45	14.20
Comp. Insurance...........	8.21	8.21	2.63	2.03	3.43	3.82	5.88	9.91	2.53	2.48
Other Req'd Benefits.....	2.55	2.55	4.39	3.87	1.91	5.36	3.42	2.12	2.38	2.34
Total Contract Rate..... .	26.21	26.21	33.64	29.35	16.95	41.65	30.05	24.88	19.36	19.03
Sioux City, IA										
Basic Wage & Fringe.....	21.50	21.50	32.63	36.82	20.26	35.08	20.93	15.90	28.70	27.32
Comp. Insurance...........	2.47	2.47	1.29	1.79	1.67	1.28	1.63	2.84	1.81	1.72
Other Req'd Benefits.....	3.55	3.55	5.38	6.08	3.34	5.79	3.45	2.62	4.74	4.51
Total Contract Rate.......	27.52	27.52	39.31	44.68	25.28	42.15	26.02	21.36	35.24	33.55
Sioux Falls, SD										
Basic Wage & Fringe.....	13.97	13.97	30.46	14.65	12.50	35.08	24.60	12.00	17.12	15.31
Comp. Insurance..........	2.10	2.10	1.72	1.25	1.64	1.55	2.04	2.48	1.96	1.75
Other Req'd Benefits.....	2.31	2.31	5.03	2.42	2.06	5.79	4.06	1.98	2.82	2.53
Total Contract Rate.......	18.38	18.38	37.20	18.32	16.20	42.42	30.70	16.46	21.90	19.58
South Bend, IN										
Basic Wage & Fringe.....	29.51	29.51	35.52	33.20	26.86	34.62	29.83	29.15	33.15	31.80
Comp. Insurance...........	2.05	2.05	.95	.99	1.30	2.79	1.53	3.41	1.16	1.11
Other Req'd Benefits.....	4.87	4.87	5.86	5.48	4.43	5.71	4.92	4.81	5.47	5.25
Total Contract Rate.......	36.43	36.43	42.33	39.66	32.59	43.12	36.28	37.37	39.78	38.16
Spokane, WA										
Basic Wage & Fringe.....	29.83	29.83	35.42	36.05	24.29	41.46	32.37	26.30	28.76	28.44
Comp. Insurance...........	2.58	2.58	1.01	1.69	2.20	1.03	4.15	5.16	2.39	2.36
Other Req'd Benefits.....	4.92	4.92	5.84	5.95	4.01	6.84	5.34	4.34	4.75	4.69
Total Contract Rate.......	37.34	37.34	42.27	43.69	30.50	49.33	41.86	35.80	35.90	35.50
Springfield, IL										
Basic Wage & Fringe.....	32.95	32.95	35.67	38.55	28.37	36.21	26.57	30.35	33.85	33.85
Comp. Insurance...........	6.18	6.18	2.94	3.80	2.86	1.11	4.47	8.92	3.19	3.19
Other Req'd Benefits.....	5.44	5.44	5.89	6.36	4.68	5.97	4.38	5.01	5.59	5.59
Total Contract Rate.......	44.57	44.57	44.49	48.72	35.91	43.29	35.43	44.27	42.63	42.63
Springfield, MA										
Basic Wage & Fringe.....	38.16	41.06	36.80	39.46	30.10	41.87	35.86	31.02	33.63	33.37
Comp. Insurance...........	6.33	6.82	1.36	2.01	2.49	3.57	5.18	10.33	2.16	2.14
Other Req'd Benefits.....	6.30	6.77	6.07	6.51	4.97	6.91	5.92	5.12	5.55	5.51
Total Contract Rate.......	50.79	54.65	44.23	47.98	37.55	52.35	46.96	46.46	41.33	41.02

Note: Contract labor rates are defined as prevailing hourly rates.

3.3 Contract Labor Rates, Selected Metro Areas

Area	Carpenter	Carpet Layer	Electrician	Plumber	Painter	Elevator Mechanic	Brick Layer	Roofer	Crane Operator	Fork Lift Driver
Springfield, MO										
Basic Wage & Fringe....	23.73	23.73	28.31	29.47	20.61	38.05	26.00	23.03	23.43	22.35
Comp. Insurance..........	4.09	4.09	2.23	2.85	3.10	3.62	4.78	8.49	2.70	2.58
Other Req'd Benefits.....	3.92	3.92	4.67	4.86	3.40	6.28	4.29	3.80	3.87	3.69
Total Contract Rate.......	31.73	31.73	35.21	37.18	27.11	47.95	35.07	35.32	30.00	28.61
Springfield, OH										
Basic Wage & Fringe....	28.79	28.79	33.79	34.42	26.42	42.24	29.34	24.77	35.20	35.10
Comp. Insurance..........	2.79	2.79	1.88	2.20	3.97	1.87	3.36	5.37	2.97	2.96
Other Req'd Benefits.....	4.75	4.75	5.58	5.68	4.36	6.97	4.84	4.09	5.81	5.79
Total Contract Rate.......	36.33	36.33	41.24	42.30	34.75	51.08	37.54	34.22	43.98	43.85
St. Louis, MO										
Basic Wage & Fringe....	36.47	30.60	44.84	40.43	33.76	39.59	37.18	33.00	37.16	37.16
Comp. Insurance..........	6.28	5.27	3.53	3.91	5.08	3.77	6.84	12.16	4.28	4.28
Other Req'd Benefits.....	6.02	5.05	7.40	6.67	5.57	6.53	6.13	5.45	6.13	6.13
Total Contract Rate.......	48.77	40.92	55.77	51.01	44.41	49.89	50.15	50.61	47.58	47.58
Stamford, CT										
Basic Wage & Fringe....	33.18	33.18	55.76	40.03	32.62	45.23	38.93	34.85	38.12	37.25
Comp. Insurance..........	9.92	9.92	5.33	4.59	6.14	4.88	10.95	18.31	4.91	4.80
Other Req'd Benefits.....	5.47	5.47	9.20	6.60	5.38	7.46	6.42	5.75	6.29	6.15
Total Contract Rate.......	48.58	48.58	70.29	51.23	44.14	57.57	56.30	58.91	49.32	48.19
Stockton, CA										
Basic Wage & Fringe....	41.73	33.87	43.77	38.23	29.83	53.74	36.00	28.29	45.51	42.86
Comp. Insurance..........	11.80	9.58	4.35	4.43	5.80	3.10	5.24	11.26	3.59	3.38
Other Req'd Benefits.....	6.89	5.59	7.22	6.31	4.92	8.87	5.94	4.67	7.51	7.07
Total Contract Rate.......	60.42	49.04	55.34	48.97	40.55	65.71	47.18	44.21	56.61	53.31
Syracuse, NY										
Basic Wage & Fringe....	29.60	29.60	34.10	34.46	26.45	40.58	32.51	30.17	36.15	36.15
Comp. Insurance..........	4.40	4.40	2.00	2.57	3.27	1.98	5.75	8.73	2.98	2.98
Other Req'd Benefits.....	4.88	4.88	5.63	5.69	4.36	6.70	5.36	4.98	5.96	5.96
Total Contract Rate.......	38.89	38.89	41.72	42.72	34.09	49.25	43.63	43.88	45.10	45.10
Tacoma, WA										
Basic Wage & Fringe....	37.90	32.83	41.68	39.52	30.18	46.13	36.87	34.53	39.19	38.74
Comp. Insurance..........	3.28	2.84	1.19	1.85	2.74	1.15	4.73	6.78	3.26	3.22
Other Req'd Benefits.....	6.25	5.42	6.88	6.52	4.98	7.61	6.08	5.70	6.47	6.39
Total Contract Rate.......	47.44	41.09	49.75	47.89	37.90	54.89	47.68	47.01	48.91	48.35
Tallahassee, FL										
Basic Wage & Fringe....	11.53	11.53	16.33	15.90	11.26	41.98	12.56	11.04	23.43	22.59
Comp. Insurance..........	4.17	4.17	2.04	1.98	2.80	3.19	3.39	5.93	3.76	3.62
Other Req'd Benefits.....	1.90	1.90	2.69	2.62	1.86	5.22	2.07	1.82	3.87	3.73
Total Contract Rate.......	17.60	17.60	21.06	20.50	15.92	40.02	18.02	18.79	31.05	29.94
Tampa, FL										
Basic Wage & Fringe....	24.65	16.88	19.07	30.23	13.23	41.98	25.96	11.04	18.51	17.08
Comp. Insurance..........	8.91	6.10	2.38	3.76	3.29	3.96	7.01	5.93	2.97	2.74
Other Req'd Benefits.....	4.07	2.79	3.15	4.99	2.18	6.48	4.28	1.82	3.05	2.82
Total Contract Rate.......	37.63	25.77	24.60	38.98	18.70	49.70	37.25	18.79	24.53	22.64

Note: Contract labor rates are defined as prevailing hourly rates.

Area	Carpenter	Carpet Layer	Electrician	Plumber	Painter	Elevator Mechanic	Brick Layer	Roofer	Crane Operator	Fork Lift Driver
Terre Haute, IN										
Basic Wage & Fringe.....	28.65	25.30	34.32	33.96	29.10	37.71	30.53	27.05	33.90	33.90
Comp. Insurance..........	1.99	1.76	.92	1.01	1.41	3.04	1.56	3.16	1.19	1.19
Other Req'd Benefits.....	4.73	4.17	5.66	5.60	4.80	6.22	5.04	4.46	5.59	5.59
Total Contract Rate.......	35.37	31.24	40.90	40.57	35.31	46.97	37.13	34.68	40.68	40.68
Toledo, OH										
Basic Wage & Fringe.....	34.01	33.16	39.97	40.30	31.64	39.13	36.18	34.39	35.20	35.10
Comp. Insurance..........	3.29	3.21	2.22	2.58	4.76	1.73	4.14	7.45	2.97	2.96
Other Req'd Benefits.....	5.61	5.47	6.60	6.65	5.22	6.46	5.97	5.67	5.81	5.79
Total Contract Rate.......	42.91	41.84	48.78	49.53	41.62	47.32	46.29	47.52	43.98	43.85
Topeka, KS										
Basic Wage & Fringe.....	14.75	24.47	31.68	33.09	22.95	35.57	29.71	31.34	28.62	27.82
Comp. Insurance..........	1.50	2.48	1.16	1.62	1.48	1.15	2.15	5.89	1.49	1.45
Other Req'd Benefits.....	2.43	4.04	5.23	5.46	3.79	5.87	4.90	5.17	4.72	4.59
Total Contract Rate.......	18.68	30.99	38.07	40.17	28.22	42.58	36.76	42.40	34.84	33.86
Trenton, NJ										
Basic Wage & Fringe.....	45.65	45.65	55.06	48.38	39.06	48.72	40.55	39.05	50.02	48.43
Comp. Insurance..........	5.11	5.11	2.12	2.78	3.50	2.28	4.87	11.07	3.49	3.38
Other Req'd Benefits.....	7.53	7.53	9.08	7.98	6.44	8.04	6.69	6.44	8.25	7.99
Total Contract Rate.......	58.29	58.29	66.26	59.14	49.01	59.03	52.11	56.57	61.76	59.80
Tucson, AZ										
Basic Wage & Fringe.....	24.42	24.42	25.41	29.80	16.22	33.80	19.24	14.00	27.10	26.07
Comp. Insurance..........	3.21	3.21	.90	1.08	.69	1.82	1.19	1.57	1.40	1.35
Other Req'd Benefits.....	4.03	4.03	4.19	4.92	2.68	5.58	3.17	2.31	4.47	4.30
Total Contract Rate.......	31.66	31.66	30.51	35.80	19.59	41.20	23.60	17.88	32.97	31.72
Tulsa, OK										
Basic Wage & Fringe.....	16.10	17.04	18.67	28.32	16.08	31.88	26.96	19.50	24.10	23.60
Comp. Insurance..........	3.11	3.29	1.38	2.47	2.75	3.91	4.53	8.24	4.45	4.35
Other Req'd Benefits.....	2.66	2.81	3.08	4.67	2.65	5.26	4.45	3.22	3.98	3.89
Total Contract Rate.......	21.87	23.15	23.13	35.46	21.48	41.05	35.94	30.96	32.52	31.85
Tuscaloosa, AL										
Basic Wage & Fringe.....	11.54	11.54	25.21	15.08	13.31	38.05	14.04	17.96	20.52	19.80
Comp. Insurance..........	3.86	3.86	2.31	1.79	3.84	3.47	4.35	12.35	3.11	3.00
Other Req'd Benefits.....	1.90	1.90	4.16	2.49	2.20	6.28	2.32	2.96	3.39	3.27
Total Contract Rate.......	17.31	17.31	31.68	19.36	19.34	47.79	20.70	33.28	27.01	26.06
Utica, NY										
Basic Wage & Fringe.....	28.12	28.12	33.75	33.10	26.34	35.57	31.71	30.17	36.15	36.15
Comp. Insurance..........	4.18	4.18	1.98	2.47	3.26	1.73	5.61	8.73	2.98	2.98
Other Req'd Benefits.....	4.64	4.64	5.57	5.46	4.35	5.87	5.23	4.98	5.96	5.96
Total Contract Rate.......	36.94	36.94	41.30	41.03	33.95	43.17	42.55	43.88	45.10	45.10
Virginia Beach, VA										
Basic Wage & Fringe.....	21.78	18.81	25.37	28.74	18.81	32.24	23.74	13.24	24.81	21.93
Comp. Insurance..........	2.47	2.13	1.05	1.74	1.91	2.03	1.90	2.97	1.84	1.63
Other Req'd Benefits.....	3.59	3.10	4.19	4.74	3.10	5.32	3.92	2.18	4.09	3.62
Total Contract Rate.......	27.85	24.05	30.61	35.22	23.82	39.59	29.56	18.39	30.74	27.18

Note: Contract labor rates are defined as prevailing hourly rates.

Area	Carpenter	Carpet Layer	Electrician	Plumber	Painter	Elevator Mechanic	Brick Layer	Roofer	Crane Operator	Fork Lift Driver
Waco, TX										
Basic Wage & Fringe....	13.38	13.38	25.62	24.14	11.54	38.59	23.29	13.58	25.30	24.30
Comp. Insurance..........	1.77	1.77	1.89	1.95	1.14	2.70	3.19	3.10	2.65	2.55
Other Req'd Benefits.....	2.21	2.21	4.23	3.98	1.90	6.37	3.84	2.24	4.17	4.01
Total Contract Rate.......	17.36	17.36	31.73	30.07	14.59	47.65	30.32	18.92	32.13	30.86
Washington DC										
Basic Wage & Fringe....	26.08	26.08	38.64	38.41	17.30	42.08	30.37	25.56	24.10	23.39
Comp. Insurance..........	4.38	4.38	2.80	4.47	2.58	2.56	5.79	6.43	2.10	2.04
Other Req'd Benefits.....	4.30	4.30	6.38	6.34	4.55	6.94	5.01	4.22	3.98	3.86
Total Contract Rate.......	34.76	34.76	47.82	49.22	34.73	51.58	41.17	36.21	30.18	29.29
Waterbury, CT										
Basic Wage & Fringe....	33.18	33.18	42.81	40.03	32.62	45.23	38.03	34.85	38.12	37.25
Comp. Insurance..........	9.92	9.92	4.09	4.59	6.14	4.88	10.69	18.31	4.91	4.80
Other Req'd Benefits.....	5.47	5.47	7.06	6.60	5.38	7.46	6.27	5.75	6.29	6.15
Total Contract Rate.......	48.58	48.58	53.96	51.23	44.14	57.57	55.00	58.91	49.32	48.19
Watertown, NY										
Basic Wage & Fringe....	31.46	31.46	32.63	30.61	22.86	34.08	33.53	30.17	36.15	36.15
Comp. Insurance..........	4.68	4.68	1.91	2.28	2.83	1.66	5.93	8.73	2.98	2.98
Other Req'd Benefits.....	5.19	5.19	5.38	5.05	3.77	5.62	5.53	4.98	5.96	5.96
Total Contract Rate.......	41.33	41.33	39.93	37.94	29.46	41.36	45.00	43.88	45.10	45.10
Wichita Falls, TX										
Basic Wage & Fringe....	15.23	15.00	23.28	22.65	18.29	32.17	25.40	13.90	25.30	24.30
Comp. Insurance..........	2.02	1.99	1.71	1.83	1.81	2.25	3.47	3.17	2.65	2.55
Other Req'd Benefits.....	2.51	2.48	3.84	3.74	3.02	5.31	4.19	2.29	4.17	4.01
Total Contract Rate.......	19.76	19.46	28.83	28.21	23.12	39.73	33.07	19.36	32.13	30.86
Wichita, KS										
Basic Wage & Fringe....	20.70	20.70	30.62	30.60	20.95	33.20	27.60	19.35	24.67	24.42
Comp. Insurance..........	2.10	2.10	1.12	1.50	1.35	1.07	2.00	3.64	1.29	1.27
Other Req'd Benefits.....	3.42	3.42	5.05	5.05	3.46	5.48	4.55	3.19	4.07	4.03
Total Contract Rate.......	26.22	26.22	36.80	37.15	25.76	39.75	34.15	26.18	30.03	29.72
Wilmington, DE										
Basic Wage & Fringe....	36.39	36.39	41.60	51.99	29.51	44.22	29.73	38.05	35.74	32.99
Comp. Insurance..........	3.60	3.60	2.10	3.15	3.75	6.77	2.75	8.25	2.79	2.58
Other Req'd Benefits.....	6.00	6.00	6.86	8.58	4.87	7.30	4.91	6.28	5.90	5.44
Total Contract Rate.......	45.99	45.99	50.57	63.71	38.13	58.28	37.39	52.58	44.43	41.01
Winston-Salem										
Basic Wage & Fringe....	14.91	14.91	15.58	19.12	12.50	29.87	14.84	10.00	18.75	18.50
Comp. Insurance..........	2.69	2.69	1.20	1.54	1.20	1.85	1.40	2.57	1.58	1.56
Other Req'd Benefits.....	2.46	2.46	2.57	3.15	2.06	4.93	2.45	1.65	3.09	3.05
Total Contract Rate.......	20.06	20.06	19.35	23.82	15.76	36.65	18.69	14.22	23.43	23.11
Worcester, MA										
Basic Wage & Fringe....	39.66	41.89	41.27	42.12	37.43	41.87	44.58	34.61	43.44	43.23
Comp. Insurance..........	6.58	6.95	1.52	2.14	3.09	3.57	6.44	11.52	2.78	2.77
Other Req'd Benefits.....	6.54	6.91	6.81	6.95	6.18	6.91	7.36	5.71	7.17	7.13
Total Contract Rate.......	52.79	55.76	49.60	51.21	46.70	52.35	58.37	51.84	53.39	53.13

Note: Contract labor rates are defined as prevailing hourly rates.

.3 Contract Labor Rates, Selected Metro Areas

Area	Carpenter	Carpet Layer	Electrician	Plumber	Painter	Elevator Mechanic	Brick Layer	Roofer	Crane Operator	Fork Lift Driver
Yonkers, NY										
Basic Wage & Fringe	48.79	65.18	59.24	51.67	45.67	58.47	46.25	48.86	50.34	49.14
Comp. Insurance...........	7.26	9.69	3.47	3.85	5.65	2.85	8.19	14.14	4.15	4.05
Other Req'd Benefits.....	8.05	10.75	9.77	8.53	7.54	9.65	7.63	8.06	8.31	8.11
Total Contract Rate... ...	64.10	85.63	72.49	64.05	58.86	70.97	62.07	71.06	62.80	61.30
Youngstown, OH										
Basic Wage & Fringe.....	30.38	29.27	38.39	35.98	28.90	38.91	33.09	29.92	36.30	34.55
Comp. Insurance....... ...	2.94	2.83	2.13	2.30	4.35	1.72	3.79	6.48	3.06	2.92
Other Req'd Benefits.....	5.01	4.83	6.33	5.94	4.77	6.42	5.46	4.94	5.99	5.70
Total Contract Rate.......	38.33	36.93	46.85	44.22	38.02	47.05	42.34	41.34	45.35	43.17
Canadian Cities										
Calgary, Alberta										
Basic Wage & Fringe.....	37.45	31.61	41.68	41.16	35.83	44.44	29.28	32.81	38.77	28.57
Comp. Insurance........ ..	1.02	.86	1.24	.82	1.46	1.22	2.24	3.18	1.00	.73
Other Req'd Benefits.....	3.15	2.66	3.50	3.46	3.01	3.73	2.46	2.76	3.26	2.40
Total Contract Rate.......	41.61	35.13	46.42	45.44	40.30	49.40	33.98	38.75	43.02	31.70
Edmonton, Alberta										
Basic Wage & Fringe.....	37.45	31.61	41.68	41.16	35.83	44.44	29.28	32.81	38.77	28.57
Comp. Insurance...........	1.02	.86	1.24	.82	1.46	1.22	2.24	3.18	1.00	.73
Other Req'd Benefits.....	3.15	2.66	3.50	3.46	3.01	3.73	2.46	2.76	3.26	2.40
Total Contract Rate..... .	41.61	35.13	46.42	45.44	40.30	49.40	33.98	38.75	43.02	31.70
Hamilton, Ontario										
Basic Wage & Fringe.....	38.50	32.27	42.40	41.18	34.16	41.47	38.85	34.44	38.95	33.59
Comp. Insurance......... .	1.93	1.61	1.28	1.63	2.42	4.27	4.80	4.26	1.64	1.41
Other Req'd Benefits.....	3.23	2.71	3.56	3.46	2.87	3.48	3.26	2.89	3.27	2.82
Total Contract Rate.......	43.66	36.59	47.25	46.27	39.45	49.22	46.92	41.59	43.86	37.83
London, Ontario										
Basic Wage & Fringe.....	36.10	31.91	41.17	40.33	34.16	41.63	38.15	32.82	37.74	32.87
Comp. Insurance...........	1.81	1.60	1.25	1.60	2.42	4.29	4.72	4.06	1.59	1.38
Other Req'd Benefits.....	3.03	2.68	3.46	3.39	2.87	3.50	3.20	2.76	3.17	2.76
Total Contract Rate......	40.94	36.19	45.88	45.31	39.45	49.41	46.07	39.63	42.50	37.01
Montreal, Quebec										
Basic Wage & Fringe.....	33.81	26.82	35.86	34.68	31.54	39.23	33.76	34.76	33.08	32.83
Comp. Insurance..........	4.29	3.40	2.04	2.47	4.00	2.76	4.69	7.71	2.55	2.53
Other Req'd Benefits.....	2.84	2.25	3.01	2.91	2.65	3.30	2.84	2.92	2.78	2.76
Total Contract Rate.......	40.94	32.48	40.91	40.07	38.19	45.28	41.29	45.39	38.41	38.12
Ottawa, Quebec										
Basic Wage & Fringe....	37.20	30.49	41.60	39.94	32.07	40.23	38.23	32.91	37.85	32.97
Comp. Insurance...........	1.86	1.52	1.26	1.58	2.27	4.14	4.73	4.07	1.59	1.39
Other Req'd Benefits.....	3.12	2.56	3.49	3.35	2.69	3.38	3.21	2.76	3.18	2.77
Total Contract Rate.......	42.18	34.58	46.35	44.88	37.04	47.75	46.17	39.74	42.62	37.13

Note: Contract labor rates are defined as prevailing hourly rates. Values for Canadian cities are reported in Canadian dollars.

3.3 Contract Labor Rates, Selected Metro Areas

Area	Carpenter	Carpet Layer	Electrician	Plumber	Painter	Elevator Mechanic	Brick Layer	Roofer	Crane Operator	Fork Lift Driver
Quebec, Quebec										
Basic Wage & Fringe.....	33.81	26.82	35.86	34.68	31.54	39.23	33.76	34.76	33.08	32.83
Comp. Insurance..........	4.29	3.40	2.04	2.47	4.00	2.76	4.69	7.71	2.55	2.53
Other Req'd Benefits.....	2.84	2.25	3.01	2.91	2.65	3.30	2.84	2.92	2.78	2.76
Total Contract Rate.......	40.94	32.48	40.91	40.07	38.19	45.28	41.29	45.39	38.41	38.12
Toronto, Ontario										
Basic Wage & Fringe.....	39.97	35.00	42.76	42.39	35.80	41.55	39.96	36.70	39.66	37.48
Comp. Insurance..........	2.00	1.75	1.30	1.68	2.54	4.28	4.94	4.54	1.67	1.58
Other Req'd Benefits.....	3.36	2.94	3.59	3.56	3.01	3.49	3.36	3.08	3.33	3.15
Total Contract Rate.......	45.33	39.69	47.65	47.63	41.35	49.32	48.26	44.32	44.66	42.20
Vancouver, B.C.										
Basic Wage & Fringe.....	33.43	30.91	35.28	35.16	33.91	38.25	33.30	31.68	35.22	31.11
Comp. Insurance..........	1.55	1.43	1.02	1.51	1.97	2.35	2.76	2.77	1.20	1.06
Other Req'd Benefits.....	2.81	2.60	2.96	2.95	2.85	3.21	2.80	2.66	2.96	2.61
Total Contract Rate.......	37.79	34.94	39.27	39.63	38.73	43.81	38.86	37.11	39.38	34.79
Winnipeg, Manitoba										
Basic Wage & Fringe.....	27.74	21.16	31.82	32.51	18.64	27.00	28.17	24.92	31.54	24.53
Comp. Insurance..........	1.05	.80	.65	.77	.78	1.23	1.07	1.35	1.20	.93
Other Req'd Benefits.....	2.33	1.78	2.67	2.73	1.57	2.27	2.37	2.09	2.65	2.06
Total Contract Rate.......	31.12	23.74	35.14	36.01	20.99	30.50	31.61	28.36	35.39	27.52

Note: Contract labor rates are defined as prevailing hourly rates. Values for Canadian cities are reported in Canadian dollars.

Chapter 4
M&R Construction Indexes

National M&R Cost Indexes

Selected M&R-Related Commodity Indexes

4. M&R Construction Indexes

This chapter presents a collection of national-level economic indicators relevant for the maintenance and repair (M&R) construction industry. Three of these measures are unique to this publication—indexes for residential, non-residential (commercial and public buildings), and military facilities M&R construction—while others are included for products and services common in building M&R activities.

Table 4-1
Maintenance and New Construction Costs
Change in average from previous year

	Whitestone M&R Indexes				
	Non-Residential Buildings	Residential Buildings	Military Facilities	New Construction[1]	CPI[2]
1994	2.1%	2.0%	2.0%	2.7%	2.6%
1995	3.5%	4.1%	3.4%	3.1%	2.8%
1996	1.7%	1.4%	2.2%	2.4%	3.0%
1997	2.1%	1.3%	2.1%	2.4%	2.3%
1998	0.9%	-0.6%	0.8%	2.0%	1.6%
1999	2.1%	1.6%	2.5%	2.2%	2.2%
2000	4.3%	4.9%	5.1%	2.8%	3.4%
2001	1.4%	0.6%	1.4%	3.5%	2.8%
2002	1.1%	0.1%	1.2%	2.9%	1.6%
2003	2.8%	3.6%	3.0%	2.6%	2.3%
2004e	2.2%	3.0%	2.5%	0.8%	1.8%
Avg.	2.2%	2.0%	2.4%	2.5%	2.4%

[1]R.S. Means Historical Cost Index.
[2]Consumer Price Index for all items, urban consumers.
e=Estimated.

Industry Outlook

The U.S. economy has grown at an average 3.3 percent following the brief 2001 recession. Construction activity has been a major contributor to this expansion. Fueled by historically low capital costs, expenditures on commercial and residential structures account for almost a third of the recent economic growth.

Implications of the new construction activity for M&R spending are not clear. The demand for M&R construction is relatively stable compared with the often-steep oscillations for new construction. In the short run, we assume demand to be buoyed by low interest costs and high real estate values, particularly in the residential sector. But in the longer run, we expect M&R demand to conform to the historic average of roughly 60 percent of the demand for new construction. A recent study by Whitestone estimated that the domestic U.S. demand for non-residential M&R activities was $210 billion, and found at the country-level this demand ebbs and flows with net changes in building inventory.

As shown in Table 4.1, maintenance and repair construction costs are expected to continue to increase, relative to 2003. Major factors include record high petroleum costs and run-ups in scrap metal and lumber costs (note that the recent sharp price increases are muted in our annualized forecasts of commodity prices).

M&R Labor and Commodity Costs

A view of specific factors affecting costs is provided in Table 4-2. The first eleven items are general product groupings that are inputs to Whitestone indexes, while the other items are commodities common to M&R construction. We follow the price changes of these particular commodities because manufacturers and distributors often use them in making price adjustments.

Table 4-2
M&R-Related Service and Commodity Costs
Change in average from previous year

Index	1999	2000	2001	2002	2003	2004e
Architect & Engineering Services*	3.1%	4.5%	2.8%	2.5%	4.2%	3.2%
Petroleum Refining & Related Products*	15.8%	46.9%	-6.7%	-6.1%	23.5%	13.1%
Lumber and Wood Products*	-2.9%	-2.8%	-5.6%	-1.5%	2.2%	3.4%
Fabricated Metal Products*	0.3%	1.0%	0.5%	0.5%	0.9%	3.2%
Stone, Clay, Glass, & Concrete*	2.5%	1.6%	1.0%	0.7%	0.8%	1.6%
Rubber & Miscellaneous Plastics*	0.0%	2.0%	1.0%	-0.4%	2.3%	1.0%
Chemicals & Allied Products*	0.7%	4.7%	1.1%	-0.7%	4.6%	2.4%
Electrical Equipment & Supplies*	-1.3%	-0.7%	-0.3%	-0.9%	-1.5%	-0.4%
Stone & Clay Mining*	1.3%	2.3%	2.9%	1.8%	2.0%	2.0%
Primary Metal Industries*	-4.3%	3.5%	-3.1%	0.1%	1.9%	11.9%
Construction Labor*	3.5%	4.1%	3.0%	2.9%	2.3%	0.8%
Prepared Paints	1.6%	2.2%	2.2%	1.5%	2.3%	1.8%
Vitreous China Fixtures	1.2%	0.7%	4.6%	0.7%	-0.4%	0.0%
Elevators, Escalators & Lifts	1.3%	1.0%	0.5%	0.5%	-1.0%	1.3%
Air Conditioning & Refrigeration Equip.	0.6%	-0.1%	0.5%	0.7%	0.3%	0.4%
Floor Coverings	-0.9%	1.9%	0.5%	0.2%	2.4%	-0.8%
Asphalt & Tar Roofing Products	-0.3%	4.8%	3.3%	3.1%	5.0%	-0.6%

*Component of Whitestone M&R Indexes

With regards to labor costs, projected wage increases for A&E workers are expected to fall short of the 4.2 percent growth seen in 2003, while wage rates for general construction are anticipated to be only slightly higher than those seen in 2003. Among organized workers, we expect an average wage increase of 4.0 percent, with the greatest increase—4.8 percent—in the New England. Wage changes by region are discussed in Chapter 6.

Using M&R Cost Indexes

Change in costs over time can be expressed in various ways. Two common methods are shown in Box 4-1. Focusing on the Whitestone Military Facility M&R Index, for example, we find costs for military facilities rose around 27 percent between 1994 and 2004. The average annual (or compound) rate of increase was 2.4 percent during the same period. By comparison, the average annual rate of increase for new construction was 2.5 percent from 1994 to 2004.

Box 4-1
Calculating Rates of Change: DoD Maintenance Expenditures

Simple percentage change:

$$\text{change (\%)} = ((\text{index}_{t+n} / \text{index}_{t}) - 1)$$

Where t and t+n are different points in time

Example: Calculate the inflation in M&R costs for military facilities during the period 1994-2004:

$$26.9\% = .269 = ((151.4_{2004} / 119.3_{1994}) - 1)$$

Average rate of change:

$$\text{average annual rate of change (\%)} = ((\text{index}_{t+n} / \text{index}_{t})^{(1/n)} - 1)$$

Example: Calculate the annual inflation rate in M&R costs for military facilities during the period 1994-2004:

$$2.4\% = .024 = ((151.4_{2004} / 119.3_{1994})^{(.10)} - 1)$$

The choice of indexes is important, particularly when used for adjusting cost trends. Consider the example of the M&R budget of the U.S. Department of Defense (DoD), one of the world's largest property owners. In 1994, DoD spent roughly $4.3 billion for facility maintenance. Starting from this point, we can calculate how much funding would be required in 2004 for the same level of M&R goods and services. Based on the Whitestone Military Facilities M&R Index, the 2004 cost in current dollars would be approximately $5.4 billion. Using an alternative inflation measure such as that for new construction—a common practice—suggests a requirement of $5.5 billion, an overstatement of $100 million in the DoD budget for a single year.

The appropriate M&R indexes can also be used to correct timeworn funding benchmarks. Funding rules established in the past may no longer reflect current cost trends. For example, a non-residential organization may have found ten years ago that an M&R budget of 3 percent of replacement value—a common rule of thumb—was adequate for maintaining their physical plant. Today that rule of thumb is incorrect because of the different inflation rates of maintenance construction and new construction, the two components in the "percent of replacement value" ratio. Using the appropriate inflation rates, it is shown—3 percent multiplied by ($1.245_{\text{Non-Residential}} / 1.274_{\text{New Construction}}$)—that the M&R goods and services purchased with 3% of replacement value in 1994 would now require only 2.9 percent.

A simpler approach is to use a funding rule expressed in terms of square feet rather than replacement cost. For example, a benchmark of $3 per gsft. in 1994 could be adjusted just by using the M&R cost index (a multiplier of 1.269 in the case of military facilities) to determine that the equivalent amount would be $3.81 in 2004. *No matter how they are expressed, funding benchmarks are not static over time and must be adjusted to reflect prevailing cost trends.*

The choice of indexes is also important for adjusting location-specific costs. The local cost index shown in Chapter 3 specifically reflects place-to-place variations in M&R costs. Use of an alternative index, such as one designed for new construction, will not measure the true change in the "market basket" of goods and services specific to M&R construction. Is the difference meaningful? The 50-year average M&R cost for a two-story office building in Washington DC is $2.13, as shown in Chapter 2. The same building would cost $2.42 in Detroit, according to the Whitestone local cost index (the index value for Detroit is 113.4). Alternatively, the use of a well-known commercially published index for new construction would show an adjusted cost of $2.36.

Monthly updates of Whitestone indexes are available at www.whitestoneresearch.com.

4.1 National M&R Cost Indexes

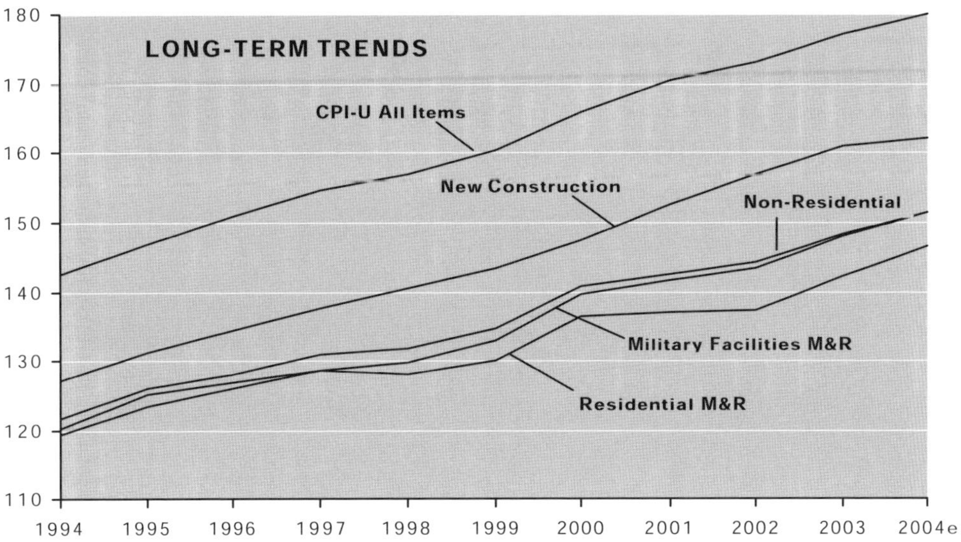

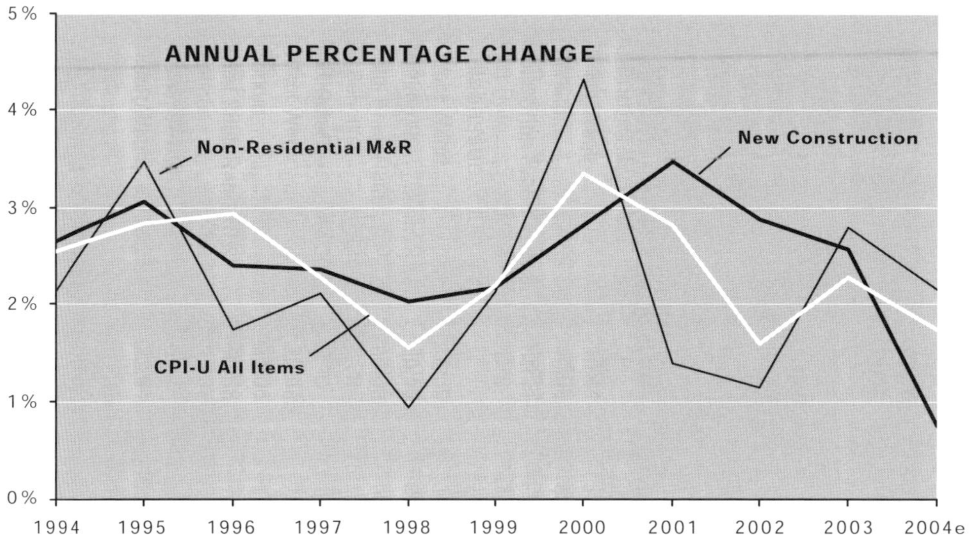

1984=100

Indexes	Annual average 1994	1995	1996	1997	1998	1999	2000	2001	2002	2003	2004e	Rate of change 10-year average	2003-04 est.
Whitestone M&R Indexes													
Non-Residential Buildings M&R	121.7	125.9	128.1	130.8	132.0	134.9	140.7	142.6	144.3	148.3	151.5	2.2%	2.2%
Residential Buildings M&R	120.2	125.1	126.9	128.6	127.9	129.9	136.3	137.1	137.2	142.2	146.5	2.0%	3.0%
Military Facilities M&R	119.3	123.3	126.0	128.7	129.7	133.0	139.8	141.7	143.5	147.8	151.4	2.4%	2.5%
Index Components													
Architect & Engineering Services	136.8	141.5	147.1	154.6	159.9	164.8	172.3	177.1	181.6	189.2	195.3	3.6%	3.2%
Petroleum Refining & Related Products	74.8	77.2	87.4	85.6	66.3	76.8	112.8	105.3	98.8	122.0	138.1	6.3%	13.1%
Lumber & Wood Products	201.1	202.8	193.9	199.6	196.4	190.7	185.3	174.9	172.3	176.1	182.1	-1.0%	3.4%
Fabricated Metal Products	120.3	124.8	126.2	127.6	128.7	129.1	130.3	131.0	131.7	132.9	137.2	1.3%	3.2%
Stone, Clay, Glass & Concrete	119.6	124.3	125.8	127.4	129.3	132.6	134.7	136.0	137.0	138.1	140.3	1.6%	1.6%
Rubber & Miscellaneous Plastics	117.1	123.3	123.1	122.8	122.1	122.2	124.6	125.9	125.5	128.4	129.7	1.0%	1.0%
Chemicals & Allied Products	130.0	143.4	145.8	147.1	148.7	149.7	156.7	158.4	157.3	164.6	168.6	2.6%	2.4%
Electrical Equipment & Supplies	106.8	108.1	108.7	108.8	109.7	108.3	107.5	107.2	106.2	104.6	104.2	-0.2%	-0.4%
Stone & Clay Mining	120.5	123.8	127.1	128.8	132.2	133.9	137.0	141.0	143.5	146.4	149.4	2.2%	2.0%
Primary Metal Industries	117.0	128.2	123.7	124.7	120.9	115.7	119.8	116.1	116.2	118.4	132.5	1.3%	11.9%
Construction Labor	124.2	127.3	130.6	135.4	140.2	145.1	151.1	155.6	160.1	163.9	165.1	2.9%	0.8%
Other Indexes													
New Construction	127.3	131.2	134.4	137.6	140.4	143.4	147.4	152.6	157.0	161.0	162.2	2.5%	0.8%
CPI-Urban Consumers, All Items	142.6	146.7	151.0	154.5	156.9	160.3	165.7	170.4	173.1	177.1	180.2	2.4%	1.8%

e = estimate based on average change, 2nd quarter 2003 to 2nd quarter 2004

Sources: U.S. Department of Commerce, Department of Labor, RS Means Co. Inc., and Whitestone Research

4.2 Selected M&R-Related Commodity Indexes

Prepared Paints, 1994-2004

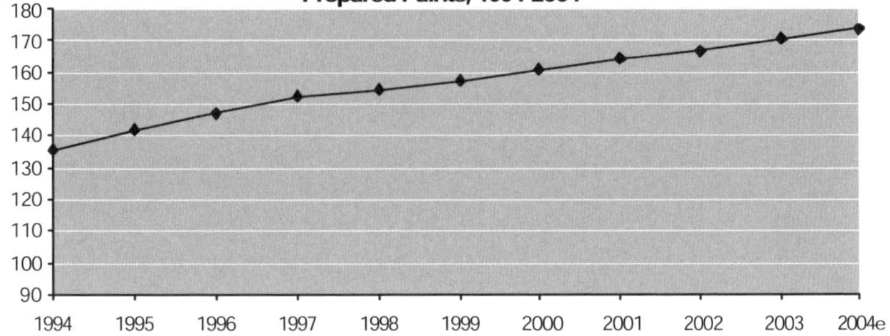

Vitreous China Fixtures, 1994-2004

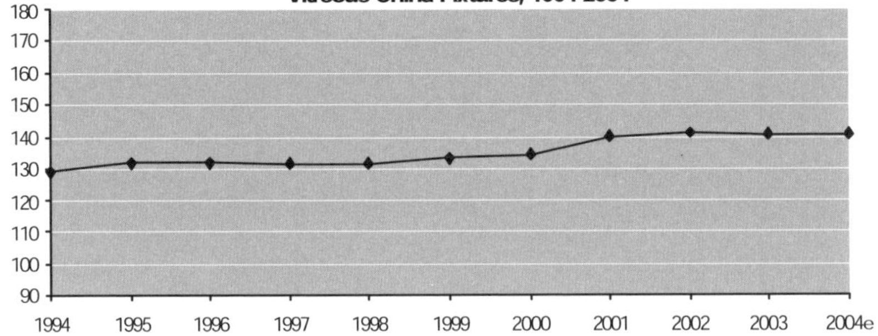

Elevators, Escaltors and Lifts, 1994-2004

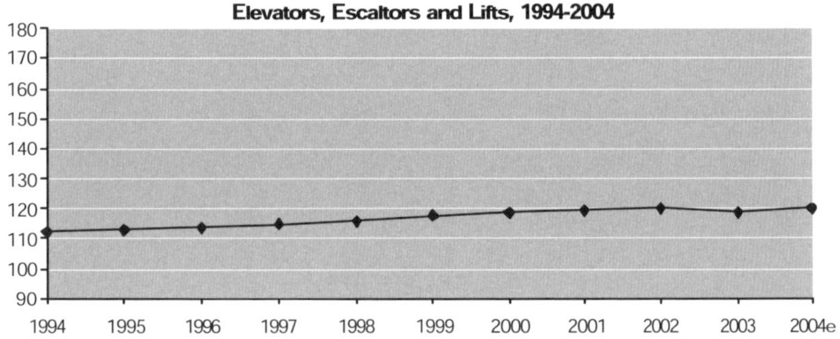

Air Conditioning & Refrigeration Equipment, 1994-2004

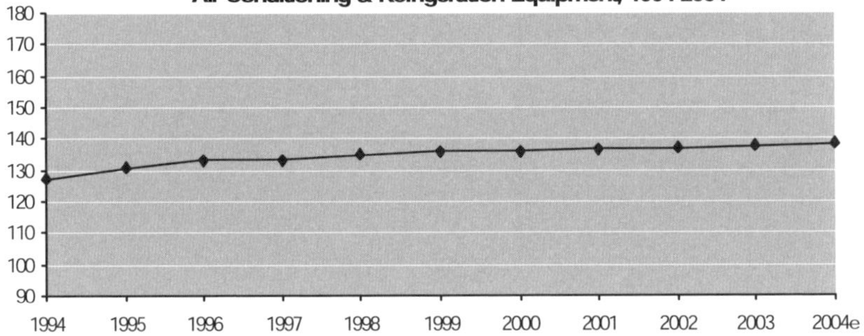

Floor Coverings, 1994-2004

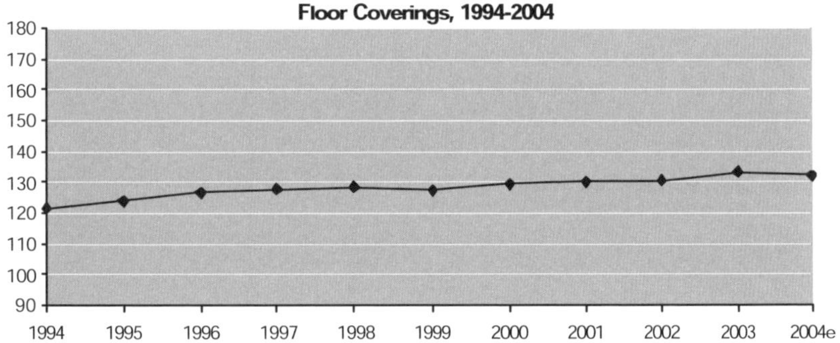

Asphalt & Tar Roofing Products, 1994-2004

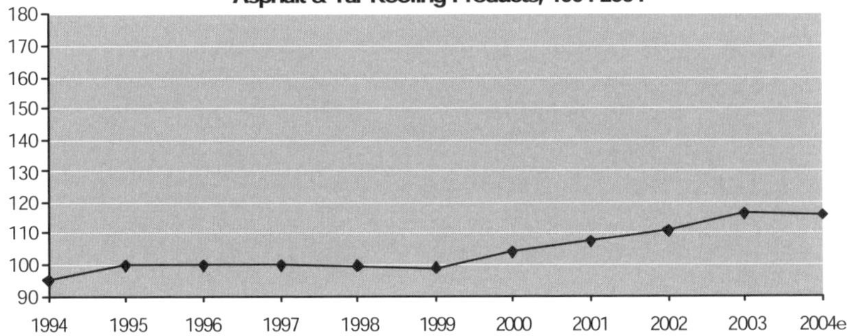

Dec. 1984=100

	Annual average											Rate of change	
Indexes	1994	1995	1996	1997	1998	1999	2000	2001	2002	2003	2004e	10-year average	2003-04 est.
Prepared Paints	135.3	142.1	147.2	152.1	154.9	157.4	160.8	164.4	166.9	170.8	173.9	156.9	1.8%
Vitreous China Fixtures	128.9	131.9	131.8	131.5	131.4	133.0	134.0	140.2	141.2	140.7	140.6	135.0	-0.1%
Elevators, Escalators & Lifts	112.4	113.0	113.7	114.8	116.0	117.5	118.7	119.3	119.9	118.7	120.2	116.8	1.3%
Air Conditioning & Refrigeration Equip.	127.0	130.2	132.7	132.6	134.6	135.5	135.3	136.0	136.9	137.2	137.6	134.1	0.3%
Floor Coverings	121.5	123.7	126.6	128.0	128.3	127.2	129.6	130.2	130.5	133.6	132.1	128.3	-1.1%
Asphalt & Tar Roofing Products	95.3	100.0	100.0	99.9	99.5	99.3	104.1	107.5	110.9	116.4	115.8	104.4	-0.6%

e = estimate based on average change, 1st quarter 2003 to 2nd quarter 2004   Sources: Department of Labor and Whitestone Research

Chapter 5
Component Data

Building Components and Repair Tasks

Repair Task Frequencies of HVAC Components by Climate Zone

5. Component Data

This chapter provides M&R data for over 450 components and more than 1,200 associated maintenance tasks. The tasks are the basis for the Building M&R Cost Profiles shown in Chapter 2. All of the components used in generating the profiles are included. Detailed information including labor hours, material and equipment costs, and task frequencies are listed in the first section of this chapter; the second section lists different task frequencies for HVAC components across climate zones. Components and tasks are listed according to Uniformat II, a 12-division classification system commonly used by architects, engineers, and cost analysts (shown in the Appendix).

Box 5-1.
Calculating Maintenance Task Costs

Cost of an individual maintenance task:

Task cost = quantity x [(labor x um multiplier x trade rate x markup) + (material x markup) + (equip x equip rate x markup)]

where: quantity = expressed in specified units; e.g. square foot, 1000 linear feet, etc.
labor = hours per unit
um (unscheduled maintenance) multiplier = 1.0 for renewal and replacement tasks or > 1.0 for all other tasks
trade rate = contract rate or in-house shop rate
markup = contract markup, including overhead and fee
material = material cost per unit
equip = special equipment required, in hours per unit
equip rate = hourly rate for equipment and operator

Example: Place new membrane over 41,500 Sq. Ft. existing built-up roof (Washington DC):

$72,426 = 41,500 x [(.018571 x 1.0 x $36.21 x 1.3) + ($.67 x 1.3) + (no special equip)]

Major repair costs for a selected component over a given lifetime:

Component cost = sum (task cost * (lifetime / frequency))

where: sum = sum for all tasks associated with the component
task cost = explained above
lifetime = arbitrary period beginning with new construction (50 yrs. was used in Ch. 2)
frequency = frequency of task (for HVAC units this varies by region)

note: results of (lifetime/frequency) are rounded to the nearest integer; frequencies are reset upon component replacement.

Example: M&R costs of a 41,500 Sq. Ft. built-up roof, over 50 years (Washington DC):

Maintain built up roof	$16,093 x 50 =	$804,650
Non-destructive moisture inspection	2,149 x 16 =	$34,384
Place new membrane over existing roof	72,426 x 2 =	144,852
Remove and replace old membrane	143,290 x 1 =	143,290
		$1,127,176

The primary reason for presenting detailed component and task data is to allow the reader to both understand the source of the estimates presented, and to make it simple to alter the cost profiles in Chapter 2. With this information one can substitute alternative labor, material, and frequency for specific tasks—or add and substitute alternative components.

Uniformat / Component / Repair Task	Units	Trade	Labor Type	Labor Hours	Material Costs	Equip Hours	Equipment Type	Freq. Yrs.*	Task Type	Repl. Task
B20 Exterior Enclosure										
B2010 Exterior Walls										
Aluminum Siding, Exterior, 1st Floor										
Refinish Aluminum Siding, Exterior, 1st Floor	Sq Ft	Painter	Contract	.0370	.10	0		20	Major	No
Finish Replaced Aluminum Siding, Exterior, 1st Floor	Sq Ft	Painter	Contract	.0250	.10	0		35	Major	No
Replace Aluminum Siding, Exterior, 1st Floor	Sq Ft	Carpenter	Contract	.0730	.86	0		35	Major	Yes
Aluminum Siding, Exterior, 2nd Floor										
Refinish Aluminum Siding, Exterior, 2nd Floor	Sq Ft	Painter	Contract	.0490	.10	0		20	Major	No
Replace Aluminum Siding, Exterior, 2nd Floor	Sq Ft	Carpenter	Contract	.0930	.86	0		35	Major	Yes
Finish Replaced Aluminum Siding, Exterior, 2nd Floor	Sq Ft	Painter	Contract	.0370	.10	0		35	Major	No
Aluminum Siding, Exterior, 3rd+ Floor										
Refinish Aluminum Siding, Exterior, 3rd+ Floor	Sq Ft	Painter	Contract	.0610	.10	0		20	Major	No
Finish Replaced Aluminum Siding, Exterior, 3rd+ Floor	Sq Ft	Painter	Contract	.0500	.10	0		35	Major	No
Replace Aluminum Siding, Exterior, 3rd+ Floor	Sq Ft	Carpenter	Contract	.1100	.86	0		35	Major	Yes
Clay Brick, Exterior, 1st Floor										
Clean & Reseal Clay Brick, Exterior, 1st Floor	Sq Ft	Maint Painter	Staff	.0320	.27	0		10	PM	No
Repair Clay Brick, Exterior, 1st Floor	Sq Ft	Brick Layer	Contract	.0163	.06	0		25	Major	No
Repoint (50% surface) Clay Brick, Exterior, 1st Floor	Sq Ft	Brick Layer	Contract	.0780	.20	0		25	Major	No
Replace Clay Brick, Exterior, 1st Floor	Sq Ft	Brick Layer	Contract	.3700	2.87	0		75	Major	Yes
Clay Brick, Exterior, 2nd Floor										
Clean & Reseal Clay Brick, Exterior, 2nd Floor	Sq Ft	Maint Painter	Staff	.0408	.27	0		10	PM	No
Repair Clay Brick, Exterior, 2nd Floor	Sq Ft	Brick Layer	Contract	.0163	.06	0		25	Major	No
Repoint (50% Surface) Clay Brick, Exterior, 2nd Floor	Sq Ft	Brick Layer	Contract	.0820	.20	0		25	Major	No
Replace Clay Brick, Exterior, 2nd Floor	Sq Ft	Brick Layer	Contract	.3788	2.87	0		75	Major	Yes
Clay Brick, Exterior, 3rd+ Floor										
Clean & Reseal Clay Brick, Exterior, 3rd+ Floor	Sq Ft	Maint Painter	Staff	.0496	.27	0		10	PM	No
Repair Clay Brick, Exterior, 3rd+ Floor	Sq Ft	Brick Layer	Contract	.0163	.06	0		25	Major	No
Repoint (50% surface) Clay Brick, Exterior, 3rd+ Floor	Sq Ft	Brick Layer	Contract	.0864	.20	0		25	Major	No
Replace Clay Brick, Exterior, 3rd+ Floor	Sq Ft	Brick Layer	Contract	.3876	2.87	0		75	Major	Yes

*Task frequency for Washington D.C.

Components ordered by UNIFORMAT II Classifications. All costs expressed in ($) 2004.

Sources: U.S. Army Corps of Engineers, Department of Defense Engineering Performance Standards, and Whitestone Research interviews.

5.1 Building Components and Repair Tasks

Uniformat / Component / Repair Task	Units	Trade	Labor Type	Labor Hours	Material Costs	Equip Hours	Equipment Type	Freq. Yrs.*	Task Type	Repl. Task
Clay Brick, Painted, Exterior, 1st Floor										
Refinish Clay Brick, Painted, Exterior, 1st Floor	Sq Ft	Painter	Contract	.0340	.09	0		10	Major	No
Repair Clay Brick, Painted, Exterior, 1st Floor	Sq Ft	Brick Layer	Contract	.0163	.06	0		25	Major	No
Repoint (50% surface) Clay Brick, Painted, Exterior, 1st Floor	Sq Ft	Brick Layer	Contract	.0818	.16	0		25	Major	No
Replace Clay Brick, Painted, Exterior, 1st Floor	Sq Ft	Brick Layer	Contract	.4100	3.07	0		75	Major	Yes
Clay Brick, Painted, Exterior, 2nd Floor										
Refinish Clay Brick, Painted, Exterior, 2nd Floor	Sq Ft	Painter	Contract	.0420	.09	0		10	Major	No
Repoint (50% surface) Clay Brick, Painted, Exterior, 2nd Floor	Sq Ft	Brick Layer	Contract	.0858	.16	0		25	Major	No
Repair Clay Brick, Painted, Exterior, 2nd Floor	Sq Ft	Brick Layer	Contract	.0163	.06	0		25	Major	No
Replace Clay Brick, Painted, Exterior, 2nd Floor	Sq Ft	Brick Layer	Contract	.4180	3.07	0		75	Major	Yes
Clay Brick, Painted, Exterior, 3rd+ Floor										
Refinish Clay Brick, Painted, Exterior, 3rd+ Floor	Sq Ft	Painter	Contract	.0428	.09	0		10	Major	No
Repoint (50% surface) Clay Brick, Painted, Exterior, 3rd+ Floor	Sq Ft	Brick Layer	Contract	.0898	.16	0		25	Major	No
Repair Clay Brick, Painted, Exterior, 3rd+ Floor	Sq Ft	Brick Layer	Contract	.0163	.06	0		25	Major	No
Replace Clay Brick, Painted, Exterior, 3rd+ Floor	Sq Ft	Brick Layer	Contract	.4260	3.07	0		75	Major	Yes
Concrete Block, Exterior, 1st Floor										
Clean & Seal Concrete Block, Exterior, 1st Floor	Sq Ft	Maint Painter	Staff	.0320	.28	0		10	PM	No
Repair Concrete Block, Exterior, 1st Floor	Sq Ft	Brick Layer	Contract	.0186	.04	0		25	Major	No
Repoint (50% surface) Concrete Block, Exterior, 1st Floor	Sq Ft	Brick Layer	Contract	.0380	.09	0		25	Major	No
Replace Concrete Block, Exterior, 1st Floor	Sq Ft	Brick Layer	Contract	.2032	1.95	0		75	Major	Yes
Concrete Block, Exterior, 2nd Floor										
Clean & Seal Concrete Block, Exterior, 2nd Floor	Sq Ft	Maint Painter	Staff	.0408	.28	0		10	PM	No
Repair Concrete Block, Exterior, 2nd Floor	Sq Ft	Brick Layer	Contract	.0186	.04	0		25	Major	No
Repoint (50% surface) Concrete Block, Exterior, 2nd Floor	Sq Ft	Brick Layer	Contract	.0420	.09	0		25	Major	No
Replace Concrete Block, Exterior, 2nd Floor	Sq Ft	Brick Layer	Contract	.2112	1.95	0		75	Major	Yes
Concrete Block, Exterior, 3rd+ Floor										
Clean & Seal Concrete Block, Exterior, 3rd+ Floor	Sq Ft	Maint Painter	Staff	.0496	.28	0		10	PM	No
Repair Concrete Block, Exterior, 3rd+ Floor	Sq Ft	Brick Layer	Contract	.0186	.05	0		25	Major	No
Repoint (50% surface) Concrete Block, Exterior, 3rd+ Floor	Sq Ft	Brick Layer	Contract	.0460	.09	0		25	Major	No
Replace Concrete Block, Exterior, 3rd+ Floor	Sq Ft	Brick Layer	Contract	.2192	1.95	0		75	Major	Yes

*Task frequency for Washington D.C.

Components ordered by UNIFORMAT II Classifications. All costs expressed in ($) 2004.

Sources: U.S. Army Corps of Engineers, Department of Defense Engineering Performance Standards, and Whitestone Research

5.1 Building Components and Repair Tasks

Uniformat / Component / Repair Task	Units	Trade	Labor Type	Labor Hours	Material Costs	Equip Hours	Equipment Type	Freq. Yrs.*	Task Type	Repl. Task
Concrete Block, Painted, Exterior, 1st Floor										
Refinish Concrete Block, Painted, Exterior, 1st Floor	Sq Ft	Painter	Contract	.0470	.20	0		10	Major	No
Repoint (50% surface) Concrete Block, Painted, Exterior, 1st Floor	Sq Ft	Brick Layer	Contract	.0460	.17	0		25	Major	No
Repair Concrete Block, Painted, Exterior, 1st Floor	Sq Ft	Brick Layer	Contract	.0040	.04	0		25	Major	No
Replace Concrete Block, Painted, Exterior, 1st Floor	Sq Ft	Brick Layer	Contract	.2100	2.19	0		75	Major	Yes
Concrete Block, Painted, Exterior, 2nd Floor										
Refinish Concrete Block, Painted, Exterior, 2nd Floor	Sq Ft	Painter	Contract	.0590	.20	0		10	Major	No
Repair Concrete Block, Painted, Exterior, 2nd Floor	Sq Ft	Brick Layer	Contract	.0040	.04	0		25	Major	No
Repoint (50% surface) Concrete Block, Painted, Exterior, 2nd Floor	Sq Ft	Brick Layer	Contract	.0500	.17	0		25	Major	No
Replace Concrete Block, Painted, Exterior, 2nd Floor	Sq Ft	Brick Layer	Contract	.2188	2.19	0		75	Major	Yes
Concrete Block, Painted, Exterior, 3rd+ Floor										
Refinish Concrete Block, Painted, Exterior, 3rd+ Floor	Sq Ft	Painter	Contract	.0710	.08	0		10	Major	No
Repair Concrete Block, Painted, Exterior, 3rd+ Floor	Sq Ft	Brick Layer	Contract	.0040	.04	0		25	Major	No
Repoint (50% surface) Concrete Block, Painted, Exterior, 3rd+ Floor	Sq Ft	Brick Layer	Contract	.0548	.17	0		25	Major	No
Replace Concrete Block, Painted, Exterior, 3rd+ Floor	Sq Ft	Brick Layer	Contract	.2276	2.19	0		75	Major	Yes
Concrete, Painted, Exterior, 1st Floor										
Refinish Concrete, Painted, Exterior, 1st Floor	Sq Ft	Painter	Contract	.0470	.07	0		10	Major	No
Concrete, Painted, Exterior, 2nd Floor										
Refinish Concrete, Painted, Exterior, 2nd Floor	Sq Ft	Painter	Contract	.0590	.07	0		10	Major	No
Concrete, Painted, Exterior, 3rd+ Floor										
Refinish Concrete, Painted, Exterior, 3rd+ Floor	Sq Ft	Painter	Contract	.0710	.07	0		10	Major	No
Concrete Wall, Exterior										
Repair Concrete Wall, Exterior	Sq Ft	Brick Layer	Contract	.0186	.03	0		10	Major	No
Clean & Seal Concrete Wall, Exterior	Sq Ft	Maint Painter	Staff	.0217	.27	0		10	PM	No
Stucco, Exterior, 1st Floor										
Refinish Stucco, Exterior, 1st Floor	Sq Ft	Painter	Contract	.0470	.09	0		10	Major	No
Stucco, Exterior, 2nd Floor										
Refinish Stucco, Exterior, 2nd Floor	Sq Ft	Painter	Contract	.0590	.09	0		10	Major	No
Stucco, Exterior, 3rd+ Floor										
Refinish Stucco, Exterior, 3rd+ Floor	Sq Ft	Painter	Contract	.0590	.09	0		10	Major	No

*Task frequency for Washington D.C.
Components ordered by UNIFORMAT II Classifications. All costs expressed in ($) 2004.
Sources: U.S. Army Corps of Engineers, Department of Defense Engineering Performance Standards, and Whitestone Research interviews.

5.1 Building Components and Repair Tasks

Uniformat / Component / Repair Task	Units	Trade	Labor Type	Labor Hours	Material Costs	Equip Hours	Equipment Type	Freq. Yrs.*	Task Type	Repl. Task
Glass Curtain Wall										
Minor Repair and Maintenance of Glass Curtain Wall	Sq Ft	Maint Carpenter	Staff	.1500	13.59	0		5	Major	No
Replace Glass Curtain Wall	Sq Ft	Carpenter	Contract	.1640	40.00	0		30	Major	Yes
Steel, Painted, Exterior Finish, 1st Floor										
Refinish Steel, Painted, Exterior Finish, 1st Floor	Sq Ft	Painter	Contract	.0360	.10	0		10	Major	No
Replace Steel, Painted, Exterior Finish, 1st Floor	Sq Ft	Carpenter	Contract	.0730	5.93	0		150	Major	Yes
Finish Replaced Steel, Painted, Exterior Finish, 1st Floor	Sq Ft	Painter	Contract	.0250	.10	0		150	Major	No
Steel, Painted, Exterior Finish, 2nd Floor										
Refinish Steel, Painted, Exterior Finish, 2nd Floor	Sq Ft	Painter	Contract	.0480	.10	0		10	Major	No
Replace Steel, Painted, Exterior Finish, 2nd Floor	Sq Ft	Carpenter	Contract	.0930	5.93	0		150	Major	Yes
Finish Replaced Steel, Painted, Exterior Finish, 2nd Floor	Sq Ft	Painter	Contract	.0370	.10	0		150	Major	No
Steel, Painted, Exterior Finish, 3rd+ Floor										
Refinish Steel, Painted, Exterior Finish, 3rd+ Floor	Sq Ft	Painter	Contract	.0610	.10	0		10	Major	No
Replace Steel, Painted, Exterior Finish, 3rd+ Floor	Sq Ft	Carpenter	Contract	.1050	5.93	0		150	Major	Yes
Finish Replaced Steel, Painted, Exterior Finish, 3rd+ Floor	Sq Ft	Painter	Contract	.0490	.10	0		150	Major	No
Steel Dampler										
Refinish Steel Damper	Each	Gen Maint Worker	Staff	.5000	.25	0		1	PM	No
Replace Steel Damper	Each	Carpenter	Contract	.7270	168.00	0		60	Major	Yes
B2020 Exterior Windows										
Aluminum Operable Window, 12 sf, 1st Floor										
Repair Aluminum Operable Window, 12 sf, 1st Floor	Each	Carpenter	Contract	.1600	3.73	0		20	Major	No
Replace Aluminum Operable Window, 12 sf, 1st Floor	Each	Carpenter	Contract	2.4800	181.21	0		75	Major	Yes
Aluminum Operable Window, 12 sf, 2nd Floor										
Repair Aluminum Operable Window, 12 sf, 2nd Floor	Each	Carpenter	Contract	.2100	3.73	0		20	Major	No
Replace Aluminum Operable Window, 12 sf, 2nd Floor	Each	Carpenter	Contract	3.2400	181.21	0		75	Major	Yes
Aluminum Operable Window, 12 sf, 3rd+ Floor										
Repair Aluminum Operable Window, 12 sf, 3rd+ Floor	Each	Carpenter	Contract	.2500	3.73	0		20	Major	No
Replace Aluminum Operable Window, 12 sf, 3rd+ Floor	Each	Carpenter	Contract	3.9450	181.21	0		75	Major	Yes
Aluminum Operable Window, 24 sf, 1st Floor										
Repair Aluminum Operable Window, 24 sf, 1st Floor	Each	Carpenter	Contract	.3000	3.73	0		20	Major	No
Replace Aluminum Operable Window, 24 sf, 1st Floor	Each	Carpenter	Contract	2.4800	403.27	0		75	Major	Yes

*Task frequency for Washington D.C.

Components ordered by UNIFORMAT II Classifications. All costs expressed in ($) 2004.

Sources: U.S. Army Corps of Engineers, Department of Defense Engineering [illegible]

5.1 Building Components and Repair Tasks

Uniformat / Component / Repair Task	Units	Trade	Labor Type	Labor Hours	Material Costs	Equip Hours	Equipment Type	Freq. Yrs.*	Task Type	Repl. Task
Aluminum Operable Window, 24 sf, 2nd Floor										
Repair Aluminum Operable Window, 24 sf, 2nd Floor	Each	Carpenter	Contract	.4000	3.73	0		20	Major	No
Replace Aluminum Operable Window, 24 sf, 2nd Floor	Each	Carpenter	Contract	3.2445	403.27	0		75	Major	Yes
Aluminum Operable Window, 24 sf, 3rd+ Floor										
Repair Aluminum Operable Window, 24 sf, 3rd+ Floor	Each	Carpenter	Contract	.5000	3.73	0		20	Major	No
Replace Aluminum Operable Window, 24 sf, 3rd+ Floor	Each	Carpenter	Contract	3.9456	403.27	0		75	Major	Yes
Aluminum Fixed Window, 12 sf, 1st Floor										
Repair Aluminum Fixed Window, 12 sf, 1st Floor	Each	Carpenter	Contract	.1600	1.86	0		20	Major	No
Replace Aluminum Fixed Window, 12 sf, 1st Floor	Each	Carpenter	Contract	2.4800	140.36	0		75	Major	Yes
Aluminum Fixed Window, 12 sf, 2nd Floor										
Repair Aluminum Fixed Window, 12 sf, 2nd Floor	Each	Carpenter	Contract	.2100	1.86	0		20	Major	No
Replace Aluminum Fixed Window, 12 sf, 2nd Floor	Each	Carpenter	Contract	3.2400	140.36	0		75	Major	Yes
Aluminum Fixed Window, 12 sf, 3rd+ Floor										
Repair Aluminum Fixed Window, 12 sf, 3rd+ Floor	Each	Carpenter	Contract	.2500	1.86	0		20	Major	No
Replace Aluminum Fixed Window, 12 sf, 3rd+ Floor	Each	Carpenter	Contract	3.9450	140.36	0		75	Major	Yes
Aluminum Fixed Window, 24 sf, 1st Floor										
Repair Aluminum Fixed Window, 24 sf, 1st Floor	Each	Carpenter	Contract	.3000	1.86	0		20	Major	No
Replace Aluminum Fixed Window, 24 sf, 1st Floor	Each	Carpenter	Contract	2.4800	181.21	0		75	Major	Yes
Aluminum Fixed Window, 24 sf, 2nd Floor										
Repair Aluminum Fixed Window, 24 sf, 2nd Floor	Each	Carpenter	Contract	.4000	1.86	0		20	Major	No
Replace Aluminum Fixed Window, 24 sf, 2nd Floor	Each	Carpenter	Contract	3.2445	181.21	0		75	Major	Yes
Aluminum Fixed Window, 24 sf, 3rd+ Floor										
Repair Aluminum Fixed Window, 24 sf, 3rd+ Floor	Each	Carpenter	Contract	.5000	1.86	0		20	Major	No
Replace Aluminum Fixed Window, 24 sf, 3rd+ Floor	Each	Carpenter	Contract	3.9456	181.21	0		75	Major	Yes
Steel Frame, Painted, Operable Window, 12 sf, 1st Floor										
Refinish Steel Frame, Painted, Operable Window, 12 sf, 1st Floor	Each	Painter	Contract	.2200	1.61	0		5	Major	No
Repair Steel Frame, Painted, Operable Window, 12 sf, 1st Floor	Each	Carpenter	Contract	.1600	1.86	0		20	Major	No
Finish Replaced Steel Frame, Painted, Operable Window, 12 sf, 1st F	Each	Painter	Contract	.2200	1.61	0		80	Major	No
Replace Steel Frame, Painted, Operable Window, 12 sf, 1st Floor	Each	Carpenter	Contract	2.5400	960.50	0		80	Major	Yes

*Task frequency for Washington D.C.
Components ordered by UNIFORMAT II Classifications. All costs expressed in ($) 2004.
Sources: U.S. Army Corps of Engineers, Department of Defense Engineering Performance Standards, and Whitestone Research interviews.

5.1 Building Components and Repair Tasks

Uniformat / Component / Repair Task	Units	Trade	Labor Type	Labor Hours	Material Costs	Equip Hours	Equipment Type	Freq. Yrs.*	Task Type	Repl. Task
Steel Frame, Painted, Operable Window, 12 sf, 2nd Floor										
Refinish Steel Frame, Painted, Operable Window, 12 sf, 2nd Floor	Each	Painter	Contract	1.3340	1.61	0		5	Major	No
Repair Steel Frame, Painted, Operable Window, 12 sf, 2nd Floor	Each	Carpenter	Contract	.2100	1.86	0		20	Major	No
Finish Replaced Steel Frame, Painted, Operable Window, 12 sf, 2nd	Each	Painter	Contract	1.3340	1.61	0		80	Major	No
Replace Steel Frame, Painted, Operable Window, 12 sf, 2nd Floor	Each	Carpenter	Contract	3.2440	960.50	0		80	Major	Yes
Steel Frame, Painted, Operable Window, 12 sf, 3rd+ Floor										
Refinish Steel Frame, Painted, Operable Window, 12 sf, 3rd+ Floor	Each	Painter	Contract	1.9825	1.61	0		5	Major	No
Repair Steel Frame, Painted, Operable Window, 12 sf, 3rd+ Floor	Each	Carpenter	Contract	.2500	1.86	0		20	Major	No
Replace Steel Frame, Painted, Operable Window, 12 sf, 3rd+ Floor	Each	Carpenter	Contract	3.9500	960.50	0		80	Major	Yes
Finish Replaced Steel Frame, Painted, Operable Window, 12 sf, 3rd+	Each	Painter	Contract	1.9825	1.61	0		80	Major	No
B2030 Exterior Doors										
Steel w/ Safety Glass, Painted, Exterior Door										
Maintain Steel w/ Safety Glass, Painted, Exterior Door, Locks	Each	Gen Maint Worker	Staff	.2160	.00	0		1	PM	No
Refinish Steel w/ Safety Glass, Painted, Exterior Door	Each	Painter	Contract	.3200	2.13	0		4	Major	No
Repair Steel w/ Safety Glass, Painted, Exterior Door	Each	Carpenter	Contract	.8400	222.00	0		14	Major	No
Finish Replaced Steel w/ Safety Glass, Painted, Exterior Door	Each	Painter	Contract	.3206	2.13	0		80	Major	No
Replace Steel w/ Safety Glass, Painted, Exterior Door	Each	Carpenter	Contract	2.1000	181.00	0		80	Major	Yes
Steel, Painted, Exterior Door										
Maintain Steel, Painted, Exterior Door, Locks	Each	Gen Maint Worker	Staff	.2160	.00	0		1	PM	No
Refinish Steel, Painted, Exterior Door	Each	Painter	Contract	.3200	2.13	0		4	Major	No
Repair Steel, Painted, Exterior Door	Each	Carpenter	Contract	.8400	222.00	0		14	Major	No
Finish Replaced Steel, Painted, Exterior Door	Each	Painter	Contract	.3200	2.13	0		80	Major	No
Replace Steel, Painted, Exterior Door	Each	Carpenter	Contract	1.7438	181.00	0		80	Major	Yes
Wood, Solid Core, Painted, Exterior Door										
Maintain Wood, Solid Core, Painted, Exterior Door Locks	Each	Gen Maint Worker	Staff	.2160	.00	0		1	PM	No
Refinish Wood, Solid Core, Painted, Exterior Door	Each	Painter	Contract	.2410	9.20	0		4	Major	No
Repair Wood, Solid Core, Painted, Exterior Door	Each	Carpenter	Contract	.6000	97.00	0		7	Major	No
Finish Replaced Wood, Solid Core, Painted, Exterior Door	Each	Painter	Contract	.2480	9.20	0		40	Major	No
Replace Wood, Solid Core, Painted, Exterior Door	Each	Carpenter	Contract	1.7385	510.15	0		40	Major	Yes
Aluminum Frame, Fully Glazed, Exterior Door										
Maintain Aluminum Frame, Fully Glazed, Exterior Door Locks	Each	Gen Maint Worker	Staff	.2160	.00	0		1	PM	No
Repair Aluminum Frame, Fully Glazed, Exterior Door	Each	Carpenter	Contract	.8400	126.50	0		12	Major	No
Replace Aluminum Frame, Fully Glazed, Exterior Door	Each	Carpenter	Contract	1.7198	574.46	0		65	Major	Yes

*Task frequency for Washington D.C.

Components ordered by UNIFORMAT II Classifications. All costs expressed in ($) 2004.

Sources: U.S. Army Corps of Engineers, Department of Defense Engineering Performance Standards, and Whitestone Research

Uniformat / Component / Repair Task	Units	Trade	Labor Type	Labor Hours	Material Costs	Equip Hours	Equipment Type	Freq. Yrs.*	Task Type	Repl. Task
Aluminum Frame, Fully Glazed, Sliding Exterior Door										
Repair Aluminum Frame, Fully Glazed, Sliding Exterior Door	Each	Carpenter	Contract	.6800	46.25	0		12	Major	No
Replace Aluminum Frame, Fully Glazed, Sliding Exterior Door	Each	Carpenter	Contract	5.6322	938.00	0		65	Major	Yes
Aluminum & Glass Revolving Door										
Maintain Aluminum & Glass Revolving Door	Each	Gen Maint Worker	Staff	1.0000	.00	0		1	PM	No
Replace Aluminum & Glass Revolving Door	Each	Carpenter	Contract	50.0000	19,865.20	0		15	Major	Yes
Steel Single 12'x12', Painted, Roll-up Door										
Maintain Steel Single 12'x12', Painted, Roll-up Door	Each	Gen Maint Worker	Staff	1.2400	.00	0		1	PM	No
Refinish Steel Single 12'x12', Painted, Roll-up Door	Each	Painter	Contract	2.0000	21.22	0		5	Major	No
Repair Steel Single 12'x12', Painted, Roll-up Door	Each	Carpenter	Contract	.8000	156.00	0		10	Major	No
Replace Steel Single 12'x12', Painted, Roll-up Door	Each	Carpenter	Contract	13.3600	1,250.00	0		35	Major	Yes
Finish Replaced Steel Single 12'x12', Painted, Roll-up Door	Each	Painter	Contract	2.0000	20.16	0		35	Major	No
Steel Double 12'x24', Painted, Roll-up Door										
Maintain Steel Double 12'x24', Painted, Roll-up Door	Each	Gen Maint Worker	Staff	1.2800	.00	0		1	PM	No
Refinish Steel Double 12'x24', Painted, Roll-up Door	Each	Painter	Contract	4.0000	42.42	0		5	Major	No
Repair Steel Double 12'x24', Painted, Roll-up Door	Each	Carpenter	Contract	.8000	312.50	0		10	Major	No
Finish Replaced Steel Double 12'x24', Painted, Roll-up Door	Each	Painter	Contract	4.0000	42.42	0		35	Major	No
Replace Steel Double 12'x24', Painted, Roll-up Door	Each	Carpenter	Contract	26.0000	2,500.00	0		35	Major	Yes
Steel 20'x20', Painted, Roll-up Door										
Maintain Steel 20'x20', Painted, Roll-up Door	Each	Gen Maint Worker	Staff	2.0000	.00	0		1	PM	No
Refinish Steel 20'x20', Painted, Roll-up Door	Each	Painter	Contract	6.4000	56.00	0		5	Major	No
Repair Steel 20'x20', Painted, Roll-up Door	Each	Carpenter	Contract	.8000	532.11	0		10	Major	No
Finish Replaced Steel 20'x20', Painted, Roll-up Door	Each	Painter	Contract	6.4000	58.18	0		35	Major	No
Replace Steel 20'x20', Painted, Roll-up Door	Each	Carpenter	Contract	24.0000	4,300.00	0		35	Major	Yes
Swinging Doors, Motor-Operated										
Maintain Swinging Doors, Motor-Operated	Each	Gen Maint Worker	Staff	8.0000	.00	0		1	PM	No
Replace Swinging Doors, Motor-Operated	Each	Carpenter	Contract	8.0000	5,325.00	0		15	Major	Yes
Sliding Doors, Motor-Operated										
Maintain Sliding Doors, Motor-Operated	Each	Gen Maint Worker	Staff	8.0000	.00	0		1	PM	No
Replace Sliding Doors, Motor-Operated	Each	Carpenter	Contract	9.0000	5,900.00	0		15	Major	Yes

B30 Roofing

B3010 Roof Coverings

*Task frequency for Washington D.C.

Components ordered by UNIFORMAT II Classifications. All costs expressed in ($) 2004.

Sources: U.S. Army Corps of Engineers, Department of Defense Engineering Performance Standards, and Whitestone Research interviews.

5.1 Building Components and Repair Tasks

Uniformat / Component / Repair Task	Units	Trade	Labor Type	Labor Hours	Material Costs	Equip Hours	Equipment Type	Freq. Yrs.*	Task Type	Repl. Task
Asphalt Shingle Roof										
Maintain Asphalt Shingle Roof	Sq Ft	Gen Maint Worker	Staff	.0019	.02	0		1	PM	No
Minor Replacement, Asphalt Shingle Roof	Sq Ft	Roofer	Contract	.0006	.43	0		5	Major	No
Place New Over Existing, Asphalt Shingle Roof	Sq Ft	Roofer	Contract	.0230	.45	0		20	Major	No
Remove & Replace Asphalt Shingle Roof	Sq Ft	Roofer	Contract	.0317	.49	0		40	Major	Yes
Built-up Roof										
Maintain Built-up Roof	Sq Ft	Gen Maint Worker	Staff	.0034	.03	0		1	PM	No
Non-Destructive Moisture Inspection	Sq Ft	Roofer	Contract	.0011	.00	0		3	Major	No
Place New Membrane Over Existing, Built-up Roof	Sq Ft	Roofer	Contract	.0186	.67	0		14	Major	No
Remove & Replace Membrane, Built-up Roof	Sq Ft	Roofer	Contract	.0380	1.28	0		28	Major	Yes
Single-Ply Thermosetting Roof										
Maintain Single-Ply Thermosetting Roof	Sq Ft	Gen Maint Worker	Staff	.0010	.03	0		1	Major	No
Non-Destructive Moisture Inspection	Sq Ft	Roofer	Contract	.0011	.00	0		3	PM	No
Minor Replacement, Single-Ply Thermosetting Roof	Sq Ft	Roofer	Contract	.0008	.50	0		10	Major	No
Remove & Replace Membrane, Single-Ply Thermosetting Roof	Sq Ft	Roofer	Contract	.0284	1.25	0		20	Major	Yes
Single-Ply Modified Bituminous/Thermoplastic Roof										
Maintain Single-Ply Modified Bituminous/Thermoplastic Roof	Sq Ft	Gen Maint Worker	Staff	.0016	.03	0		1	PM	No
Non-Destructive Moisture Inspection	Sq Ft	Roofer	Contract	.0011	.00	0		3	Major	No
Repair Single-Ply Modified Bituminous/Thermoplastic Roof	Sq Ft	Roofer	Contract	.0003	.63	0		10	Major	No
Remove & Replace Membrane, Single-Ply Modified Bituminous/Ther	Sq Ft	Roofer	Contract	.0435	1.12	0		20	Major	Yes
Clay Tile Roof										
Maintain Clay Tile Roof	Sq Ft	Gen Maint Worker	Staff	.0013	.00	0		1	PM	No
Minor Replacement, Clay Tile Roof	Sq Ft	Roofer	Contract	.0023	.11	0		20	Major	No
Remove & Replace Clay Tile Roof	Sq Ft	Roofer	Contract	.0782	5.48	0		70	Major	Yes
Metal Roof										
Maintain Metal Roof	Sq Ft	Gen Maint Worker	Staff	.0010	.00	0		1	PM	No
Minor Repair, Metal Roof	Sq Ft	Roofer	Contract	.0005	.45	0		5	Major	No
Minor Replacement, Metal Roof	Sq Ft	Roofer	Contract	.0052	.07	0		20	Major	No
Remove & Replace Metal Roof	Sq Ft	Roofer	Contract	.2790	1.06	0		30	Major	Yes
Concrete Decking										
Refinish Concrete Decking	Sq Ft	Painter	Contract	.0165	.08	0		7	PM	No
Repair Concrete Decking	Sq Ft	Gen Maint Worker	Staff	.0338	.03	0		20	PM	No
Refinish Repaired Concrete Decking	Sq Ft	Maint Painter	Staff	.0005	.06	0		20	PM	No

*Task frequency for Washington D.C.

Components ordered by UNIFORMAT II Classifications. All costs expressed in ($) 2004.

Sources: U.S. Army Corps of Engineers, [illegible]

5.1 Building Components and Repair Tasks

Uniformat / Component / Repair Task	Units	Trade	Labor Type	Labor Hours	Material Costs	Equip Hours	Equipment Type	Freq. Yrs.*	Task Type	Repl. Task
Concrete Steps										
Refinish Concrete Steps	Sq Ft	Painter	Contract	.0165	.06	0		7	PM	No
Repair Concrete Steps	Sq Ft	Gen Maint Worker	Staff	.0338	.03	0		20	PM	No
Refinish Repaired Concrete Steps	Sq Ft	Maint Painter	Staff	.0005	.06	0		20	PM	No
Metal Decking										
Refinish Metal Decking	Sq Ft	Painter	Contract	.0165	.12	0		7	PM	No
Repair Metal Decking	Sq Ft	Maint Carpenter	Staff	.0287	.02	0		20	PM	No
Metal Steps										
Refinish Metal Steps	Sq Ft	Painter	Contract	.0165	.06	0		7	PM	No
Repair Metal Steps	Sq Ft	Maint Carpenter	Staff	.0287	28.46	0		20	PM	No
Wood Decking										
Refinish Wood Decking	Sq Ft	Painter	Contract	.0122	.06	0		5	Major	No
Repair Wood Decking	Sq Ft	Carpenter	Contract	.0008	.05	0		20	Major	No
Wooden Steps										
Refinish Wood Steps	Sq Ft	Painter	Contract	.0122	.06	0		5	Major	No
Repair Wood Steps	Sq Ft	Carpenter	Contract	.0008	.05	0		20	Major	No
Fiberglass Dome Roof										
Debris Removal and Visual Inspection on Roof	Sq Ft	Gen Maint Worker	Staff	.0040	.02	0		5	PM	No
Fiberglass Roof Replacement	Sq Ft	Roofer	Contract	.0941	28.46	0		20	Major	Yes
Roof Hatch, Aluminum										
Annual Maintenance, Aluminum Roof Hatch	Each	Maint Carpenter	Staff	1.0000	.05	0		1	PM	No
Replace Aluminum, Roof Hatch	Each	Roofer	Contract	3.2000	3.16	0		50	Major	Yes
B3020 Roof Openings										
Roof Skylight, 20 sf										
Repair Roof Skylight, 20 sf	Each	Carpenter	Contract	.5000	3.78	0		20	Major	No
Replace Roof Skylight	Each	Carpenter	Contract	3.9456	324.00	0		75	Major	Yes
C10 Interior Construction										
C1020 Interior Doors										

*Task frequency for Washington D.C.

Components ordered by UNIFORMAT II Classifications. All costs expressed in ($) 2004.

Sources: U.S. Army Corps of Engineers, Department of Defense Engineering Performance Standards, and Whitestone Research interviews.

5.1 Building Components and Repair Tasks

Uniformat / Component / Repair Task	Units	Trade	Labor Type	Labor Hours	Material Costs	Equip Hours	Equipment Type	Freq. Yrs.*	Task Type	Repl. Task
Wood Frame, Fully Glazed, Painted, Interior Door										
Maintain Wood Frame, Fully Glazed, Painted, Door Locks	Each	Gen Maint Worker	Staff	.2160	.00	0		1	PM	No
Refinish Wood Frame, Fully Glazed, Painted, Interior Door	Each	Painter	Contract	.1280	1.61	0		4	Major	No
Replace Wood Frame, Fully Glazed, Painted, Door Locks	Each	Carpenter	Contract	.8210	68.50	0		10	Major	No
Finish Replaced Wood Frame, Fully Glazed, Painted, Interior Door	Each	Painter	Contract	.1280	2.08	0		50	Major	No
Replace Wood Frame, Fully Glazed, Painted, Interior Door	Each	Carpenter	Contract	2.1000	492.50	0		50	Major	Yes
Wood Frame, Painted, Fully Glazed, Sliding Interior Door										
Maintain Wood Frame, Painted, Fully Glazed, Sliding Door Locks	Each	Gen Maint Worker	Staff	.2160	.00	0		1	PM	No
Refinish Wood Frame, Painted, Fully Glazed, Sliding Interior Door	Each	Painter	Contract	.3200	1.61	0		4	Major	No
Replace Wood Frame, Painted, Fully Glazed, Sliding Door Locks	Each	Carpenter	Contract	.8880	68.50	0		10	Major	No
Replace Wood Frame, Painted, Fully Glazed, Sliding Interior Door	Each	Carpenter	Contract	4.3300	492.84	0		40	Major	Yes
Refinish Replaced Wood Frame, Painted, Fully Glazed, Sliding Interio	Each	Painter	Contract	.3240	2.13	0		40	Major	No
Wood, Solid Core, Painted, Interior Door										
Maintain Wood, Solid Core, Painted, Door Locks	Each	Gen Maint Worker	Staff	.2160	.00	0		1	PM	No
Refinish Wood, Solid Core, Painted, Interior Door	Each	Painter	Contract	.2410	3.60	0		4	Major	No
Replace Solid Core, Painted, Door Locks	Each	Carpenter	Contract	.0821	68.50	0		10	Major	No
Replace Wood, Solid Core, Painted, Interior Door	Each	Carpenter	Contract	2.1000	214.00	0		30	Major	Yes
Finish Replaced Wood, Solid Core, Painted, Interior Door	Each	Painter	Contract	.2480	4.04	0		30	Major	No
Wood, Solid Core, Painted, Sliding Interior Door										
Maintain Wood, Solid Core, Painted, Sliding Door Locks	Each	Gen Maint Worker	Staff	.2160	.00	0		1	PM	No
Refinish Wood, Solid Core, Painted, Sliding Interior Door	Each	Painter	Contract	.3200	3.60	0		4	Major	No
Replace Wood, Solid Core, Painted, Sliding Door Locks	Each	Carpenter	Contract	.8880	68.50	0		10	Major	No
Replace Wood, Solid Core, Painted, Sliding Interior Door	Each	Carpenter	Contract	4.3300	243.50	0		40	Major	Yes
Finish Replaced Wood, Solid Core, Painted, Sliding Interior Door	Each	Painter	Contract	.3240	3.60	0		40	Major	No
Wood, Solid Core w/ Safety Glass, Painted, Interior Door										
Maintain Wood, Solid Core w/ Safety Glass, Painted, Door Locks	Each	Gen Maint Worker	Staff	.2160	.00	0		1	PM	No
Refinish Wood, Solid Core w/ Safety Glass, Painted, Interior Door	Each	Painter	Contract	.2410	1.65	0		4	Major	No
Replace Wood, Solid Core w/ Safety Glass, Painted, Door Locks	Each	Carpenter	Contract	.8210	68.50	0		10	Major	No
Replace Wood, Solid Core w/ Safety Glass, Painted, Interior Door	Each	Carpenter	Contract	4.3320	221.73	0		40	Major	Yes
Finish Replaced Wood, Solid Core w/ Safety Glass, Painted, Interior	Each	Painter	Contract	.2480	2.19	0		40	Major	No

*Task frequency for Washington D.C.
Components ordered by UNIFORMAT II Classifications. All costs expressed in ($) 2004.
Sources: U.S. Army Corps of Engineers, Department of Defense [illegible]

Uniformat / Component / Repair Task	Units	Trade	Labor Type	Labor Hours	Material Costs	Equip Hours	Equipment Type	Freq. Yrs.*	Task Type	Repl. Task
Wood, Hollow Core, Painted, Interior Door										
Maintain Wood, Hollow Core, Painted, Door Locks	Each	Gen Maint Worker	Staff	.2160	.00	0		1	PM	No
Refinish Wood, Hollow Core, Painted, Interior Door	Each	Painter	Contract	.2410	3.60	0		4	Major	No
Replace Wood, Hollow Core, Painted, Door Locks	Each	Carpenter	Contract	.8210	68.50	0		10	Major	No
Finish Replaced Wood, Hollow Core, Painted, Interior Door	Each	Painter	Contract	.2480	2.13	0		30	Major	No
Replace Wood, Hollow Core, Painted, Interior Door	Each	Carpenter	Contract	2.1000	157.06	0		30	Major	Yes
Wood, Hollow Core, Painted, Sliding Interior Door										
Maintain Wood, Hollow Core, Painted, Sliding Door Locks	Each	Gen Maint Worker	Staff	.2160	.00	0		1	PM	No
Refinish Wood, Hollow Core, Painted, Sliding Interior Door	Each	Painter	Contract	.3200	3.60	0		4	Major	No
Replace Wood, Hollow Core, Painted, Sliding Door Locks	Each	Carpenter	Contract	.8800	68.50	0		10	Major	No
Finish Replaced Wood, Hollow Core, Painted, Sliding Interior Door	Each	Painter	Contract	.3240	2.13	0		40	Major	No
Replace Wood, Hollow Core, Painted, Sliding Interior Door	Each	Carpenter	Contract	2.1000	153.00	0		40	Major	Yes
Steel, Painted, Interior Door										
Maintain Steel, Painted, Door Locks	Each	Gen Maint Worker	Staff	.2160	.00	0		1	PM	No
Refinish Steel, Painted, Interior Door	Each	Painter	Contract	.3195	2.13	0		4	Major	No
Replace Steel, Painted, Door Locks	Each	Carpenter	Contract	.8800	165.95	0		10	Major	No
Replace Steel, Painted, Interior Door	Each	Carpenter	Contract	2.1000	344.47	0		80	Major	Yes
Finish Replaced Steel, Painted, Interior Door	Each	Painter	Contract	.3206	2.13	0		80	Major	No
Aluminum, w/ Safety Glass, Plain or Anodized, Interior Door										
Maintain Aluminum, w/ Safety Glass, Plain or Anodized, Door Locks	Each	Gen Maint Worker	Staff	.2160	.00	0		1	PM	No
Replace Aluminum, w/ Safety Glass, Plain or Anodized, Door Locks	Each	Carpenter	Contract	.8210	2.11	0		10	Major	No
Replace Aluminum, w/ Safety Glass, Plain or Anodized, Interior Door	Each	Carpenter	Contract	2.1000	562.16	0		65	Major	Yes
Aluminum Frame, Fully Glazed, Interior Door										
Maintain Aluminum Frame, Fully Glazed, Door Locks	Each	Gen Maint Worker	Staff	.2160	.00	0		1	PM	No
Replace Aluminum Frame, Fully Glazed, Door Locks	Each	Carpenter	Contract	.8210	165.95	0		10	Major	No
Replace Aluminum Frame, Fully Glazed, Interior Door	Each	Carpenter	Contract	2.1000	568.82	0		65	Major	Yes
Wood Lead-Lined Medical Interior Door										
Maintain Wood Lead-Lined Medical Door Locks	Each	Gen Maint Worker	Staff	.2160	.00	0		1	PM	No
Repair Wood Lead-Lined Medical Interior Door	Each	Carpenter	Contract	1.0680	96.83	0		9	Major	No
Replace Wood Lead-Lined Medical Door Locks	Each	Carpenter	Contract	.8210	161.81	0		10	Major	No
Replace Wood Lead-Lined Medical Interior Door	Each	Carpenter	Contract	2.7300	522.74	0		80	Major	Yes

C20 Stairs

C2020 Stair Finishes

*Task frequency for Washington D.C.

Components ordered by UNIFORMAT II Classifications. All costs expressed in ($) 2004.

Sources: U.S. Army Corps of Engineers, Department of Defense Engineering Performance Standards, and Whitestone Research interviews.

5.1 Building Components and Repair Tasks

Uniformat / Component / Repair Task	Units	Trade	Labor Type	Labor Hours	Material Costs	Equip Hours	Equipment Type	Freq. Yrs.*	Task Type	Repl. Task
Concrete Interior Steps										
Repair Concrete Interior Steps	Sq Ft	Brick Layer	Contract	.0310	.04	0		15	Major	No
Masonry Interior Steps, Painted										
Refinish Masonry Interior Steps, Painted	Sq Ft	Painter	Contract	.0164	.15	0		7	Major	No
Repair Masonry Interior Steps, Painted	Sq Ft	Brick Layer	Contract	.0240	.05	0		27	Major	No
Refinish Repaired Masonry Interior Steps, Painted	Sq Ft	Painter	Contract	.0005	.15	0		27	Major	No
Masonry Interior Steps, Unpainted										
Repair Masonry Interior Steps, Unpainted	Sq Ft	Brick Layer	Contract	.2450	.05	0		27	Major	No
Metal Interior Steps, Painted										
Refinish Metal Interior Steps, Painted	Sq Ft	Painter	Contract	.0165	.05	0		9	Major	No
Repair Metal Interior Steps, Painted	Sq Ft	Carpenter	Contract	.0086	.05	0		15	Major	No
Refinish Repaired Metal Interior Steps, Painted	Sq Ft	Painter	Contract	.0005	.05	0		15	Major	No
Metal Interior Railing, Painted										
Refinish Metal Interior Railing, Painted	Ln Ft	Painter	Contract	.0165	.05	0		7	Major	No
Repair Metal Interior Railing, Painted	Ln Ft	Carpenter	Contract	.0023	.05	0		15	Major	No
Refinish Repaired Metal Interior Railing, Painted	Ln Ft	Painter	Contract	.0005	.05	0		15	Major	No
C30 Interior Finishes										
C3010 Wall Finishes										
Clay Brick Interior Wall Finish										
Minor Repair, Clay Brick Interior Wall Finish	Sq Ft	Brick Layer	Contract	.0189	.02	0		25	Major	No
Sheetrock, Unstippled, Interior Wall Finish										
Refinish Sheetrock, Unstippled, Interior Wall Finish	Sq Ft	Painter	Contract	.0143	.04	0		4	Major	No
Minor Repair, Sheetrock, Unstippled, Interior Wall Finish	Sq Ft	Carpenter	Contract	.0063	.01	0		20	Major	No
Finish Repaired Sheetrock, Unstippled, Interior Wall Finish	Sq Ft	Painter	Contract	.0024	.04	0		20	Major	No
Sheetrock, Stippled, Interior Wall Finish										
Refinish Sheetrock, Stippled, Interior Wall Finish	Sq Ft	Painter	Contract	.0143	.04	0		4	Major	No
Minor Repair, Sheetrock, Stippled, Interior Wall Finish	Sq Ft	Carpenter	Contract	.0063	.01	0		20	Major	No
Finish Repaired Sheetrock, Stippled, Interior Wall Finish	Sq Ft	Painter	Contract	.0024	.04	0		20	Major	No

*Task frequency for Washington D.C.

Components ordered by UNIFORMAT II Classifications. All costs expressed in ($) 2004.

Sources: U.S. Army Corps of Engineers, Department [illegible]

5.1 Building Components and Repair Tasks

Uniformat / Component / Repair Task	Units	Trade	Labor Type	Labor Hours	Material Costs	Equip Hours	Equipment Type	Freq. Yrs.*	Task Type	Repl. Task
Plaster Interior Wall Finish										
Refinish Plaster Interior Wall Finish	Sq Ft	Painter	Contract	.0143	.04	0		4	Major	No
Minor Repair, Plaster Interior Wall Finish	Sq Ft	Carpenter	Contract	.0066	.05	0		12	Major	No
Finish Repaired Plaster Interior Wall Finish	Sq Ft	Painter	Contract	.0024	.04	0		12	Major	No
Concrete, Painted, Interior Wall Finish										
Refinish Concrete, Painted, Interior Wall Finish	Sq Ft	Painter	Contract	.0143	.14	0		4	Major	No
Minor Repair, Concrete, Painted, Interior Wall Finish	Sq Ft	Brick Layer	Contract	.0359	.02	0		15	Major	No
Finish Repaired Concrete, Painted, Interior Wall Finish	Sq Ft	Painter	Contract	.0024	.07	0		15	Major	No
Concrete Block, Painted, Interior Wall Finish										
Refinish Concrete Block, Painted, Interior Wall Finish	Sq Ft	Painter	Contract	.0143	.14	0		4	Major	No
Finish Repaired Concrete Block, Painted, Interior Wall Finish	Sq Ft	Painter	Contract	.0024	.07	0		24	Major	No
Minor Repair, Concrete Block, Painted, Interior Wall Finish	Sq Ft	Brick Layer	Contract	.0049	.03	0		24	Major	No
Fabric Interior Wall Finish										
Minor Repair, Fabric Interior Wall Finish	Sq Ft	Carpenter	Contract	.0023	.03	0		5	PM	No
Replace Fabric Interior Wall Finish	Sq Ft	Painter	Contract	.0225	.72	0		10	Major	Yes
Wallpaper Interior Wall Finish										
Minor Repair, Wallpaper Interior Wall Finish	Sq Ft	Carpenter	Contract	.0023	.04	0		5	PM	No
Replace Wallpaper Interior Wall Finish	Sq Ft	Painter	Contract	.0225	.64	0		10	Major	Yes
Vinyl Interior Wall Finish										
Minor Repair, Vinyl Interior Wall Finish	Sq Ft	Carpenter	Contract	.0023	.08	0		5	PM	No
Replace Vinyl Interior Wall Finish	Sq Ft	Painter	Contract	.0225	.72	0		10	Major	Yes
Ceramic Tile Wall Finish										
Minor Repair, Ceramic Tile Wall Finish	Sq Ft	Brick Layer	Contract	.0205	.04	0		10	Major	No
Replace Ceramic Tile Wall Finish	Sq Ft	Brick Layer	Contract	.3200	2.17	0		100	Major	Yes
Steel, Painted, Interior Wall Finish										
Refinish Steel, Painted, Interior Wall Finish	Sq Ft	Painter	Contract	.0139	.04	0		4	Major	No
Finish Replaced Steel, Painted, Interior Wall Finish	Sq Ft	Painter	Contract	.0252	.04	0		200	Major	No
Replace Steel, Painted, Interior Wall Finish	Sq Ft	Carpenter	Contract	.4138	3.25	0		200	Major	Yes
Acoustical Interior Wall Tile										
Refinish Acoustical Interior Wall Tile	Sq Ft	Painter	Contract	.0164	.20	0		4	Major	No
Minor Repair, Acoustical Interior Wall Tile	Sq Ft	Carpenter	Contract	.0073	.03	0		24	Major	No

*Task frequency for Washington D.C.
Components ordered by UNIFORMAT II Classifications. All costs expressed in ($) 2004.
Sources: U.S. Army Corps of Engineers, Department of Defense Engineering Performance Standards, and Whitestone Research interviews.

5.1 Building Components and Repair Tasks

Uniformat / Component / Repair Task	Units	Trade	Labor Type	Labor Hours	Material Costs	Equip Hours	Equipment Type	Freq. Yrs.*	Task Type	Repl. Task
Wall & Corner Guards										
Annual PM, Wall & Corner Guards	Ln Ft	Gen Maint Worker	Staff	.0250	.00	0		1	PM	No
Replace Wall & Corner Guards	Ln Ft	Carpenter	Contract	.1000	9.35	0		4	Major	Yes
Protective Chair Rails										
Refinish Protective Chair Rails	Ln Ft	Painter	Contract	.0165	.06	0		7	Major	No
Minor Repair, Protective Chair Rails	Ln Ft	Carpenter	Contract	.0023	.01	0		15	Major	No
Finish Repaired Protective Chair Rails	Ln Ft	Painter	Contract	.0005	.03	0		15	Major	No
C3020 Floor Finishes										
Access Flooring (Computer Room Raised Floor)										
Minor Repair, Access Flooring	Sq Ft	Carpenter	Contract	.0066	.06	0		15	Major	No
Replace Access Flooring	Sq Ft	Carpenter	Contract	.0320	10.60	0		75	Major	Yes
Concrete, Finished Flooring										
Minor Repair, Concrete, Finished Flooring	Sq Ft	Brick Layer	Contract	.0066	.01	0		15	Major	No
Refinish Concrete, Finished Flooring	Sq Ft	Brick Layer	Contract	.0046	.34	0		25	Major	No
Replace Concrete, Finished Flooring	Sq Ft	Brick Layer	Contract	.0759	.93	0		75	Major	Yes
Concrete, Unfinished Flooring										
Minor Repair, Concrete, Unfinished Flooring	Sq Ft	Brick Layer	Contract	.0066	.01	0		15	Major	No
Replace Concrete, Unfinished Flooring	Sq Ft	Brick Layer	Contract	.0725	.83	0		75	Major	Yes
Ceramic Tile Flooring										
Minor Repair, Ceramic Tile Flooring	Sq Ft	Brick Layer	Contract	.0003	.01	0		15	Major	No
Replace Ceramic Tile Flooring	Sq Ft	Brick Layer	Contract	.1404	4.01	0		50	Major	Yes
Quarry Tile Flooring										
Minor Repair, Quarry Tile Flooring	Sq Ft	Brick Layer	Contract	.0003	.14	0		15	Major	No
Replace Quarry Tile Flooring	Sq Ft	Brick Layer	Contract	.6940	4.01	0		50	Major	Yes
Terrazzo Flooring										
Minor Repair, Terrazzo Flooring	Sq Ft	Brick Layer	Contract	.0066	.14	0		15	Major	No
Replace Terrazzo Flooring	Sq Ft	Brick Layer	Contract	.1404	2.22	0		75	Major	Yes
Wood Parquetry Flooring										
Sand & Refinish Wood Parquetry Flooring	Sq Ft	Carpenter	Contract	.0132	.12	0		10	Major	No
Replace Wood Parquetry Floor	Sq Ft	Carpenter	Contract	.1537	3.62	0		40	Major	Yes

*Task frequency for Washington D.C.

Components ordered by UNIFORMAT II Classifications. All costs expressed in ($) 2004.

Sources: U.S. Army Corps of Engineers, [illegible]

5.1 Building Components and Repair Tasks

Uniformat / Component / Repair Task	Units	Trade	Labor Type	Labor Hours	Material Costs	Equip Hours	Equipment Type	Freq. Yrs.*	Task Type	Repl. Task
Maple Sports Floor										
Screen & Recoat Maple Sports Floor	Sq Ft	Carpenter	Contract	.0048	.13	0		1	PM	No
Sand & Refinish Maple Sports Floor	Sq Ft	Carpenter	Contract	.0132	.12	0		10	Major	No
Replace Maple Sports Floor	Sq Ft	Carpenter	Contract	.0292	5.42	0		40	Major	Yes
Vinyl Tile Flooring										
Minor Repair, Vinyl Tile Flooring	Sq Ft	Carpet Layer	Contract	.0004	.02	0		9	Major	No
Replace Vinyl Tile Flooring	Sq Ft	Carpet Layer	Contract	.0249	.75	0		18	Major	Yes
Rubber Tile Flooring										
Minor Repair, Rubber Tile Flooring	Sq Ft	Carpet Layer	Contract	.0004	.13	0		9	Major	No
Replace Rubber Tile Flooring	Sq Ft	Carpet Layer	Contract	.0249	5.55	0		18	Major	Yes
Vinyl Sheet Flooring										
Minor Repair, Vinyl Sheet Flooring	Sq Ft	Carpet Layer	Contract	.0038	.02	0		9	Major	No
Replace Vinyl Sheet Flooring	Sq Ft	Carpet Layer	Contract	.2924	2.12	0		18	Major	Yes
Linoleum Flooring										
Minor Repair, Linoleum Flooring	Sq Ft	Carpet Layer	Contract	.0005	.04	0		9	Major	No
Replace Linoleum Flooring	Sq Ft	Carpet Layer	Contract	.0229	2.12	0		18	Major	Yes
Carpet, Nylon 20 oz., Low Traffic										
Minor Repair, Carpet, Nylon 20 oz., Low Traffic	Sq Ft	Carpet Layer	Contract	.0002	.01	0		5	PM	No
Replace Carpet, Nylon 20 oz., Low Traffic	Sq Ft	Carpet Layer	Contract	.0400	2.33	0		15	Major	Yes
Carpet, Nylon 20 oz., High Traffic										
Minor Repair, Carpet, Nylon 20 oz., High Traffic	Sq Ft	Carpet Layer	Contract	.0002	.01	0		2	PM	No
Replace Carpet, Nylon 20 oz., High Traffic	Sq Ft	Carpet Layer	Contract	.0400	2.33	0		8	Major	Yes
Carpet, Wool 45 oz., Premium										
Minor Repair, Wool 45 oz., Premium	Sq Ft	Carpet Layer	Contract	.0002	.04	0		5	PM	No
Replace Carpet, Wool 45 oz., Premium	Sq Ft	Carpet Layer	Contract	.0400	7.55	0		25	Major	Yes
Metal Grating										
Metal Floor Grating Repairs - (2% of Grating)	Sq Ft	Maint Carpenter	Staff	.0700	.15	0		10	PM	No
Replace Metal Floor Grating	Sq Ft	Carpenter	Contract	.3480	8.80	0		30	PM	No
C3030 Ceiling Finishes										

*Task frequency for Washington D.C.
Components ordered by UNIFORMAT II Classifications. All costs expressed in ($) 2004.
Sources: U.S. Army Corps of Engineers, Department of Defense Engineering Performance Standards, and Whitestone Research interviews.

5.1 Building Components and Repair Tasks

Uniformat / Component / Repair Task	Units	Trade	Labor Type	Labor Hours	Material Costs	Equip Hours	Equipment Type	Freq. Yrs.*	Task Type	Repl. Task
Concrete, Finished Ceiling										
Minor Repair, Concrete, Finished Ceiling	Sq Ft	Brick Layer	Contract	.0066	.05	0		15	Major	No
Finish Repaired Concrete, Finished Ceiling	Sq Ft	Painter	Contract	.0010	.10	0		15	Major	No
Refinish Concrete, Finished Ceiling	Sq Ft	Brick Layer	Contract	.0046	.14	0		25	Major	No
Plaster Ceiling										
Refinish Plaster Ceiling	Sq Ft	Painter	Contract	.0143	.04	0		4	Major	No
Minor Repair, Plaster Ceiling	Sq Ft	Carpenter	Contract	.0066	.72	0		12	Major	No
Finish Repaired Plaster Ceiling	Sq Ft	Painter	Contract	.0024	.04	0		12	Major	No
Sheetrock, Stippled, Ceiling										
Refinish Sheetrock, Stippled, Ceiling	Sq Ft	Painter	Contract	.0143	.04	0		4	Major	No
Finish Repaired Sheetrock, Stippled, Ceiling	Sq Ft	Painter	Contract	.0024	.04	0		20	Major	No
Minor Repair, Sheetrock, Stippled, Ceiling	Sq Ft	Carpenter	Contract	.0063	.28	0		20	Major	No
Sheetrock, Unstippled, Ceiling										
Refinish Sheetrock, Unstippled, Ceiling	Sq Ft	Painter	Contract	.0143	.04	0		4	Major	No
Minor Repair, Sheetrock, Unstippled, Ceiling	Sq Ft	Carpenter	Contract	.0063	.28	0		20	Major	No
Finish Repaired Sheetrock, Unstippled, Ceiling	Sq Ft	Painter	Contract	.0024	.04	0		20	Major	No
Textured, Spray, Ceiling										
Refinish Textured, Spray, Ceiling	Sq Ft	Painter	Contract	.0040	.04	0		5	Major	No
Finished Repaired Textured, Spray, Ceiling	Sq Ft	Carpenter	Contract	.0024	.04	0		20	Major	No
Minor Repair, Textured, Spray, Ceiling	Sq Ft	Carpenter	Contract	.0063	.01	0		20	Major	No
Acoustical Tile Ceiling										
Minor Repair, Acoustic Tile Ceiling	Sq Ft	Carpenter	Contract	.0070	.03	0		9	Major	No
Replace Acoustic Tile Ceiling	Sq Ft	Carpenter	Contract	.0180	1.68	0		70	Major	Yes
Acoustical Tile, Dropped, Ceiling										
Minor Repair, Acoustic Tile, Dropped, Ceiling	Sq Ft	Carpenter	Contract	.0022	.02	0		10	Major	No
Replace Acoustical Tile, Dropped, Ceiling	Sq Ft	Carpenter	Contract	.0058	1.11	0		70	Major	Yes
Metal Ceiling, Painted										
Refinish Metal Ceiling, Painted	Sq Ft	Painter	Contract	.0039	.04	0		5	Major	No
Minor Repair, Metal Ceiling, Painted	Sq Ft	Carpenter	Contract	.0030	.05	0		8	Major	No
Finish Repaired Metal Ceiling, Painted	Sq Ft	Painter	Contract	.0021	.02	0		8	Major	No
Replace Metal Ceiling, Painted	Sq Ft	Carpenter	Contract	.0466	3.86	0		100	Major	Yes

*Task frequency for Washington D.C.
Components ordered by UNIFORMAT II Classifications. All costs expressed in ($) 2004.
Sources: U.S. Army Corps of Engineers, Department of [illegible]

Uniformat / Component / Repair Task	Units	Trade	Labor Type	Labor Hours	Material Costs	Equip Hours	Equipment Type	Freq. Yrs.*	Task Type	Repl. Task
Metal Unpainted Interior Ceiling										
Repair Metal Unpainted Interior Ceiling	Sq Ft	Carpenter	Contract	.0078	.12	0		24	Major	No
Replace Metal Unpainted Interior Ceiling	Sq Ft	Carpenter	Contract	.4421	3.86	0		200	Major	Yes
Wood Finished Ceiling										
Refinish Wood Finished Ceiling	Sq Ft	Painter	Contract	.0143	.04	0		6	Major	No
Replace Wood Finished Ceiling	Sq Ft	Carpenter	Contract	.2310	2.28	0		80	Major	Yes
Finish Replaced Wood Finished Ceiling	Sq Ft	Painter	Contract	.0143	.04	0		80	Major	No
D10 Conveying										
D1010 Elevators and Lifts										
Elevator, Hydraulic, 2500 lbs, 3-5 Floor, 200 fpm										
Maintain Elevator, Hydraulic, 2500 lbs, 3-5 Floor, 200 fpm	Each	Elevator Mechanic	Contract	58.0000	998.73	0		1	Major	No
Renovate Elevator, Hydraulic, 2500 lbs, 3-5 Floor, 200 fpm	Each	Elevator Mechanic	Contract	175.0000	8,903.03	0		25	Major	No
Elevator, Hydraulic, 6000 lbs, 3-5 Floor, 150 fpm										
Maintain Elevator, Hydraulic, 6000 lbs, 3-5 Floor, 150 fpm	Each	Elevator Mechanic	Contract	168.0000	998.73	0		1	Major	No
Renovate Elevator, Hydraulic, 6000 lbs, 3-5 Floor, 150 fpm	Each	Elevator Mechanic	Contract	136.0000	15,241.77	0		25	Major	No
Elevator, Geared, 3000 lbs, 3-5 Floor, 300 fpm										
Maintain Elevator, Geared, 3000 lbs, 3-5 Floor, 300 fpm	Each	Elevator Mechanic	Contract	58.0000	1,681.81	0		1	Major	No
Renovate Elevator, Geared, 3000 lbs, 3-5 Floor, 300 fpm	Each	Elevator Mechanic	Contract	175.0000	31,176.94	0		25	Major	No
Elevator, Geared, 3500 lbs, 6-15 Floor, 350 fpm										
Maintain Elevator, Geared, 3500 lbs, 6-15 Floor, 350 fpm	Each	Elevator Mechanic	Contract	134.0000	1,681.81	0		1	Major	No
Renovate Elevator, Geared, 3500 lbs, 6-15 Floor, 350 fpm	Each	Elevator Mechanic	Contract	290.0000	42,744.35	0		25	Major	No
Elevator, Geared, 3500 lbs, 16-25 Floor, 350 fpm										
Maintain Elevator, Geared, 3500 lbs, 16-25 Floor, 350 fpm	Each	Elevator Mechanic	Contract	154.0000	1,681.81	0		1	Major	No
Renovate Elevator, Geared, 3500 lbs, 16-25 Floor, 350 fpm	Each	Elevator Mechanic	Contract	840.0000	42,744.35	0		25	Major	No
Elevator, Geared, 3500 lbs, 16-25 Floor, 800 fpm										
Maintain Elevator, Geared, 3500 lbs, 16-25 Floor, 800 fpm	Each	Elevator Mechanic	Contract	185.0000	1,681.81	0		1	Major	No
Renovate Elevator, Geared, 3500 lbs, 16-25 Floor, 800 fpm	Each	Elevator Mechanic	Contract	840.0000	42,744.35	0		25	Major	No
Elevator, Geared, 5000 lbs										
Maintain Elevator, Geared, 5000 lbs	Each	Elevator Mechanic	Contract	120.0000	1,300.46	0		1	PM	No
Replace Elevator, Geared, 5000 lbs	Each	Elevator Mechanic	Contract	411.0000	93,952.60	0		25	Major	Yes

*Task frequency for Washington D.C.
Components ordered by UNIFORMAT II Classifications. All costs expressed in ($) 2004.
Sources: U.S. Army Corps of Engineers, Department of Defense Engineering Performance Standards, and Whitestone Research interviews.

5.1 Building Components and Repair Tasks

Uniformat / Component / Repair Task	Units	Trade	Labor Type	Labor Hours	Material Costs	Equip Hours	Equipment Type	Freq. Yrs.*	Task Type	Repl. Task
Cartlift										
Maintain Cartlift	Each	Gen Maint Worker	Staff	24.0000	5.18	0		1	Major	No
Dumbwaiter, Electric, 5 Floor										
Maintain Dumbwaiter, Electric, 5 Floor	Each	Elevator Mechanic	Contract	.5000	36.37	0		1	Major	No
Renovate Dumbwaiter, Electric, 5 Floor	Each	Elevator Mechanic	Contract	40.0000	4,796.68	0		25	Major	No
Platform Lifter, 5 Ton										
Maintain Platform Lifter, 5 Ton	Each	Elevator Mechanic	Contract	18.0000	62.36	0		1	Major	No
Renovate Platform Lifter, 5 Ton	Each	Elevator Mechanic	Contract	5.0000	1,672.77	0		25	Major	Yes
Wheel Chair Lift										
Maintain Wheel Chair Lift	Each	Elevator Mechanic	Contract	12.0000	42.43	0		1	Major	No
Renovate Wheel Chair Lift	Each	Elevator Mechanic	Contract	63.0000	5,229.68	0		25	Major	No
Crane, Jib 1/2 Ton										
Maintain Crane, Jib Electric, Chain, 1/2 Ton	Each	Elevator Mechanic	Contract	16.0000	12.83	0		1	PM	No
Replace Crane Jib Electric, 1/2 Ton	Each	Elevator Mechanic	Contract	96.0000	1,378.65	0		25	Major	Yes
Crane, Jib 1 Ton										
Maintain Crane, Jib Electric, Chain, 1 Ton	Each	Elevator Mechanic	Contract	16.0000	12.83	0		1	PM	No
Replace Crane Jib Electric, 1 Ton	Each	Elevator Mechanic	Contract	96.0000	2,297.76	0		25	Major	Yes
Hoist Electric, Overhead, Chain, 3 Ton										
Maintain Hoist Electric, Chain, 3 Ton	Each	Elevator Mechanic	Contract	1.6490	11.87	0		1	PM	No
Replace Hoist Electric, Chain, 3 Ton	Each	Elevator Mechanic	Contract	8.0000	2,350.00	0		25	Major	Yes
D1020 Escalators & Moving Walks										
Escalator, 16 ft rise, 100 fpm										
Maintain Escalator, 16 ft rise, 100 fpm	Each	Elevator Mechanic	Contract	230.0000	1,811.17	0		1	Major	No
D1090 Other Conveying Systems										
Pneumatic Tube Station										
Maintain Pneumatic Tube Station	Each	Gen Maint Worker	Staff	7.0000	.00	0		1	PM	No
Dock Ramp Leveler										
Maintain Dock Ramp Leveler	Each	Gen Maint Worker	Staff	7.6000	.00	0		1	PM	No
Automatic Cart Washer										
Maintain Automatic Cart Washer	Each	Gen Maint Worker	Staff	126.0000	.00	0		1	PM	No

*Task frequency for Washington D.C.
Components ordered by UNIFORMAT II Classifications. All costs expressed in ($) 2004.
Sources: U.S. Army Corps of Engineers, Department of Defense Engineering R...

5.1 Building Components and Repair Tasks

Uniformat / Component / Repair Task	Units	Trade	Labor Type	Labor Hours	Material Costs	Equip Hours	Equipment Type	Freq. Yrs.*	Task Type	Repl. Task
D20 Plumbing										
D2010 Plumbing Fixtures										
Emergency Eye Wash										
Inspect & Clean Spray Heads, Emergency Eye Wash	Each	Maint Plumber	Staff	.0890	.18	0		3	PM	No
Repack Valve Glands, Emergency Eye Wash	Each	Plumber	Contract	.2720	.54	0		10	Major	No
Replace Emergency Eye Wash Station	Each	Plumber	Contract	2.0000	134.00	0		25	Major	Yes
Emergency Shower Station										
Inspect & Clean Emergency Shower Station	Each	Maint Plumber	Staff	.0890	.18	0		3	PM	No
Repack Valve Glands, Emergency Shower Station	Each	Plumber	Contract	.2720	.54	0		10	Major	No
Replace Emergency Shower Station	Each	Plumber	Contract	2.0000	255.00	0		25	Major	Yes
Tankless Water Closet										
Replace Flush Valve	Each	Plumber	Contract	.0850	109.00	0		10	Major	No
Replace Tankless Water Closet	Each	Plumber	Contract	1.5000	348.38	0		35	Major	Yes
Flush Tank Water Closet										
Replace Washer in Ball Cock, Flush Tank Water Closet	Each	Maint Plumber	Staff	.1440	3.34	0		5	PM	No
Replace Worn Parts, Flush Tank Water Closet	Each	Plumber	Contract	.9310	11.15	0		15	Major	No
Install Gasket in Spud Connection, Flush Tank Water Closet	Each	Plumber	Contract	.1030	1.24	0		20	Major	No
Replace Flush Tank Water Closet	Each	Plumber	Contract	3.1600	403.92	0		35	Major	Yes
Urinal, Vitreous China										
Replace Flush Valve, Urinal, Vitreous China	Each	Plumber	Contract	.0900	24.00	0		7	Major	No
Replace Urinal, Vitreous China	Each	Plumber	Contract	1.6400	291.00	0		35	Major	Yes
Toilet & Wash Basin, Stainless Steel, Detention Fixture										
Replace Flush Valve, Toilet & Wash Basin, Stainless Steel	Each	Plumber	Contract	.0850	11.37	0		10	Major	No
Replace Toilet & Wash Basin, Stainless Steel, Detention Fixture	Each	Plumber	Contract	2.0000	1,683.10	0		40	Major	Yes
Lavatory, Iron, Enamel										
Replace Faucet Washer & Clean Trap, Lavatory, Iron, Enamel	Each	Maint Plumber	Staff	.4810	.26	0		2	PM	No
Replace Washer & Spud Connection, Lavatory, Iron, Enamel	Each	Plumber	Contract	.1030	9.80	0		7	Major	No
Replace Valve Set, Lavatory, Iron, Enamel	Each	Plumber	Contract	.8177	37.50	0		10	Major	No
Replace Lavatory, Iron, Enamel	Each	Plumber	Contract	1.5350	282.74	0		40	Major	Yes

*Task frequency for Washington D.C.

Components ordered by UNIFORMAT II Classifications. All costs expressed in ($) 2004.

Sources: U.S. Army Corps of Engineers, Department of Defense Engineering Performance Standards, and Whitestone Research interviews.

5.1 Building Components and Repair Tasks

Uniformat / Component / Repair Task	Units	Trade	Labor Type	Labor Hours	Material Costs	Equip Hours	Equipment Type	Freq. Yrs.*	Task Type	Repl. Task
Lavatory, Vitreous China										
Replace Faucet Washer & Clean Trap, Lavatory, Vitreous China	Each	Maint Plumber	Staff	.4820	.26	0		2	PM	No
Replace Washer & Spud Connection, Lavatory, Vitreous China	Each	Plumber	Contract	.1030	9.80	0		7	Major	No
Replace Valve Set, Lavatory, Vitreous China	Each	Plumber	Contract	.8177	37.50	0		10	Major	No
Replace Lavatory, Vitreous China	Each	Plumber	Contract	1.9968	184.50	0		35	Major	Yes
Sink, Enameled Steel										
Replace Faucet Washer & Clean Trap, Sink, Enameled Steel	Each	Maint Plumber	Staff	.2400	.26	0		2	PM	No
Repair Strainer, Sink, Enameled Steel	Each	Plumber	Contract	.9000	1.37	0		8	Major	No
Replace Valve Set, Sink, Enameled Steel	Each	Plumber	Contract	.8177	37.50	0		10	Major	No
Replace Sink, Enameled Steel	Each	Plumber	Contract	3.0680	107.00	0		35	Major	Yes
Sink, Stainless Steel										
Replace Faucet Washer & Clean Trap, Sink, Stainless Steel	Each	Maint Plumber	Staff	.2400	.26	0		2	PM	No
Repair Strainer, Sink, Stainless Steel	Each	Plumber	Contract	.9000	1.37	0		8	Major	No
Replace Valve Set, Sink, Stainless Steel	Each	Plumber	Contract	.8177	37.50	0		10	Major	No
Replace Sink, Stainless Steel	Each	Plumber	Contract	2.3600	328.19	0		40	Major	Yes
Sink, Plastic										
Replace Faucet Washer & Clean Trap, Sink, Plastic	Each	Maint Plumber	Staff	.2400	.26	0		2	PM	No
Repair Strainer, Sink, Plastic	Each	Plumber	Contract	.9000	1.37	0		8	Major	No
Replace Valve Set, Sink, Plastic	Each	Plumber	Contract	.8177	37.50	0		10	Major	No
Replace Sink, Plastic	Each	Plumber	Contract	2.1800	90.88	0		15	Major	Yes
Service Sink, Iron Enamel										
Replace Faucet Washer & Clean Trap, Sink, Iron Enamel	Each	Maint Plumber	Staff	.2400	.26	0		2	PM	No
Repair Strainer, Sink, Iron Enamel	Each	Plumber	Contract	.9000	1.37	0		8	Major	No
Replace Valve Set, Service Sink, Iron Enamel	Each	Plumber	Contract	.8177	37.50	0		10	Major	No
Replace Service Sink, Iron Enamel	Each	Plumber	Contract	2.3600	201.00	0		35	Major	Yes
Shower, Terrazzo										
Replace Washer & Clean Shower Head, Terrazzo	Each	Maint Plumber	Staff	.2000	.26	0		2	PM	No
Reseal Shower, Terrazzo	Each	Maint Plumber	Staff	.5900	1.45	0		5	PM	No
Replace Valve Set, Shower, Terrazzo	Each	Plumber	Contract	3.0000	101.00	0		10	Major	No
Replace Terrazzo Surfaces	Each	Brick Layer	Contract	16.0000	324.65	0		30	Major	Yes

*Task frequency for Washington D.C.

Components ordered by UNIFORMAT II Classifications. All costs expressed in ($) 2004.

Sources: U.S. Army Corps of Engineers, Department of Defense Engineering Re[illegible]

5.1 Building Components and Repair Tasks

Uniformat / Component / Repair Task	Units	Trade	Labor Type	Labor Hours	Material Costs	Equip Hours	Equipment Type	Freq. Yrs.*	Task Type	Repl. Task
Shower, Fiberglass										
Replace Faucet Washer & Clean Shower Head, Fiberglass	Each	Maint Plumber	Staff	.2000	.26	0		2	PM	No
Reseal Shower, Fiberglass	Each	Maint Plumber	Staff	.5900	1.49	0		5	PM	No
Replace Valve Set, Shower, Fiberglass	Each	Plumber	Contract	3.0000	101.00	0		10	Major	No
Replace Shower, Fiberglass	Each	Plumber	Contract	2.5900	395.00	0		20	Major	Yes
Shower, Ceramic Tile										
Replace Faucet Washer & Clean Shower Head, Ceramic Tile	Each	Maint Plumber	Staff	.2000	.26	0		2	PM	No
Reseal Shower, Ceramic Tile	Each	Maint Plumber	Staff	.5900	1.45	0		5	PM	No
Replace Valve Set, Shower, Ceramic Tile	Each	Plumber	Contract	3.0000	101.00	0		10	Major	No
Replace Ceramic Tile Surfaces	Each	Brick Layer	Contract	16.0000	206.00	0		30	Major	Yes
Bath Tub, Enameled Steel										
Replace Faucet Washer & Clean Shower Head, Bath Tub, Enameled	Each	Maint Plumber	Staff	.2000	.26	0		2	PM	No
Replace Diverter Valve, Bath Tub, Enameled Steel	Each	Maint Plumber	Staff	.7900	100.98	0		4	PM	No
Reseal Bath Tub, Enameled Steel	Each	Maint Plumber	Staff	.5900	1.49	0		5	PM	No
Replace Valve Set, Bath Tub, Enameled Steel	Each	Plumber	Contract	3.0000	101.00	0		10	Major	No
Replace Bath Tub, Enameled Steel	Each	Plumber	Contract	2.5700	286.00	0		35	Major	Yes
Bathtub & Shower Enclosure, Fiberglass										
Replace Faucet Washer & Clean Shower Head, Bathtub Shower Encl	Each	Maint Plumber	Staff	.2000	.27	0		2	PM	No
Repair Diverter Valve, Bathtub & Shower Enclosure, Fiberglass	Each	Maint Plumber	Staff	.7900	45.44	0		4	PM	No
Reseal Bathtub & Shower Enclosure, Fiberglass	Each	Maint Plumber	Staff	.5900	1.49	0		5	PM	No
Replace Valve Set, Bathtub & Shower Enclosure, Fiberglass	Each	Plumber	Contract	3.0000	101.00	0		10	Major	No
Replace Bathtub & Shower Enclosure, Fiberglass	Each	Plumber	Contract	10.0000	490.00	0		20	Major	Yes
Drinking Fountain, Vitreous China										
Repack Valve Gland, Drinking Fountain, Vitreous China	Each	Maint Plumber	Staff	.2590	.54	0		2	PM	No
Repair Valve, Drinking Fountain, Vitreous China	Each	Plumber	Contract	.2000	.00	0		10	Major	No
Replace Drinking Fountain, Vitreous China	Each	Plumber	Contract	1.0600	450.00	0		35	Major	Yes
Drinking Fountain, Refrigerated										
Replace Coolant & Adjust Drinking Fountain, Refrigerated	Each	Maint Plumber	Staff	1.3000	3.40	0		2	PM	No
Replace Drinking Fountain, Refrigerated	Each	Plumber	Contract	1.0600	500.00	0		10	Major	Yes
D2020 Domestic Water Distribution										

*Task frequency for Washington D.C.
Components ordered by UNIFORMAT II Classifications. All costs expressed in ($) 2004.
Sources: U.S. Army Corps of Engineers, Department of Defense Engineering Performance Standards, and Whitestone Research interviews.

5.1 Building Components and Repair Tasks

Uniformat / Component / Repair Task	Units	Trade	Labor Type	Labor Hours	Material Costs	Equip Hours	Equipment Type	Freq. Yrs.*	Task Type	Repl. Task
Pipe & Fittings, 3/4" Copper, Cold Water										
Resolder Joint, Pipe & Fittings, 3/4" Copper, Cold Water	K Ln Ft	Plumber	Contract	.1420	.26	0		10	Major	No
Install 10' Section, Pipe & Fittings, 3/4" Copper, Cold Water	K Ln Ft	Plumber	Contract	4.2700	18.98	0		20	Major	No
Replace Pipe & Fittings, 3/4" Copper, Cold Water	K Ln Ft	Plumber	Contract	427.0000	1,898.70	0		25	Major	Yes
Pipe & Fittings, 3/4" Copper, Hot Water										
Install 10' Section, Pipe & Fittings, 3/4" Copper, Hot Water	K Ln Ft	Plumber	Contract	4.2700	17.60	0		13	Major	No
Resolder Joint, Pipe & Fittings, 3/4" Copper, Hot Water	K Ln Ft	Plumber	Contract	.1420	.26	0		13	Major	No
Replace Pipe & Fittings, 3/4" Copper, Hot Water	K Ln Ft	Plumber	Contract	427.0000	1,760.00	0		25	Major	Yes
Pipe & Fittings, 2" Copper, Cold Water										
Resolder Joint, Pipe & Fittings, 2" Copper, Cold Water	K Ln Ft	Plumber	Contract	.1420	.26	0		10	Major	No
Install 10' Section, Pipe & Fittings, 2" Copper, Cold Water	K Ln Ft	Plumber	Contract	4.2700	50.19	0		20	Major	No
Replace Pipe & Fittings, 2" Copper, Cold Water	K Ln Ft	Plumber	Contract	427.0000	5,018.71	0		25	Major	Yes
Pipe & Fittings, 4" Steel										
Replace 10' Section, Pipe & Fittings, 4" Steel	K Ln Ft	Plumber	Contract	8.2650	126.00	0		13	Major	No
Install New Gasket & Bolts, Pipe & Fittings, 4" Steel	K Ln Ft	Plumber	Contract	48.8000	817.07	0		25	Major	No
Replace Pipe & Fittings, 4" Steel	K Ln Ft	Plumber	Contract	826.0000	12,600.00	0		75	Major	Yes
Pipe Insulation, Cold Water										
Re-tape Pipe Insulation, Cold Water	K Ln Ft	Maint Plumber	Staff	.2600	134.00	0		5	PM	No
Replace Pipe Insulation, Cold Water	K Ln Ft	Plumber	Contract	24.2300	1,364.36	0		25	Major	Yes
Pipe Insulation, Hot Water										
Re-tape Pipe Insulation, Hot Water	K Ln Ft	Maint Plumber	Staff	.2600	134.00	0		2	PM	No
Replace Pipe Insulation, Hot Water	K Ln Ft	Plumber	Contract	24.2300	1,364.36	0		25	Major	Yes
Booster Pump, 1.5 HP										
Inspect & Lubricate Booster Pump, 1.5 HP	Each	Maint Plumber	Staff	.1250	.75	0		1	PM	No
Overhaul Booster Pump, 1.5 HP	Each	Maint Plumber	Staff	.6820	32.50	0		5	PM	No
Replace Assembly, Booster Pump, 1.5 HP	Each	Plumber	Contract	12.8000	3,675.00	0		20	Major	Yes
Booster Pump, 7.5 HP										
Inspect & Lubricate Booster Pump, 7.5 HP	Each	Maint Plumber	Staff	.1250	1.72	0		1	PM	No
Overhaul Booster Pump, 7.5 HP	Each	Maint Plumber	Staff	.6820	32.50	0		5	PM	No
Replace Assembly, Booster Pump, 7.5 HP	Each	Plumber	Contract	16.0000	4,725.00	0		20	Major	Yes

*Task frequency for Washington D.C.
Components ordered by UNIFORMAT II Classifications. All costs expressed in ($) 2004.
Sources: U.S. Army Corps of Engineers, Department of Defense Engineering

5.1 Building Components and Repair Tasks

Uniformat / Component / Repair Task	Units	Trade	Labor Type	Labor Hours	Material Costs	Equip Hours	Equipment Type	Freq. Yrs.*	Task Type	Repl. Task
Booster Pump, 10 HP										
Inspect & Lubricate Booster Pump, 10 HP	Each	Maint Plumber	Staff	.1250	1.72	0		1	PM	No
Overhaul Booster Pump, 10 HP	Each	Maint Plumber	Staff	.6820	32.50	0		5	PM	No
Replace Assembly, Booster Pump, 10 HP	Each	Plumber	Contract	16.0000	4,925.00	4	2T Forklift	20	Major	Yes
Circulator Pump 1/12 HP, Cold Water										
Inspect & Lubricate Circulator Pump 1/12 HP, Cold Water	Each	Maint Plumber	Staff	.1250	1.75	0		1	PM	No
Overhaul Circulator Pump 1/12 HP, Cold Water	Each	Maint Plumber	Staff	.6820	3.65	0		5	PM	No
Replace Circulator Pump 1/12 HP, Cold Water	Each	Plumber	Contract	.8300	326.42	0		20	Major	Yes
Circulator Pump 1/6 HP, Cold Water										
Inspect & Lubricate Circulator Pump 1/6 HP, Cold Water	Each	Maint Plumber	Staff	.1250	1.72	0		1	PM	No
Overhaul Circulator Pump 1/6 HP, Cold Water	Each	Maint Plumber	Staff	.6820	3.65	0		5	PM	No
Replace Circulator Pump 1/6 HP, Cold Water	Each	Plumber	Contract	1.4200	563.47	0		20	Major	Yes
Circulator Pump 1/2 HP, Cold Water										
Inspect & Lubricate Circulator Pump 1/2 HP, Cold Water	Each	Maint Plumber	Staff	.1250	1.72	0		1	PM	No
Overhaul Circulator Pump 1/2 HP, Cold Water	Each	Maint Plumber	Staff	.6820	3.61	0		5	PM	No
Replace Circulator Pump 1/2 HP, Cold Water	Each	Plumber	Contract	2.2100	1,083.01	0		20	Major	Yes
Circulator Pump, 1/12 HP, Hot Water										
Inspect & Lubricate Circulator Pump 1/12 HP, Hot Water	Each	Maint Plumber	Staff	.1250	1.72	0		1	PM	No
Overhaul Circulator Pump 1/12 HP, Hot Water	Each	Maint Plumber	Staff	.6820	3.65	0		5	PM	No
Replace Circulator Pump, 1/12 HP, Hot Water	Each	Plumber	Contract	.8300	326.42	0		10	Major	Yes
Circulator Pump, 1/6 HP, Hot Water										
Inspect & Lubricate Circulator Pump 1/6 HP, Hot Water	Each	Maint Plumber	Staff	.1250	1.72	0		1	PM	No
Overhaul Circulator Pump 1/6 HP, Hot Water	Each	Maint Plumber	Staff	.6820	3.65	0		5	PM	No
Replace Circulator Pump, 1/6 HP, Hot Water	Each	Plumber	Contract	1.4200	805.00	0		10	Major	Yes
Circulator Pump, 1/2 HP, Hot Water										
Inspect & Lubricate Circulator Pump 1/2 HP, Hot Water	Each	Maint Plumber	Staff	.1250	1.72	0		1	PM	No
Overhaul Circulator Pump 1/2 HP, Hot Water	Each	Maint Plumber	Staff	.6820	3.65	0		5	PM	No
Replace Circulator Pump, 1/2 HP, Hot Water	Each	Plumber	Contract	2.2100	1,375.00	0		10	Major	Yes
Circulator Pump, 1 HP, Hot Water										
Inspect & Lubricate Circulator Pump, 1 HP, Hot Water	Each	Maint Plumber	Staff	.1250	1.72	0		1	PM	No
Overhaul Circulator Pump 1 HP, Hot Water	Each	Maint Plumber	Staff	.6820	3.65	0		5	PM	No
Replace Circulator Pump, 1 HP, Hot Water	Each	Plumber	Contract	3.2100	1,855.51	0		10	Major	Yes

*Task frequency for Washington D.C.
Components ordered by UNIFORMAT II Classifications. All costs expressed in ($) 2004.
Sources: U.S. Army Corps of Engineers, Department of Defense Engineering Performance Standards, and Whitestone Research interviews.

5.1 Building Components and Repair Tasks

Uniformat / Component / Repair Task	Units	Trade	Labor Type	Labor Hours	Material Costs	Equip Hours	Equipment Type	Freq. Yrs.*	Task Type	Repl. Task
Steam Converter, Domestic Hot Water, 10 Gpm										
Inspect Steam Converter, Domestic Hot Water, 10 Gpm	Each	Maint Plumber	Staff	.0710	.00	0		1	PM	No
Replace Steam Converter, Domestic Hot Water, 10 Gpm	Each	Plumber	Contract	5.6600	954.26	0		20	Major	Yes
Water Heater, Instant Hot										
Check Operation, Water Heater, Instant Hot	Each	Maint Plumber	Staff	.0350	.00	0		1	PM	No
Replace Water Heater, Instant Hot	Each	Plumber	Contract	.5710	160.56	0		15	Major	Yes
Water Heater Gas/Oil 25 Gph										
Clean Water Heater, Gas/Oil 25 Gph	Each	Maint Plumber	Staff	2.3100	.00	0		1	PM	No
Overhaul Water Heater, Gas/Oil 25 Gph	Each	Maint Plumber	Staff	1.8420	7.01	0		5	PM	No
Replace Water Heater, Gas/Oil 25 Gph	Each	Plumber	Contract	3.0000	385.00	0		8	Major	Yes
Water Heater, Gas/Oil 65 Gph										
Clean Water Heater, Gas/Oil 65 Gph	Each	Maint Plumber	Staff	2.3100	.00	0		1	PM	No
Overhaul Water Heater, Gas/Oil 65 Gph	Each	Maint Plumber	Staff	1.8420	7.01	0		5	PM	No
Replace Water Heater, Gas/Oil 65 Gph	Each	Plumber	Contract	3.0000	1,375.00	0		8	Major	Yes
Water Heater, Gas/Oil 75 Gph										
Clean Water Heater, Gas/Oil 75 Gph	Each	Maint Plumber	Staff	2.3100	.00	0		1	PM	No
Overhaul Water Heater, Gas/Oil 75 Gph	Each	Maint Plumber	Staff	1.8420	7.01	0		5	PM	No
Replace Water Heater, Gas/Oil 75 Gph	Each	Plumber	Contract	3.0000	1,375.00	0		8	Major	Yes
Water Heater, Gas/Oil 130 Gph										
Clean Water Heater, Gas/Oil 130 Gph	Each	Maint Plumber	Staff	2.3100	.00	0		1	PM	No
Overhaul Water Heater, Gas/Oil 130 Gph	Each	Maint Plumber	Staff	1.8420	7.01	0		5	PM	No
Replace Water Heater, Gas/Oil 130 Gph	Each	Plumber	Contract	8.0000	4,150.00	0		8	Major	Yes
Water Heater, Gas/Oil 175 Gph										
Clean Water Heater, Gas/Oil 175 Gph	Each	Maint Plumber	Staff	2.3100	.00	0		1	PM	No
Overhaul Water Heater, Gas/Oil 175 Gph	Each	Maint Plumber	Staff	1.8420	7.01	0		5	PM	No
Replace Water Heater, Gas/Oil 175 Gph	Each	Plumber	Contract	8.0000	4,275.00	0		8	Major	Yes
Water Heater, Gas/Oil 275 Gph										
Clean Water Heater, Gas/Oil 275 Gph	Each	Maint Plumber	Staff	2.3100	.00	0		1	PM	No
Overhaul Water Heater, Gas/Oil 275 Gph	Each	Maint Plumber	Staff	1.8420	7.01	0		5	PM	No
Replace Water Heater, Gas/Oil 275 Gph	Each	Plumber	Contract	8.0000	5,900.00	0		8	Major	Yes

*Task frequency for Washington D.C.

Components ordered by UNIFORMAT II Classifications. All costs expressed in ($) 2004.

Sources: U.S. Army Corps of Engineers, Department of

Uniformat / Component / Repair Task	Units	Trade	Labor Type	Labor Hours	Material Costs	Equip Hours	Equipment Type	Freq. Yrs.*	Task Type	Repl. Task
Water Heater, Gas/Oil 575 Gph										
Clean & Service Water Heater, Gas/Oil 575 Gph	Each	Maint Plumber	Staff	26.0200	.00	0		2	PM	No
Overhaul Water Heater, Gas/Oil 575 Gph	Each	Maint Plumber	Staff	1.8420	7.01	0		5	PM	No
Replace Water Heater, Gas/Oil 575 Gph	Each	Plumber	Contract	12.0000	8,200.00	4	2T Forklift	8	Major	Yes
Water Heater, Gas/Oil 850 Gph										
Clean & Service Water Heater, Gas/Oil 850 Gph	Each	Maint Plumber	Staff	26.0200	.00	0		2	PM	No
Overhaul Water Heater, Gas/Oil 850 Gph	Each	Maint Painter	Staff	1.8420	7.01	0		5	PM	No
Replace Water Heater, Gas/Oil 850 Gph	Each	Plumber	Contract	24.0000	9,425.00	4	2T Forklift	8	Major	Yes
Water Heater, Electric 52 Gal.										
Check Operation, Water Heater, Electric 52 Gal.	Each	Maint Plumber	Staff	.0440	.00	0		3	PM	No
Drain & Flush Water Heater, Electric 52 Gal.	Each	Plumber	Contract	3.5800	.00	0		8	Major	No
Replace Water Heater, Electric 52 Gal.	Each	Plumber	Contract	3.0000	345.00	0		15	Major	Yes
Water Heater, Electric 120 Gal.										
Check Operation, Water Heater, Electric 120 Gal.	Each	Maint Plumber	Staff	.0440	.00	0		3	PM	No
Drain & Flush Water Heater, Electric 120 Gal.	Each	Plumber	Contract	3.5800	.00	0		8	Major	No
Replace Water Heater, Electric 120 Gal.	Each	Plumber	Contract	3.0000	770.00	0		15	Major	Yes
Water Heater, Electric 300 Gal., 738 Gph										
Check Operation, Water Heater, Electric 300 Gal., 738 Gph	Each	Maint Plumber	Staff	.0440	.00	0		3	PM	No
Drain & Flush Water Heater, Electric 300 Gal., 738 Gph	Each	Plumber	Contract	3.5800	.00	0		8	Major	No
Replace Water Heater, Electric 300 Gal., 738 Gph	Each	Plumber	Contract	3.0000	17,000.00	0		15	Major	Yes
Water Heater, Electric 1500 Gal., 1970 Gph										
Check Operation, Water Heater, Electric 1500 Gal., 1970 Gph	Each	Maint Plumber	Staff	.0440	.00	0		3	PM	No
Drain & Flush Water Heater, Electric 1500 Gal., 1970 Gph	Each	Plumber	Contract	3.5800	.00	0		8	Major	No
Replace Water Heater, Electric 1500 Gal., 1970 Gph	Each	Plumber	Contract	3.6000	48,300.00	4	2T Forklift	15	Major	Yes
Water Storage Tank, 150 Gal.										
Maintain Water Storage Tank, 150 Gal.	Each	Maint Plumber	Staff	1.0000	.00	0		1	PM	No
Clean & Scrape Water Storage Tank, 150 Gal.	Each	Maint Plumber	Staff	.4307	7.53	0		2	PM	No
Replace Water Storage Tank, 150 Gal.	Each	Plumber	Contract	3.0000	965.00	0		50	Major	Yes
Water Storage Tank, 250 Gal.										
Maintain Water Storage Tank, 250 Gal.	Each	Maint Plumber	Staff	1.0000	.00	0		1	PM	No
Clean & Scrape Water Storage Tank, 250 Gal.	Each	Maint Plumber	Staff	.7179	7.53	0		2	PM	No
Replace Water Storage Tank, 250 Gal.	Each	Plumber	Contract	4.0000	1,125.00	0		50	Major	Yes

*Task frequency for Washington D.C.
Components ordered by UNIFORMAT II Classifications. All costs expressed in ($) 2004.
Sources: U.S. Army Corps of Engineers, Department of Defense Engineering Performance Standards, and Whitestone Research interviews.

5.1 Building Components and Repair Tasks

Uniformat / Component / Repair Task	Units	Trade	Labor Type	Labor Hours	Material Costs	Equip Hours	Equipment Type	Freq. Yrs.*	Task Type	Repl. Task
Water Storage Tank, 500 Gal.										
Maintain Water Storage Tank, 500 Gal.	Each	Maint Plumber	Staff	1.0000	.00	0		1	PM	No
Clean & Scrape Water Storage Tank, 500 Gal.	Each	Maint Plumber	Staff	1.4357	7.53	0		2	PM	No
Replace Water Storage Tank, 500 Gal.	Each	Plumber	Contract	6.0000	1,161.27	4	2T Forklift	50	Major	Yes
Water Storage Tank, 1000 Gal.										
Maintain Water Storage Tank, 1000 Gal.	Each	Maint Plumber	Staff	1.9120	.00	0		1	PM	No
Clean & Scrape Water Storage Tank, 1000 Gal.	Each	Maint Plumber	Staff	2.8715	7.53	0		2	PM	No
Replace Water Storage Tank, 1000 Gal.	Each	Plumber	Contract	6.0000	1,817.64	4	2T Forklift	50	Major	Yes
Water Storage Tank, 2500 Gal.										
Maintain Water Storage Tank, 2500 Gal.	Each	Maint Plumber	Staff	1.9120	.00	0		1	PM	No
Clean & Scrape Water Storage Tank, 2500 Gal.	Each	Maint Plumber	Staff	7.1790	7.53	0		2	PM	No
Replace Water Storage Tank, 2500 Gal.	Each	Plumber	Contract	8.0000	4,089.69	4	2T Forklift	50	Major	Yes
Water Softener										
Maintain Water Softener	Each	Maint Plumber	Staff	8.1080	.00	0		1	PM	No
Replace Water Softener	Each	Plumber	Contract	4.0000	600.00	0		15	Major	Yes
Water Flow Meter, 3/4"										
Maintain Water Flow Meter	Each	Maint Plumber	Staff	.3000	24.00	0		1	PM	No
Remove Old Meter, Install New, 3/4"	Each	Plumber	Contract	1.1140	67.50	0		25	Major	Yes
Water Flow Meter, 2"										
Maintain Water Flow Meter	Each	Maint Plumber	Staff	.3000	24.00	0		1	PM	No
Remove Old Meter, Install New, 2"	Each	Plumber	Contract	2.6000	415.00	0		25	Major	Yes
Water Flow Meter, 4"										
Maintain Water Flow Meter	Each	Maint Plumber	Staff	.3000	24.00	0		1	PM	No
Remove Old Meter, Install New, 4"	Each	Plumber	Contract	20.7970	3,025.00	0		25	Major	Yes
D2030 Sanitary Waste										
Pipe & Fittings, 6" Cast Iron										
Replace 10' Section, Pipe & Fittings, 6" Cast Iron	K Ln Ft	Plumber	Contract	8.2650	141.28	0		13	Major	No
Install New Gasket & Bolts, Pipe & Fittings, 6" Cast Iron	K Ln Ft	Plumber	Contract	48.8000	768.00	0		25	Major	No
Replace Pipe & Fittings, 6" Cast Iron	K Ln Ft	Plumber	Contract	826.0000	14,127.86	0		75	Major	Yes

*Task frequency for Washington D.C.
Components ordered by UNIFORMAT II Classifications. All costs expressed in ($) 2004.
Sources: U.S. Army Corps of Engineers, Department of Defense Engi[illegible]

5.1 Building Components and Repair Tasks

Uniformat / Component / Repair Task	Units	Trade	Labor Type	Labor Hours	Material Costs	Equip Hours	Equipment Type	Freq. Yrs.*	Task Type	Repl. Task
Pipe & Fittings, 10" Cast Iron										
Replace 10' Section, Pipe & Fittings, 10" Cast Iron	K Ln Ft	Plumber	Contract	8.2650	357.02	0		13	Major	No
Install New Gasket & Bolts, Pipe & Fittings, 10" Cast Iron	K Ln Ft	Plumber	Contract	55.0000	1,685.00	0		25	Major	No
Replace Pipe & Fittings, 10" Cast Iron	K Ln Ft	Plumber	Contract	826.0000	35,702.18	0		75	Major	Yes
Pipe & Fittings, 12" Cast Iron										
Replace 12" Section, Pipe & Fittings, 12" Cast Iron	K Ln Ft	Plumber	Contract	6.6700	504.93	0		13	Major	No
Install New Gasket and Bolts, Pipe & Fittings, 12" Cast Iron	K Ln Ft	Plumber	Contract	5.5000	2,524.65	0		25	Major	No
Replace Pipe & Fittings, 12" Cast Iron	K Ln Ft	Plumber	Contract	667.0000	50,493.53	0		75	Major	No
Sewage Ejector Pump, 1 HP										
Inspect & Lubricate Sewage Ejector Pump, 1 HP	Each	Maint Plumber	Staff	.9650	.75	0		1	PM	No
Replace Sewage Ejector Pump, 1 HP	Each	Plumber	Contract	8.0000	943.56	0		20	Major	Yes
Backflow Preventer, 2"										
Maintain Backflow Preventer, 2"	Each	Maint Plumber	Staff	1.7100	.00	0		1	PM	No
Replace Backflow Preventer, 2"	Each	Plumber	Contract	2.0000	958.86	0		10	Major	Yes
Backflow Preventer, 4"										
Maintain Backflow Preventer, 4"	Each	Maint Plumber	Staff	1.7100	.00	0		1	PM	No
Replace Backflow Preventer, 4"	Each	Plumber	Contract	5.0000	2,175.00	0		10	Major	Yes
Backflow Preventer, 8"										
Maintain Backflow Preventer, 8"	Each	Maint Plumber	Staff	1.7100	.00	0		1	PM	No
Replace Backflow Preventer, 8"	Each	Plumber	Contract	13.0000	6,477.39	0		10	Major	Yes
D2040 Rain Water Drainage										
Aluminum Gutter, Downspouts, Fittings										
Maintain Aluminum Gutter, Downspouts, Fittings	K Ln Ft	Gen Maint Worker	Staff	4.0000	2.14	0		1	PM	No
Replace Aluminum Gutter, Downspouts, Fittings	K Ln Ft	Carpenter	Contract	159.0000	1,231.96	0		20	Major	Yes
Roof Drain, 2"										
Maintain Roof Drain, 2"	Each	Maint Plumber	Staff	.3900	.40	0		1	PM	No
Replace Roof Drain, 2"	Each	Plumber	Contract	1.0100	120.09	0		40	Major	Yes
Roof Drain, 4-6"										
Maintain Roof Drain, 4-6"	Each	Maint Plumber	Staff	.3900	.40	0		1	PM	No
Replace Roof Drain, 4-6"	Each	Plumber	Contract	1.0000	151.00	0		40	Major	Yes

*Task frequency for Washington D.C.
Components ordered by UNIFORMAT II Classifications. All costs expressed in ($) 2004.
Sources: U.S. Army Corps of Engineers, Department of Defense Engineering Performance Standards, and Whitestone Research interviews.

5.1 Building Components and Repair Tasks

Uniformat / Component / Repair Task	Units	Trade	Labor Type	Labor Hours	Material Costs	Equip Hours	Equipment Type	Freq. Yrs.*	Task Type	Repl. Task
Pipe & Fittings, 4" DWV PVC										
Maintain Pipe & Fittings, 4" DWV PVC	K Ln Ft	Maint Plumber	Staff	4.0000	2.03	0		1	PM	No
Replace Pipe & Fittings, 4" DWV PVC	K Ln Ft	Plumber	Contract	32.0810	570.54	0		25	Major	Yes
Rain Water Drainage										
Inspect & Lubricate Sump Pump, 1/2 HP	Each	Maint Plumber	Staff	.1620	.37	0		1	PM	No
Overhaul Sump Pump, 1/2 HP	Each	Maint Plumber	Staff	.6820	3.72	0		5	PM	No
Replace Pump & Motor Assembly, Sump Pump, 1/2 HP	Each	Plumber	Contract	2.0000	176.00	0		20	Major	Yes
Sump Pump, 3 HP										
Inspect & Lubricate Sump Pump, 3 HP	Each	Maint Plumber	Staff	1.4180	.37	0		1	PM	No
Overhaul Sump Pump, 3 HP	Each	Maint Plumber	Staff	.6820	3.69	0		5	PM	No
Replace Pump & Motor Assembly, Sump Pump, 3HP	Each	Plumber	Contract	2.2100	176.00	0		20	Major	Yes
D2090 Other Plumbing Systems										
Pipe & Fittings, 3/4" Copper, Oxygen										
Replace 10' Section, Pipe & Fittings, 3/4" Copper, Oxygen	K Ln Ft	Plumber	Contract	5.5510	22.00	0		13	Major	No
Resolder Joint, Pipe & Fittings, 3/4" Copper, Oxygen	K Ln Ft	Plumber	Contract	.1420	.26	0		13	Major	No
Replace Pipe & Fittings, 3/4" Copper, Oxygen	K Ln Ft	Plumber	Contract	555.1000	1,971.00	0		25	Major	Yes
Air Compressor, 3/4 HP										
Check & Adjust, Air Compressor, 3/4 HP	Each	Maint Plumber	Staff	1.1300	.00	0		1	PM	No
Repair Air Compressor, 3/4 HP	Each	Maint Plumber	Staff	10.2100	30.73	0		4	PM	No
Replace Air Compressor, 3/4 HP	Each	Plumber	Contract	3.0780	955.00	0		25	Major	Yes
Air Compressor, 2.0 HP										
Check & Adjust, Air Compressor, 2.0 HP	Each	Maint Plumber	Staff	1.1520	.00	0		1	PM	No
Repair Air Compressor, 2.0 HP	Each	Maint Plumber	Staff	10.2100	21.92	0		4	PM	No
Replace Air Compressor, 2.0 HP	Each	Plumber	Contract	13.3700	2,600.24	0		25	Major	Yes
Air Compressor, 10.0 HP										
Check & Adjust, Air Compressor, 10.0 HP	Each	Maint Plumber	Staff	1.1520	.00	0		1	PM	No
Repair Air Compressor, 10.0 HP	Each	Maint Plumber	Staff	6.0000	32.26	0		4	PM	No
Replace Air Compressor, 10.0 HP	Each	Plumber	Contract	51.9000	4,973.27	0		25	Major	Yes
Air Compressor, Gas, 7.5 HP										
Check & Adjust, Air Compressor, Gas, 7.5 HP	Each	Maint Plumber	Staff	1.1300	.00	0		1	PM	No
Repair Air Compressor, Gas, 7.5 HP	Each	Maint Plumber	Staff	10.2100	32.17	0		4	PM	No
Replace Air Compressor, Gas, 7.5 HP	Each	Plumber	Contract	3.0000	4,550.00	0		25	Major	Yes

*Task frequency for Washington D.C.
Components ordered by UNIFORMAT II Classifications. All costs expressed in ($) 2004.
Sources: U.S. Army Corps of Engineers, Department of Defense [illegible]

Uniformat / Component / Repair Task	Units	Trade	Labor Type	Labor Hours	Material Costs	Equip Hours	Equipment Type	Freq. Yrs.*	Task Type	Repl. Task
Compressed Air Dryer										
Check & Adjust Compressed Air Dryer	Each	Maint Plumber	Staff	.2500	.00	0		1	PM	No
Recharge Compressed Air Dryer	Each	Maint Plumber	Staff	.5000	.00	0		5	PM	No
Replace Compressor, Compressed Air Dryer	Each	Plumber	Contract	1.7500	693.64	0		10	Major	No
Replace Compressed Air Dryer	Each	Plumber	Contract	1.5000	2,248.66	0		15	Major	Yes
Anesthesia Vacuum Pump, <16 HP										
Annual PM, Anesthesia Vacuum Pump, <16 HP	Each	Maint Plumber	Staff	9.0000	.00	0		1	PM	No
Replace Anesthesia Vacuum Pump, <16 HP	Each	Plumber	Contract	20.0000	24,481.49	0		10	Major	Yes
Dental Vacuum Pump, <11 HP										
Annual PM, Dental Vacuum Pump, <11 HP	Each	Maint Plumber	Staff	16.0000	.00	0		1	PM	No
Replace Dental Vacuum Pump, <11 HP	Each	Plumber	Contract	18.5000	14,280.59	0		10	Major	Yes
Medical Gas Alarm Panel										
Annual PM, Medical Gas Alarm Panel	Each	Maint Plumber	Staff	.5000	.00	0		1	PM	No
Replace Medical Gas Alarm Panel	Each	Plumber	Contract	2.5000	650.14	0		15	Major	Yes
Medical Gas Compressor/Dryer										
Annual PM, Medical Gas Compressor/Dryer	Each	Maint Plumber	Staff	8.5000	.00	0		1	PM	No
Replace Medical Gas Compressor/Dryer	Each	Plumber	Contract	1.5000	693.64	0		15	Major	Yes
Medical Gas Outlet										
Annual PM, Medical Gas Outlet	Each	Maint Plumber	Staff	.1600	.00	0		1	PM	No
Replace Medical Gas Outlet	Each	Plumber	Contract	2.5000	51.00	0		25	Major	Yes
Reverse Osmosis/Deionized Water System										
Maintain Reverse Osmosis/Deionized Water System	Each	Maint Plumber	Staff	4.0000	.00	0		1	PM	No
Replace Reverse Osmosis/Deionized Water System	Each	Plumber	Contract	32.0000	7,370.96	0		15	Major	Yes
D30 HVAC										
D3010 Energy Supply										
Leak Detection, Panel & Float Assembly										
Replace Leak Detection, Panel & Float Assembly	Each	Plumber	Contract	16.0000	2,223.98	0		75	Major	Yes
Fuel Oil Pump, 1.5 HP										
Annual PM, Fuel Oil Pump, 1.5 HP	Each	Gen Maint Worker	Staff	.4000	.00	0		1	PM	No
Repair Fuel Oil Pump, 1.5 HP	Each	Plumber	Contract	3.5300	31.95	0		10	Major	No
Replace Fuel Oil Pump, 1.5 HP	Each	Plumber	Contract	2.0000	795.00	0		15	Major	Yes

*Task frequency for Washington D.C.
Components ordered by UNIFORMAT II Classifications. All costs expressed in ($) 2004.
Sources: U.S. Army Corps of Engineers, Department of Defense Engineering Performance Standards, and Whitestone Research interviews.

5.1 Building Components and Repair Tasks

Uniformat / Component / Repair Task	Units	Trade	Labor Type	Labor Hours	Material Costs	Equip Hours	Equipment Type	Freq. Yrs.*	Task Type	Repl. Task
Generator Day Tank, 275 Gal.										
Maintain Generator Day Tank, 275 Gal.	Each	Maint HVAC Tech	Staff	3.9100	.00	0		1	PM	No
Replace Generator Day Tank, 275 Gal.	Each	Plumber	Contract	4.0000	251.00	0		30	Major	Yes
Oil Storage Tank, 275 Gal.										
Maintain Oil Storage Tank, 275 Gal.	Each	Maint Plumber	Staff	3.9100	.00	0		1	PM	No
Replace Oil Storage Tank, 275 Gal.	Each	Plumber	Contract	2.0000	251.00	0		30	Major	Yes
Oil Storage Tank, Fiberglass, 500 Gal										
Maintain Oil Storage Tank, Fiberglass, 500 Gal	Each	Maint HVAC Tech	Staff	3.9100	.00	0		1	PM	No
Replace Oil Storage Tank, Fiberglass, 500 Gal	Each	Plumber	Contract	5.9930	1,475.00	0		100	Major	Yes
Oil Storage Tank, 25,000 Gal.										
Maintain Oil Storage Tank, 25,000 Gal.	Each	Maint Plumber	Staff	3.9100	.00	0		1	PM	No
Replace Oil Storage Tank, 25,000 Gal.	Each	Plumber	Contract	.0000	30,407.18	0		30	Major	Yes
Pipe & Fittings, 2" Copper, Fuel Oil										
Resolder Joint, Pipe & Fittings, 2" Copper, Fuel Oil	K Ln Ft	Plumber	Contract	.1420	.25	0		10	Major	No
Install 10' Section, Pipe & Fittings, 2" Copper, Fuel Oil	K Ln Ft	Plumber	Contract	4.2700	53.00	0		20	Major	No
Replace Pipe & Fittings, 2" Copper, Fuel Oil	K Ln Ft	Plumber	Contract	427.0000	5,300.00	0		25	Major	Yes
Pipe & Fittings, 4" Steel, Gas										
Replace 10 Foot Section, Pipe & Fittings, 4" Steel, Gas	K Ln Ft	Plumber	Contract	8.2650	90.98	0		12	Major	No
Install New Gasket & Bolts, Pipe & Fittings, 4" Steel, Gas	K Ln Ft	Plumber	Contract	48.8000	879.48	0		30	Major	No
Replace Pipe & Fittings, 4" Steel, Gas	K Ln Ft	Plumber	Contract	826.0000	9,098.10	0		75	Major	Yes
D3020 Heat Generating Systems										
Fuel Oil Pump 1/2 HP										
Inspect & Lubricate Fuel Oil Pump 1/2 HP	Each	Maint Plumber	Staff	.1250	.70	0		1	PM	No
Overhaul Fuel Oil Pump 1/2 HP	Each	Maint Plumber	Staff	.6820	3.53	0		5	PM	No
Replace Fuel Oil Pump 1/2 HP	Each	Plumber	Contract	4.0000	970.71	0		20	Major	Yes
Gate Valve, 2-3"										
Repack Gland, Gate Valve, 2-3"	Each	Maint Plumber	Staff	.2730	9.31	0		10	PM	No
Replace Gate Valve, 2-3"	Each	Plumber	Contract	.4400	296.00	0		20	Major	Yes
Pressure Reducer Valve, 2"										
Annual PM, Pressure Reducer Valve, 2"	Each	Maint HVAC Tech	Staff	.4000	.00	0		1	PM	No
Replace Pressure Reducer Valve, 2"	Each	Plumber	Contract	6.0000	835.00	0		6	Major	Yes

*Task frequency for Washington D.C.

Components ordered by UNIFORMAT II Classifications. All costs expressed in ($) 2004.

Sources: U.S. Army Corps of Engineers, [illegible]

Uniformat / Component / Repair Task	Units	Trade	Labor Type	Labor Hours	Material Costs	Equip Hours	Equipment Type	Freq. Yrs.*	Task Type	Repl. Task
Air Separator, 2"										
Annual PM, Air Separator, 2"	Each	Maint HVAC Tech	Staff	1.0000	.00	0		1	PM	No
Replace Air Separator, 2"	Each	Plumber	Contract	1.0000	710.66	0		15	Major	Yes
AIR Separator, 8"										
Annual PM, Air Separator, 8"	Each	Maint HVAC Tech	Staff	.2100	.00	0		1	PM	No
Replace Air Separator, 8"	Each	Plumber	Contract	5.4550	3,000.00	0		15	Major	Yes
Boiler Blowoff System										
Repair Boiler Blowoff System	Each	Maint HVAC Tech	Staff	1.0000	.00	0		7	PM	No
Replace Boiler Blowoff System	Each	Plumber	Contract	2.0000	2,750.00	0		30	Major	Yes
Chemical Feed System										
Maintain Chemical Feed System	Each	Maint HVAC Tech	Staff	.5000	.00	0		1	PM	No
Replace Chemical Feed System	Each	Plumber	Contract	6.4300	420.00	0	2T Forklift	15	Major	Yes
Deaerator										
Maintain Deaerator	Each	Maint HVAC Tech	Staff	15.0000	.00	0		1	PM	No
Repair Deaerator	Each	Maint HVAC Tech	Staff	.5000	.00	0		10	Major	No
Replace Deaerator	Each	Plumber	Contract	200.0000	21,700.00	0		20	Major	Yes
Circulation Pump 5 HP, Hot Water										
Maintain Circulation Pump 5 HP, Hot Water	Each	Maint HVAC Tech	Staff	.0840	10.17	0		1	PM	No
Repair Circulation Pump 5 HP, Hot Water	Each	Maint HVAC Tech	Staff	.4640	32.85	0		5	PM	No
Replace Circulation Pump 5 HP, Hot Water	Each	Plumber	Contract	3.2300	1,475.00	0		15	Major	Yes
Circulation Pump 7.5 HP, Hot Water										
Maintain Circulation Pump 7.5 HP, Hot Water	Each	Maint HVAC Tech	Staff	.1000	10.17	0		1	PM	No
Repair Circulation Pump 7.5 HP, Hot Water	Each	Maint HVAC Tech	Staff	.5700	32.85	0		5	PM	No
Replace Circulation Pump 7.5 HP, Hot Water	Each	Plumber	Contract	4.0000	2,050.00	0		15	Major	Yes
Circulation Pump 10 HP, Hot Water										
Maintain Circulation Pump 10 HP, Hot Water	Each	Maint HVAC Tech	Staff	.1200	10.17	0		1	PM	No
Repair Circulation Pump 10 HP, Hot Water	Each	Maint HVAC Tech	Staff	.6500	56.30	0		5	PM	No
Replace Circulation Pump 10 HP, Hot Water	Each	Plumber	Contract	4.5000	2,525.00	0		15	Major	Yes
Circulation Pump 25 HP, Hot Water										
Maintain Circulation Pump 25 HP, Hot Water	Each	Maint HVAC Tech	Staff	.2100	12.20	0		1	PM	No
Repair Circulation Pump 25 HP, Hot Water	Each	Maint HVAC Tech	Staff	1.1500	65.71	0		5	PM	No
Replace Circulation Pump 25 HP, Hot Water	Each	Plumber	Contract	8.0000	3,975.00	4	2T Forklift	15	Major	Yes

*Task frequency for Washington D.C.
Components ordered by UNIFORMAT II Classifications. All costs expressed in ($) 2004.
Sources: U.S. Army Corps of Engineers, Department of Defense Engineering Performance Standards, and Whitestone Research interviews.

5.1 Building Components and Repair Tasks

Uniformat / Component / Repair Task	Units	Trade	Labor Type	Labor Hours	Material Costs	Equip Hours	Equipment Type	Freq. Yrs.*	Task Type	Repl. Task
Circulation Pump 75 HP, Hot Water										
Maintain Circulation Pump 75 HP, Hot Water	Each	Maint HVAC Tech	Staff	.2600	13.22	0		1	PM	No
Repair Circulation Pump 75 HP, Hot Water	Each	Maint HVAC Tech	Staff	1.4400	210.85	0		5	PM	No
Replace Circulation Pump 75 HP, Hot Water	Each	Plumber	Contract	10.0000	12,535.16	4	2T Forklift	15	Major	Yes
Circulation Pump 100 HP, Hot Water										
Maintain Circulation Pump 100 HP, Hot Water	Each	Maint HVAC Tech	Staff	.2600	15.25	0		1	PM	No
Repair Circulation Pump 100 HP, Hot Water	Each	Maint HVAC Tech	Staff	1.4400	238.99	0		5	PM	No
Replace Circulation Pump 100 HP, Hot Water	Each	Plumber	Contract	10.0000	11,726.44	4	2T Forklift	15	Major	Yes
Circulation Pump 150 HP, Hot Water										
Maintain Circulation Pump 150 HP, Hot Water	Each	Maint HVAC Tech	Staff	.4200	16.18	0		1	PM	No
Repair Circulation Pump 150 HP, Hot Water	Each	Maint HVAC Tech	Staff	2.3000	384.14	0		5	PM	No
Replace Circulation Pump 150 HP, Hot Water	Each	Plumber	Contract	16.0000	16,882.03	4	2T Forklift	15	Major	Yes
Condensing Heat Exchanger, Gas/Oil, 10 Mbh										
Maintain Condensing Heat Exchanger, Gas, 10 Mbh	Each	Maint HVAC Tech	Staff	31.3000	.00	0		1	PM	No
Replace Condensing Heat Exchanger, Gas, 10 Mbh	Each	Plumber	Contract	336.0000	35,405.76	0		30	Major	Yes
Condensing Heat Exchanger, Gas/Oil, 20 Mbh										
Maintain Condensing Heat Exchanger, Gas/Oil, 20 Mbh	Each	Maint HVAC Tech	Staff	31.3000	.00	0		1	PM	No
Replace Condensing Heat Exchanger, Gas/Oil, 20 Mbh	Each	Plumber	Contract	643.0000	67,687.85	0		30	Major	Yes
Condensate Receiver Station, 10-15 Gal.										
Maintain Condensate Receiver Station, 10-15 Gal.	Each	Maint HVAC Tech	Staff	1.0000	.00	0		1	PM	No
Repair Condensate Receiver Station, 10-15 Gal.	Each	Plumber	Contract	1.5750	641.92	0		3	Major	No
Repair Condensate Receiver Station, Motor, 10-15 Gal.	Each	Plumber	Contract	.4990	1,122.10	0		5	Major	No
Replace Condensate Receiver Station, 10-15 Gal.	Each	Plumber	Contract	12.0000	3,184.33	0		15	Major	Yes
Expansion Tank, 30 Gal.										
Annual PM, Expansion Tank, 30 Gal.	Each	Maint HVAC Tech	Staff	1.3000	.00	0		1	PM	No
Replace Expansion Tank, 30 Gal.	Each	Plumber	Contract	2.0000	405.00	0		50	Major	Yes
Expansion Tank, 60 Gal.										
Annual PM, Expansion Tank, 60 Gal.	Each	Maint HVAC Tech	Staff	1.3000	.00	0		1	PM	No
Replace Expansion Tank, 60 Gal.	Each	Plumber	Contract	2.0000	540.39	0		50	Major	Yes
Expansion Tank, 100 Gal.										
Annual PM, Expansion Tank, 100 Gal.	Each	Maint HVAC Tech	Staff	1.3000	.00	0		1	PM	No
Replace Expansion Tank, 100 Gal.	Each	Plumber	Contract	4.0000	735.00	0		50	Major	Yes

*Task frequency for Washington D.C.
Components ordered by UNIFORMAT II Classifications. All costs expressed in ($) 2004.
Sources: U.S. Army Corps of Engineers, Department of Defense Engineering Ref...

5.1 Building Components and Repair Tasks

Uniformat / Component / Repair Task	Units	Trade	Labor Type	Labor Hours	Material Costs	Equip Hours	Equipment Type	Freq. Yrs.*	Task Type	Repl. Task
Expansion Tank, 250 Gal.										
Annual PM, Expansion Tank, 250 Gal.	Each	Maint HVAC Tech	Staff	1.3000	.00	0		1	PM	No
Replace Expansion Tank, 250 Gal.	Each	Plumber	Contract	4.0000	1,500.00	0		50	Major	Yes
Surge Tank, 1000 Gal.										
Replace Surge Tank, 1000 Gal.	Each	Plumber	Contract	4.0000	1,837.43	0		15	Major	Yes
Flash Tank, 24 Gal.										
Maintain Flash Tank, 24 Gal.	Each	Maint HVAC Tech	Staff	.3000	.00	0		1	PM	No
Repair Flash Tank, 24 Gal.	Each	Maint HVAC Tech	Staff	4.5700	.00	0		5	PM	No
Replace Flash Tank, 24 Gal.	Each	Plumber	Contract	5.0000	370.00	0		15	Major	Yes
Feed Water Pump, <1000 kbtu Boiler										
Maintain Feed Water Pump, <1000 kbtu Boiler	Each	Maint HVAC Tech	Staff	15.0000	444.79	0		1	PM	No
Replace Feed Water Pump, <1000 kbtu Boiler	Each	Plumber	Contract	22.0000	4,275.00	4	2T Forklift	15	Major	Yes
Feed Water Pump, 1000-3000 kbtu Boiler										
Maintain Feed Water Pump, 1000-3000 kbtu Boiler	Each	Maint HVAC Tech	Staff	15.0000	1,187.81	0		1	PM	No
Replace Feed Water Pump, 1000-3000 kbtu Boiler	Each	Plumber	Contract	22.0000	4,275.00	4	2T Forklift	15	Major	Yes
Gas Boiler, 250 Mbh										
Inspect & Test Gas Boiler, 250 Mbh	Each	Maint HVAC Tech	Staff	3.5400	.00	0		1	PM	No
Clean Fire Box, Gas Boiler, 250 Mbh	Each	Maint HVAC Tech	Staff	1.1000	.00	0		2	PM	No
Repair Gas Boiler, 250 Mbh	Each	Plumber	Contract	4.7000	375.46	0		7	Major	No
Replace Gas Boiler, 250 Mbh	Each	Plumber	Contract	50.0000	1,975.00	4	2T Lift/45T Crane	30	Major	Yes
Gas Boiler, 750 Mbh										
Inspect & Test Gas Boiler, 750 Mbh	Each	Maint HVAC Tech	Staff	10.6200	.00	0		1	PM	No
Clean Fire Box, Gas Boiler, 750 Mbh	Each	Maint HVAC Tech	Staff	1.1000	.00	0		2	PM	No
Repair Gas Boiler, 750 Mbh	Each	Plumber	Contract	4.7000	1,115.00	0		7	Major	No
Replace Gas Boiler, 750 Mbh	Each	Plumber	Contract	90.0000	7,175.00	4	2T Lift/45T Crane	30	Major	Yes
Gas Boiler, 1000 Mbh										
Inspect & Test Gas Boiler, 1000 Mbh	Each	Maint HVAC Tech	Staff	14.1500	.00	0		1	PM	No
Clean Fire Box, Gas Boiler, 1000 Mbh	Each	Maint HVAC Tech	Staff	1.1000	.00	0		2	PM	No
Repair Gas Boiler, 1000 Mbh	Each	Plumber	Contract	4.7000	1,350.99	0		7	Major	No
Replace Gas Boiler, 1000 Mbh	Each	Plumber	Contract	125.0000	8,275.00	4	2T Lift/45T Crane	30	Major	Yes

*Task frequency for Washington D.C.

Components ordered by UNIFORMAT II Classifications. All costs expressed in ($) 2004.

Sources: U.S. Army Corps of Engineers, Department of Defense Engineering Performance Standards, and Whitestone Research interviews.

5.1 Building Components and Repair Tasks

Uniformat / Component / Repair Task	Units	Trade	Labor Type	Labor Hours	Material Costs	Equip Hours	Equipment Type	Freq. Yrs.*	Task Type	Repl. Task
Gas Boiler, 2000 Mbh										
Inspect & Test Gas Boiler, 2000 Mbh	Each	Maint HVAC Tech	Staff	28.3000	.00	0		1	PM	No
Clean Fire Box, Gas Boiler, 2000 Mbh	Each	Maint HVAC Tech	Staff	1.1000	.00	0		2	PM	No
Repair Gas Boiler, 2000 Mbh	Each	Plumber	Contract	4.7000	2,380.00	0		7	Major	No
Replace Gas Boiler, 2000 Mbh	Each	Plumber	Contract	142.0000	11,900.00	4	2T Lift/45T Crane	30	Major	Yes
Gas Boiler, 2500 Mbh										
Inspect & Test Gas Boiler, 2500 Mbh	Each	Maint HVAC Tech	Staff	28.3000	.00	0		1	PM	No
Clean Fire Box, Gas Boiler, 2500 Mbh	Each	Maint HVAC Tech	Staff	1.1000	.00	0		2	PM	No
Repair Gas Boiler, 2500 Mbh	Each	Plumber	Contract	4.7000	2,380.00	0		7	Major	No
Replace Gas Boiler, 2500 Mbh	Each	Plumber	Contract	153.0000	14,800.00	4	2T Lift/45T Crane	30	Major	Yes
Gas Boiler, 4200 Mbh										
Inspect & Test Gas Boiler, 4200 Mbh	Each	Maint HVAC Tech	Staff	28.3000	.00	0		1	PM	No
Clean Fire Box, Gas Boiler, 4200 Mbh	Each	Maint HVAC Tech	Staff	1.1000	.00	0		2	PM	No
Repair Gas Boiler, 4200 Mbh	Each	Plumber	Contract	4.7000	3,831.80	0		7	PM	No
Replace Gas Boiler, 4200 Mbh	Each	Plumber	Contract	205.0000	19,900.00	0		30	Major	Yes
Gas Boiler, Steam, 24,000 Mbh										
Inspect & Test Gas Boiler, Steam, 24,000 Mbh	Each	Maint HVAC Tech	Staff	28.3000	.00	0		1	PM	No
Clean Fire Box, Gas Boiler, Steam, 24,000 Mbh	Each	Maint HVAC Tech	Staff	1.1000	.00	0		2	PM	No
Repair Gas Boiler, Steam, 24,000 Mbh	Each	Plumber	Contract	48.0000	20,000.00	0		7	Major	No
Replace Gas Boiler, Steam, 24,000 Mbh	Each	Plumber	Contract	2400.0000	155,000.00	4	4T Lift/90T Crane	30	Major	Yes
House Furnace, Gas/Oil, 25 Mbh										
Maintain House Furnace, Gas/Oil, 25 Mbh	Each	Maint HVAC Tech	Staff	4.1000	.00	0		1	PM	No
Repair House Furnace, Gas/Oil, 25 Mbh	Each	Plumber	Contract	2.5600	186.46	0		10	Major	No
Replace House Furnace, Gas/Oil, 25 Mbh	Each	Plumber	Contract	8.0000	435.00	0		15	Major	Yes
House Furnace, Gas/Oil, 100 Mbh										
Maintain House Furnace, Gas/Oil, 100 Mbh	Each	Maint HVAC Tech	Staff	4.1000	.00	0		1	PM	No
Repair House Furnace, Gas/Oil, 100 Mbh	Each	Plumber	Contract	2.6000	477.85	0		10	Major	No
Replace House Furnace, Gas/Oil, 100 Mbh	Each	Plumber	Contract	16.0000	655.48	0		15	Major	Yes
House Furnace, Electric 25 Mbh, 8 kw										
Maintain House Furnace, Electric 25 Mbh, 8 kw	Each	Maint HVAC Tech	Staff	1.5000	.00	0		1	PM	No
Repair House Furnace, Electric 25 Mbh, 8 kw	Each	Plumber	Contract	3.3000	230.66	0		10	Major	No
Replace House Furnace, Electric 25 Mbh, 8 kw	Each	Plumber	Contract	8.0000	395.00	0		15	Major	Yes

*Task frequency for Washington D.C.

Components ordered by UNIFORMAT II Classifications. All costs expressed in ($) 2004.

Sources: U.S. Army Corps of Engineers, Department of Defense Engineering [illegible]

Uniformat / Component / Repair Task	Units	Trade	Labor Type	Labor Hours	Material Costs	Equip Hours	Equipment Type	Freq. Yrs.*	Task Type	Repl. Task
House Furnace, Electric 100 Mbh, 30 kw										
Maintain House Furnace, Electric 100 Mbh, 30 kw	Each	Maint HVAC Tech	Staff	1.5000	.00	0		1	PM	No
Repair House Furnace, Electric 100 Mbh, 30 kw	Each	Plumber	Contract	3.3000	1,840.00	0		10	Major	No
Replace House Furnace, Electric 100 Mbh, 30 kw	Each	Plumber	Contract	16.0000	4,225.00	0		15	Major	Yes
Heat Exchanger, Steam-to-Water, 130 Gph										
Annual PM, Heat Exchanger, Steam-to-Water, 130 Gph	Each	Maint Plumber	Staff	1.0000	.00	0		1	PM	No
Replace Heat Exchanger, Steam-to-Water, 130 Gph	Each	Plumber	Contract	24.0000	1,175.86	0		28	Major	Yes
Steam Converter, 200 Gph										
Inspect & Maintain Steam Converter, 200 Gph	Each	Maint HVAC Tech	Staff	1.2140	.00	0		2	PM	No
Repair Steam Converter, 200 Gph	Each	Maint HVAC Tech	Staff	4.5700	.00	0		5	PM	No
Replace Steam Converter, 200 Gph	Each	Plumber	Contract	36.0000	7,989.55	0		30	Major	Yes
Steam Trap, F&T, 2"										
Maintain Steam Trap, F&T, 2"	Each	Maint HVAC Tech	Staff	.5000	.00	0		1	PM	No
Repair Steam Trap, F&T, 2"	Each	Plumber	Contract	.4270	235.00	0		3	Major	No
Replace Steam Trap, F&T, 2"	Each	Plumber	Contract	1.0000	330.00	0		10	Major	Yes
D3030 Cooling Generating Systems										
Flow Control Valve, Motorized, 2"										
Maintain Flow Control Valve & Actuator, 2"	Each	Maint Plumber	Staff	1.0000	10.18	0		1	PM	No
Replace Valve Actuator, 2"	Each	Plumber	Contract	17.5800	530.39	0		15	Major	Yes
Replace Valve, Flow Control Motorized, 2"	Each	Plumber	Contract	17.5800	770.00	0		17	Major	Yes
Flow Control Valve, Motorized, 4"										
Maintain Flow Control Valve & Actuator, 4"	Each	Maint Plumber	Staff	1.0000	10.18	0		1	PM	No
Replace Valve Actuator, 4"	Each	Plumber	Contract	17.5800	3,048.61	0		15	Major	Yes
Replace Valve, Flow Control Motorized, 4"	Each	Plumber	Contract	17.5800	1,200.88	0		17	Major	Yes
Condensor, Air Cooled, 7.5 Ton										
Maintain Condenser, Air-Cooled, 7.5 Ton	Each	Maint HVAC Tech	Staff	1.8000	.00	0		1	PM	No
Repair Condenser, Air-Cooled, 7.5 Ton	Each	Plumber	Contract	5.0000	451.90	0		10	Major	No
Replace Condenser, Air-Cooled, 7.5 Ton	Each	Plumber	Contract	7.0000	1,725.00	4	2T Lift/45T Crane	15	Major	Yes
Condenser, Air-Cooled, 20 Ton										
Maintain Condenser, Air-Cooled, 20 Ton	Each	Maint HVAC Tech	Staff	1.8000	26.28	0		1	PM	No
Repair Condenser, Air-Cooled, 20 Ton	Each	Plumber	Contract	8.0000	456.00	0		10	Major	No
Replace Condenser, Air-Cooled, 20 Ton	Each	Plumber	Contract	16.0000	3,450.00	4	2T Lift/45T Crane	15	Major	Yes

*Task frequency for Washington D.C.

Components ordered by UNIFORMAT II Classifications. All costs expressed in ($) 2004.

Sources: U.S. Army Corps of Engineers, Department of Defense Engineering Performance Standards, and Whitestone Research interviews.

5.1 Building Components and Repair Tasks

Uniformat / Component / Repair Task	Units	Trade	Labor Type	Labor Hours	Material Costs	Equip Hours	Equipment Type	Freq. Yrs.*	Task Type	Repl. Task
Condenser, Air-Cooled, 40 Ton										
Maintain Condenser, Air-Cooled, 40 Ton	Each	Maint HVAC Tech	Staff	3.2000	50.00	0		1	PM	No
Repair Condenser, Air-Cooled, 40 Ton	Each	Plumber	Contract	14.0000	470.23	0		10	Major	No
Replace Condenser, Air-Cooled, 40 Ton	Each	Plumber	Contract	24.0000	10,100.00	4	2T Lift/45T Crane	15	Major	Yes
Condenser, Air-Cooled, 60 Ton										
Maintain Condenser, Air-Cooled, 60 Ton	Each	Maint HVAC Tech	Staff	3.2000	50.00	0		1	PM	No
Repair Condenser, Air-Cooled, 60 Ton	Each	Plumber	Contract	19.6000	747.22	0		10	Major	No
Replace Condenser, Air-Cooled, 60 Ton	Each	Plumber	Contract	28.0000	11,200.00	4	2T Lift/45T Crane	15	Major	Yes
Circulation Pump, 5 HP, Chiller & Condenser Water										
Maintain Circulation Pump, 5 HP, Chiller & Condenser Water	Each	Maint HVAC Tech	Staff	3.2000	11.09	0		1	PM	No
Repair Circulation Pump, 5 HP, Chiller & Condenser Water	Each	Plumber	Contract	.4640	40.31	0		5	Major	No
Replace Circulation Pump, 5 HP, Chiller & Condenser Water	Each	Plumber	Contract	3.2300	1,475.00	0		15	Major	Yes
Circulation Pump, 7.5 HP, Chiller & Condenser Water										
Maintain Circulation Pump, 7.5 HP, Chiller & Condenser Water	Each	Maint HVAC Tech	Staff	.1000	11.09	0		1	PM	No
Repair Circulation Pump, 7.5 HP, Chiller & Condenser Water	Each	Plumber	Contract	.5700	47.24	0		5	Major	No
Replace Circulation Pump, 7.5 HP, Chiller & Condenser Water	Each	Plumber	Contract	4.0000	2,050.00	0		15	Major	Yes
Circulation Pump, 10 HP, Chiller & Condenser Water										
Maintain Circulation Pump, 10 HP, Chiller & Condenser Water	Each	Maint HVAC Tech	Staff	.1200	11.09	0		1	PM	No
Repair Circulation Pump, 10 HP, Chiller & Condenser Water	Each	Plumber	Contract	.6500	55.77	0		5	Major	No
Replace Circulation Pump, 10 HP, Chiller & Condenser Water	Each	Plumber	Contract	4.5000	2,525.00	0		15	Major	Yes
Circulation Pump, 25 HP, Chiller & Condenser Water										
Maintain Circulation Pump, 25 HP, Chiller & Condenser Water	Each	Maint HVAC Tech	Staff	.2100	13.10	0		1	PM	No
Repair Circulation Pump, 25 HP, Chiller & Condenser Water	Each	Plumber	Contract	1.1500	80.82	0		5	Major	No
Replace Circulation Pump, 25 HP, Chiller & Condenser Water	Each	Plumber	Contract	8.0000	3,975.00	4	2T Forklift	15	Major	Yes
Circulation Pump, 75 HP, Chiller & Condenser Water										
Maintain Circulation Pump, 75 HP, Chiller & Condenser Water	Each	Maint HVAC Tech	Staff	.2600	14.15	0		1	PM	No
Repair Circulation Pump, 75 HP, Chiller & Condenser Water	Each	Plumber	Contract	1.4400	208.89	0		5	Major	No
Replace Circulation Pump, 75 HP, Chiller & Condenser Water	Each	Plumber	Contract	10.0000	12,494.11	4	2T Forklift	15	Major	Yes
Circulation Pump, 100 HP, Chiller & Condenser Water										
Maintain Circulation Pump, 100 HP, Chiller & Condenser Water	Each	Maint HVAC Tech	Staff	.2600	16.12	0		1	PM	No
Repair Circulation Pump, 100 HP, Chiller & Condenser Water	Each	Plumber	Contract	1.4400	241.83	0		5	Major	No
Replace Circulation Pump, 100 HP, Chiller & Condenser Water	Each	Plumber	Contract	10.0000	13,700.00	4	2T Forklift	15	Major	Yes

*Task frequency for Washington D.C.

Components ordered by UNIFORMAT II Classifications. All costs expressed in ($) 2004.

Sources: U.S. Army Corps of Engineers, Department of Defense Engineering [illegible]

5.1 Building Components and Repair Tasks

Uniformat / Component / Repair Task	Units	Trade	Labor Type	Labor Hours	Material Costs	Equip Hours	Equipment Type	Freq. Yrs.*	Task Type	Repl. Task
Circulation Pump, 150 HP, Chiller & Condenser Water										
Maintain Circulation Pump, 150 HP, Chiller & Condenser Water	Each	Maint HVAC Tech	Staff	.4200	17.13	0		1	PM	No
Repair Circulation Pump, 150 HP, Chiller & Condenser Water	Each	Plumber	Contract	2.3000	290.74	0		5	Major	No
Replace Circulation Pump, 150 HP, Chiller & Condenser Water	Each	Plumber	Contract	16.0000	18,500.00	4	2T Forklift	15	Major	Yes
Chiller, Reciprocal Air-Cooled Hermetic, 20 Ton										
Maintain Chiller, Reciprocal Air Cooled Hermetic, 20 Ton	Each	Maint HVAC Tech	Staff	17.3000	40.31	0		1	PM	No
Repair Chiller, Reciprocal Air Cooled Hermetic, 20 Ton	Each	Plumber	Contract	15.3000	6,524.36	0		10	Major	No
Replace Chiller, Reciprocal Air Cooled Hermetic, 20 Ton	Each	Plumber	Contract	22.0000	17,600.00	4	2T Lift/45T Crane	20	Major	Yes
Chiller, Reciprocal Air-Cooled Hermetic, 40 Ton										
Maintain Chiller, Reciprocal Air Cooled Hermetic, 40 Ton	Each	Maint HVAC Tech	Staff	20.1000	55.42	0		1	PM	No
Repair Chiller, Reciprocal Air Cooled Hermetic, 40 Ton	Each	Plumber	Contract	25.0000	15,643.35	0		10	Major	No
Replace Chiller, Reciprocal Air Cooled Hermetic, 40 Ton	Each	Plumber	Contract	24.0000	23,900.00	4	2T Lift/45T Crane	20	Major	Yes
Chiller, Reciprocal Air-Cooled Hermetic, 60 Ton										
Maintain Chiller, Reciprocal Air Cooled Hermetic, 60 Ton	Each	Maint HVAC Tech	Staff	21.5000	55.42	0		1	PM	No
Repair Chiller, Reciprocal Air Cooled Hermetic, 60 Ton	Each	Plumber	Contract	32.3000	16,625.78	0		10	Major	No
Replace Chiller, Reciprocal Air Cooled Hermetic, 60 Ton	Each	Plumber	Contract	31.0000	39,600.00	4	2T Lift/45T Crane	20	Major	Yes
Chiller, Reciprocal Air-Cooled Hermetic, 200 Ton										
Maintain Chiller, Reciprocal Air Cooled Hermetic, 200 Ton	Each	Maint HVAC Tech	Staff	21.0000	250.00	0		1	PM	No
Repair Chiller, Reciprocal Air Cooled Hermetic, 200 Ton	Each	Plumber	Contract	25.0000	45,000.00	0		10	Major	No
Replace Chiller, Reciprocal Air Cooled Hermetic, 200 Ton	Each	Plumber	Contract	83.0000	113,356.69	4	2T Lift/45T Crane	20	Major	Yes
Chiller, Reciprocal Water-Cooled Hermetic, 100 Ton										
Maintain Chiller, Reciprocal Water-Cooled Hermetic, 100 Ton	Each	Maint HVAC Tech	Staff	21.5000	42.32	0		1	PM	No
Repair Chiller, Reciprocal Water-Cooled Hermetic, 100 Ton	Each	Plumber	Contract	33.3000	19,850.17	0		10	Major	No
Replace Chiller, Reciprocal Water-Cooled Hermetic, 100 Ton	Each	Plumber	Contract	74.1000	51,500.00	4	2T Lift/45T Crane	20	Major	Yes
Chiller, Reciprocal Water-Cooled Hermetic, 165 Ton										
Maintain Chiller, Reciprocal Water-Cooled Hermetic, 165 Ton	Each	Maint HVAC Tech	Staff	21.5000	42.32	0		1	PM	No
Repair Chiller, Reciprocal Water-Cooled Hermetic, 165 Ton	Each	Plumber	Contract	37.1000	23,981.43	0		10	Major	No
Replace Chiller, Reciprocal Water-Cooled Hermetic, 165 Ton	Each	Plumber	Contract	74.1000	76,500.00	4	2T Lift/45T Crane	20	Major	Yes
Chiller, Reciprocal Water-Cooled Hermetic, 300 Ton										
Maintain Chiller, Reciprocal Water-Cooled Hermetic, 300 Ton	Each	Maint HVAC Tech	Staff	21.5000	42.32	0		1	PM	No
Repair Chiller, Reciprocal Water-Cooled Hermetic, 300 Ton	Each	Plumber	Contract	22.0000	30,228.69	0		10	Major	No
Replace Chiller, Reciprocal Water-Cooled Hermetic, 300 Ton	Each	Plumber	Contract	74.1000	102,248.56	4	2T Lift/45T Crane	20	Major	Yes

*Task frequency for Washington D.C.
Components ordered by UNIFORMAT II Classifications. All costs expressed in ($) 2004.
Sources: U.S. Army Corps of Engineers, Department of Defense Engineering Performance Standards, and Whitestone Research interviews.

5.1 Building Components and Repair Tasks

Uniformat / Component / Repair Task	Units	Trade	Labor Type	Labor Hours	Material Costs	Equip Hours	Equipment Type	Freq. Yrs.*	Task Type	Repl. Task
Chiller, Centrifugal Hermetic, Bundled w/ Condenser, 200 Ton										
Maintain Chiller, Centrifugal Hermetic, Bnd'd w/ Condenser, 200 Ton	Each	Maint HVAC Tech	Staff	29.5000	191.45	0		1	PM	No
Repair Chiller, Centrifugal Hermetic, Bnd'd w/ Condenser, 200 Ton	Each	Plumber	Contract	23.1000	27,508.11	0		10	Major	No
Replace Chiller, Centrifugal Hermetic, Bnd'd w/ Condenser, 200 Ton	Each	Plumber	Contract	83.0000	91,316.61	4	4T Lift/90T Crane	20	Major	Yes
Chiller, Centrifugal Hermetic, Bundled w/ Condenser, 250 Ton										
Maintain Chiller, Centrifugal Hermetic, Bnd'd w/ Condenser, 250 Ton	Each	Maint HVAC Tech	Staff	29.5000	191.45	0		1	PM	No
Repair Chiller, Centrifugal Hermetic, Bnd'd w/ Condenser, 250 Ton	Each	Plumber	Contract	23.1000	30,732.51	0		10	Major	No
Replace Chiller, Centrifugal Hermetic, Bnd'd w/ Condenser, 250 Ton	Each	Plumber	Contract	83.0000	102,777.55	4	4T Lift/90T Crane	20	Major	Yes
Chiller, Centrifugal Hermetic, Bundled w/ Condenser, 300 Ton										
Maintain Chiller, Centrifugal Hermetic, Bnd'd w/ Condenser, 300 Ton	Each	Maint HVAC Tech	Staff	29.5000	191.45	0		1	PM	No
Repair Chiller, Centrifugal Hermetic, Bnd'd w/ Condenser, 300 Ton	Each	Plumber	Contract	37.3000	79,970.00	0		10	Major	No
Replace Chiller, Centrifugal Hermetic, Bnd'd w/ Condenser, 300 Ton	Each	Plumber	Contract	83.0000	138,000.00	4	4T Lift/90T Crane	20	Major	Yes
Chiller, Centrifugal Hermetic, Bundled w/ Condenser, 900 Ton										
Maintain Chiller, Centrifugal Hermetic, Bnd'd w/ Condenser, 900 Ton	Each	Maint HVAC Tech	Staff	29.5000	191.45	0		1	PM	No
Repair Chiller, Centrifugal Hermetic, Bnd'd w/ Condenser, 900 Ton	Each	Plumber	Contract	97.0000	133,006.25	0		10	Major	No
Replace Chiller, Centrifugal Hermetic, Bnd'd w/ Condenser, 900 Ton	Each	Plumber	Contract	178.1000	315,500.00	4	4T Lift/90T Crane	20	Major	Yes
Cooling Tower, 50 Ton										
Maintain Cooling Tower, 50 Ton	Each	Maint HVAC Tech	Staff	10.0000	27.00	0		1	PM	No
Repair Cooling Tower, 50 Ton	Each	Plumber	Contract	24.0000	289.85	0		10	Major	No
Replace Cooling Tower, 50 Ton	Each	Plumber	Contract	36.0000	4,753.47	4	2T Lift/45T Crane	15	Major	Yes
Cooling Tower, 100 Ton										
Maintain Cooling Tower, 100 Ton	Each	Maint HVAC Tech	Staff	10.9000	50.00	0		1	PM	No
Repair Cooling Tower, 100 Ton	Each	Maint HVAC Tech	Staff	28.0000	1,112.80	0		10	Major	No
Replace Cooling Tower, 100 Ton	Each	Plumber	Contract	40.0000	6,254.56	4	2T Lift/45T Crane	15	Major	Yes
Cooling Tower, 200 Ton										
Maintain Cooling Tower, 200 Ton	Each	Maint HVAC Tech	Staff	11.8000	50.00	0		1	PM	No
Repair Cooling Tower, 200 Ton	Each	Plumber	Contract	34.0000	1,297.00	0		10	Major	No
Replace Cooling Tower, 200 Ton	Each	Plumber	Contract	50.0000	10,400.00	4	2T Lift/45T Crane	15	Major	Yes
Cooling Tower, 250 Ton										
Maintain Cooling Tower, 250 Ton	Each	Maint HVAC Tech	Staff	12.2500	50.00	0		1	PM	No
Repair Cooling Tower, 250 Ton	Each	Plumber	Contract	34.0000	1,621.00	0		10	Major	No
Replace Cooling Tower, 250 Ton	Each	Plumber	Contract	60.0000	12,667.00	4	2T Lift/45T Crane	15	Major	Yes

*Task frequency for Washington D.C.

Components ordered by UNIFORMAT II Classifications. All costs expressed in ($) 2004.

Sources: U.S. Army Corps of Engineers, Department of Defense Engineering Performance Standards, and Whitestone R

5.1 Building Components and Repair Tasks

Uniformat / Component / Repair Task	Units	Trade	Labor Type	Labor Hours	Material Costs	Equip Hours	Equipment Type	Freq. Yrs.*	Task Type	Repl. Task
Cooling Tower, 300 Ton										
Maintain Cooling Tower, 300 Ton	Each	Maint HVAC Tech	Staff	12.7000	50.00	0		1	PM	No
Repair Cooling Tower, 300 Ton	Each	Plumber	Contract	34.0000	2,041.00	0		10	Major	No
Replace Cooling Tower, 300 Ton	Each	Plumber	Contract	60.0000	15,200.00	4	4T Lift/90T Crane	15	Major	Yes
Cooling Tower, 500 Ton										
Maintain Cooling Tower, 500 Ton	Each	Maint HVAC Tech	Staff	13.2500	50.00	0		1	PM	No
Repair Cooling Tower, 500 Ton	Each	Plumber	Contract	34.0000	3,244.00	0		10	Major	No
Replace Cooling Tower, 500 Ton	Each	Plumber	Contract	70.0000	22,833.00	4	4T Lift/90T Crane	15	Major	Yes
Cooling Tower, 1000 Ton										
Maintain Cooling Tower, 1000 Ton	Each	Maint HVAC Tech	Staff	14.1000	110.00	0		1	PM	No
Repair Cooling Tower, 1000 Ton	Each	Plumber	Contract	80.0000	6,426.85	0		10	Major	No
Replace Cooling Tower, 1000 Ton	Each	Plumber	Contract	200.0000	47,000.00	4	4T Lift/90T Crane	15	Major	Yes
Cooling Tower, 1400 Ton										
Maintain Cooling Tower, 1400 Ton	Each	Maint HVAC Tech	Staff	15.0000	136.03	0		1	PM	No
Repair Cooling Tower, 1400 Ton	Each	Plumber	Contract	80.0000	8,336.08	0		10	Major	No
Replace Cooling Tower, 1400 Ton	Each	Plumber	Contract	224.0000	65,800.00	4	4T Lift/90T Crane	15	Major	Yes
Evaporative Cooler, Direct, Packaged w/ Heat, 4 Ton										
Maintain Evaporative Cooler, Direct, Packaged w/ Heat, 4 Ton	Each	Maint HVAC Tech	Staff	3.0000	22.17	0		1	PM	No
Replace Evaporative Cooler, Direct, Packaged w/ Heat, 4 Ton	Each	Plumber	Contract	71.0000	9,238.51	0		20	Major	Yes
Evaporative Cooler, Indirect & Direct, Packaged w/ Heat, 4 Ton										
Maintain Evaporative Cooler, Indirect & Direct, Pkg'd w/ Heat, 4 Ton	Each	Maint HVAC Tech	Staff	4.0000	22.17	0		1	PM	No
Replace Evaporative Cooler, Indirect & Direct, Pkg'd w/ Heat, 4 Ton	Each	Plumber	Contract	65.0000	8,407.60	0		20	Major	Yes
Evaporative Cooler, Indirect & Direct, Packaged w/ Heat, 20 Ton										
Maintain Evaporative Cooler, Indirect & Direct, Pkg'd w/ Heat, 20 Ton	Each	Maint HVAC Tech	Staff	12.0000	22.17	0		1	PM	No
Replace Evaporative Cooler, Indirect & Direct, Pkg'd w/ Heat, 20 Ton	Each	Plumber	Contract	181.0000	23,458.30	0		20	Major	Yes
Evaporative Cooler, Indirect, 2000 Scfm										
Maintain Evaporative Cooler, Indirect, 2000 Scfm	Each	Maint HVAC Tech	Staff	4.0000	27.21	0		1	PM	No
Replace Evaporative Cooler, Indirect, 2000 Scfm	Each	Plumber	Contract	44.0000	5,709.38	0		20	Major	Yes
Evaporative Cooler, Indirect, 5000 Scfm										
Maintain Evaporative Cooler, Indirect, 5000 Scfm	Each	Maint HVAC Tech	Staff	4.0000	27.21	0		1	PM	No
Replace Evaporative Cooler, Indirect, 5000 Scfm	Each	Plumber	Contract	68.0000	8,874.96	0		20	Major	Yes

*Task frequency for Washington D.C.

Components ordered by UNIFORMAT II Classifications. All costs expressed in ($) 2004.

Sources: U.S. Army Corps of Engineers, Department of Defense Engineering Performance Standards, and Whitestone Research interviews.

5.1 Building Components and Repair Tasks

Uniformat / Component / Repair Task	Units	Trade	Labor Type	Labor Hours	Material Costs	Equip Hours	Equipment Type	Freq. Yrs.*	Task Type	Repl. Task
Evaporative Cooler, Indirect, 15,000 Scfm										
Maintain Evaporative Cooler, Indirect, 15,000 Scfm	Each	Maint HVAC Tech	Staff	12.0000	27.21	0		1	PM	No
Replace Evaporative Cooler, Indirect, 15,000 Scfm	Each	Plumber	Contract	181.0000	23,873.49	0		20	Major	Yes
Chiller, Centrifugal Hermetic, Bundled w/ Condenser, 1000 Ton										
Maintain Chiller, Centrifugal Hermetic, Bnd'd w/ Condenser, 1000 Ton	Each	Maint HVAC Tech	Staff	29.5000	191.45	0		1	PM	No
Repair Chiller, Centrifugal Hermetic, Bnd'd w/ Condenser, 1000 Ton	Each	Plumber	Contract	100.0000	94,054.61	0		10	Major	No
Replace Chiller, Centrifugal Hermetic, Bnd'd w/ Condenser, 1000 Ton	Each	Plumber	Contract	190.0000	315,500.00	4	4T Lift/90T Crane	20	Major	Yes
D3040 Distribution Systems										
Air Curtain, 1000 Cfm										
Maintain Air Curtain, 1000 Cfm	Each	Maint HVAC Tech	Staff	.2500	.00	0		1	PM	No
Replace Fan & Motor, Air Curtain, 1000 Cfm	Each	Plumber	Contract	1.2000	78.30	0		15	Major	No
Replace Air Curtain, 1000 Cfm	Each	Plumber	Contract	2.5000	531.65	0		20	Major	Yes
Ceiling Panel, Radiant, Hydronic										
Replace Ceiling Panel, Radiant, Hydronic	Sq Ft	Plumber	Contract	.2500	8.37	0		50	Major	Yes
Clean Room (Air Shower) Class 10,000										
Maintain, Clean Room (Air Shower) Class 10,000	Each	Maint HVAC Tech	Staff	1.0000	758.17	0		1	PM	No
Repair, Clean Room (Air Shower) Class 10,000	Each	Maint Plumber	Staff	2.4000	2,350.34	0		5	Major	No
Replace, Clean Room (Air Shower) Class 10,000	Each	Plumber	Contract	44.0000	9,401.37	0		20	Major	Yes
Dehumidifier & Control 120-150 lbs, 4500 Cfm										
Annual PM, Dehumidifier & Control 120-150 lbs, 4500 Cfm	Each	Maint HVAC Tech	Staff	6.5000	151.14	0		1	PM	No
Replace Dehumidifier & Control 120-150 lbs, 4500 Cfm	Each	Plumber	Contract	12.0000	49,000.00	0		20	Major	Yes
Humidifer, Steam, Duct w/ Controls										
Annual PM, Humidifier, Steam, Duct w/ Controls	Each	Maint HVAC Tech	Staff	2.5000	32.24	0		1	PM	No
Replace Humidifier, Steam, Duct w/ Controls	Each	Plumber	Contract	4.0000	3,155.87	0		20	Major	Yes
Pipe & Fittings, 3/4" Copper, High Temp.										
Install 10' Section, Pipe & Fittings, 3/4" Copper, High Temp.	K Ln Ft	Plumber	Contract	4.2700	17.60	0		13	Major	No
Resolder Joint, Pipe & Fittings, 3/4" Copper, High Temp.	K Ln Ft	Plumber	Contract	.1420	.25	0		13	Major	No
Replace Pipe & Fittings, 3/4" Copper, High Temp.	K Ln Ft	Plumber	Contract	427.0000	1,760.00	0		25	Major	Yes
Variable Air Volume Box										
Annual PM, Variable Air Volume Box	Each	Maint HVAC Tech	Staff	.6600	5.03	0		1	PM	No

*Task frequency for Washington D.C.

Components ordered by UNIFORMAT II Classifications. All costs expressed in ($) 2004.

Sources: U.S. Army Corps of Engineers, Department of Defense [illegible]

5.1 Building Components and Repair Tasks

Uniformat / Component / Repair Task	Units	Trade	Labor Type	Labor Hours	Material Costs	Equip Hours	Equipment Type	Freq. Yrs.*	Task Type	Repl. Task
Air Handler, Singlezone, 1300 Cfm										
Maintain Air Handler, Singlezone, 1300 Cfm	Each	Maint HVAC Tech	Staff	2.1000	76.77	0		1	PM	No
Repair Air Handler, Singlezone, 1300 Cfm	Each	Plumber	Contract	3.0000	26.28	0		10	Major	No
Replace Air Handler, Singlezone, 1300 Cfm	Each	Plumber	Contract	13.3000	2,401.75	4	2T Lift/45T Crane	15	Major	Yes
Air Handler, Singlezone, 6500 Cfm										
Maintain Air Handler, Singlezone, 6500 Cfm	Each	Maint HVAC Tech	Staff	5.4000	70.53	0		1	PM	No
Repair Air Handler, Singlezone, 6500 Cfm	Each	Plumber	Contract	4.0000	268.03	0		10	Major	No
Replace Air Handler, Singlezone, 6,500 Cfm	Each	Plumber	Contract	25.0000	4,125.00	4	2T Lift/45T Crane	15	Major	Yes
Air Handler, Singlezone, 20,000 Cfm										
Maintain Air Handler, Singlezone, 20,000 Cfm	Each	Maint HVAC Tech	Staff	4.8000	167.07	0		1	PM	No
Repair Air Handler, Singlezone, 20,000 Cfm	Each	Plumber	Contract	3.4500	25.77	0		10	Major	No
Replace Air Handler, Singlezone, 20,000 Cfm	Each	Plumber	Contract	26.0000	7,859.46	4	2T Lift/45T Crane	15	Major	Yes
Air Handler, Singlezone, 50,000 Cfm										
Maintain Air Handler, Singlezone, 50,000 Cfm	Each	Maint HVAC Tech	Staff	6.0000	337.56	0		1	PM	No
Repair Air Handler, Singlezone, 50,000 Cfm	Each	Plumber	Contract	6.0000	2,443.48	0		10	Major	No
Replace Air Handler, Singlezone, 50,000 Cfm	Each	Plumber	Contract	105.3000	31,883.21	4	2T Lift/45T Crane	15	Major	Yes
Air Handler, Singlezone, 100,000 Cfm										
Maintain Air Handler, Singlezone, 100,000 Cfm	Each	Maint HVAC Tech	Staff	6.0000	337.56	0		1	PM	No
Repair Air Handler, Singlezone, 100,000 Cfm	Each	Plumber	Contract	8.0000	4,398.28	0		10	Major	No
Replace Air Handler, Singlezone, 100,000 Cfm	Each	Plumber	Contract	250.0000	65,735.32	4	2T Lift/45T Crane	15	Major	Yes
Air Handler, Singlezone, 120,000 Cfm										
Maintain Air Handler, Singlezone, 120,000 Cfm	Each	Maint HVAC Tech	Staff	6.0000	370.80	0		1	PM	No
Repair Air Handler, Singlezone, 120,000 Cfm	Each	Plumber	Contract	8.0000	3,865.24	0		10	Major	No
Replace Air Handler, Singlezone, 120,000 Cfm	Each	Plumber	Contract	495.0000	56,696.94	4	2T Lift/45T Crane	15	Major	Yes
Air Handler, Singlezone, 135,000 Cfm										
Maintain Air Handler, Singlezone, 135,000 Cfm	Each	Maint HVAC Tech	Staff	6.0000	403.05	0		1	PM	No
Repair Air Handler, Singlezone, 135,000 Cfm	Each	Plumber	Contract	8.0000	4,098.00	0		10	Major	No
Replace Air Handler, Singlezone, 135,000 Cfm	Each	Plumber	Contract	557.0000	60,476.54	4	2T Lift/45T Crane	15	Major	Yes
Air Handler, Singlezone, 150,000 Cfm										
Maintain Air Handler, Singlezone, 150,000 Cfm	Each	Maint HVAC Tech	Staff	6.0000	437.31	0		1	PM	No
Repair Air Handler, Singlezone, 150,000 Cfm	Each	Plumber	Contract	8.0000	4,381.15	0		10	Major	No
Replace Air Handler, Singlezone, 150,000 Cfm	Each	Plumber	Contract	619.0000	64,256.12	4	2T Lift/45T Crane	15	Major	Yes

*Task frequency for Washington D.C.
Components ordered by UNIFORMAT II Classifications. All costs expressed in ($) 2004.
Sources: U.S. Army Corps of Engineers, Department of Defense Engineering Performance Standards, and Whitestone Research interviews.

5.1 Building Components and Repair Tasks

Uniformat / Component / Repair Task	Units	Trade	Labor Type	Labor Hours	Material Costs	Equip Hours	Equipment Type	Freq. Yrs.*	Task Type	Repl. Task
Air Handler, Multizone, 2500 Cfm										
Maintain Air Handler, Multizone, 2,500 Cfm	Each	Maint HVAC Tech	Staff	5.5000	65.50	0		1	PM	No
Repair Air Handler, Multizone, 2500 Cfm	Each	Plumber	Contract	4.0000	247.87	0		10	Major	No
Replace Air Handler, Multizone, 2,500 Cfm	Each	Plumber	Contract	20.0000	3,778.59	4	2T Lift/45T Crane	15	Major	Yes
Air Handler, Multizone, 5200 Cfm										
Maintain Air Handler, Multizone, 5,200 Cfm	Each	Maint HVAC Tech	Staff	5.5000	285.21	0		1	PM	No
Repair Air Handler, Multizone, 5200 Cfm	Each	Plumber	Contract	4.0000	26.28	0		10	Major	No
Replace Air Handler, Multizone, 5,200 Cfm	Each	Plumber	Contract	21.6000	4,978.63	4	2T Lift/45T Crane	15	Major	Yes
Air Handler, Multizone, 6500 Cfm										
Maintain Air Handler, Multizone, 6500 Cfm	Each	Maint HVAC Tech	Staff	5.5000	73.55	0		1	PM	No
Repair Air Handler, Multizone, 6500 Cfm	Each	Plumber	Contract	4.0000	268.03	0		10	Major	No
Replace Air Handler, Multizone, 6,500 Cfm	Each	Plumber	Contract	28.0000	6,185.80	4	2T Lift/45T Crane	15	Major	Yes
Air Handler, Multizone, 10,000 Cfm										
Maintain Air Handler, Multizone, 10,000 Cfm	Each	Maint HVAC Tech	Staff	5.5000	287.17	0		1	PM	No
Repair Air Handler, Multizone, 10,000 Cfm	Each	Plumber	Contract	4.0000	358.71	0		10	Major	No
Replace Air Handler, Multizone, 10,000 Cfm	Each	Plumber	Contract	33.0000	7,494.20	4	2T Lift/45T Crane	15	Major	Yes
Air Handler, Multizone, 15,000 Cfm										
Maintain Air Handler, Multizone, 15,000 Cfm	Each	Maint HVAC Tech	Staff	4.8000	287.17	0		1	PM	No
Repair Air Handler, Multizone, 15,000 Cfm	Each	Plumber	Contract	4.0000	472.57	0		10	Major	No
Replace Air Handler, Multizone, 15,000 Cfm	Each	Plumber	Contract	40.0000	9,233.86	4	2T Lift/45T Crane	15	Major	Yes
Air Handler, Multizone, 20,000 Cfm										
Maintain Air Handler, Multizone, 20,000 Cfm	Each	Maint HVAC Tech	Staff	4.8000	287.17	0		1	PM	No
Repair Air Handler, Multizone, 20,000 Cfm	Each	Plumber	Contract	4.0000	629.77	0		10	Major	No
Replace Air Handler, Multizone, 20,000 Cfm	Each	Plumber	Contract	40.0000	9,925.00	4	2T Lift/45T Crane	15	Major	Yes
Air Handler, Multizone, 25,000 Cfm										
Maintain Air Handler, Multizone, 25,000 Cfm	Each	Maint HVAC Tech	Staff	4.8000	372.82	0		1	PM	No
Repair Air Handler, Multizone, 25,000 Cfm	Each	Plumber	Contract	6.0000	670.07	0		10	Major	No
Replace Air Handler, Multizone, 25,000 Cfm	Each	Plumber	Contract	58.0000	16,182.42	4	2T Lift/45T Crane	15	Major	Yes
Air Handler, Multizone, 50,000 Cfm										
Maintain Air Handler, Multizone, 50,000 Cfm	Each	Maint HVAC Tech	Staff	6.1000	372.82	0		1	PM	No
Repair Air Handler, Multizone, 50,000 Cfm	Each	Plumber	Contract	6.0000	2,198.63	0		10	Major	No
Replace Air Handler, Multizone, 50,000 Cfm	Each	Plumber	Contract	105.3000	32,867.66	4	2T Lift/45T Crane	15	Major	Yes

*Task frequency for Washington D.C.

Components ordered by UNIFORMAT II Classifications. All costs expressed in ($) 2004.

Sources: U.S. Army Corps of Engineers, Department of [illegible]

5.1 Building Components and Repair Tasks

Uniformat / Component / Repair Task	Units	Trade	Labor Type	Labor Hours	Material Costs	Equip Hours	Equipment Type	Freq. Yrs.*	Task Type	Repl. Task
Air Handler, Exhaust, 15,000 Cfm										
Maintain Air Handler, Exhaust, 15,000 Cfm	Each	Maint HVAC Tech	Staff	3.4000	261.98	0		1	PM	No
Repair Air Handler, Exhaust, 15,000 Cfm	Each	Plumber	Contract	4.0000	503.81	0		10	Major	No
Replace Air Handler, Exhaust, 15,000 Cfm	Each	Plumber	Contract	20.0000	9,849.51	0		20	Major	Yes
Air Handler, Exhaust, 20,000 Cfm										
Maintain Air Handler, Exhaust, 20,000 Cfm	Each	Maint HVAC Tech	Staff	3.4000	261.98	0		1	PM	No
Repair Air Handler, Exhaust, 20,000 Cfm	Each	Plumber	Contract	4.0000	629.77	0		10	Major	No
Replace Air Handler, Exhaust, 20,000 Cfm	Each	Plumber	Contract	25.0000	12,311.98	0		20	Major	Yes
Exhaust Fan, Ceiling, 200-500 Cfm										
Maintain Fan & Motor, Exhaust Fan, Ceiling 200-500 Cfm	Each	Maint HVAC Tech	Staff	.2500	.00	0		1	PM	No
Replace Fan & Motor, Exhaust Fan, Ceiling 200-500 Cfm	Each	Plumber	Contract	2.5000	233.00	0		20	Major	Yes
Exhaust Fan, Roof Mounted, 800 Cfm										
Maintain Exhaust Fan, Roof Mounted, 800 Cfm	Each	Maint HVAC Tech	Staff	1.0000	.00	0		1	PM	No
Repair Exhaust Fan, Roof Mounted, 800 Cfm	Each	Plumber	Contract	2.4000	26.15	0		10	Major	No
Replace Exhaust Fan, Roof Mounted, 800 Cfm	Each	Plumber	Contract	5.0000	325.00	0		20	Major	Yes
Exhaust Fan, Roof Mounted, 1000 Cfm										
Maintain Exhaust Fan, Roof Mounted, 1000 Cfm	Each	Maint HVAC Tech	Staff	1.0000	5.54	0		1	PM	No
Repair Exhaust Fan, Roof Mounted, 1000 Cfm	Each	Plumber	Contract	2.4000	26.49	0		10	Major	No
Replace Exhaust Fan, Roof Mounted, 1000 Cfm	Each	Plumber	Contract	4.0000	360.00	0		20	Major	Yes
Exhaust Fan, Roof Mounted, 5000 Cfm										
Maintain Exhaust Fan, Roof Mounted, 5000 Cfm	Each	Maint HVAC Tech	Staff	1.0000	5.54	0		1	PM	No
Repair Exhaust Fan, Roof Mounted, 5000 Cfm	Each	Plumber	Contract	4.0000	26.41	0		10	Major	No
Replace Exhaust Fan, Roof Mounted, 5000 Cfm	Each	Plumber	Contract	12.0000	580.00	0		20	Major	Yes
Exhaust Fan, Roof Mounted, 10,000 Cfm										
Maintain Exhaust Fan, Roof Mounted, 10,000 Cfm	Each	Maint HVAC Tech	Staff	3.4000	5.54	0		1	PM	No
Repair Exhaust Fan, Roof Mounted, 10,000 Cfm	Each	Plumber	Contract	4.0000	26.49	0		10	Major	No
Replace Exhaust Fan, Roof Mounted, 10,000 Cfm	Each	Plumber	Contract	26.0000	1,748.23	0		20	Major	Yes
Exhaust Fan, Roof Mounted, 20,000 Cfm										
Maintain Exhaust Fan, Roof Mounted, 20,000 Cfm	Each	Maint HVAC Tech	Staff	3.4000	5.54	0		1	PM	No
Repair Exhaust Fan, Roof Mounted, 20,000 Cfm	Each	Plumber	Contract	4.0000	26.41	0		10	Major	No
Replace Exhaust Fan, Roof Mounted, 20,000 Cfm	Each	Plumber	Contract	26.0000	4,307.59	0		20	Major	Yes

*Task frequency for Washington D.C.

Components ordered by UNIFORMAT II Classifications. All costs expressed in ($) 2004.

Sources: U.S. Army Corps of Engineers, Department of Defense Engineering Performance Standards, and Whitestone Research interviews.

5.1 Building Components and Repair Tasks

Uniformat / Component / Repair Task	Units	Trade	Labor Type	Labor Hours	Material Costs	Equip Hours	Equipment Type	Freq. Yrs.*	Task Type	Repl. Task
Fan, Axial Flow, 5000 Cfm										
Maintain Fan, Axial Flow, 5000 Cfm	Each	Maint HVAC Tech	Staff	3.0000	5.54	0		1	PM	No
Repair Fan, Axial Flow, 5000 Cfm	Each	Plumber	Contract	4.0000	26.41	0		10	Major	No
Replace Fan, Axial Flow, 5000 Cfm	Each	Plumber	Contract	12.0000	4,125.00	0		20	Major	Yes
Fan, Axial Flow, 10,000 Cfm										
Annual PM, Fan, Axial Flow, 10,000 Cfm	Each	Maint HVAC Tech	Staff	3.0000	5.54	0		1	PM	No
Repair Fan, Axial Flow, 10,000 Cfm	Each	Plumber	Contract	4.0000	26.41	0		10	Major	No
Replace Fan, Axial Flow, 10,000 Cfm	Each	Plumber	Contract	15.0000	6,336.94	0		20	Major	Yes
Fan, Axial Flow, 15,000 Cfm										
Maintain Fan, Axial Flow, 15,000 Cfm	Each	Maint HVAC Tech	Staff	3.0000	5.54	0		1	PM	No
Repair Fan, Axial Flow, 15,000 Cfm	Each	Plumber	Contract	4.0000	26.41	0		10	Major	No
Replace Fan, Axial Flow, 15,000 Cfm	Each	Plumber	Contract	24.0000	6,475.00	0		20	Major	Yes
Fan, Axial Flow, 30,000 Cfm										
Annual PM, Fan, Axial Flow, 30,000 Cfm	Each	Maint HVAC Tech	Staff	1.0000	100.76	0		1	PM	No
Repair Fan, Axial Flow, 30,000 Cfm	Each	Plumber	Contract	5.0000	906.66	0		10	Major	No
Replace Fan, Axial Flow, 30,000 Cfm	Each	Plumber	Contract	20.0000	18,288.36	0		20	Major	Yes
Supply Fan, Roof Mounted, 1000 Cfm										
Maintain Supply Fan, Roof Mounted, 1000 Cfm	Each	Maint HVAC Tech	Staff	1.0000	5.54	0		1	PM	No
Repair Supply Fan, Roof Mounted, 1000 Cfm	Each	Plumber	Contract	2.4000	156.18	0		10	Major	No
Replace Supply Fan, Roof Mounted, 1000 Cfm	Each	Plumber	Contract	4.0000	415.00	0		20	Major	Yes
Supply Fan, Roof Mounted, 5000 Cfm										
Maintain Supply Fan, Roof Mounted, 5000 Cfm	Each	Maint HVAC Tech	Staff	1.0000	5.54	0		1	PM	No
Repair Supply Fan, Roof Mounted, 5000 Cfm	Each	Plumber	Contract	4.0000	251.90	0		10	Major	No
Replace Supply Fan, Roof Mounted, 5000 Cfm	Each	Plumber	Contract	12.0000	1,150.00	0		20	Major	Yes
Supply Fan, Roof Mounted, 10,000 Cfm										
Maintain Supply Fan, Roof Mounted, 10,000 Cfm	Each	Maint HVAC Tech	Staff	3.4000	5.54	0		1	PM	No
Repair Supply Fan, Roof Mounted, 10,000 Cfm	Each	Plumber	Contract	4.0000	523.96	0		10	Major	No
Replace Supply Fan, Roof Mounted, 10,000 Cfm	Each	Plumber	Contract	20.0000	1,975.00	0		20	Major	Yes
D3050 Terminal & Package Units										

*Task frequency for Washington D.C.
Components ordered by UNIFORMAT II Classifications. All costs expressed in ($) 2004.
Sources: U.S. Army Corps of Engineers, Department of [illegible]

Uniformat / Component / Repair Task	Units	Trade	Labor Type	Labor Hours	Material Costs	Equip Hours	Equipment Type	Freq. Yrs.*	Task Type	Repl. Task
Air Conditioner, Rooftop, 5 Ton										
Maintain Air Conditioner, Rooftop, 5 Ton	Each	Maint HVAC Tech	Staff	2.8000	53.41	0		1	PM	No
Repair Air Conditioner, Rooftop, 5 Ton	Each	Plumber	Contract	16.0000	2,093.84	0		10	Major	No
Replace Air Conditioner, Rooftop, 5 Ton	Each	Plumber	Contract	6.4500	6,121.32	0		15	Major	Yes
Air Conditioner, Rooftop, 10 Ton										
Maintain Air Conditioner, Rooftop, 10 Ton	Each	Maint HVAC Tech	Staff	2.8000	53.41	0		1	PM	No
Repair Air Conditioner, Rooftop, 10 Ton	Each	Plumber	Contract	16.0000	2,035.48	0		10	Major	No
Replace Air Conditioner, Rooftop, 10 Ton	Each	Plumber	Contract	35.0000	9,565.37	0		15	Major	Yes
Air Conditioner, Rooftop, 20 Ton										
Maintain Air Conditioner, Rooftop, 20 Ton	Each	Maint HVAC Tech	Staff	7.1000	53.41	0		1	PM	No
Repair Air Conditioner, Rooftop, 20 Ton	Each	Plumber	Contract	16.0000	8,584.95	0		10	Major	No
Replace Air Conditioner, Rooftop, 20 Ton	Each	Plumber	Contract	15.7000	21,260.85	4	2T Lift/45T Crane	15	Major	Yes
Air Conditioner, Rooftop, 50 Ton										
Maintain Air Conditioner, Rooftop, 50 Ton	Each	Maint HVAC Tech	Staff	9.1000	75.58	0		1	PM	No
Repair Air Conditioner, Rooftop, 50 Ton	Each	Plumber	Contract	18.0000	23,326.48	0		10	Major	No
Replace Air Conditioner, Rooftop, 50 Ton	Each	Plumber	Contract	36.7000	36,375.19	4	2T Lift/45T Crane	15	Major	Yes
Air Conditioner, Window, 1 Ton										
Maintain Air Conditioner, Window, 1 Ton	Each	Maint HVAC Tech	Staff	1.3000	20.15	0		1	PM	No
Replace Air Conditioner, Window, 1 Ton	Each	Plumber	Contract	4.6000	549.15	0		10	Major	Yes
Air Conditioner, Window, 2 Ton										
Maintain Air Conditioner, Window, 2 Ton	Each	Maint HVAC Tech	Staff	1.3000	20.15	0		1	PM	No
Replace Air Conditioner, Window, 2 Ton	Each	Plumber	Contract	5.0000	1,133.58	0		10	Major	Yes
Air Conditioner, Indoor DX Packaged, Air Cooled, 20 Ton										
Maintain Air Conditioner, Indoor DX Packaged, Air Cooled, 20 Ton	Each	Maint HVAC Tech	Staff	7.1000	52.40	0		1	PM	No
Repair Air Conditioner, Indoor DX Packaged, Air Cooled, 20 Ton	Each	Plumber	Contract	12.5000	5,155.00	0		10	Major	No
Replace Air Conditioner, Indoor DX Packaged, Air Cooled, 20 Ton	Each	Plumber	Contract	15.7000	14,318.32	4	2T Lift/45T Crane	20	Major	Yes
Air Conditioner, Indoor DX Packaged, Air Cooled, 5 Ton										
Maintain Air Conditioner, Indoor DX Packaged, Air Cooled, 5 Ton	Each	Maint HVAC Tech	Staff	2.8000	52.40	0		1	PM	No
Repair Air Conditioner, Indoor DX Packaged, Air Cooled, 5 Ton	Each	Plumber	Contract	12.5000	1,985.02	0		10	Major	No
Replace Air Conditioner, Indoor DX Packaged, Air Cooled, 5 Ton	Each	Plumber	Contract	6.4500	6,121.32	0		20	Major	Yes

*Task frequency for Washington D.C.

Components ordered by UNIFORMAT II Classifications. All costs expressed in ($) 2004.

Sources: U.S. Army Corps of Engineers, Department of Defense Engineering Performance Standards, and Whitestone Research interviews.

5.1 Building Components and Repair Tasks

Uniformat / Component / Repair Task	Units	Trade	Labor Type	Labor Hours	Material Costs	Equip Hours	Equipment Type	Freq. Yrs.*	Task Type	Repl. Task
Air Conditioner, Indoor DX Packaged, Air Cooled, 50 Ton										
Maintain Air Conditioner, Indoor DX Packaged, Air Cooled, 50 Ton	Each	Maint HVAC Tech	Staff	9.1000	75.58	0		1	PM	No
Repair Air Conditioner, Indoor DX Packaged, Air Cooled, 50 Ton	Each	Plumber	Contract	12.5000	12,595.29	0		10	Major	No
Replace Air Conditioner, Indoor DX Packaged, Air Cooled, 50 Ton	Each	Plumber	Contract	36.7000	36,375.19	4	2T Lift/45T Crane	20	Major	Yes
Baseboard Radiator 10 ft.										
Replace Baseboard Radiator, 10 ft.	Each	Plumber	Contract	4.0000	221.68	0		20	Major	Yes
Electric Cabinet Heater										
Inspect & Maintain Electric Cabinet Heater	Each	Maint Electrician	Staff	1.0000	.00	0		1	PM	No
Replace Electric Cabinet Heater	Each	Electrician	Contract	3.3340	1,495.31	0		15	Major	Yes
Explosion Proof Suspended Heater										
Inspect & Maintain Explosion Proof Suspended Heater	Each	Maint Electrician	Staff	1.6920	.00	0		1	PM	No
Replace Contactor, Explosion Proof Suspended Heater	Each	Maint Electrician	Staff	.4620	70.50	0		2	PM	No
Replace Explosion Proof Suspended Heater	Each	Electrician	Contract	2.4999	1,926.41	0		15	Major	Yes
Finned Radiator										
Replace Finned Radiator, 10 ft.	Each	Plumber	Contract	4.0000	347.63	0		20	Major	Yes
Radiant Heating Panel, Electric, 8 sf										
Maintain Radiant Heating Panel, Electric, 8 sf	Each	Maint Electrician	Staff	1.6920	.00	0		1	PM	No
Repair Wiring, Radiant Heating Panel, Electric, 8 sf	Each	Maint Electrician	Staff	.4620	.00	0		2	PM	No
Replace Radiant Heating Panel, Electric, 8 sf	Each	Electrician	Contract	1.9230	195.00	0		15	Major	Yes
Fan Coil, Two-Pipe, 400 Cfm										
Maintain Fan Coil, Two-Pipe, 400 Cfm	Each	Maint HVAC Tech	Staff	.8000	22.17	0		1	PM	No
Repair Fan Coil, Two-Pipe, 400 Cfm	Each	Plumber	Contract	1.2000	140.00	0		10	Major	No
Replace Fan Coil, Two-Pipe, 400 Cfm	Each	Plumber	Contract	2.0000	927.02	0		15	Major	Yes
Fan Coil, Two-Pipe, 1800 Cfm										
Maintain Fan Coil, Two-Pipe, 1800 Cfm	Each	Maint HVAC Tech	Staff	3.3000	43.72	0		1	PM	No
Repair Fan Coil, Two-Pipe, 1800 Cfm	Each	Plumber	Contract	3.5000	194.00	0		10	Major	No
Replace Fan Coil, Two-Pipe, 1800 Cfm	Each	Plumber	Contract	7.7920	2,176.59	0		15	Major	Yes
Heat Pump, 1 Ton										
Maintain Heat Pump, 1 Ton	Each	Maint HVAC Tech	Staff	2.8000	33.25	0		1	PM	No
Repair Heat Pump, 1 Ton	Each	Plumber	Contract	13.6000	677.06	0		10	Major	No
Replace Heat Pump, 1 Ton	Each	Plumber	Contract	4.2600	1,234.34	0		20	Major	Yes

*Task frequency for Washington D.C.

Components ordered by UNIFORMAT II Classifications. All costs expressed in ($) 2004.

Sources: U.S. Army Corps of Engineers, Department of D[illegible]

Uniformat / Component / Repair Task	Units	Trade	Labor Type	Labor Hours	Material Costs	Equip Hours	Equipment Type	Freq. Yrs.*	Task Type	Repl. Task
Heat Pump, 5 Ton										
Maintain Heat Pump, 5 Ton	Each	Maint HVAC Tech	Staff	2.8000	33.25	0		1	PM	No
Repair Heat Pump, 5 Ton	Each	Plumber	Contract	23.4000	1,103.04	0		10	Major	No
Replace Heat Pump, 5 Ton	Each	Plumber	Contract	8.3850	4,307.59	0		20	Major	Yes
Heat Pump, 10 Ton										
Maintain Heat Pump, 10 Ton	Each	Maint HVAC Tech	Staff	7.1000	53.41	0		1	PM	No
Repair Heat Pump, 10 Ton	Each	Plumber	Contract	16.0000	2,046.08	0		10	Major	No
Replace Heat Pump, 10 Ton	Each	Plumber	Contract	11.0000	11,184.62	4	2T Lift/45T Crane	20	Major	Yes
Heat Pump, 25 Ton										
Maintain Heat Pump, 25 Ton	Each	Maint HVAC Tech	Staff	9.1000	53.41	0		1	PM	No
Repair Heat Pump, 25 Ton	Each	Plumber	Contract	16.0000	4,569.15	0		10	Major	No
Replace Heat Pump, 25 Ton	Each	Plumber	Contract	15.0000	23,578.38	4	2T Lift/45T Crane	20	Major	Yes
Suspended Heater, Hot Water, 1200 Cfm, 55 Mbh										
Maintain Suspended Heater, Hot Water, 1200 Cfm, 55 Mbh	Each	Maint HVAC Tech	Staff	1.4000	22.17	0		1	PM	No
Repair Suspended Heater, Hot Water, 1200 Cfm, 55 Mbh	Each	Plumber	Contract	.8000	176.34	0		10	Major	No
Replace Suspended Heater, Hot Water, 1200 Cfm, 55 Mbh	Each	Plumber	Contract	2.5000	975.71	0		15	Major	Yes
Suspended Heater, Hot Water, 4000 Cfm, 180 Mbh										
Maintain Suspended Heater, Hot Water, 4000 Cfm, 180 Mbh	Each	Maint HVAC Tech	Staff	1.4000	22.17	0		1	PM	No
Repair Suspended Heater, Hot Water, 4000 Cfm, 180 Mbh	Each	Plumber	Contract	.8000	179.36	0		10	Major	No
Replace Suspended Heater, Hot Water, 4000 Cfm, 180 Mbh	Each	Plumber	Contract	2.5000	1,486.24	0		15	Major	Yes
Suspended Heater, Hot Water, 8000 Cfm, 360 Mbh										
Maintain Suspended Heater, Hot Water, 8000 Cfm, 360 Mbh	Each	Maint HVAC Tech	Staff	1.4000	22.17	0		1	PM	No
Repair Suspended Heater, Hot Water, 8000 Cfm, 360 Mbh	Each	Plumber	Contract	.8000	179.36	0		10	Major	No
Replace Suspended Heater, Hot Water, 8000 Cfm, 360 Mbh	Each	Plumber	Contract	2.5000	2,090.82	0		15	Major	Yes
Secondary Coil, 30" x 72"										
Replace Secondary Coil, 30" x 72"	Each	Plumber	Contract	24.0000	1,175.00	0		15	Major	Yes
Terminal Reheat Unit, 10,000 Cfm										
Maintain Terminal Reheat Unit, 10,000 Cfm	Each	Maint HVAC Tech	Staff	5.4000	47.50	0		1	PM	No
Repair Terminal Reheat Unit, 10,000 Cfm	Each	Plumber	Contract	1.2000	.00	0		10	Major	No
Replace Terminal Reheat Unit, 10,000 Cfm	Each	Plumber	Contract	29.0000	695.26	0		15	Major	Yes

*Task frequency for Washington D.C.
Components ordered by UNIFORMAT II Classifications. All costs expressed in ($) 2004.
Sources: U.S. Army Corps of Engineers, Department of Defense Engineering Performance Standards, and Whitestone Research interviews.

5.1 Building Components and Repair Tasks

Uniformat / Component / Repair Task	Units	Trade	Labor Type	Labor Hours	Material Costs	Equip Hours	Equipment Type	Freq. Yrs.*	Task Type	Repl. Task
Terminal Reheat Unit, 25,000 Cfm										
Maintain Terminal Reheat Unit, 25,000 Cfm	Each	Maint HVAC Tech	Staff	5.7000	117.74	0		1	PM	No
Repair Terminal Reheat Unit, 25,000 Cfm	Each	Plumber	Contract	1.2000	.00	0		10	Major	No
Replace Terminal Reheat Unit, 25,000 Cfm	Each	Plumber	Contract	29.0000	1,225.00	0		15	Major	Yes
D3060 Controls & Instrumentation										
Direct Digital Controls, System Points										
Annual PM, Direct Digital Controls, System Points	Each	Maint HVAC Tech	Staff	.1500	32.00	0		1	PM	No
Replace Direct Digital Controls, System Points	Each	Electrician	Contract	.0000	353.73	0		10	Major	Yes
Pneumatic Thermostat										
Annual PM, Pneumatic Thermostat	Each	Maint HVAC Tech	Staff	1.0000	.00	0		1	PM	No
Replace Pneumatic Thermostat	Each	Electrician	Contract	6.6000	560.00	0		10	Major	Yes
D40 Fire Protection										
D4010 Sprinklers										
Fire Standpipe System										
Fire Standpipe System, Flow & Valve Check	Each	Plumber	Contract	8.0000	.00	0		5	Major	No
Fire Sprinkler System										
Fire Sprinkler System, Annual PM	Each	Plumber	Contract	27.0000	.00	0		1	Major	No
Fire Sprinkler System, Test Gages & Valves	Each	Plumber	Contract	11.0000	.00	0		5	Major	No
Fire Sprinkler Head										
Inspect & Test, Sprinkler Head	Each	Maint Plumber	Staff	.0103	.00	0		1	PM	No
Repair Fire Sprinkler Head	Each	Plumber	Contract	.5000	.00	0		10	Major	No
Replace Fire Sprinkler Head	Each	Plumber	Contract	.3250	5.45	0		20	Major	Yes
Fire Suppression System Water Pump										
Annual PM, Fire Suppression System Water Pump	Each	Maint Plumber	Staff	4.0000	86.64	0		1	PM	No
Annual PM, Fire Suppression System Water Pump Testing	Each	Maint Plumber	Staff	3.0000	86.64	0		1	PM	No
Annual PM, Fire Suppression System Water Pump Motor	Each	Maint Plumber	Staff	5.0000	86.64	0		1	PM	No
Kitchen Fire Suppression System (CO2)										
Annual PM, Kitchen Fire Suppression System (CO2)	Each	Maint Plumber	Staff	1.0000	.00	0		1	PM	No
Replace Kitchen Fire Supression System Extinguisher (CO2)	Each			2.6670	968.33	0		10		No
D4030 Fire Protection Specialties										

*Task frequency for Washington D.C.

Components ordered by UNIFORMAT II Classifications. All costs expressed in ($) 2004.

Sources: U.S. Army Corps of Engineers, Department of Defense Engineering Performance Standards, and Whitestone Research.

5.1 Building Components and Repair Tasks

Uniformat / Component / Repair Task	Units	Trade	Labor Type	Labor Hours	Material Costs	Equip Hours	Equipment Type	Freq. Yrs.*	Task Type	Repl. Task
Fire Damper										
Annual PM, Fire Damper	Each	Gen Maint Worker	Staff	1.0000	.00	0		1	PM	No
Fire Shutter, Motor-Operated										
Maintain Fire Shutter, Motor-Operated	Each	Gen Maint Worker	Staff	2.0000	.00	0		1	PM	No
D50 Electrical										
D5010 Electrical Service & Distribution										
Circuit Breaker, 600v, 125-400 Amp										
Annual PM, Circuit Breaker, 600 V, 125-400 Amp., 3 Ph.	Each	Maint Electrician	Staff	.4000	.00	0		1	PM	No
Repair Circuit Breaker, 600 V, 125-400 Amp., 3 Ph.	Each	Electrician	Contract	2.1000	26.73	0		10	Major	No
Replace Circuit Breaker, 600 V, 125-400 Amp., 3 Ph.	Each	Electrician	Contract	5.3000	1,120.64	0		200	Major	Yes
Circuit Breaker, Main, 600 V, 1200 Amp., 1 Ph.										
Annual PM, Circuit Breaker, Main, 600 V, 1200 Amp., 1 Ph.	Each	Maint Electrician	Staff	.4000	.00	0		1	PM	No
Repair Circuit Breaker, Main, 600 V, 1200 Amp., 1 Ph.	Each	Electrician	Contract	2.1000	89.15	0		10	Major	No
Replace Circuit Breaker, Main, 600 V, 1200 Amp., 1 Ph.	Each	Electrician	Contract	8.8890	4,731.57	0		200	Major	Yes
Circuit Breaker, Main, 600 V, 1600 Amp., 1 Ph.										
Annual PM, Circuit Breaker, Main, 600 V, 1600 Amp., 1 Ph.	Each	Maint Electrician	Staff	.9300	.00	0		1	PM	No
Repair Circuit Breaker, Main, 600 V, 1600 Amp., 1 Ph.	Each	Electrician	Contract	1.6150	89.25	0		10	Major	No
Replace Circuit Breaker, Main, 600 V, 1600 Amp., 1 Ph.	Each	Electrician	Contract	19.0480	9,338.63	0		200	Major	Yes
Circuit Breaker, Main, 240 V, 15-60 Amp., 3 Ph.										
Annual PM, Circuit Breaker, 240 V, 15-60 Amp., 3 Ph.	Each	Maint Electrician	Staff	.5000	.00	0		1	PM	No
Repair Circuit Breaker, 240 V, 15-60 Amp., 3 Ph.	Each	Electrician	Contract	.4000	4.14	0		10	Major	No
Replace Circuit Breaker, 240 V, 15-60 Amp., 3 Ph.	Each	Electrician	Contract	2.9600	293.00	0		200	Major	Yes
Circuit Breaker, 600 V, 70-100 Amp., 3 Ph.										
Annual PM, Circuit Breaker, 600 V, 70-100 Amp., 3 Ph.	Each	Maint Electrician	Staff	.9300	.00	0		1	PM	No
Repair Circuit Breaker, 600 V, 70-100 Amp., 3 Ph.	Each	Electrician	Contract	2.1000	8.32	0		10	Major	No
Replace Circuit Breaker, 600 V, 70-100 Amp., 3 Ph.	Each	Electrician	Contract	2.2800	348.64	0		200	Major	Yes
Circuit Breaker, Main, 600 V, 800 Amp., 3 Ph.										
Annual PM, Circuit Breaker, Main, 600 V, 800 Amp., 3 Ph.	Each	Maint Electrician	Staff	.9300	.00	0		1	PM	No
Repair Circuit Breaker, Main, 600 V, 800 Amp., 3 Ph.	Each	Electrician	Contract	1.6150	83.47	0		10	Major	No
Replace Circuit Breaker, Main, 600 V, 800 Amp., 3 Ph.	Each	Electrician	Contract	.7480	3,500.00	0		200	Major	Yes

*Task frequency for Washington D.C.
Components ordered by UNIFORMAT II Classifications. All costs expressed in ($) 2004.
Sources: U.S. Army Corps of Engineers, Department of Defense Engineering Performance Standards, and Whitestone Research interviews.

5.1 Building Components and Repair Tasks

Uniformat / Component / Repair Task	Units	Trade	Labor Type	Labor Hours	Material Costs	Equip Hours	Equipment Type	Freq. Yrs.*	Task Type	Repl. Task
Distribution Switch, Fused, <600 V										
Maintain Distribution Switch, Fused, <600 V	Each	Maint Electrician	Staff	.0100	.00	0		3	PM	No
Repair Distribution Switch, Fused, <600 V	Each	Electrician	Contract	.1040	3.92	0		10	Major	No
Replace Distribution Switch, Fused, <600 V	Each	Electrician	Contract	.8600	1,725.00	0		30	Major	Yes
Safety Switch, Fused, 400 Amp., 3 Ph.										
Annual PM, Safety Switch, Fused, 400 Amp., 3 Ph.	Each	Maint Electrician	Staff	.6100	.00	0		1	PM	No
Repair Safety Switch, Fused, 400 Amp., 3 Ph.	Each	Electrician	Contract	.3850	.00	0		8	Major	No
Replace Safety Switch, Fused, 400 Amp., 3 Ph.	Each	Electrician	Contract	.4350	2,050.00	0		25	Major	Yes
Transfer Switch, Auto, 600 V, 400 Amp.										
Inspect & Clean Transfer Switch, Auto, 600 V, 400 Amp.	Each	Maint Electrician	Staff	.3200	.00	0		1	PM	No
Replace Coil, Transfer Switch, Auto, 600 V, 400 Amp.	Each	Maint Electrician	Staff	2.0000	37.00	0		5	PM	No
Replace Transfer Switch, Auto, 600 V, 400 Amp.	Each	Electrician	Contract	10.0000	6,284.50	0		18	Major	Yes
Transfer Switch, Auto, 600 V, 800 Amp.										
Inspect & Clean Transfer Switch, Auto, 600 V, 800 Amp.	Each	Maint Electrician	Staff	.3200	.00	0		1	PM	No
Repair Transfer Switch, Auto, 600 V, 800 Amp.	Each	Maint Electrician	Staff	2.0000	46.50	0		5	PM	No
Replace Transfer Switch, Auto, 600 V, 800 Amp.	Each	Electrician	Contract	24.0000	9,450.00			18	Major	Yes
Distribution Panel Board										
Annual PM, Distribution Panel Board	Each	Maint Electrician	Staff	1.0000	.00	0		1	PM	No
Primary Transformer, Dry, >500 KVA										
Maintain Primary Transformer, Dry, >500 KVA	Each	Maint Electrician	Staff	.5916	.00	0		1	PM	No
Repair Primary Transformer, Dry, >500 KVA	Each	Electrician	Contract	1.5000	37.00	0		10	Major	No
Replace Primary Transformer, Dry, >500 KVA	Each	Electrician	Contract	23.8000	29,818.00	4	2T Lift/45T Crane	30	Major	Yes
Primary Transformer, Oil Filled, <500 KVA										
Maintain Primary Transformer, Oil Filled, <500 KVA	Each	Maint Electrician	Staff	1.0000	.00	0		1	PM	No
Repair Primary Transformer, Oil Filled, <500 KVA	Each	Electrician	Contract	3.0000	2,216.37	0		10	Major	No
Replace Primary Transformer, Oil Filled, <500 KVA	Each	Electrician	Contract	12.2300	12,600.00	4	2T Lift/45T Crane	30	Major	Yes
Primary Transformer, Dry, <500 KVA										
Maintain Primary Transformer, Dry, <500 KVA	Each	Maint Electrician	Staff	.5916	.00	0		1	PM	No
Repair Primary Transformer, Dry, <500 KVA	Each	Electrician	Contract	1.5000	37.00	0		10	Major	No
Replace Primary Transformer, Dry, <500 KVA	Each	Electrician	Contract	23.8000	26,200.00	4	2T Lift/45T Crane	30	Major	Yes

*Task frequency for Washington D.C.

Components ordered by UNIFORMAT II Classifications. All costs expressed in ($) 2004.

Sources: U.S. Army Corps of Engineers, Department of Defense Engi[illegible]

5.1 Building Components and Repair Tasks

Uniformat / Component / Repair Task	Units	Trade	Labor Type	Labor Hours	Material Costs	Equip Hours	Equipment Type	Freq. Yrs.*	Task Type	Repl. Task
Primary Transformer, Oil Filled, >500 KVA										
Maintain Primary Transformer, Oil Filled, >500 KVA	Each	Maint Electrician	Staff	.5916	.00	0		1	PM	No
Repair Primary Transformer, Oil Filled, >500 KVA	Each	Electrician	Contract	3.0000	2,225.00	0		10	Major	No
Replace Primary Transformer, Oil Filled, >500 KVA	Each	Electrician	Contract	12.2300	15,060.42	4	2T Lift/45T Crane	30	Major	Yes
Secondary Transformer, Dry, 50 KVA										
Maintain Secondary Transformer, Dry, 50 KVA	Each	Maint Electrician	Staff	.0600	.00	0		1	PM	No
Repair Secondary Transformer, Dry, 50 KVA	Each	Electrician	Contract	1.0000	112.00	0		10	Major	No
Replace Secondary Transformer, Dry, 50 KVA	Each	Electrician	Contract	2.0000	2,189.72	0	2T Forklift	30	Major	Yes
Secondary Transformer, Dry, 150 KVA										
Maintain Secondary Transformer, Dry, 150 KVA	Each	Maint Electrician	Staff	.0600	.00	0		1	PM	No
Repair Secondary Transformer, Dry, 150 KVA	Each	Electrician	Contract	1.0000	112.00	0		10	Major	No
Replace Secondary Transformer, Dry, 150 KVA	Each	Electrician	Contract	2.0000	3,795.52	0	2T Forklift	30	Major	Yes
Secondary Transformer, Oil Filled, >600 V										
Maintain Secondary Transformer, Oil Filled, >600 V	Each	Maint Electrician	Staff	1.0000	.00	0		1	PM	No
Repair Secondary Transformer, Oil Filled, >600 V	Each	Electrician	Contract	1.0000	37.00	0		25	Major	No
Replace Secondary Transformer, Oil Filled, >600 V	Each	Electrician	Contract	2.0000	5,200.00	0		50	Major	Yes
Main Switchgear, >1200 Amp.										
Maintain Main Switchgear, >1200 Amp.	Each	Maint Electrician	Staff	.6000	.00	0		1	PM	No
Repair Main Switchgear, >1200 Amp.	Each	Maint Electrician	Staff	5.1500	500.00	0		5	PM	No
Replace Main Switchgear, >1200 Amp.	Each	Electrician	Contract	1.5000	3,412.07	0		20	Major	Yes
Main Switchgear, <1200 Amp.										
Maintain Main Switchgear, <1200 Amp.	Each	Maint Electrician	Staff	.6000	.00	0		1	PM	No
Repair Main Switchgear, <1200 Amp.	Each	Maint Electrician	Staff	5.1500	500.00	0		5	PM	No
Replace Main Switchgear, <1200 Amp.	Each	Electrician	Contract	1.5000	2,050.00	0		20	Major	Yes
Motor Control Center										
Annual PM, Motor Control Center	Each	Maint Electrician	Staff	1.5000	.00	0		1	PM	No
Motor Starter, 5-20 HP, <600 V										
Inspect & Clean Motor Starter, 5-20 HP, <600 V	Each	Maint Electrician	Staff	1.0000	.00	0		1	PM	No
Replace Coil, Motor Starter, 5-20 HP, <600 V	Each	Maint Electrician	Staff	1.5000	38.50	0		5	PM	No
Replace Motor Starter, 5-20 HP, <600 V	Each	Electrician	Contract	4.0000	201.46	0		18	Major	Yes

*Task frequency for Washington D.C.
Components ordered by UNIFORMAT II Classifications. All costs expressed in ($) 2004.
Sources: U.S. Army Corps of Engineers, Department of Defense Engineering Performance Standards, and Whitestone Research interviews.

5.1 Building Components and Repair Tasks

Uniformat / Component / Repair Task	Units	Trade	Labor Type	Labor Hours	Material Costs	Equip Hours	Equipment Type	Freq. Yrs.*	Task Type	Repl. Task
Motor Starter, 21-50 HP, <600 V										
Inspect & Clean Motor Starter, 21-50 HP, <600 V	Each	Maint Electrician	Staff	1.0000	.00	0		1	PM	No
Replace Coil, Motor Starter, 21-50 HP, <600 V	Each	Maint Electrician	Staff	1.5000	32.67	0		5	PM	No
Replace Motor Starter, 21-50 HP, <600 V	Each	Electrician	Contract	8.0000	720.00	0		18	Major	Yes
Motor Starter, 50-75 HP, <600 V										
Inspect & Clean Motor Starter, 51-75 HP, <600 V	Each	Maint Electrician	Staff	1.0000	.00	0		1	PM	No
Replace Coil, Motor Starter, 51-75 HP, <600 V	Each	Maint Electrician	Staff	1.5000	115.00	0		5	PM	No
Replace Motor Starter, 51-75 HP, <600 V	Each	Electrician	Contract	12.0000	1,600.00	0		18	Major	Yes
D5020 Lighting & Branch Wiring										
Exit Lighting Fixture, w/ Battery										
Replace Lamp, Exit Lighting Fixture, w/ Battery	Each	Maint Electrician	Staff	.0528	34.00	0		5	PM	No
Replace Exit Lighting Fixture, w/ Battery	Each	Electrician	Contract	1.3000	107.00	0		20	Major	Yes
Emergency Horn & Strobe										
Annual PM, Emergency Horn & Strobe	Each	Maint Electrician	Staff	.1500	.00	0		1	PM	No
Replace Emergency Horn & Strobe	Each	Electrician	Contract	1.5000	95.00	0		20	Major	Yes
Emergency Lighting Pack, 2 Light w/ Battery										
Replace Lamp, Replace Emergency Lighting Pack, 2 Light w/ Battery	Each	Maint Electrician	Staff	.0528	23.00	0		2	PM	No
Replace Lens, Replace Emergency Lighting Pack, 2 Light w/ Battery	Each	Electrician	Contract	.3000	8.05	0		10	Major	No
Replace Emergency Lighting Pack, 2 Light w/ Battery	Each	Electrician	Contract	2.5000	550.00	0		20	Major	Yes
Light Dimming Panel										
Light Dimming Panel, Maint & Inspection	Each	Maint Electrician	Staff	.5330	.00	0		1	PM	No
Light Dimming Panel, Replace Dimmer Switch	Each	Maint Electrician	Staff	1.2890	49.00	0		5	PM	No
Replace Light Dimming Panel	Each	Electrician	Contract	9.8000	435.00	0		15	Major	Yes
Incandescent Lighting Fixture, Basic, 100w										
Replace Lamp, Incandescent Lighting Fixture, Basic, 100w	Each	Maint Electrician	Staff	.0528	2.01	0		5	PM	No
Replace Incandescent Lighting Fixture, Basic, 100w	Each	Electrician	Contract	.4485	46.00	0		20	Major	Yes
Incandescent Lighting Fixture, Basic, 200w										
Replace Lamp, Incandescent Lighting Fixture, Basic, 200w	Each	Maint Electrician	Staff	.0528	2.01	0		5	PM	No
Replace Incandescent Lighting Fixture, Basic, 200w	Each	Electrician	Contract	.4485	46.00	0		20	Major	Yes
Fluorescent Lighting Fixture, 80w										
Replace Ballast & Lamps, Fluorescent Lighting Fixture, 80w	Each	Electrician	Contract	2.0000	50.30	0		10	Major	Yes
Replace Fluorescent Lighting Fixture, 80w	Each	Electrician	Contract	1.3000	38.35	0		20	Major	Yes

*Task frequency for Washington D.C.

Components ordered by UNIFORMAT II Classifications. All costs expressed in ($) 2004.

Sources: U.S. Army Corps of Engineers, Department of Defence Engineering [illegible]

5.1 Building Components and Repair Tasks

Uniformat / Component / Repair Task	Units	Trade	Labor Type	Labor Hours	Material Costs	Equip Hours	Equipment Type	Freq. Yrs.*	Task Type	Repl. Task
Fluorescent Lighting Fixture, 160w										
Replace Ballast & Lamps, Fluorescent Lighting Fixture, 160w	Each	Electrician	Contract	.6757	48.00	0		10	Major	No
Replace Fluorescent Lighting Fixture, 160w	Each	Electrician	Contract	.5700	38.35	0		20	Major	Yes
Metal Halide Lighting Fixture, Wall Mount, 150w										
Replace Lamp, Metal Halide Lighting Fixture, Wall Mount, 150w	Each	Maint Electrician	Staff	.0528	52.30	0		5	PM	No
Replace Ballast, Metal Halide Lighting Fixture, Wall Mount, 150w	Each	Electrician	Contract	.4734	71.00	0		10	Major	No
Replace Metal Halide Lighting Fixture, Wall Mount, 150w	Each	Electrician	Contract	.3350	330.00	0		20	Major	Yes
Metal Halide Lighting Fixture, Low Bay, 250w										
Replace Lamp, Metal Halide Lighting Fixture, Low Bay, 250w	Each	Maint Electrician	Staff	.0528	44.83	0		5	PM	No
Replace Ballast, Metal Halide Lighting Fixture, Low Bay, 250w	Each	Electrician	Contract	.4734	91.64	0		10	Major	No
Replace Metal Halide Lighting Fixture, Low Bay, 250w	Each	Electrician	Contract	.3350	256.00	0		20	Major	Yes
Metal Halide Lighting Fixture, High Bay, 400w										
Replace Lamp, Metal Halide Lighting Fixture, High Bay, 400w	Each	Maint Electrician	Staff	.0528	47.81	0		5	PM	No
Replace Ballast, Metal Halide Lighting Fixture, High Bay, 400w	Each	Electrician	Contract	.4734	113.56	0		10	Major	No
Replace Metal Halide Lighting Fixture, High Bay, 400w	Each	Electrician	Contract	.3350	297.84	0		20	Major	Yes
Metal Halide Lighting Fixture, High Bay, 1000w										
Replace Lamp, Metal Halide Lighting Fixture, High Bay, 1000w	Each	Maint Electrician	Staff	.0528	137.46	0		5	PM	No
Replace Ballast, Metal Halide Lighting Fixture, High Bay, 1000w	Each	Electrician	Contract	.4734	190.75	0		10	Major	No
Replace Metal Halide Lighting Fixture, High Bay, 1000w	Each	Electrician	Contract	.3350	508.02	0		20	Major	Yes
High Pressure Sodium Lighting Fixture, 250w										
Replace Ballast & Lamp, HP Sodium Lighting Fixture, 250w	Each	Electrician	Contract	.6276	253.01	0		10	Major	No
Replace HP Sodium Lighting Fixture, 250w	Each	Electrician	Contract	.9720	535.00	0		20	Major	Yes
Wiring Device, Switch										
Repair Wiring Device, Switch	Each	Electrician	Contract	.3850	1.80	0		10	Major	No
Replace Wiring Device, Switch	Each	Electrician	Contract	.3600	7.50	0		15	Major	Yes
Grounded Fault Circuit Interruptor, 15 Amps										
Replace Receptacle	Each	Electrician	Contract	.3600	23.42	0		20	Major	Yes
Receptacle, 120V, 15 Amp										
Replace Receptacle	Each	Electrician	Contract	.3600	7.05	0		20	Major	Yes
Receptacle, 120v, 20 Amp										
Replace Receptacle	Each	Electrician	Contract	.3600	9.24	0		20	Major	Yes

*Task frequency for Washington D.C.
Components ordered by UNIFORMAT II Classifications. All costs expressed in ($) 2004.
Sources: U.S. Army Corps of Engineers, Department of Defense Engineering Performance Standards, and Whitestone Research interviews.

5.1 Building Components and Repair Tasks

Uniformat / Component / Repair Task	Units	Trade	Labor Type	Labor Hours	Material Costs	Equip Hours	Equipment Type	Freq. Yrs.*	Task Type	Repl. Task
Receptacle, 208v, 3 phase										
Replace Receptacle	Each	Electrician	Contract	.3600	35.36	0		20	Major	Yes
Welding Receptacle										
Clean and Inspect Welding Receptacle	Each	Maint Electrician	Staff	.2500	.00	0		1	PM	No
Replace Welding Receptacle	Each	Electrician	Contract	2.0000	291.96	0		30	Major	Yes
TV Cable Outlet										
Repair TV Cable Outlet	Each	Electrician	Contract	.4620	1.98	0		10	Major	No
Replace TV Cable Outlet	Each	Electrician	Contract	.6500	14.00	0		20	Major	Yes
D5030 Communications & Security										
Annunciation Panel										
Inspect & Maintain Annunciation Panel	Each	Maint Electrician	Staff	.8000	.00	0		1	PM	No
Replace Annunciation Panel	Each	Electrician	Contract	1.3500	300.00	0		15	Major	Yes
Nurse Call Station										
Annual PM, Nurse Call Station	Each	Maint Electrician	Staff	.2150	.00	0		1	PM	No
Replace Nurse Call Station	Each	Electrician	Contract	1.3200	178.00	0		15	Major	Yes
Fire Alarm Bell, 6"										
Replace Fire Alarm Bell, 6"	Each	Electrician	Contract	.2200	49.50	0		20	Major	Yes
Fire Alarm Control Panel										
Inspect & Maintain Fire Alarm Control Panel	Each	Maint Electrician	Staff	.8000	.00	0		1	PM	No
Minor Repair, Fire Alarm Control Panel	Each	Maint Electrician	Staff	.3140	26.00	0		5	PM	No
Replace Fire Alarm Control Panel	Each	Electrician	Contract	1.3500	1,400.00	0		15	Major	Yes
Manual Pull Station										
Check & Repair Manual Pull Station	Each	Electrician	Contract	.4000	17.55	0		10	Major	No
Replace Manual Pull Station	Each	Electrician	Contract	.2000	37.00	0		15	Major	Yes
Heat Detector										
Check Operation, Heat Detector	Each	Maint Electrician	Staff	.1500	.00	0		1	PM	No
Repair Heat Detector	Each	Electrician	Contract	.4000	6.43	0		10	Major	No
Replace Heat Detector	Each	Electrician	Contract	.3100	28.00	0		15	Major	Yes

*Task frequency for Washington D.C.
Components ordered by UNIFORMAT II Classifications. All costs expressed in ($) 2004.
Sources: U.S. Army Corps of Engineers, Department of Defense Engineering Performance Standards [illegible]

5.1 Building Components and Repair Tasks

Uniformat / Component / Repair Task	Units	Trade	Labor Type	Labor Hours	Material Costs	Equip Hours	Equipment Type	Freq. Yrs.*	Task Type	Repl. Task
Smoke Detector										
Check Operation, Smoke Detector	Each	Maint Electrician	Staff	.1500	.00	0		1	PM	No
Repair Smoke Detector	Each	Electrician	Contract	.4000	3.42	0		10	Major	No
Replace Smoke Detector	Each	Electrician	Contract	.3100	75.00	0		15	Major	Yes
Security System Panel										
Replace Standby Batteries	Each	Maint Electrician	Staff	1.0000	20.00	0		4	PM	No
Replace Security System Panel	Each	Electrician	Contract	1.3550	330.00	0		10	Major	Yes
Closed Circuit Surveillance Camera										
Maintain, Closed Circuit Surveillance Camera	Camera	Maint Electrician	Staff	.2500	87.93	0		1	PM	No
Repair, Closed Circuit Surveillance Camera	Camera	Electrician	Contract	2.9000	480.00	0		15	Major	Yes
Replace Closed Circuit Surveillance Camera	Camera	Electrician	Contract	6.1540	1,494.18	0		20	Major	Yes
Security Camera System										
Annual PM, Security Camera System	Camera	Maint Electrician	Staff	.2500	.00	0		1	PM	No
Replace Security Camera System	Camera	Electrician	Contract	2.9000	985.00	0		15	Major	Yes
Sound System										
Annual PM, Sound System	Each	Maint Electrician	Staff	.2150	.00	0		1	PM	No
Replace Sound System	Each	Electrician	Contract	1.4500	1,475.00	0		15	Major	Yes
Public Address Speaker										
Annual PM, Public Address Speaker	Each	Maint Electrician	Staff	.2150	.00	0		1	PM	No
Replace Public Address Speaker	Each	Electrician	Contract	1.4500	97.50	0		15	Major	Yes
Time Control Clock										
Maintain Time Control Clock	Each	Maint Electrician	Staff	.2500	.00	0		1	PM	No
Repair Time Control Clock	Each	Electrician	Contract	1.0000	16.27	0		10	Major	No
Replace Time Control Clock	Each	Electrician	Contract	1.2000	75.50	0		15	Major	Yes
Chart Recorder										
Annual PM, Chart Recorder	Each	Maint Electrician	Staff	18.0000	.00	0		1	PM	No
D5090 Other Electrical Systems										
Generator Battery, Wet										
Clean Terminals, Check Water & Test Generator Battery, Wet	Each	Maint Electrician	Staff	19.2500	.00	0		1	PM	No
Replace Generator Battery, Wet	Each	Electrician	Contract	.3000	296.00	0		10	Major	Yes

*Task frequency for Washington D.C.
Components ordered by UNIFORMAT II Classifications. All costs expressed in ($) 2004.
Sources: U.S. Army Corps of Engineers, Department of Defense Engineering Performance Standards, and Whitestone Research interviews.

5.1 Building Components and Repair Tasks

Uniformat / Component / Repair Task	Units	Trade	Labor Type	Labor Hours	Material Costs	Equip Hours	Equipment Type	Freq. Yrs.*	Task Type	Repl. Task
Generator, Propane, 30kw										
Electrical Testing, Generator, Propane, 30kw	Each	Maint Electrician	Staff	7.6800	.00	0		1	PM	No
Replace Generator, Propane, 30kw	Each	Electrician	Contract	36.3400	12,000.00	0		25	Major	Yes
Generator, Diesel, 30kw										
Electrical Testing, Generator, Diesel, 30kw	Each	Maint Electrician	Staff	7.6875	.00	0		1	PM	No
Replace Generator, Diesel, 30kw	Each	Electrician	Contract	48.0000	16,000.00	8	2T Lift/45T Crane	25	Major	Yes
Generator, Diesel, 500kw										
Electrical Testing, Generator, Diesel, 500kw	Each	Maint Electrician	Staff	7.6875	.00	0		1	PM	No
Replace Generator, Diesel, 500kw	Each	Electrician	Contract	120.0000	86,500.00	8	4T Lift/90T Crane	25	Major	Yes
Generator, Diesel, 1000kw										
Electrical Testing, Generator, Diesel, 1000kw	Each	Maint Electrician	Staff	7.6875	.00	0		1	PM	No
Replace Generator, Diesel, 1000kw	Each	Electrician	Contract	180.0000	176,000.00	8	4T Lift/90T Crane	25	Major	Yes
Generator Switchgear										
Annual PM, Generator Switchgear	Each	Maint Electrician	Staff	6.0000	.00	0		1	PM	No
Replace Generator Switchgear	Each	Electrician	Contract	.6000	5,887.92	0		18	Major	Yes
Uninterruptable Power Supply										
Annual PM, Uninterruptable Power Supply	Each	Maint Electrician	Staff	3.0000	.00	0		1	PM	No
Replace Uninterruptable Power Supply Battery	Each	Electrician	Contract	.3070	294.85	0		10	Major	No
Replace Uninterruptable Power Supply Motor Generator	Each	Electrician	Contract	4.1000	46,000.00	0		15	Major	Yes
Skylight Shade System Program & Controls										
Annual PM, Skylight Shade System Program & Controls	Each	Maint Electrician	Staff	1.0000	.00	0		1	PM	No
Lightning Protection System										
Maintenance & Repair of General Wiring, Lightning Protection System	Each	Electrician	Contract	1.5000	5.60	0		1	PM	No
Replace Lightning Protection General Wiring	Each	Electrician	Contract	62.1000	1,549.80	0		25	Major	Yes
E10 Equipment										
E1010 Commercial Equipment										
Trash Compactor, 600 lbs										
Annual PM, Trash Compactor, 600 lbs	Each	Maint Electrician	Staff	23.0000	.00	0		1	PM	No
Replace Trash Compactor, 600 lbs	Each	Electrician	Contract	24.0000	19,500.00	0		10	Major	Yes

*Task frequency for Washington D.C.

Components ordered by UNIFORMAT II Classifications. All costs expressed in ($) 2004.

Sources: U.S. Army Corps of Engineers, Departm[illegible]

Uniformat / Component / Repair Task	Units	Trade	Labor Type	Labor Hours	Material Costs	Equip Hours	Equipment Type	Freq. Yrs.*	Task Type	Repl. Task
Commercial Dishwasher										
Annual PM, Commercial Dishwasher	Each	Maint Plumber	Staff	9.0000	.00	0		1	PM	No
Replace Commercial Dishwasher	Each	Plumber	Contract	35.0000	11,077.23	0		10	Major	Yes
Commercial Icemaker										
Annual PM, Commercial Icemaker	Each	Maint Plumber	Staff	14.0000	.00	0		1	PM	No
Replace Commercial Icemaker	Each	Plumber	Contract	16.0000	8,854.32	0		10	Major	Yes
Commercial Walk-In Freezer/Cooler										
Annual PM, Commercial Walk-In Freezer/Cooler	Each	Maint Plumber	Staff	7.0000	.00	0		1	PM	No
Replace Commercial Walk-In Freezer/Cooler	Each	Plumber	Contract	64.0000	9,125.00	0		10	Major	Yes
Commercial Kitchen Exhaust Hood										
Annual PM, Commercial Kitchen Exhaust Hood	Each	Gen Maint Worker	Staff	1.0000	.00	0		1	PM	No
Replace Commercial Kitchen Exhaust Hood	Each	Carpenter	Contract	4.4000	4,675.00	0		15	Major	Yes
E1020 Institutional Equipment										
Detention Doors & Hardware										
Maintain & Repair Detention Doors & Hardware	Each	Carpenter	Contract	.3300	.00	0		1	PM	No
Nourishment Station										
Annual PM, Nourishment Station	Each	Gen Maint Worker	Staff	4.5000	.00	0		1	PM	No
Dust Collection System										
Annual PM, Dust Collection System	Each	Maint Electrician	Staff	2.5000	.00	0		1	PM	No
Replace Dust Collection System	Each	Electrician	Contract	40.0000	3,567.40	0		30	Major	Yes
Mortuary Unit Cooler										
Annual PM, Mortuary Unit Cooler	Each	Maint Plumber	Staff	3.5000	.00	0		1	PM	No
Replace Mortuary Unit Cooler	Each	Plumber	Contract	.0000	21,400.00	0		10	Major	Yes
Laboratory Exhaust Hood										
Annual PM, Laboratory Exhaust Hood	Each	Gen Maint Worker	Staff	1.0000	.00	0		1	PM	No
Replace Laboratory Exhaust Hood	Each	Carpenter	Contract	6.6000	1,225.00	0		15	Major	Yes
Silver Recovery System										
Annual PM, Silver Recovery System	Each	Maint Plumber	Staff	32.0000	.00	0		1	PM	No
Waste Acid Neutralization System										
Annual PM, Waste Acid Neutralization System	Each	Maint Plumber	Staff	8.0000	.00	0		1	PM	No

*Task frequency for Washington D.C.
Components ordered by UNIFORMAT II Classifications. All costs expressed in ($) 2004.
Sources: U.S. Army Corps of Engineers, Department of Defense Engineering Performance Standards, and Whitestone Research interviews.

5.1 Building Components and Repair Tasks

Uniformat / Component / Repair Task	Units	Trade	Labor Type	Labor Hours	Material Costs	Equip Hours	Equipment Type	Freq. Yrs.*	Task Type	Repl. Task
Medical Waste Incinerator										
Replace Medical Waste Incinerator	Each	Plumber	Contract	48.0000	18,100.00	0		10	Major	Yes
G20 Site Improvements										
G2020 Parking Lots										
Parking control equipment, barrier gate & controller										
Replace barrier gate and controller	Each	Electrician	Contract	5.3333	3,050.00	0		10	Major	Yes
Parking control equipment, card reader										
Replace parking control equipment, card reader	Each	Electrician	Contract	4.0000	1,675.00	0		10	Major	Yes
Parking stall										
Paint parking stall	Each	Painter	Contract	.1090	3.34	0		10	Major	No

*Task frequency for Washington D.C.
Components ordered by UNIFORMAT II Classifications. All costs expressed in ($) 2004.
Sources: U.S. Army Corps of Engineers, Department of Defense [illegible]

The frequency of repair tasks for HVAC components is dependent on local climate conditions. These conditions are summarized in terms of 11 climate zones. Climate zone designations for 330 metro areas are listed in table 5-1. Broad climate zones for the U. S. are shown in figure 5.1.

Table 5-1.
Metro Areas by Climate Zone

City, State	Zone
Aberdeen, SD	9
Abilene, TX	3
Akron, OH	7
Alamogordo, NM	4
Alamosa, CO	9
Albany, NY	8
Albuquerque, NM	5
Allentown, PA	7
Alpena, MI	9
Altus, OK	3
Amarillo, TX	5
Anaheim, CA	4
Anchorage, AK	11
Ann Arbor, MI	8
Annette, AK	8
Apalachicola, FL	2
Asheville, NC	6
Astoria, OR	7
Athens, GA	4
Atlanta, GA	3
Atlantic City, NJ	6
Augusta, GA	4
Austin, TX	2
Avoca, PA	7
Bakersfield, CA	4
Baltimore, MD	6
Barrow, AK	11
Baton Rouge, LA	2
Battle Creek, MI	8
Beaufort, SC	1
Beaumont, TX	1
Beckley, WV	7
Bethel, AK	11
Bettles, AK	11
Big Delta, AK	11
Billings, MT	8
Biloxi, MS	2
Binghamton, NY	8
Birmingham, AL	3
Bishop, CA	5
Bismarck, ND	9
Blue Hill, MA	7
Boise, ID	7
Boston, MA	7
Boulder, CO	8
Bowling Green, KY	5
Bridgeport, CT	7
Bristol, TN	6
Brockton, MA	7
Brownsville, TX	1
Buffalo, NY	8
Burlington, VT	9
Burns, OR	9
Camden, NJ	6
Cape Hatteras, NC	4
Caribou, ME	10
Carson City NV	7
Casper, WY	8
Cedar Rapids, IA	6
Charleston, SC	3
Charleston, WV	6
Charlotte, NC	4
Chattanooga, TN	5
Cheyenne, WY	8
Chicago, IL	7
Cincinnati, OH	6
Clayton, NM	6
Cleveland, OH	7
Cold Bay, AK	10
Colorado Springs, CO	7
Columbia, MO	6
Columbia, SC	4
Columbus, GA	3
Columbus, OH	6
Concord, NH	8
Concordia, KS	6
Corpus Christi, TX	1
Cumberland, MD	6
Dallas, TX	3
Danbury, CT	6
Davenport, IA	7
Dayton, OH	6
Daytona Beach, FL	1
Del Rio, TX	2
Denver, CO	7
Des Moines, IA	7
Detroit, MI	8
Dodge City, KS	6
Dubuque, IA	8
Duluth, MN	10
Eau Claire, WI	9
El Paso, TX	3
Elkins, WV	7
Elko, NV	8
Ely, NV	8
Erie, PA	7
Eugene, OR	6
Eureka, CA	6
Evansville, IN	5
Fairbanks, AK	11
Fall River, MA	11
Fargo, ND	9
Flagstaff, AZ	8
Flint, MI	8
Fort Lauderdale, FL	1
Fort Myers, FL	1
Fort Smith, AR	4
Fort Wayne, IN	7
Fort Worth, TX	3
Fresno, CA	4
Gainesville, FL	2
Galveston, TX	2
Gary, IN	7
Glasgow, MT	9
Goodland, KS	7
Grand Island, NE	7
Grand Junction, CO	6
Grand Rapids, MI	8
Great Falls, MT	8
Green Bay, WI	9
Greensboro, NC	5
Greenville, SC	4
Gulkana, AK	11
Hampton, VA	5
Harrisburg, PA	6
Hartford, CT	7
Helena, MT	9
Hilo, HI	1
Homer, AK	10
Honolulu, HI	1
Houghton Lake, MI	9
Houston, TX	2
Huntington, WV	6
Huntsville, AL	3
Huron, SD	8
Indianapolis, IN	6
International Falls, MN	11
Islip, NY	7
Jackson, KY	6
Jackson, MS	3
Jacksonville, FL	2
Jersey City, NJ	6
Juneau, AK	10
Kahului, HI	1
Kalamazoo, MI	8
Kalispell, MT	9
Kansas City, MO	6
Key West, FL	1
King Salmon, AK	11
Knoxville, TN	5
Kodiak, AK	9
Kokomo, IN	6
Kotzebue, AK	11
La Crosse, WI	8
Lake Charles, LA	2
Lander, WY	8
Lansing, MI	8
Las Cruces, NM	3
Las Vegas, NV	3
Lawton, OK	3
Lewiston, ID	6
Lewiston, ME	8
Lexington, KY	6
Lihue, HI	1
Lincoln, NE	7
Little Rock, AR	3
Los Angeles, CA	4
Louisville, KY	6
Lowell, MA	7
Lubbock, TX	5
Lynchburg, VA	6
Macon, GA	3
Madison, WI	8
Manchester, NH	7
Mansfield, OH	7
Marquette, MI	9
McGrath, AK	11
Medford, OR	6
Memphis, TN	3
Meridian, MS	3
Miami, FL	1
Midland-Odessa, TX	4
Milford, UT	7
Milwaukee, WI	8
Minneapolis, MN	9
Missoula, MT	9
Mobile, AL	2
Moline, IL	7
Montgomery, AL	3
Mount Shasta, CA	7
Mt. Washington, NH	11
Muncie, IN	6
Muskegon, MI	8
Nashville, TN	5
New Brunswick, NJ	6
New Haven, CT	7
New Orleans, LA	2
New York, NY	6
Newark, NJ	6
Newport News, VA	5
Nome, AK	11
Norfolk, NE	7
Norfolk, VA	5
North Little Rock, AR	4
North Platte, NE	8
Norwalk, CT	6
Norwich, CT	7
Oak Ridge, TN	5
Oakland, CA	5
Ogden, UT	7
Oklahoma City, OK	5
Olympia, WA	8
Omaha, NE	7
Orlando, FL	1
Owensboro, KY	5
Oxnard, CA	4
Paducah KY	5
Parkersburg, WV	6
Pendleton, OR	6
Pensacola, FL	2
Peoria, IL	7
Philadelphia, PA	6
Phoenix, AZ	2
Pittsburgh, PA	6
Pocatello, ID	8
Port Arthur, TX	2
Portland, ME	8
Portland, OR	5
Providence, RI	7
Pueblo, CO	6
Quillayute, WA	7
Raleigh, NC	5
Raleigh-Durham, NC	4
Rapid City, SD	8
Reading, PA	6
Redding, CA	4
Reno, NV	7
Richland, WA	5
Richmond, VA	5
Riverside, CA	4
Roanoke, VA	6
Rochester, MN	8
Rochester, NY	8
Rockford, IL	8
Roswell, NM	4
Rutland, VT	8
Sacramento, CA	4
Saginaw, MI	8
Salem, OR	6
Salt Lake City, UT	7
San Angelo, TX	3
San Antonio, TX	2
San Diego, CA	4
San Francisco, CA	5
San Jose, CA	5
Santa Barbara, CA	4
Santa Maria, CA	5
Sault Ste. Marie, MI	10
Savannah, GA	3
Scottsbluff, NE	7
Scranton, PA	6
Seattle, WA	8
Sexton Summit, OR	8
Sheridan, WY	8
Shreveport, LA	3
Sioux City, IA	7
Sioux Falls, SD	8
South Bend, IN	7
Spokane, WA	8
Springfield, IL	6
Springfield, MA	7
Springfield, MO	6
Springfield, OH	6
St. Cloud, MN	9
St. Louis, MO	5
St. Paul Island, AK	11
Stamford, CT	6
Stockton, CA	4
Syracuse, NY	8
Tacoma, WA	8
Talkeetna, AK	11
Tallahassee, FL	3
Tampa, FL	1
Terre Haute, IN	6
Toledo, OH	7
Topeka, KS	6
Trenton, NJ	6
Tucson, AZ	2
Tulsa, OK	5
Tupelo, MS	4
Tuscaloosa, AL	3
Unalakleet, AK	11
Utica, NY	8
Valdez, AK	10
Valentine, NE	8
Vero Beach, FL	1
Victoria, TX	2
Virginia Beach, VA	5
Waco, TX	2
Wallops Island, VA	5
Washington, DC	6
Waterbury, CT	6
Waterloo, IA	8
Watertown, NY	8
West Palm Beach, FL	1
Wichita Falls, TX	3
Wichita, KS	5
Williamsport, PA	7
Williston, ND	9
Wilmington, DE	6
Wilmington, NC	4
Winnemucca, NV	7
Winslow, AZ	6
Winston-Salem, NC	6
Worcester, MA	8
Yakima, WA	7
Yakutat, AK	10
Yonkers, NY	6
Youngstown, OH	7
Yuma, AZ	1

Canadian Cities

City	Zone
Calgary, Alberta	10
Edmonton, Alberta	11
Hamilton, Ontario	8
London, Ontario	8
Montreal, Quebec	9
Ottawa, Quebec	9
Quebec, Quebec	10
Toronto, Ontario	7
Vancouver, B.C.	7
Winnipeg, Manitoba	11

Figure 5-1
HVAC Climate Zones

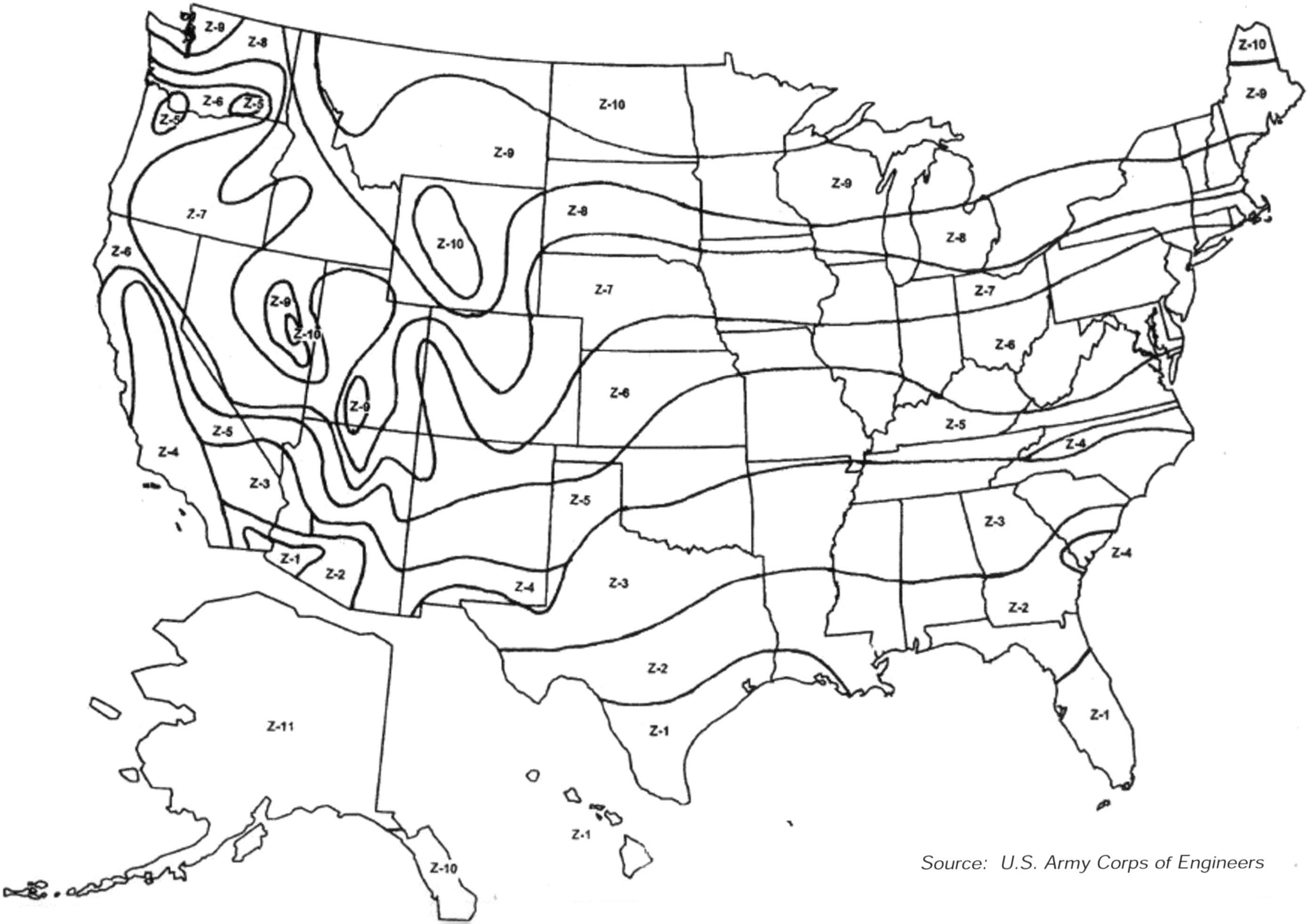

Source: U.S. Army Corps of Engineers

5.2 Repair Task Frequencies of HVAC Components by Climate Zone

Uniformat / Component / Repair Task	Climate Zone										
	1	2	3	4	5	6	7	8	9	10	11
D30 HVAC											
D3010 Energy Supply											
Leak Detection, Panel & Float Assembly											
Replace Leak Detection, Panel & Float Assembly	>50	>50	>50	>50	>50	>50	>50	>50	>50	>50	>50
Fuel Oil Pump, 1.5 HP											
Annual PM, Fuel Oil Pump, 1.5 HP	1	1	1	1	1	1	1	1	1	1	1
Repair Fuel Oil Pump, 1.5 HP	42	19	13	12	11	10	9	8	8	7	7
Replace Fuel Oil Pump, 1.5 HP	>50	28	20	19	16	15	13	13	12	11	11
Generator Day Tank, 275 Gal.											
Maintain Generator Day Tank, 275 Gal.	1	1	1	1	1	1	1	1	1	1	1
Replace Generator Day Tank, 275 Gal.	>50	>50	40	37	32	30	26	25	23	22	22
Oil Storage Tank, 275 Gal.											
Maintain Oil Storage Tank, 275 Gal.	1	1	1	1	1	1	1	1	1	1	1
Replace Oil Storage Tank, 275 Gal.	>50	>50	40	37	32	30	26	25	23	22	22
Oil Storage Tank, Fiberglass, 500 Gal											
Maintain Oil Storage Tank, Fiberglass, 500 Gal	1	1	1	1	1	1	1	1	1	1	1
Replace Oil Storage Tank, Fiberglass, 500 Gal	>50	>50	>50	>50	>50	>50	>50	>50	>50	>50	>50
Oil Storage Tank, 25,000 Gal.											
Maintain Oil Storage Tank, 25,000 Gal.	1	1	1	1	1	1	1	1	1	1	1
Replace Oil Storage Tank, 25,000 Gal.	>50	>50	40	37	32	30	26	25	23	22	22
Pipe & Fittings, 2" Copper, Fuel Oil											
Resolder Joint, Pipe & Fittings, 2" Copper, Fuel Oil	10	10	10	10	10	10	10	10	10	10	10
Install 10' Section, Pipe & Fittings, 2" Copper, Fuel Oil	20	20	20	20	20	20	20	20	20	20	20
Replace Pipe & Fittings, 2" Copper, Fuel Oil	25	25	25	25	25	25	25	25	25	25	25
Pipe & Fittings, 4" Steel, Gas											
Replace 10 Foot Section, Pipe & Fittings, 4" Steel, Gas	50	23	16	15	13	12	11	10	9	9	9
Install New Gasket & Bolts, Pipe & Fittings, 4" Steel, Gas	>50	>50	40	37	32	30	26	25	23	22	22
Replace Pipe & Fittings, 4" Steel, Gas	>50	>50	>50	>50	>50	>50	>50	>50	>50	>50	>50
D3020 Heat Generating Systems											

Components ordered by Uniformat II Classifications.

Sources: U.S. Army Corps of Engineers, Department of Defense Engineering Performance Standards, and Whitestone Research interviews.

5.2 Repair Task Frequencies of HVAC Components by Climate Zone

Uniformat / Component / Repair Task	Climate Zone										
	1	2	3	4	5	6	7	8	9	10	11
Fuel Oil Pump 1/2 HP											
Inspect & Lubricate Fuel Oil Pump 1/2 HP	1	1	1	1	1	1	1	1	1	1	1
Overhaul Fuel Oil Pump 1/2 HP	5	5	5	5	5	5	5	5	5	5	5
Replace Fuel Oil Pump 1/2 HP	20	20	20	20	20	20	20	20	20	20	20
Gate Valve, 2-3"											
Repack Gland, Gate Valve, 2-3"	38	16	12	11	9	10	8	8	7	6	6
Replace Gate Valve, 2-3"	>50	32	24	22	19	20	15	15	14	12	12
Pressure Reducer Valve, 2"											
Annual PM, Pressure Reducer Valve, 2"	1	1	1	1	1	1	1	1	1	1	1
Replace Pressure Reducer Valve, 2"	23	10	7	7	6	6	5	5	4	4	4
Air Separator, 2"											
Annual PM, Air Separator, 2"	1	1	1	1	1	1	1	1	1	1	1
Replace Air Separator, 2"	>50	24	18	17	14	15	12	12	10	9	9
AIR Separator, 8"											
Annual PM, Air Separator, 8"	1	1	1	1	1	1	1	1	1	1	1
Replace Air Separator, 8"	15	15	15	15	15	15	15	15	15	15	15
Boiler Blowoff System											
Repair Boiler Blowoff System	26	11	8	8	7	7	5	5	5	4	4
Replace Boiler Blowoff System	>50	49	36	33	28	30	23	23	20	18	18
Chemical Feed System											
Maintain Chemical Feed System	1	1	1	1	1	1	1	1	1	1	1
Replace Chemical Feed System	>50	24	18	17	14	15	12	12	10	9	9
Deaerator											
Maintain Deaerator	1	1	1	1	1	1	1	1	1	1	1
Repair Deaerator	38	16	12	11	9	10	8	8	7	6	6
Replace Deaerator	>50	32	24	22	19	20	15	15	14	12	12

Components ordered by Uniformat II Classifications.

Sources: U.S. Army Corps of Engineers, Department of Defense Engineering Performance Standards, and Whitestone Research interviews

5.2 Repair Task Frequencies of HVAC Components by Climate Zone

Uniformat / Component / Repair Task	Climate Zone										
	1	2	3	4	5	6	7	8	9	10	11
Circulation Pump 5 HP, Hot Water											
Maintain Circulation Pump 5 HP, Hot Water	1	1	1	1	1	1	1	1	1	1	1
Repair Circulation Pump 5 HP, Hot Water	19	8	6	6	5	5	4	4	3	3	3
Replace Circulation Pump 5 HP, Hot Water	>50	24	18	17	14	15	12	12	10	9	9
Circulation Pump 7.5 HP, Hot Water											
Maintain Circulation Pump 7.5 HP, Hot Water	1	1	1	1	1	1	1	1	1	1	1
Repair Circulation Pump 7.5 HP, Hot Water	19	8	6	6	5	5	4	4	3	3	3
Replace Circulation Pump 7.5 HP, Hot Water	>50	24	18	17	14	15	12	12	10	9	9
Circulation Pump 10 HP, Hot Water											
Maintain Circulation Pump 10 HP, Hot Water	1	1	1	1	1	1	1	1	1	1	1
Repair Circulation Pump 10 HP, Hot Water	19	8	6	6	5	5	4	4	3	3	3
Replace Circulation Pump 10 HP, Hot Water	>50	24	18	17	14	15	12	12	10	9	9
Circulation Pump 25 HP, Hot Water											
Maintain Circulation Pump 25 HP, Hot Water	1	1	1	1	1	1	1	1	1	1	1
Repair Circulation Pump 25 HP, Hot Water	19	8	6	6	5	5	4	4	3	3	3
Replace Circulation Pump 25 HP, Hot Water	>50	24	18	17	14	15	12	12	10	9	9
Circulation Pump 75 HP, Hot Water											
Maintain Circulation Pump 75 HP, Hot Water	1	1	1	1	1	1	1	1	1	1	1
Repair Circulation Pump 75 HP, Hot Water	19	8	6	6	5	5	4	4	3	3	3
Replace Circulation Pump 75 HP, Hot Water	>50	24	18	17	14	15	12	12	10	9	9
Circulation Pump 100 HP, Hot Water											
Maintain Circulation Pump 100 HP, Hot Water	1	1	1	1	1	1	1	1	1	1	1
Repair Circulation Pump 100 HP, Hot Water	19	8	6	6	5	5	4	4	3	3	3
Replace Circulation Pump 100 HP, Hot Water	>50	24	18	17	14	15	12	12	10	9	9
Circulation Pump 150 HP, Hot Water											
Maintain Circulation Pump 150 HP, Hot Water	1	1	1	1	1	1	1	1	1	1	1
Repair Circulation Pump 150 HP, Hot Water	19	8	6	6	5	5	4	4	3	3	3
Replace Circulation Pump 150 HP, Hot Water	>50	24	18	17	14	15	12	12	10	9	9

Components ordered by Uniformat II Classifications.

Sources: U.S. Army Corps of Engineers, Department of Defense Engineering Performance Standards, and Whitestone Research interviews.

5.2 Repair Task Frequencies of HVAC Components by Climate Zone

Uniformat / Component / Repair Task	Climate Zone										
	1	2	3	4	5	6	7	8	9	10	11
Condensing Heat Exchanger, Gas/Oil, 10 Mbh											
Maintain Condensing Heat Exchanger, Gas, 10 Mbh	1	1	1	1	1	1	1	1	1	1	1
Replace Condensing Heat Exchanger, Gas, 10 Mbh	>50	49	36	33	28	30	23	23	20	18	18
Condensing Heat Exchanger, Gas/Oil, 20 Mbh											
Maintain Condensing Heat Exchanger, Gas/Oil, 20 Mbh	1	1	1	1	1	1	1	1	1	1	1
Replace Condensing Heat Exchanger, Gas/Oil, 20 Mbh	>50	49	36	33	28	30	23	23	20	18	18
Condensate Receiver Station, 10-15 Gal.											
Maintain Condensate Receiver Station, 10-15 Gal.	1	1	1	1	1	1	1	1	1	1	1
Repair Condensate Receiver Station, 10-15 Gal.	11	5	4	3	3	3	2	2	2	2	2
Repair Condensate Receiver Station, Motor, 10-15 Gal.	19	8	6	6	5	5	4	4	3	3	3
Replace Condensate Receiver Station, 10-15 Gal.	>50	24	18	17	14	15	12	12	10	9	9
Expansion Tank, 30 Gal.											
Annual PM, Expansion Tank, 30 Gal.	1	1	1	1	1	1	1	1	1	1	1
Replace Expansion Tank, 30 Gal.	>50	>50	>50	>50	47	50	39	39	34	30	30
Expansion Tank, 60 Gal.											
Annual PM, Expansion Tank, 60 Gal.	1	1	1	1	1	1	1	1	1	1	1
Replace Expansion Tank, 60 Gal.	>50	>50	>50	>50	47	50	39	39	34	30	30
Expansion Tank, 100 Gal.											
Annual PM, Expansion Tank, 100 Gal.	1	1	1	1	1	1	1	1	1	1	1
Replace Expansion Tank, 100 Gal.	>50	>50	>50	>50	47	50	39	39	34	30	30
Expansion Tank, 250 Gal.											
Annual PM, Expansion Tank, 250 Gal.	1	1	1	1	1	1	1	1	1	1	1
Replace Expansion Tank, 250 Gal.	>50	>50	>50	>50	47	50	39	39	34	30	30
Surge Tank, 1000 Gal.											
Replace Surge Tank, 1000 Gal.	>50	24	18	17	14	15	12	12	10	9	9
Flash Tank, 24 Gal.											
Maintain Flash Tank, 24 Gal.	1	1	1	1	1	1	1	1	1	1	1
Repair Flash Tank, 24 Gal.	19	8	6	6	5	5	4	4	3	3	3
Replace Flash Tank, 24 Gal.	>50	24	18	17	14	15	12	12	10	9	9

Components ordered by Uniformat II Classifications.

Sources: U.S. Army Corps of Engineers, Department of Defense Engineering Performance Standards, and Whitestone Research interviews.

5.2 Repair Task Frequencies of HVAC Components by Climate Zone

Uniformat / Component / Repair Task	Climate Zone										
	1	2	3	4	5	6	7	8	9	10	11
Feed Water Pump, <1000 kbtu Boiler											
Maintain Feed Water Pump, <1000 kbtu Boiler	1	1	1	1	1	1	1	1	1	1	1
Replace Feed Water Pump, <1000 kbtu Boiler	>50	24	18	17	14	15	12	12	10	9	9
Feed Water Pump, 1000-3000 kbtu Boiler											
Maintain Feed Water Pump, 1000-3000 kbtu Boiler	1	1	1	1	1	1	1	1	1	1	1
Replace Feed Water Pump, 1000-3000 kbtu Boiler	>50	24	18	17	14	15	12	12	10	9	9
Gas Boiler, 250 Mbh											
Inspect & Test Gas Boiler, 250 Mbh	1	1	1	1	1	1	1	1	1	1	1
Clean Fire Box, Gas Boiler, 250 Mbh	2	2	2	2	2	2	2	2	2	2	2
Repair Gas Boiler, 250 Mbh	26	11	8	8	7	7	5	5	5	4	4
Replace Gas Boiler, 250 Mbh	>50	49	36	33	28	30	23	23	20	18	18
Gas Boiler, 750 Mbh											
Inspect & Test Gas Boiler, 750 Mbh	1	1	1	1	1	1	1	1	1	1	1
Clean Fire Box, Gas Boiler, 750 Mbh	2	2	2	2	2	2	2	2	2	2	2
Repair Gas Boiler, 750 Mbh	26	11	8	8	7	7	5	5	5	4	4
Replace Gas Boiler, 750 Mbh	>50	49	36	33	28	30	23	23	20	18	18
Gas Boiler, 1000 Mbh											
Inspect & Test Gas Boiler, 1000 Mbh	1	1	1	1	1	1	1	1	1	1	1
Clean Fire Box, Gas Boiler, 1000 Mbh	2	2	2	2	2	2	2	2	2	2	2
Repair Gas Boiler, 1000 Mbh	26	11	8	8	7	7	5	5	5	4	4
Replace Gas Boiler, 1000 Mbh	>50	49	36	33	28	30	23	23	20	18	18
Gas Boiler, 2000 Mbh											
Inspect & Test Gas Boiler, 2000 Mbh	1	1	1	1	1	1	1	1	1	1	1
Clean Fire Box, Gas Boiler, 2000 Mbh	2	2	2	2	2	2	2	2	2	2	2
Repair Gas Boiler, 2000 Mbh	26	11	8	8	7	7	5	5	5	4	4
Replace Gas Boiler, 2000 Mbh	>50	49	36	33	28	30	23	23	20	18	18
Gas Boiler, 2500 Mbh											
Inspect & Test Gas Boiler, 2500 Mbh	1	1	1	1	1	1	1	1	1	1	1
Clean Fire Box, Gas Boiler, 2500 Mbh	2	2	2	2	2	2	2	2	2	2	2
Repair Gas Boiler, 2500 Mbh	26	11	8	8	7	7	5	5	5	4	4
Replace Gas Boiler, 2500 Mbh	>50	49	36	33	28	30	23	23	20	18	18

Components ordered by Uniformat II Classifications.

Sources: U.S. Army Corps of Engineers, Department of Defense Engineering Performance Standards, and Whitestone Research interviews.

5.2 Repair Task Frequencies of HVAC Components by Climate Zone

Uniformat / Component / Repair Task	Climate Zone										
	1	2	3	4	5	6	7	8	9	10	11
Gas Boiler, 4200 Mbh											
Inspect & Test Gas Boiler, 4200 Mbh	1	1	1	1	1	1	1	1	1	1	1
Clean Fire Box, Gas Boiler, 4200 Mbh	2	2	2	2	2	2	2	2	2	2	2
Repair Gas Boiler, 4200 Mbh	7	7	7	7	7	7	7	7	7	7	7
Replace Gas Boiler, 4200 Mbh	30	30	30	30	30	30	30	30	30	30	30
Gas Boiler, Steam, 24,000 Mbh											
Inspect & Test Gas Boiler, Steam, 24,000 Mbh	1	1	1	1	1	1	1	1	1	1	1
Clean Fire Box, Gas Boiler, Steam, 24,000 Mbh	2	2	2	2	2	2	2	2	2	2	2
Repair Gas Boiler, Steam, 24,000 Mbh	26	11	8	8	7	7	5	5	5	4	4
Replace Gas Boiler, Steam, 24,000 Mbh	>50	49	36	33	28	30	23	23	20	18	18
House Furnace, Gas/Oil, 25 Mbh											
Maintain House Furnace, Gas/Oil, 25 Mbh	1	1	1	1	1	1	1	1	1	1	1
Repair House Furnace, Gas/Oil, 25 Mbh	38	16	12	11	9	10	8	8	7	6	6
Replace House Furnace, Gas/Oil, 25 Mbh	>50	24	18	17	14	15	12	12	10	9	9
House Furnace, Gas/Oil, 100 Mbh											
Maintain House Furnace, Gas/Oil, 100 Mbh	1	1	1	1	1	1	1	1	1	1	1
Repair House Furnace, Gas/Oil, 100 Mbh	38	16	12	11	9	10	8	8	7	6	6
Replace House Furnace, Gas/Oil, 100 Mbh	>50	24	18	17	14	15	12	12	10	9	9
House Furnace, Electric 25 Mbh, 8 kw											
Maintain House Furnace, Electric 25 Mbh, 8 kw	1	1	1	1	1	1	1	1	1	1	1
Repair House Furnace, Electric 25 Mbh, 8 kw	38	16	12	11	9	10	8	8	7	6	6
Replace House Furnace, Electric 25 Mbh, 8 kw	>50	24	18	17	14	15	12	12	10	9	9
House Furnace, Electric 100 Mbh, 30 kw											
Maintain House Furnace, Electric 100 Mbh, 30 kw	1	1	1	1	1	1	1	1	1	1	1
Repair House Furnace, Electric 100 Mbh, 30 kw	38	16	12	11	9	10	8	8	7	6	6
Replace House Furnace, Electric 100 Mbh, 30 kw	>50	24	18	17	14	15	12	12	10	9	9
Heat Exchanger, Steam-to-Water, 130 Gph											
Annual PM, Heat Exchanger, Steam-to-Water, 130 Gph	1	1	1	1	1	1	1	1	1	1	1
Replace Heat Exchanger, Steam-to-Water, 130 Gph	28	28	28	28	28	28	28	28	28	28	28

Components ordered by Uniformat II Classifications.

Sources: U.S. Army Corps of Engineers, Department of Defense Engineering Performance Standards, and Whitestone Research interviews.

5.2 Repair Task Frequencies of HVAC Components by Climate Zone

Uniformat / Component / Repair Task	Climate Zone										
	1	2	3	4	5	6	7	8	9	10	11
Steam Converter, 200 Gph											
Inspect & Maintain Steam Converter, 200 Gph	2	2	2	2	2	2	2	2	2	2	2
Repair Steam Converter, 200 Gph	19	8	6	6	5	5	4	4	3	3	3
Replace Steam Converter, 200 Gph	>50	49	36	33	28	30	23	23	20	18	18
Steam Trap, F&T, 2"											
Maintain Steam Trap, F&T, 2"	1	1	1	1	1	1	1	1	1	1	1
Repair Steam Trap, F&T, 2"	11	5	4	3	3	3	2	2	2	2	2
Replace Steam Trap, F&T, 2"	38	16	12	11	9	10	8	8	7	6	6
D3030 Cooling Generating Systems											
Flow Control Valve, Motorized, 2"											
Maintain Flow Control Valve & Actuator, 2"	1	1	1	1	1	1	1	1	1	1	1
Replace Valve Actuator, 2"	15	15	15	15	15	15	15	15	15	15	15
Flow Control Valve, Motorized, 4"											
Maintain Flow Control Valve & Actuator, 4"	1	1	1	1	1	1	1	1	1	1	1
Replace Valve Actuator, 4"	15	15	15	15	15	15	15	15	15	15	15
Replace Valve, Flow Control Motorized, 4"	17	17	17	17	17	17	17	17	17	17	17
Condensor, Air Cooled, 7.5 Ton											
Maintain Condenser, Air-Cooled, 7.5 Ton	1	1	1	1	1	1	1	1	1	1	1
Repair Condenser, Air-Cooled, 7.5 Ton	9	10	15	15	20	10	32	46	46	>50	>50
Replace Condenser, Air-Cooled, 7.5 Ton	13	14	23	22	29	15	48	>50	>50	>50	>50
Condenser, Air-Cooled, 20 Ton											
Maintain Condenser, Air-Cooled, 20 Ton	1	1	1	1	1	1	1	1	1	1	1
Repair Condenser, Air-Cooled, 20 Ton	9	10	15	15	20	10	32	46	46	>50	>50
Replace Condenser, Air-Cooled, 20 Ton	13	14	23	22	29	15	48	>50	>50	>50	>50
Condenser, Air-Cooled, 40 Ton											
Maintain Condenser, Air-Cooled, 40 Ton	1	1	1	1	1	1	1	1	1	1	1
Repair Condenser, Air-Cooled, 40 Ton	9	10	15	15	20	10	32	46	46	>50	>50
Replace Condenser, Air-Cooled, 40 Ton	13	14	23	22	29	15	48	>50	>50	>50	>50

Components ordered by Uniformat II Classifications.

Sources: U.S. Army Corps of Engineers, Department of Defense Engineering Performance Standards, and Whitestone Research interviews.

5.2 Repair Task Frequencies of HVAC Components by Climate Zone

Uniformat / Component / Repair Task	Climate Zone										
	1	2	3	4	5	6	7	8	9	10	11
Condenser, Air-Cooled, 60 Ton											
Maintain Condenser, Air-Cooled, 60 Ton	1	1	1	1	1	1	1	1	1	1	1
Repair Condenser, Air-Cooled, 60 Ton	9	10	15	15	20	10	32	46	46	>50	>50
Replace Condenser, Air-Cooled, 60 Ton	13	14	23	22	29	15	48	>50	>50	>50	>50
Circulation Pump, 5 HP, Chiller & Condenser Water											
Maintain Circulation Pump, 5 HP, Chiller & Condenser Water	1	1	1	1	1	1	1	1	1	1	1
Repair Circulation Pump, 5 HP, Chiller & Condenser Water	4	5	8	7	10	5	16	23	23	29	>50
Replace Circulation Pump, 5 HP, Chiller & Condenser Water	13	14	23	22	29	15	48	>50	>50	>50	>50
Circulation Pump, 7.5 HP, Chiller & Condenser Water											
Maintain Circulation Pump, 7.5 HP, Chiller & Condenser Water	1	1	1	1	1	1	1	1	1	1	1
Repair Circulation Pump, 7.5 HP, Chiller & Condenser Water	4	5	8	7	10	5	16	23	23	29	>50
Replace Circulation Pump, 7.5 HP, Chiller & Condenser Water	13	14	23	22	29	15	48	>50	>50	>50	>50
Circulation Pump, 10 HP, Chiller & Condenser Water											
Maintain Circulation Pump, 10 HP, Chiller & Condenser Water	1	1	1	1	1	1	1	1	1	1	1
Repair Circulation Pump, 10 HP, Chiller & Condenser Water	4	5	8	7	10	5	16	23	23	29	>50
Replace Circulation Pump, 10 HP, Chiller & Condenser Water	13	14	23	22	29	15	48	>50	>50	>50	>50
Circulation Pump, 25 HP, Chiller & Condenser Water											
Maintain Circulation Pump, 25 HP, Chiller & Condenser Water	1	1	1	1	1	1	1	1	1	1	1
Repair Circulation Pump, 25 HP, Chiller & Condenser Water	4	5	8	7	10	5	16	23	23	29	>50
Replace Circulation Pump, 25 HP, Chiller & Condenser Water	13	14	23	22	29	15	48	>50	>50	>50	>50
Circulation Pump, 75 HP, Chiller & Condenser Water											
Maintain Circulation Pump, 75 HP, Chiller & Condenser Water	1	1	1	1	1	1	1	1	1	1	1
Repair Circulation Pump, 75 HP, Chiller & Condenser Water	4	5	8	7	10	5	16	23	23	29	>50
Replace Circulation Pump, 75 HP, Chiller & Condenser Water	13	14	23	22	29	15	48	>50	>50	>50	>50
Circulation Pump, 100 HP, Chiller & Condenser Water											
Maintain Circulation Pump, 100 HP, Chiller & Condenser Water	1	1	1	1	1	1	1	1	1	1	1
Repair Circulation Pump, 100 HP, Chiller & Condenser Water	4	5	8	7	10	5	16	23	23	29	>50
Replace Circulation Pump, 100 HP, Chiller & Condenser Water	13	14	23	22	29	15	48	>50	>50	>50	>50

Components ordered by Uniformat II Classifications.

Sources: U.S. Army Corps of Engineers, Department of Defense Engineering Performance Standards, and Whitestone Research interviews.

5.2 Repair Task Frequencies of HVAC Components by Climate Zone

Uniformat / Component / Repair Task	Climate Zone										
	1	2	3	4	5	6	7	8	9	10	11
Circulation Pump, 150 HP, Chiller & Condenser Water											
Maintain Circulation Pump, 150 HP, Chiller & Condenser Water	1	1	1	1	1	1	1	1	1	1	1
Repair Circulation Pump, 150 HP, Chiller & Condenser Water	4	5	8	7	10	5	16	23	23	29	>50
Replace Circulation Pump, 150 HP, Chiller & Condenser Water	13	14	23	22	29	15	48	>50	>50	>50	>50
Chiller, Reciprocal Air-Cooled Hermetic, 20 Ton											
Maintain Chiller, Reciprocal Air Cooled Hermetic, 20 Ton	1	1	1	1	1	1	1	1	1	1	1
Repair Chiller, Reciprocal Air Cooled Hermetic, 20 Ton	9	10	15	15	20	10	32	46	46	>50	>50
Replace Chiller, Reciprocal Air Cooled Hermetic, 20 Ton	18	19	30	30	39	20	>50	>50	>50	>50	>50
Chiller, Reciprocal Air-Cooled Hermetic, 40 Ton											
Maintain Chiller, Reciprocal Air Cooled Hermetic, 40 Ton	1	1	1	1	1	1	1	1	1	1	1
Repair Chiller, Reciprocal Air Cooled Hermetic, 40 Ton	9	10	15	15	20	10	32	46	46	>50	>50
Replace Chiller, Reciprocal Air Cooled Hermetic, 40 Ton	18	19	30	30	39	20	>50	>50	>50	>50	>50
Chiller, Reciprocal Air-Cooled Hermetic, 60 Ton											
Maintain Chiller, Reciprocal Air Cooled Hermetic, 60 Ton	1	1	1	1	1	1	1	1	1	1	1
Repair Chiller, Reciprocal Air Cooled Hermetic, 60 Ton	9	10	15	15	20	10	32	46	46	>50	>50
Replace Chiller, Reciprocal Air Cooled Hermetic, 60 Ton	18	19	30	30	39	20	>50	>50	>50	>50	>50
Chiller, Reciprocal Air-Cooled Hermetic, 200 Ton											
Maintain Chiller, Reciprocal Air Cooled Hermetic, 200 Ton	1	1	1	1	1	1	1	1	1	1	1
Repair Chiller, Reciprocal Air Cooled Hermetic, 200 Ton	10	10	10	10	10	10	10	10	10	10	10
Replace Chiller, Reciprocal Air Cooled Hermetic, 200 Ton	20	20	20	20	20	20	20	20	20	20	20
Chiller, Reciprocal Water-Cooled Hermetic, 100 Ton											
Maintain Chiller, Reciprocal Water-Cooled Hermetic, 100 Ton	1	1	1	1	1	1	1	1	1	1	1
Repair Chiller, Reciprocal Water-Cooled Hermetic, 100 Ton	9	10	15	15	20	10	32	46	46	>50	>50
Replace Chiller, Reciprocal Water-Cooled Hermetic, 100 Ton	18	19	30	30	39	20	>50	>50	>50	>50	>50
Chiller, Reciprocal Water-Cooled Hermetic, 165 Ton											
Maintain Chiller, Reciprocal Water-Cooled Hermetic, 165 Ton	1	1	1	1	1	1	1	1	1	1	1
Repair Chiller, Reciprocal Water-Cooled Hermetic, 165 Ton	9	10	15	15	20	10	32	46	46	>50	>50
Replace Chiller, Reciprocal Water-Cooled Hermetic, 165 Ton	18	19	30	30	39	20	>50	>50	>50	>50	>50

Components ordered by Uniformat II Classifications.

Sources: U.S. Army Corps of Engineers, Department of Defense Engineering Performance Standards, and Whitestone Research interviews.

5.2 Repair Task Frequencies of HVAC Components by Climate Zone

Uniformat / Component / Repair Task	Climate Zone										
	1	2	3	4	5	6	7	8	9	10	11
Chiller, Reciprocal Water-Cooled Hermetic, 300 Ton											
Maintain Chiller, Reciprocal Water-Cooled Hermetic, 300 Ton	1	1	1	1	1	1	1	1	1	1	1
Repair Chiller, Reciprocal Water-Cooled Hermetic, 300 Ton	9	10	15	15	20	10	32	46	46	>50	>50
Replace Chiller, Reciprocal Water-Cooled Hermetic, 300 Ton	18	19	30	30	39	20	>50	>50	>50	>50	>50
Chiller, Centrifugal Hermetic, Bundled w/ Condenser, 200 Ton											
Maintain Chiller, Centrifugal Hermetic, Bnd'd w/ Condenser, 200 Ton	1	1	1	1	1	1	1	1	1	1	1
Repair Chiller, Centrifugal Hermetic, Bnd'd w/ Condenser, 200 Ton	9	10	15	15	20	10	32	46	46	>50	>50
Replace Chiller, Centrifugal Hermetic, Bnd'd w/ Condenser, 200 Ton	18	19	30	30	39	20	>50	>50	>50	>50	>50
Chiller, Centrifugal Hermetic, Bundled w/ Condenser, 250 Ton											
Maintain Chiller, Centrifugal Hermetic, Bnd'd w/ Condenser, 250 Ton	1	1	1	1	1	1	1	1	1	1	1
Repair Chiller, Centrifugal Hermetic, Bnd'd w/ Condenser, 250 Ton	9	10	15	15	20	10	32	46	46	>50	>50
Replace Chiller, Centrifugal Hermetic, Bnd'd w/ Condenser, 250 Ton	18	19	30	30	39	20	>50	>50	>50	>50	>50
Chiller, Centrifugal Hermetic, Bundled w/ Condenser, 300 Ton											
Maintain Chiller, Centrifugal Hermetic, Bnd'd w/ Condenser, 300 Ton	1	1	1	1	1	1	1	1	1	1	1
Repair Chiller, Centrifugal Hermetic, Bnd'd w/ Condenser, 300 Ton	9	10	15	15	20	10	32	46	46	>50	>50
Replace Chiller, Centrifugal Hermetic, Bnd'd w/ Condenser, 300 Ton	18	19	30	30	39	20	>50	>50	>50	>50	>50
Chiller, Centrifugal Hermetic, Bundled w/ Condenser, 900 Ton											
Maintain Chiller, Centrifugal Hermetic, Bnd'd w/ Condenser, 900 Ton	1	1	1	1	1	1	1	1	1	1	1
Repair Chiller, Centrifugal Hermetic, Bnd'd w/ Condenser, 900 Ton	9	10	15	15	20	10	32	46	46	>50	>50
Replace Chiller, Centrifugal Hermetic, Bnd'd w/ Condenser, 900 Ton	18	19	30	30	39	20	>50	>50	>50	>50	>50
Cooling Tower, 50 Ton											
Maintain Cooling Tower, 50 Ton	1	1	1	1	1	1	1	1	1	1	1
Repair Cooling Tower, 50 Ton	9	10	15	15	20	10	32	46	46	>50	>50
Replace Cooling Tower, 50 Ton	13	14	23	22	29	15	48	>50	>50	>50	>50
Cooling Tower, 100 Ton											
Maintain Cooling Tower, 100 Ton	1	1	1	1	1	1	1	1	1	1	1
Repair Cooling Tower, 100 Ton	9	10	15	15	20	10	32	46	46	>50	>50
Replace Cooling Tower, 100 Ton	13	14	23	22	29	15	48	>50	>50	>50	>50

Components ordered by Uniformat II Classifications.

Sources: U.S. Army Corps of Engineers, Department of Defense Engineering Performance Standards, and Whitestone Research interviews.

5.2 Repair Task Frequencies of HVAC Components by Climate Zone

Uniformat / Component / Repair Task	Climate Zone										
	1	2	3	4	5	6	7	8	9	10	11
Cooling Tower, 200 Ton											
Maintain Cooling Tower, 200 Ton	1	1	1	1	1	1	1	1	1	1	1
Repair Cooling Tower, 200 Ton	9	10	15	15	20	10	32	46	46	>50	>50
Replace Cooling Tower, 200 Ton	13	14	23	22	29	15	48	>50	>50	>50	>50
Cooling Tower, 250 Ton											
Maintain Cooling Tower, 250 Ton	1	1	1	1	1	1	1	1	1	1	1
Repair Cooling Tower, 250 Ton	9	10	15	15	20	10	32	46	46	>50	>50
Replace Cooling Tower, 250 Ton	13	14	23	22	29	15	48	>50	>50	>50	>50
Cooling Tower, 300 Ton											
Maintain Cooling Tower, 300 Ton	1	1	1	1	1	1	1	1	1	1	1
Repair Cooling Tower, 300 Ton	9	10	15	15	20	10	32	46	46	>50	>50
Replace Cooling Tower, 300 Ton	13	14	23	22	29	15	48	>50	>50	>50	>50
Cooling Tower, 500 Ton											
Maintain Cooling Tower, 500 Ton	1	1	1	1	1	1	1	1	1	1	1
Repair Cooling Tower, 500 Ton	9	10	15	15	20	10	32	46	46	>50	>50
Replace Cooling Tower, 500 Ton	13	14	23	22	29	15	48	>50	>50	>50	>50
Cooling Tower, 1000 Ton											
Maintain Cooling Tower, 1000 Ton	1	1	1	1	1	1	1	1	1	1	1
Repair Cooling Tower, 1000 Ton	9	10	15	15	20	10	32	46	46	>50	>50
Replace Cooling Tower, 1000 Ton	13	14	23	22	29	15	48	>50	>50	>50	>50
Cooling Tower, 1400 Ton											
Maintain Cooling Tower, 1400 Ton	1	1	1	1	1	1	1	1	1	1	1
Repair Cooling Tower, 1400 Ton	9	10	15	15	20	10	32	46	46	>50	>50
Replace Cooling Tower, 1400 Ton	13	14	23	22	29	15	48	>50	>50	>50	>50
Evaporative Cooler, Direct, Packaged w/ Heat, 4 Ton											
Maintain Evaporative Cooler, Direct, Packaged w/ Heat, 4 Ton	1	1	1	1	1	1	1	1	1	1	1
Replace Evaporative Cooler, Direct, Packaged w/ Heat, 4 Ton	18	19	30	30	39	20	>50	>50	>50	>50	>50
Evaporative Cooler, Indirect & Direct, Packaged w/ Heat, 4 Ton											
Maintain Evaporative Cooler, Indirect & Direct, Pkg'd w/ Heat, 4 Ton	1	1	1	1	1	1	1	1	1	1	1
Replace Evaporative Cooler, Indirect & Direct, Pkg'd w/ Heat, 4 Ton	18	19	30	30	39	20	>50	>50	>50	>50	>50

Components ordered by Uniformat II Classifications.

Sources: U.S. Army Corps of Engineers, Department of Defense Engineering Performance Standards, and Whitestone Research interviews.

5.2 Repair Task Frequencies of HVAC Components by Climate Zone

Uniformat / Component / Repair Task	Climate Zone										
	1	2	3	4	5	6	7	8	9	10	11
Evaporative Cooler, Indirect & Direct, Packaged w/ Heat, 20 Ton											
Maintain Evaporative Cooler, Indirect & Direct, Pkg'd w/ Heat, 20 Ton	1	1	1	1	1	1	1	1	1	1	1
Replace Evaporative Cooler, Indirect & Direct, Pkg'd w/ Heat, 20 Ton	18	19	30	30	39	20	>50	>50	>50	>50	>50
Evaporative Cooler, Indirect, 2000 Scfm											
Maintain Evaporative Cooler, Indirect, 2000 Scfm	1	1	1	1	1	1	1	1	1	1	1
Replace Evaporative Cooler, Indirect, 2000 Scfm	18	19	30	30	39	20	>50	>50	>50	>50	>50
Evaporative Cooler, Indirect, 5000 Scfm											
Maintain Evaporative Cooler, Indirect, 5000 Scfm	1	1	1	1	1	1	1	1	1	1	1
Replace Evaporative Cooler, Indirect, 5000 Scfm	18	19	30	30	39	20	>50	>50	>50	>50	>50
Evaporative Cooler, Indirect, 15,000 Scfm											
Maintain Evaporative Cooler, Indirect, 15,000 Scfm	1	1	1	1	1	1	1	1	1	1	1
Replace Evaporative Cooler, Indirect, 15,000 Scfm	18	19	30	30	39	20	>50	>50	>50	>50	>50
Chiller, Centrifugal Hermetic, Bundled w/ Condenser, 1000 Ton											
Maintain Chiller, Centrifugal Hermetic, Bnd'd w/ Condenser, 1000 Ton	1	1	1	1	1	1	1	1	1	1	1
Repair Chiller, Centrifugal Hermetic, Bnd'd w/ Condenser, 1000 Ton	9	10	15	15	20	10	32	46	46	>50	>50
Replace Chiller, Centrifugal Hermetic, Bnd'd w/ Condenser, 1000 Ton	18	19	30	30	39	20	>50	>50	>50	>50	>50
D3040 Distribution Systems											
Air Curtain, 1000 Cfm											
Maintain Air Curtain, 1000 Cfm	1	1	1	1	1	1	1	1	1	1	1
Replace Fan & Motor, Air Curtain, 1000 Cfm	13	13	13	14	15	15	16	17	18	19	26
Replace Air Curtain, 1000 Cfm	18	17	18	19	19	20	22	23	24	26	35
Ceiling Panel, Radiant, Hydronic											
Replace Ceiling Panel, Radiant, Hydronic	>50	>50	>50	>50	47	50	39	39	34	30	30
Clean Room (Air Shower) Class 10,000											
Maintain, Clean Room (Air Shower) Class 10,000	1	1	1	1	1	1	1	1	1	1	1
Repair, Clean Room (Air Shower) Class 10,000	5	5	5	5	5	5	5	5	5	5	5
Replace, Clean Room (Air Shower) Class 10,000	20	20	20	20	20	20	20	20	20	20	20

Components ordered by Uniformat II Classifications.

Sources: U.S. Army Corps of Engineers, Department of Defense Engineering Performance Standards, and Whitestone Research interviews

5.2 Repair Task Frequencies of HVAC Components by Climate Zone

Uniformat / Component / Repair Task	Climate Zone										
	1	2	3	4	5	6	7	8	9	10	11
Dehumidifier & Control 120-150 lbs, 4500 Cfm											
Annual PM, Dehumidifier & Control 120-150 lbs, 4500 Cfm	1	1	1	1	1	1	1	1	1	1	1
Replace Dehumidifier & Control 120-150 lbs, 4500 Cfm	20	20	20	20	20	20	20	20	20	20	20
Humidifer, Steam, Duct w/ Controls											
Annual PM, Humidifier, Steam, Duct w/ Controls	1	1	1	1	1	1	1	1	1	1	1
Replace Humidifier, Steam, Duct w/ Controls	18	17	18	19	19	20	22	23	24	26	35
Pipe & Fittings, 3/4" Copper, High Temp.											
Install 10' Section, Pipe & Fittings, 3/4" Copper, High Temp.	13	13	13	13	13	13	13	13	13	13	13
Resolder Joint, Pipe & Fittings, 3/4" Copper, High Temp.	13	13	13	13	13	13	13	13	13	13	13
Replace Pipe & Fittings, 3/4" Copper, High Temp.	25	25	25	25	25	25	25	25	25	25	25
Variable Air Volume Box											
Annual PM, Variable Air Volume Box	1	1	1	1	1	1	1	1	1	1	1
Air Handler, Singlezone, 1300 Cfm											
Maintain Air Handler, Singlezone, 1300 Cfm	1	1	1	1	1	1	1	1	1	1	1
Repair Air Handler, Singlezone, 1300 Cfm	9	8	9	10	10	10	11	11	12	13	18
Replace Air Handler, Singlezone, 1300 Cfm	13	13	13	14	15	15	16	17	18	19	26
Air Handler, Singlezone, 6500 Cfm											
Maintain Air Handler, Singlezone, 6500 Cfm	1	1	1	1	1	1	1	1	1	1	1
Repair Air Handler, Singlezone, 6500 Cfm	9	8	9	10	10	10	11	11	12	13	18
Replace Air Handler, Singlezone, 6500 Cfm	13	13	13	14	15	15	16	17	18	19	26
Air Handler, Singlezone, 20,000 Cfm											
Maintain Air Handler, Singlezone, 20,000 Cfm	1	1	1	1	1	1	1	1	1	1	1
Repair Air Handler, Singlezone, 20,000 Cfm	9	8	9	10	10	10	11	11	12	13	18
Replace Air Handler, Singlezone, 20,000 Cfm	13	13	13	14	15	15	16	17	18	19	26
Air Handler, Singlezone, 50,000 Cfm											
Maintain Air Handler, Singlezone, 50,000 Cfm	1	1	1	1	1	1	1	1	1	1	1
Repair Air Handler, Singlezone, 50,000 Cfm	9	8	9	10	10	10	11	11	12	13	18
Replace Air Handler, Singlezone, 50,000 Cfm	13	13	13	14	15	15	16	17	18	19	26

Components ordered by Uniformat II Classifications.

Sources: U.S. Army Corps of Engineers, Department of Defense Engineering Performance Standards, and Whitestone Research interviews.

5.2 Repair Task Frequencies of HVAC Components by Climate Zone

Uniformat / Component / Repair Task	Climate Zone										
	1	2	3	4	5	6	7	8	9	10	11
Air Handler, Singlezone, 100,000 Cfm											
Maintain Air Handler, Singlezone, 100,000 Cfm	1	1	1	1	1	1	1	1	1	1	1
Repair Air Handler, Singlezone, 100,000 Cfm	9	8	9	10	10	10	11	11	12	13	18
Replace Air Handler, Singlezone, 100,000 Cfm	13	13	13	14	15	15	16	17	18	19	26
Air Handler, Singlezone, 120,000 Cfm											
Maintain Air Handler, Singlezone, 120,000 Cfm	1	1	1	1	1	1	1	1	1	1	1
Repair Air Handler, Singlezone, 120,000 Cfm	9	8	9	10	10	10	11	11	12	13	18
Replace Air Handler, Singlezone, 120,000 Cfm	13	13	13	14	15	15	16	17	18	19	26
Air Handler, Singlezone, 135,000 Cfm											
Maintain Air Handler, Singlezone, 135,000 Cfm	1	1	1	1	1	1	1	1	1	1	1
Repair Air Handler, Singlezone, 135,000 Cfm	9	8	9	10	10	10	11	11	12	13	18
Replace Air Handler, Singlezone, 135,000 Cfm	13	13	13	14	15	15	16	17	18	19	26
Air Handler, Singlezone, 150,000 Cfm											
Maintain Air Handler, Singlezone, 150,000 Cfm	1	1	1	1	1	1	1	1	1	1	1
Repair Air Handler, Singlezone, 150,000 Cfm	9	8	9	10	10	10	11	11	12	13	18
Replace Air Handler, Singlezone, 150,000 Cfm	13	13	13	14	15	15	16	17	18	19	26
Air Handler, Multizone, 2500 Cfm											
Maintain Air Handler, Multizone, 2500 Cfm	1	1	1	1	1	1	1	1	1	1	1
Repair Air Handler, Multizone, 2500 Cfm	9	8	9	10	10	10	11	11	12	13	18
Replace Air Handler, Multizone, 2500 Cfm	13	13	13	14	15	15	16	17	18	19	26
Air Handler, Multizone, 5200 Cfm											
Maintain Air Handler, Multizone, 10,000 Cfm	1	1	1	1	1	1	1	1	1	1	1
Repair Air Handler, Multizone, 5200 Cfm	9	8	9	10	10	10	11	11	12	13	18
Replace Air Handler, Multizone, 5200 Cfm	13	13	13	14	15	15	16	17	18	19	26
Air Handler, Multizone, 6500 Cfm											
Maintain Air Handler, Multizone, 6500 Cfm	1	1	1	1	1	1	1	1	1	1	1
Repair Air Handler, Multizone, 6500 Cfm	9	8	9	10	10	10	11	11	12	13	18
Replace Air Handler, Multizone, 6500 Cfm	13	13	13	14	15	15	16	17	18	19	26

Components ordered by Uniformat II Classifications.

Sources: U.S. Army Corps of Engineers, Department of Defense Engineering Performance Standards, and Whitestone Research interviews.

5.2 Repair Task Frequencies of HVAC Components by Climate Zone

Uniformat / Component / Repair Task	Climate Zone										
	1	2	3	4	5	6	7	8	9	10	11
Air Handler, Multizone, 10,000 Cfm											
Maintain Air Handler, Multizone, 10,000 Cfm	1	1	1	1	1	1	1	1	1	1	1
Repair Air Handler, Multizone, 10,000 Cfm	9	8	9	10	10	10	11	11	12	13	18
Replace Air Handler, Multizone, 10,000 Cfm	13	13	13	14	15	15	16	17	18	19	26
Air Handler, Multizone, 15,000 Cfm											
Maintain Air Handler, Multizone, 15,000 Cfm	1	1	1	1	1	1	1	1	1	1	1
Repair Air Handler, Multizone, 15,000 Cfm	9	8	9	10	10	10	11	11	12	13	18
Replace Air Handler, Multizone, 15,000 Cfm	13	13	13	14	15	15	16	17	18	19	26
Air Handler, Multizone, 20,000 Cfm											
Maintain Air Handler, Multizone, 20,000 Cfm	1	1	1	1	1	1	1	1	1	1	1
Repair Air Handler, Multizone, 20,000 Cfm	9	8	9	10	10	10	11	11	12	13	18
Replace Air Handler, Multizone, 20,000 Cfm	13	13	13	14	15	15	16	17	18	19	26
Air Handler, Multizone, 25,000 Cfm											
Maintain Air Handler, Multizone, 25,000 Cfm	1	1	1	1	1	1	1	1	1	1	1
Repair Air Handler, Multizone, 25,000 Cfm	9	8	9	10	10	10	11	11	12	13	18
Replace Air Handler, Multizone, 25,000 Cfm	13	13	13	14	15	15	16	17	18	19	26
Air Handler, Multizone, 50,000 Cfm											
Maintain Air Handler, Multizone, 50,000 Cfm	1	1	1	1	1	1	1	1	1	1	1
Repair Air Handler, Multizone, 50,000 Cfm	9	8	9	10	10	10	11	11	12	13	18
Replace Air Handler, Multizone, 50,000 Cfm	13	13	13	14	15	15	16	17	18	19	26
Air Handler, Exhaust, 15,000 Cfm											
Maintain Air Handler, Exhaust, 15,000 Cfm	1	1	1	1	1	1	1	1	1	1	1
Repair Air Handler, Exhaust, 15,000 Cfm	9	8	9	10	10	10	11	11	12	13	18
Replace Air Handler, Exhaust, 15,000 Cfm	18	17	18	19	19	20	22	23	24	26	35
Air Handler, Exhaust, 20,000 Cfm											
Maintain Air Handler, Exhaust, 20,000 Cfm	1	1	1	1	1	1	1	1	1	1	1
Repair Air Handler, Exhaust, 20,000 Cfm	9	8	9	10	10	10	11	11	12	13	18
Replace Air Handler, Exhaust, 20,000 Cfm	18	17	18	19	19	20	22	23	24	26	35

Components ordered by Uniformat II Classifications.

Sources: U.S. Army Corps of Engineers, Department of Defense Engineering Performance Standards, and Whitestone Research interviews.

5.2 Repair Task Frequencies of HVAC Components by Climate Zone

Uniformat / Component / Repair Task	Climate Zone										
	1	2	3	4	5	6	7	8	9	10	11
Exhaust Fan, Ceiling, 200-500 Cfm											
Maintain Fan & Motor, Exhaust Fan, Ceiling 200-500 Cfm	1	1	1	1	1	1	1	1	1	1	1
Replace Fan & Motor, Exhaust Fan, Ceiling 200-500 Cfm	18	17	18	19	19	20	22	23	24	26	35
Exhaust Fan, Roof Mounted, 800 Cfm											
Maintain Exhaust Fan, Roof Mounted, 800 Cfm	1	1	1	1	1	1	1	1	1	1	1
Repair Exhaust Fan, Roof Mounted, 800 Cfm	9	8	9	10	10	10	11	11	12	13	18
Replace Exhaust Fan, Roof Mounted, 800 Cfm	18	17	18	19	19	20	22	23	24	26	35
Exhaust Fan, Roof Mounted, 1000 Cfm											
Maintain Exhaust Fan, Roof Mounted, 1000 Cfm	1	1	1	1	1	1	1	1	1	1	1
Repair Exhaust Fan, Roof Mounted, 1000 Cfm	9	8	9	10	10	10	11	11	12	13	18
Replace Exhaust Fan, Roof Mounted, 1000 Cfm	18	17	18	19	19	20	22	23	24	26	35
Exhaust Fan, Roof Mounted, 5000 Cfm											
Maintain Exhaust Fan, Roof Mounted, 5000 Cfm	1	1	1	1	1	1	1	1	1	1	1
Repair Exhaust Fan, Roof Mounted, 5000 Cfm	9	8	9	10	10	10	11	11	12	13	18
Replace Exhaust Fan, Roof Mounted, 5000 Cfm	18	17	18	19	19	20	22	23	24	26	35
Exhaust Fan, Roof Mounted, 10,000 Cfm											
Maintain Exhaust Fan, Roof Mounted, 10,000 Cfm	1	1	1	1	1	1	1	1	1	1	1
Repair Exhaust Fan, Roof Mounted, 10,000 Cfm	9	8	9	10	10	10	11	11	12	13	18
Replace Exhaust Fan, Roof Mounted, 10,000 Cfm	18	17	18	19	19	20	22	23	24	26	35
Exhaust Fan, Roof Mounted, 20,000 Cfm											
Maintain Exhaust Fan, Roof Mounted, 20,000 Cfm	1	1	1	1	1	1	1	1	1	1	1
Repair Exhaust Fan, Roof Mounted, 20,000 Cfm	9	8	9	10	10	10	11	11	12	13	18
Replace Exhaust Fan, Roof Mounted, 20,000 Cfm	18	17	18	19	19	20	22	23	24	26	35
Fan, Axial Flow, 5000 Cfm											
Maintain Fan, Axial Flow, 5000 Cfm	1	1	1	1	1	1	1	1	1	1	1
Repair Fan, Axial Flow, 5000 Cfm	9	8	9	10	10	10	11	11	12	13	18
Replace Fan, Axial Flow, 5000 Cfm	18	17	18	19	19	20	22	23	24	26	35

Components ordered by Uniformat II Classifications.

Sources: U.S. Army Corps of Engineers, Department of Defense Engineering Performance Standards, and Whitestone Research interviews

5.2 Repair Task Frequencies of HVAC Components by Climate Zone

Uniformat / Component / Repair Task	Climate Zone										
	1	2	3	4	5	6	7	8	9	10	11
Fan, Axial Flow, 10,000 Cfm											
Annual PM, Fan, Axial Flow, 10,000 Cfm	1	1	1	1	1	1	1	1	1	1	1
Repair Fan, Axial Flow, 10,000 Cfm	9	8	9	10	10	10	11	11	12	13	18
Replace Fan, Axial Flow, 10,000 Cfm	18	17	18	19	19	20	22	23	24	26	35
Fan, Axial Flow, 15,000 Cfm											
Maintain Fan, Axial Flow, 15,000 Cfm	1	1	1	1	1	1	1	1	1	1	1
Repair Fan, Axial Flow, 15,000 Cfm	9	8	9	10	10	10	11	11	12	13	18
Replace Fan, Axial Flow, 15,000 Cfm	18	17	18	19	19	20	22	23	24	26	35
Fan, Axial Flow, 30,000 Cfm											
Annual PM, Fan, Axial Flow, 30,000 Cfm	1	1	1	1	1	1	1	1	1	1	1
Repair Fan, Axial Flow, 30,000 Cfm	9	8	9	10	10	10	11	11	12	13	18
Replace Fan, Axial Flow, 30,000 Cfm	18	17	18	19	19	20	22	23	24	26	35
Supply Fan, Roof Mounted, 1000 Cfm											
Maintain Supply Fan, Roof Mounted, 1000 Cfm	1	1	1	1	1	1	1	1	1	1	1
Repair Supply Fan, Roof Mounted, 1000 Cfm	9	8	9	10	10	10	11	11	12	13	18
Replace Supply Fan, Roof Mounted, 1000 Cfm	18	17	18	19	19	20	22	23	24	26	35
Supply Fan, Roof Mounted, 5000 Cfm											
Maintain Supply Fan, Roof Mounted, 5000 Cfm	1	1	1	1	1	1	1	1	1	1	1
Repair Supply Fan, Roof Mounted, 5000 Cfm	9	8	9	10	10	10	11	11	12	13	18
Replace Supply Fan, Roof Mounted, 5000 Cfm	18	17	18	19	19	20	22	23	24	26	35
Supply Fan, Roof Mounted, 10,000 Cfm											
Maintain Supply Fan, Roof Mounted, 10,000 Cfm	1	1	1	1	1	1	1	1	1	1	1
Repair Supply Fan, Roof Mounted, 10,000 Cfm	9	8	9	10	10	10	11	11	12	13	18
Replace Supply Fan, Roof Mounted, 10,000 Cfm	18	17	18	19	19	20	22	23	24	26	35
D3050 Terminal & Package Units											
Air Conditioner, Rooftop, 5 Ton											
Maintain Air Conditioner, Rooftop, 5 Ton	1	1	1	1	1	1	1	1	1	1	1
Repair Air Conditioner, Rooftop, 5 Ton	9	10	15	15	20	10	32	46	46	>50	>50
Replace Air Conditioner, Rooftop, 5 Ton	13	14	23	22	29	15	48	>50	>50	>50	>50

Components ordered by Uniformat II Classifications.

Sources: U.S. Army Corps of Engineers, Department of Defense Engineering Performance Standards, and Whitestone Research interviews.

5.2 Repair Task Frequencies of HVAC Components by Climate Zone

Uniformat / Component / Repair Task	Climate Zone										
	1	2	3	4	5	6	7	8	9	10	11
Air Conditioner, Rooftop, 10 Ton											
Maintain Air Conditioner, Rooftop, 5 Ton	1	1	1	1	1	1	1	1	1	1	1
Repair Air Conditioner, Rooftop, 10 Ton	9	10	15	15	20	10	32	46	46	>50	>50
Replace Air Conditioner, Rooftop, 10 Ton	13	14	23	22	29	15	48	>50	>50	>50	>50
Air Conditioner, Rooftop, 20 Ton											
Maintain Air Conditioner, Rooftop, 20 Ton	1	1	1	1	1	1	1	1	1	1	1
Repair Air Conditioner, Rooftop, 20 Ton	9	10	15	15	20	10	32	46	46	>50	>50
Replace Air Conditioner, Rooftop, 20 Ton	13	14	23	22	29	15	48	>50	>50	>50	>50
Air Conditioner, Rooftop, 50 Ton											
Maintain Air Conditioner, Rooftop, 50 Ton	1	1	1	1	1	1	1	1	1	1	1
Repair Air Conditioner, Rooftop, 50 Ton	9	10	15	15	20	10	32	46	46	>50	>50
Replace Air Conditioner, Rooftop, 50 Ton	13	14	23	22	29	15	48	>50	>50	>50	>50
Air Conditioner, Window, 1 Ton											
Maintain Air Conditioner, Window, 1 Ton	1	1	1	1	1	1	1	1	1	1	1
Replace Air Conditioner, Window, 1 Ton	9	10	15	15	20	10	32	46	46	>50	>50
Air Conditioner, Window, 2 Ton											
Maintain Air Conditioner, Window, 2 Ton	1	1	1	1	1	1	1	1	1	1	1
Replace Air Conditioner, Window, 2 Ton	9	10	15	15	20	10	32	46	46	>50	>50
Air Conditioner, Indoor DX Packaged, Air Cooled, 20 Ton											
Maintain Air Conditioner, Indoor DX Packaged, Air Cooled, 20 Ton	1	1	1	1	1	1	1	1	1	1	1
Repair Air Conditioner, Indoor DX Packaged, Air Cooled, 20 Ton	9	10	15	15	20	10	32	46	46	>50	>50
Replace Air Conditioner, Indoor DX Packaged, Air Cooled, 20 Ton	18	19	30	30	39	20	>50	>50	>50	>50	>50
Air Conditioner, Indoor DX Packaged, Air Cooled, 5 Ton											
Maintain Air Conditioner, Indoor DX Packaged, Air Cooled, 5 Ton	1	1	1	1	1	1	1	1	1	1	1
Repair Air Conditioner, Indoor DX Packaged, Air Cooled, 5 Ton	9	10	15	15	20	10	32	46	46	>50	>50
Replace Air Conditioner, Indoor DX Packaged, Air Cooled, 5 Ton	18	19	30	30	39	20	>50	>50	>50	>50	>50
Air Conditioner, Indoor DX Packaged, Air Cooled, 50 Ton											
Maintain Air Conditioner, Indoor DX Packaged, Air Cooled, 50 Ton	1	1	1	1	1	1	1	1	1	1	1
Repair Air Conditioner, Indoor DX Packaged, Air Cooled, 50 Ton	9	10	15	15	20	10	32	46	46	>50	>50
Replace Air Conditioner, Indoor DX Packaged, Air Cooled, 50 Ton	18	19	30	30	39	20	>50	>50	>50	>50	>50

Components ordered by Uniformat II Classifications.

Sources: U.S. Army Corps of Engineers, Department of Defense Engineering Performance Standards, and Whitestone Research interviews

5.2 Repair Task Frequencies of HVAC Components by Climate Zone

Uniformat / Component / Repair Task	Climate Zone										
	1	2	3	4	5	6	7	8	9	10	11
Baseboard Radiator 10 ft.											
Replace Baseboard Radiator, 10 ft.	>50	32	24	22	19	20	15	15	14	12	12
Electric Cabinet Heater											
Inspect & Maintain Electric Cabinet Heater	1	1	1	1	1	1	1	1	1	1	1
Replace Electric Cabinet Heater	13	14	23	22	29	15	48	>50	>50	>50	>50
Explosion Proof Suspended Heater											
Inspect & Maintain Explosion Proof Suspended Heater	1	1	1	1	1	1	1	1	1	1	1
Replace Contactor, Explosion Proof Suspended Heater	2	2	2	2	2	2	2	2	2	2	2
Replace Explosion Proof Suspended Heater	13	14	23	22	29	15	48	>50	>50	>50	>50
Finned Radiator											
Replace Finned Radiator, 10 ft.	>50	32	24	22	19	20	15	15	14	12	12
Radiant Heating Panel, Electric, 8 sf											
Maintain Radiant Heating Panel, Electric, 8 sf	1	1	1	1	1	1	1	1	1	1	1
Repair Wiring, Radiant Heating Panel, Electric, 8 sf	2	2	2	2	2	2	2	2	2	2	2
Replace Radiant Heating Panel, Electric, 8 sf	15	15	15	15	15	15	15	15	15	15	15
Fan Coil, Two-Pipe, 400 Cfm											
Maintain Fan Coil, Two-Pipe, 400 Cfm	1	1	1	1	1	1	1	1	1	1	1
Repair Fan Coil, Two-Pipe, 400 Cfm	9	10	15	15	20	10	32	46	46	>50	>50
Replace Fan Coil, Two-Pipe, 400 Cfm	13	14	23	22	29	15	48	>50	>50	>50	>50
Fan Coil, Two-Pipe, 1800 Cfm											
Maintain Fan Coil, Two-Pipe, 1800 Cfm	1	1	1	1	1	1	1	1	1	1	1
Repair Fan Coil, Two-Pipe, 1800 Cfm	9	10	15	15	20	10	32	46	46	>50	>50
Replace Fan Coil, Two-Pipe, 1800 Cfm	13	14	23	22	29	15	48	>50	>50	>50	>50
Heat Pump, 1 Ton											
Maintain Heat Pump, 1 Ton	1	1	1	1	1	1	1	1	1	1	1
Repair Heat Pump, 1 Ton	9	10	15	15	20	10	32	46	46	>50	>50
Replace Heat Pump, 1 Ton	18	19	30	30	39	20	>50	>50	>50	>50	>50

Components ordered by Uniformat II Classifications.

Sources: U.S. Army Corps of Engineers, Department of Defense Engineering Performance Standards, and Whitestone Research interviews.

5.2 Repair Task Frequencies of HVAC Components by Climate Zone

Uniformat / Component / Repair Task	Climate Zone										
	1	2	3	4	5	6	7	8	9	10	11
Heat Pump, 5 Ton											
Maintain Heat Pump, 5 Ton	1	1	1	1	1	1	1	1	1	1	1
Repair Heat Pump, 5 Ton	9	10	15	15	20	10	32	46	46	>50	>50
Replace Heat Pump, 5 Ton	18	19	30	30	39	20	>50	>50	>50	>50	>50
Heat Pump, 10 Ton											
Maintain Heat Pump, 10 Ton	1	1	1	1	1	1	1	1	1	1	1
Repair Heat Pump, 10 Ton	9	10	15	15	20	10	32	46	46	>50	>50
Replace Heat Pump, 10 Ton	18	19	30	30	39	20	>50	>50	>50	>50	>50
Heat Pump, 25 Ton											
Maintain Heat Pump, 25 Ton	1	1	1	1	1	1	1	1	1	1	1
Repair Heat Pump, 25 Ton	9	10	15	15	20	10	32	46	46	>50	>50
Replace Heat Pump, 25 Ton	18	19	30	30	39	20	>50	>50	>50	>50	>50
Suspended Heater, Hot Water, 1200 Cfm, 55 Mbh											
Maintain Suspended Heater, Hot Water, 1200 Cfm, 55 Mbh	1	1	1	1	1	1	1	1	1	1	1
Repair Suspended Heater, Hot Water, 1200 Cfm, 55 Mbh	9	10	15	15	20	10	32	46	46	>50	>50
Replace Suspended Heater, Hot Water, 1200 Cfm, 55 Mbh	13	14	23	22	29	15	48	>50	>50	>50	>50
Suspended Heater, Hot Water, 4000 Cfm, 180 Mbh											
Maintain Suspended Heater, Hot Water, 4000 Cfm, 180 Mbh	1	1	1	1	1	1	1	1	1	1	1
Repair Suspended Heater, Hot Water, 4000 Cfm, 180 Mbh	9	10	15	15	20	10	32	46	46	>50	>50
Replace Suspended Heater, Hot Water, 4000 Cfm, 180 Mbh	13	14	23	22	29	15	48	>50	>50	>50	>50
Suspended Heater, Hot Water, 8000 Cfm, 360 Mbh											
Maintain Suspended Heater, Hot Water, 8000 Cfm, 360 Mbh	1	1	1	1	1	1	1	1	1	1	1
Repair Suspended Heater, Hot Water, 8000 Cfm, 360 Mbh	9	10	15	15	20	10	32	46	46	>50	>50
Replace Suspended Heater, Hot Water, 8000 Cfm, 360 Mbh	13	14	23	22	29	15	48	>50	>50	>50	>50
Secondary Coil, 30" x 72"											
Replace Secondary Coil, 30" x 72"	13	14	23	22	29	15	48	>50	>50	>50	>50
Terminal Reheat Unit, 10,000 Cfm											
Maintain Terminal Reheat Unit, 10,000 Cfm	1	1	1	1	1	1	1	1	1	1	1
Repair Terminal Reheat Unit, 10,000 Cfm	9	10	15	15	20	10	32	46	46	>50	>50
Replace Terminal Reheat Unit, 10,000 Cfm	13	14	23	22	29	15	48	>50	>50	>50	>50

Components ordered by Uniformat II Classifications.

Sources: U.S. Army Corps of Engineers, Department of Defense Engineering Performance Standards, and Whitestone Research interviews.

5.2 Repair Task Frequencies of HVAC Components by Climate Zone

Uniformat / Component / Repair Task	Climate Zone										
	1	2	3	4	5	6	7	8	9	10	11
Terminal Reheat Unit, 25,000 Cfm											
Maintain Terminal Reheat Unit, 25,000 Cfm	1	1	1	1	1	1	1	1	1	1	1
Repair Terminal Reheat Unit, 25,000 Cfm	9	10	15	15	20	10	32	46	46	>50	>50
Replace Terminal Reheat Unit, 25,000 Cfm	13	14	23	22	29	15	48	>50	>50	>50	>50
D3060 Controls & Instrumentation											
Direct Digital Controls, System Points											
Annual PM, Direct Digital Controls, System Points	1	1	1	1	1	1	1	1	1	1	1
Replace Direct Digital Controls, System Points	10	10	10	10	10	10	10	10	10	10	10
Pneumatic Thermostat											
Annual PM, Pneumatic Thermostat	1	1	1	1	1	1	1	1	1	1	1
Replace Pneumatic Thermostat	10	10	10	10	10	10	10	10	10	10	10

Components ordered by Uniformat II Classifications.

Sources: U.S. Army Corps of Engineers, Department of Defense Engineering Performance Standards, and Whitestone Research interviews.

Chapter 6
Definitions and Methods

6. Definitions and Methods

In this chapter we define key terms and describe the methods and assumptions behind Whitestone statistics. The discussion is loosely organized by chapter, with the exception of a number of terms that are used throughout the book and need to be defined at the outset.

Building maintenance and repair (M&R) is defined as the collection of activities necessary for keeping a building in good working order. Included in this definition are regularly scheduled adjustments and inspections, preventive maintenance tasks, emergency response and service calls for minor repairs, and the periodic major repair or replacement of building components. Other tasks associated with facilities operation such as custodial services, landscaping, waste disposal, and the provision of central utilities are not included in our definition of M&R.

For the purposes of this book, M&R activities are divided into three types:

1. ***Preventative maintenance (PM) and minor repair*** consists of scheduled tasks that sustain a component's level of service during a prescribed lifetime. These tasks are usually done frequently and require a relatively constant amount of labor and materials. In generating the cost profiles in Chapter 2, we assume that the facility staff does PM and minor repairs.

2. ***Unscheduled maintenance*** consists of service calls, emergency response, and other tasks that cannot be individually anticipated. Like PM and minor repair, these tasks tend to consume a steady level of annual facility resources. In estimating building cost profiles, we assume that the facility staff is responsible for unscheduled maintenance. The details of estimating unscheduled maintenance are discussed later in this chapter.

3. ***Renewal and replacement*** consists of component overhaul or replacement tasks. These are tasks that extend a component lifetime, and reset the schedule of PM and minor repair tasks. These tasks tend to be expensive, infrequent, and require specialized expertise and equipment. In generating building model cost profiles, we assume this work is done by contract.

Maintenance and Repair does not include a collection of related costs that are often considered capital expenditures. The distinction is summarized neatly in a taxonomy that has been adopted in Federal accounting language: **Sustainment** costs include the three types of costs presented in this book, while **Restoration and Modernization** costs are addressed elsewhere (see Additional Reading).

Table 6-1
M&R Requirements

Sustainment:	Preventative maintenance and minor repair
	Unscheduled maintenance
	Renewal & replacement
Restoration & Modernization:	Replacement due to obsolescence
	Change in use modifications
	Policy mandated retrofits
	Catastrophic damage
	Repairs from neglect
	Long-lived components

Source: Whitestone Report A Standard Terminology for Describing Facility Maintenance and Repair Requirements. Santa Barbara, CA: Whitestone Research, July 2003.

Chapter 2: Building M&R Cost Profiles

The profiles presented in Chapter 2 provide a comprehensive estimate of building M&R costs over a 50-year period, beginning with the completion of building construction. Profiles were estimated for 51 common building types, a sample intended to reflect the diversity of commercial and public building inventories, and provide cost analysts with a useful set of models. Selection of components was based on both actual building surveys and staff judgment of industry design practices. Particular care was taken to select HVAC equipment appropriate for each building's estimated heating and cooling loads.

Figure 6-1
MARS Cost Forecast Model

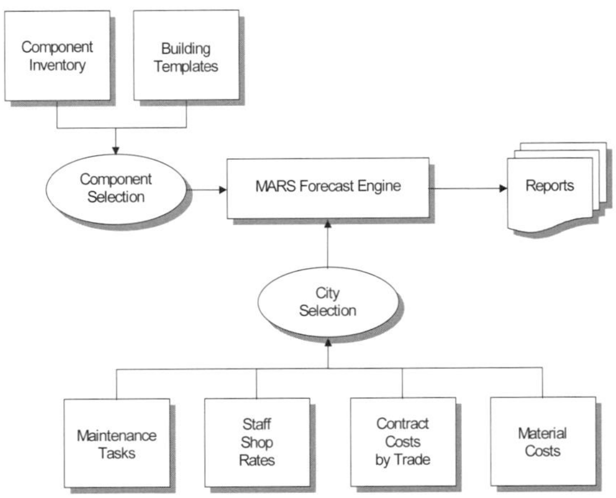

The estimation of each profile requires two basic inputs: the facility location (which establishes local labor rates and the frequency of climate-specific maintenance tasks), and an inventory of building components. The component inventory includes those components required for a safe and comfortable environment consistent with building usage. Not included in the building inventories are furniture, fixtures, appliances, or specialized equipment; notable examples include movable partitions in office buildings, restaurant equipment, gymnasium seating, bowling lanes and pin-setting equipment. Certain specialized equipment is included in component inventories for the jail and hospital models.

Also not included are components that typically require little or no major maintenance within 50 years, such as foundations and electrical wiring. No explicit provision is made for the costs of remediation measures such as asbestos or lead paint removal.

The actual estimation of M&R costs is done with the MARS Facility Maintenance Cost Forecast System, a software package developed by Whitestone Research. The estimation process (shown in Figure 6-1) draws from a relational database that includes building templates, component-related maintenance tasks, in-house shop rates and contract trade rates, and material and equipment costs. Output from the MARS system is a detailed year-by-year forecast of M&R tasks and costs presented at various levels of aggregation, as shown in the Chapter 2 profiles.

MARS is an asset management and budget tool currently used by many government agencies, corporations, and large individual facilities. Technical details, case studies, and ordering information are available at www.whitestoneresearch.com.

Accuracy of the Whitestone cost profiles. Ideally, the accuracy of Whitestone estimates could be measured by direct comparison with actual M&R costs. Unfortunately, data strictly comparable in detail and definition is not available. Information that is available from industry surveys tends to address a broad collection of facility services, rather than focusing on building M&R. Moreover, these surveys tend to emphasize operating costs and revenues, with little information on capital expenditures. This makes it difficult to get a complete view of M&R costs, as many major repairs are defined as capital improvements rather than operating expenses. To the extent this is true, M&R costs estimates derived from such surveys can understate actual costs.[2]

However, studies of deferred maintenance using the MARS system are capable of producing estimates close to those made by traditional condition inspections. A recent experiment by the Department of

[2] M&R costs reported by one survey (*2002 BOMA Experience Exchange Report*) averaged $1.82 per square foot for a sample of 55 office buildings in downtown Washington DC. The use of survey-based cost data is reviewed in *Whitestone Report: Notes on M&R Cost Benchmarks*. Santa Barbara: Whitestone Research, August 1996.

Energy found that MARS-based estimates of deferred maintenance were within 6 percent of costs determined by physical inspection, but required only a fraction of the costs.[3]

Sensitivity of the estimated profiles. In Table 6-2, the two-story office building model (shown in Chapter 2) is used to demonstrate the sensitivity of Whitestone estimates to a variety of alternative assumptions and MARS model settings.

Changing individual components can have an appreciable affect on average costs, and changing combinations of components can substantially alter M&R requirements. Substituting terrazzo flooring for carpeting, for example, can reduce M&R costs by an estimated 8 percent over a 50-year period. Type of building use can also affect M&R costs, both in terms of the number and type of components, and the necessary frequency of repair. An office building with high lighting requirements would require M&R expenditures an estimated 6 percent greater than those for the original building. The same office building with more frequent wall repainting—typical of many public buildings, for example—would require expenditures an estimated 4 percent greater than the model building.

Table 6-2
Sensitivity of Average Annual M&R Cost Estimates

Office Building, Two Story, in Wash. DC, as described in Chapter 2.	100%
Component Selection:	
• Select terrazzo flooring in place of carpet (75% of gsft.)	- 8%
• Select painted stucco exterior in place of clay brick	- 7%
• Increase lighting from 3 watts per gsft. to 6 watts	+ 6%
• Change painting frequency of interior walls from every 4 to every 2 years.	+ 4%
Location:	
• Washington. DC	100%
• Beaufort, SC	- 45%
• New York, NY	+ 59%
Average Unscheduled Maintenance Multiplier:	
• 2.10	100%
• 1.00	- 18%
• 3.00	+ 15%
Average In-house Shop Markup Rate:	
• 2.00	100%
• 1.00	- 11%
• 3.00	+ 11%
Utilization:	
• Typical Schedule	100%
• Three Shift, 24/7	+ 22%
• Minimal use	- 13%

Source: Whitestone Research

The sensitivity of estimated costs to geographic location is demonstrated clearly in the local maintenance cost index presented in Chapter 3. M&R costs for the same building model range from $1.17 per gsft. in Beaufort, SC to $3.39 in New York, NY.

Costs forecasts are affected by different assumptions about unscheduled maintenance multipliers. Setting all multiplier values to 1 (there is a separate multiplier for each building system) implies there is no unscheduled maintenance, and reduces model costs by 18 percent. Alternatively, raising all multipliers to 3 increases model costs by 15 percent. Multiplier values range from about 1.25 to 2.75, according to survey results discussed in a following section.

Shop markup is the multiplier applied to direct staff wages to establish a shop rate, the hourly rate a facility shop (carpentry, HVAC, etc.) must bill internal clients to fully recoup operating costs. For the purposes of this study, we assume a markup rate of 2 for each of eight staff trade categories. An alternative markup rate of 1 reduces forecast M&R costs by 11 percent, while a markup rate of 3 increases forecast costs by 11 percent.

Different types of utilization can affect building M&R costs. Alternative utilization scenarios can be described in terms of hours of operation, type of clientele, and volume of users.[4] Two alternative scenarios were defined based just on hours of operation. Around the clock operation increases estimated costs by 22 percent over single shift operation. Minimal operation, which extends PM schedules and involves no service calls or major repairs, would have estimated M&R costs of just 13 percent of typical usage.

---

[3] See *Parametric Estimation of Deferred Maintenance, Phase 1 Prototype at Lawrence Livermore National Laboratory.* Prepared for the National Nuclear Security Administration by Whitestone Research and Jacobs Facilities, May 2004. Also see *Whitestone Report: Current and Future M&R Funding Needs for a Historic School.* Santa Barbara: Whitestone Research, July 2002.

[4] The estimation of the impact of alternative utilization rates on M&R costs is discussed in the project report: *Implementation of the Department of Defense Sustainment Model.* Prepared for the Office of the Deputy Under Secretary of Defense (Installations) by Jacobs Facilities and Whitestone Research, April 2001.

Extraordinary work environments can also affect M&R costs. In one study, security requirements typical for military facilities were found to increase M&R labor requirements by 3 percent. The same study found that travel to and from remote work sites increased labor costs by an additional 15 percent. For nuclear related facilities, the overhead rate on labor costs can be over a third higher than typical rates due to elaborate permitting and safety requirements.[5]

Unscheduled maintenance multipliers. Unscheduled maintenance costs are often overlooked in component life-cycle based studies. For example, a major shortcoming in the Maintenance Resource Prediction Model (MRPM), a pioneering system developed by the U.S. Army Corps of Engineers in the 1980's, was that it failed to account for maintenance costs that could not be defined as scheduled tasks.[6] One study found that the MRPM underestimated costs by 24 to 40 percent over a 50-year lifetime.[7]

The MARS model is the first methodology we are aware of that explicitly recognizes unscheduled maintenance costs. Using this model, unscheduled maintenance costs are estimated using a multiplier applied to scheduled PM and minor repair labor hours. A separate multiplier is defined for each building system. For example, a facility or building may have a predicted 100 hours of PM and minor repair for interior surfaces in a given year; applying a multiplier value of 1.67, the total required for interior surfaces would be 167 hours (100 scheduled and 67 unscheduled). The values of the multipliers by building system, shown in Table 6-3, were determined by Whitestone survey.[8]

Table 6-3
Unscheduled Maintenance Multipliers by Building System

Building System	Unsched Percent[1]	Unsched Multiplier[2]
Exterior Closure	60.3	2.52
Roofing	51.7	2.07
Interior Construction	63.0	2.70
Staircases	55.0	2.22
Interior Finishes	40.1	1.67
Conveying Systems	21.3	1.27
Plumbing	62.5	2.67
HVAC	43.8	1.78
Fire Protection	24.2	1.32
Electrical	47.4	1.90
Equipment	49.7	1.99

[1]Unscheduled maintenance hours as a percentage of total (unscheduled and scheduled PM) minor repairs.
[2]Used in MARS to estimate total minor repair as a function of scheduled PM hours: M = 1 / (1-Unscheduled Maint %).
Source: Whitestone Survey

Unscheduled maintenance is expressed as a multiple of scheduled PM and minor repair, rather than all M&R tasks, because historical cost data for scheduled PM and minor repairs is readily available. Most facilities can use their work order histories to compute the simple percentage of unscheduled maintenance to their total workload by building system (not including renewal and replacement work). These percentages allow us, and other MARS users, to calibrate the model based on actual experience. The definition of building systems corresponds to level 2 group elements of the Uniformat II classification shown in the Appendix.

Building replacement costs. M&R costs are commonly expressed as a percentage of replacement costs, as in the summary section of each profile in Chapter 2. The replacement cost of each building model was adjusted for location and inflation using the City Cost Indexes from R.S. Means.

Chapter 3: Local M&R Costs

Adjusting the profiles for other metro areas. The primary use of the local maintenance cost indexes shown in Chapter 3.1 is for localizing the cost profiles shown in Chapter 2. These indexes are based on cost forecasts made for the same building model for each of the 210 metropolitan areas addressed in this book. The two-story office building described in Chapter 2 was used as the model. Each index is calculated as the simple ratio of costs for a selected area over costs for Washington DC, where the Washington DC value equals 100. These measures are also useful for inter-regional comparisons of M&R costs, as shown in our reported rankings.

---

[5] *Whitestone Report: Sensitivity of M&R Estimates to Alternative Assumptions.* Santa Barbara: Whitestone Research, January 2003.
[6] See Neely and Neathammer, *Maintenance Resource Prediction in the Facility Life-Cycle Process. Technical Report P-91/10.* Champaign: U.S. Army Corps of Engineers, May 1991.
[7] Lufkin and Desai, *The Estimation of Building Maintenance and Repair Costs.* Seattle: Science Applications International Corporation, 1992.
[8] *Whitestone Report: Estimates of Unscheduled Facility Maintenance.* Santa Barbara: Whitestone Research, April 1999.

In-House Shop Rates In estimating the cost profiles shown in Chapter 2, we assumed that the facility staff does PM, minor repair tasks, and unscheduled maintenance. The labor costs of these tasks are estimated with shop rates specific to selected trades and management positions. The shop rate includes direct wages and a shop markup. Whitestone estimates direct wages using knowledge of the typical distribution of wages across trades and their relationship to commercial contract rates. Earlier editions of this book reported in-house wage rates for maintenance workers taken directly from the Department of Labor Occupational Compensation Surveys. This data is no longer available.

The shop markup includes fringe and required benefits, non-productive time, support and supervision, and most tools and equipment costs. It is expressed as a multiplier applied to direct wages. For example, a maintenance carpenter in Washington DC has a direct wage rate of $30.68, a markup of $61.36 given a multiplier value of 2, and a total shop rate of $92.04. In actual experience, the markup rate is unique to every facility—the amount of non-productive time and support and supervision costs rolled into the rate, for example, are a function of workload and facility policy. The markup rate of 2 used in this book is an approximation based on our interviews with facility management and a review of secondary sources. Note that a facility markup rate can easily vary within a range from 1 to 3; the cost implications of this range are shown in Table 6-2.

Table 6-4
Estimated Construction Wage Changes, 2004

Region/State	
New England	
CT, MA, ME, NH, RI, VT	4.8%
Middle Atlantic	
DC, DE, MD, NJ, NY, PA	4.3%
Southeast	
AL, FL, GA, KY, MS, NC, SC, TN, VA	4.1%
East North Central	
IL, IN, MI, MN, OH, WI, WV	4.1%
West North Central	
IA, KS, MO, NE CO, MT, ND, SD, UT, WY	4.5%
South Central	
AR, LA, NM, OK, TX	3.4%
Southwest Pacific	
AZ, CA, HI, NV	3.4%
Northwest	
AK, ID, OR, WA	3.0%
National	4.0%

Source: Construction Labor Research Council

Contract Labor Rates Contract labor rates are the basis for the hourly rates used in generating cost forecasts for renewal and replacement tasks. Specified by trade for each metro area, contract labor costs include basic wage and fringe, compensation insurance, and other required benefits. When used for estimation purposes, these costs are burdened with an overhead and profit multiplier of 1.30, a rate common in light construction contracts.

Table 6-4 illustrates estimated construction wage changes, including fringe and other benefits, by geographic area as reported by the Construction Labor Research Council (CLRC). The weighted national average is expected to increase 4% for 2004. This increase is based on union wage agreements, and is much greater than the 0.8 percent increase foreseen in Chapter 4 for general construction labor.

For earlier editions of this book, ***basic wage and fringe*** rates were taken primarily from surveys of local wages conducted by the U.S. Department of Labor under the Davis-Bacon Act. These surveys are conducted regularly to determine the wage rate to be used in evaluating construction bids for federal projects. Both union and non-union wages are considered in determining prevailing rates by locality. State and local wage surveys were also used when available.

Table 6-5
Average Compensation Insurance Rates, 2004

	NCCI Code	Rate*
Carpenter	5403	$17.56
Carpet layer	9521	17.68
Electrician	5190	8.86
Plumber	5183	7.67
Painter	5474	12.86
Elevator Mechanic	5160	7.24
Brick Layer	5022	15.10
Roofer	5551	31.13
Crane Operator	6017	9.55
Fork Lift Driver	6017	9.59

*Per $100 of payroll

Source: R.S. Means

For this edition, basic wage and fringe data was provided by R.S. Means and compared with the U.S. Department of Labor surveys. The Means data was found to be very similar to the Davis-Bacon rates. The relatively small difference is attributed to the fact that the Means data is derived exclusively from union wage agreements.

Compensation insurance rates for earlier editions of this book were determined using data from the National Council on Compensation Insurance (NCCI) and interviews with the Workers Compensation Rating Board (or similar organization) from selected states. For this edition, rates were provided by R.S. Means. We believe these are reasonable approximations of prevailing insurance rates, although there is considerable variation among the state programs and differing rates for M&R contractors depending on their past claim performance. Table 6-5 shows the average distribution of insurance rates by trade for the 210 metro areas represented in this book.

Other required benefits include social security, state and federal unemployment costs, and risk and liability insurance. These costs were again provided by R.S. Means and verified with Davis-Bacon values.

Chapter 4: M&R Construction Indexes

Cost trends vary among the different divisions within the M&R construction industry. To capture these differences we estimate three specific M&R indexes: *Non-Residential Buildings M&R,* which applies to repairs and improvements to commercial and public buildings; *Residential Buildings M&R,* which applies to owner-occupied and rental units—not including major additions or alterations, or repairs to motor homes; and *Military Facilities M&R,* which applies to overall military M&R expenditures, including roads, airfields, utilities, docks, and other non-building assets.[9]

Figure 6-2
Comparison of M&R Indexes

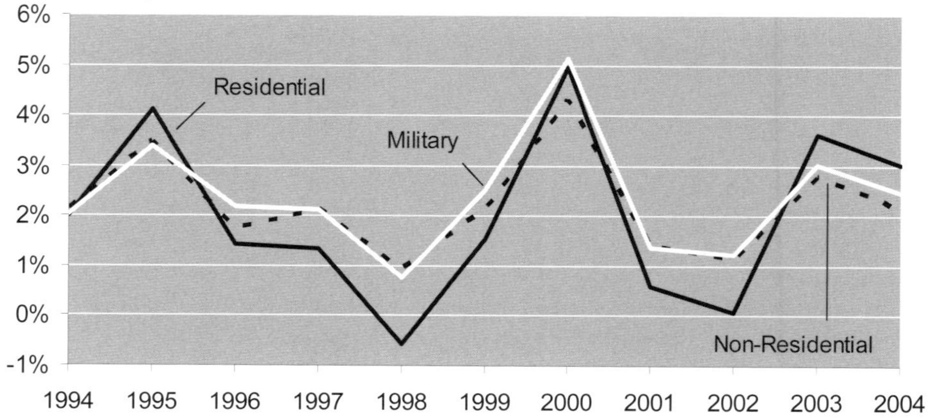

The comparison in Figure 6-2 shows only slight differences in the annual rate of change among the indexes for most years, particularly for military facilities and non-residential buildings. The similarity is due in large part to the almost identical weight given to labor, the single most important component in these measures. A number of commodities and services are inputs to the various industries, but labor accounts for 47-48 percent of total production costs, except in the residential M&R construction industry. Much of the labor in residential M&R is uncompensated work done by the owner/resident or small-scale landlord, thus labor makes up a relatively smaller portion of total costs.

The components of the index and their respective weights were defined using national input/output relationships for the selected maintenance construction industries. Sources include R.S. Means (new construction costs) and U.S. Bureau of Labor Statistics (producer and consumer price indexes). Each of the indexes reported was standardized to the base month of December 1984 (i.e., December, 1984=100). Projections of average 2004 values are based on second quarter-to-second quarter changes from 2003 to 2004.

Chapter 5: Component Data

The primary source of the component data shown in Chapter 5 is the Maintenance Task Database developed by the U.S. Army Corps of Engineers.[10] A detailed example of the task data is shown in Box 6-1. Developed in the 1980's, much of the labor information for this database is based on Engineered Performance Standards developed jointly by the U.S. Navy, Army, and Air Force.[11] For our purposes, this data was supplemented with components common to civilian construction (such as elevators, storefront

---

[9] For an overview of index use and construction, see Frumkin, Norman, *Guide to Economic Indicators.* Armonk, New York: M.E. Sharp, 1994. Pp. 3-21. Park and Sharpe-Bette, *Advanced Engineering Economics.* New York: John Wiley & Sons, 1990 is a good source on the use of indexes in financial and economic analysis.

[10] See *Maintenance Task Data Base for Buildings* [Various Systems], U.S. Army Corps of Engineers, May 1991.

[11] See *Engineered Performance Standards, Public Works Maintenance* [Various Handbooks], Naval Facilities Engineering Command *et al.*, February 1982.

doors and windows, and detailed wall and floor coverings). Labor requirements and material costs were updated through extensive interviews with manufacturers, distributors and service providers. Equipment requirements are shown only for heavy HVAC and electrical equipment. For most M&R tasks it is assumed that the necessary equipment is included in contract overhead costs.

Box 6-1
Example of a Maintenance Task:
Membrane Removal & Replacement for a Built-up Roof

Labor Requirements:

Subtask	Hours per Gsft.
Set up, secure, and take down ladder	0.000160
Sweep ballast	0.005011
Tear off existing membrane & insulation	0.016162
Replace insulation	0.002000
Install membrane	0.009769
Reinstall ballast	0.003886
Clean up	0.001000
	0.037988

Material

Description	Cost per Gsft.
Insulation, membrane & ballast	1.28
	1.28

Task frequencies shown in Chapter 5.1 are the expected incident of repair or replacement in the Washington DC area. These frequencies are assumed to hold for other regions. The major exception to this assumption is HVAC equipment, for which specific frequencies are reported in Chapter 5.2 for each climate zone. M&R task frequencies for roofing surfaces are also thought to vary by climatic zone, but we have found no data to statistically support this hypothesis.[12] Moreover, interviews with experts indicate that roof service life might vary by climate, but quality of installation is a much greater influence.

Equipment Costs. These are a function of equipment and operator labor costs in a given metro area. The rates used in estimating the profiles in Chapter 2 are shown in Table 6-6.

While the labor rates shown are applicable to the Washington DC area, they can be modified for other areas by substituting the appropriate contract labor rate shown in Chapter 3.4; equipment rental costs are similar across metro areas and require no adjustment. Note that these rates do not include contractor markup for overhead and fee (typically an additional 30 percent for light construction).

Table 6-6
Equipment Rates, 2004

Equip Type		Rental Rate	Contract Labor	Hourly Rate*
1	2 ton forklift	$51.00	$29.29	$80.29
2	45 ton crane	229.00	30.18	259.18
3	4 ton forklift	61.00	29.29	90.29
4	90 ton crane	539.00	30.18	569.18
5	2t forklift, 45t crane	280.00	59.47	339.47
6	4t forklift, 90t crane	600.00	59.47	659.47

*Hourly rate in Washington DC, including equipment rental, set-up fee, and operator (average hourly rates were calculated assuming a minimum one-day rental period). Does not include contractor markup. Labor rates can be adjusted for other areas using contract rates in Chapter 3.4.

Appendix: Reference Tables

Building M&R Labor Requirements. The MARS model used for generating M&R cost profiles can also be used to generate labor requirements. In fact, the same type of detailed building profile shown in Chapter 2 could be prepared for M&R labor requirements. This type of analysis can be used for a variety of purposes, including benchmarking staff size and composition for existing facilities, and anticipating the labor requirements of planned facilities. Table 6-7 provides an example of labor requirements computed for the office building shown in Chapter 2. A summary of average labor requirements for all building models is shown in Table A-2 in the Appendix.

[12] Our review of a large U.S. sample of home repair data showed no significant regional patterns in the incidence of roof repairs or average roof age. See NAHB Research Center, Inc., *1999 Consumer Repair and Replacement Data Files and Documentation, CD-ROM.* Upper Marlboro, MD, 2000.

Service Lives of Buildings and Infrastructure. The service life of a facility is a key variable in long-range M&R planning. Service life estimates provide an end-point for asset depreciation schedules and, similarly, are used for calculating required capital investment in facilities. For example, with a large portfolio of buildings, knowledge of facility replacement value and estimated service life can be used to approximate the annual rate of capital expenditures required to maintain the portfolio asset value (replacement cost ÷ target service life = annual recapitalization). Service lives are used to calculate the remaining service life of facilities—a common metric—and are used more broadly for estimating the value of national capital stocks. Estimated service lives for selected building and infrastructure types are shown in Table A-4 in the Appendix.

Table 6-7
Average Annual Labor Requirements: Office Building, Two Story

	In-House Staff	Contract Labor	Total Hours
PM & Minor Repair	323	5	328
Unscheduled Maintenance	373	3	376
Renewal & Replacement	0	1,119	1,119
	696	1,127	1,823

Includes only direct labor (not including supervision or support) for the Office Building, Two Story model described in Chapter 2. Average annual requirements are calculated over a 50-year building lifetime.

Source: Whitestone Research

Additional Reading

Additional Reading

A chronological list of references, organized by general topic.

Surveys of M&R Funding Policies and Planning

Organization for Economic Co-operation and Development (OECD*), Measuring Capital, A Manual on the Measurement of Capital Stocks, Consumption of Fixed Capital, and Capital Services* (Paris: OECD, 2003).

Federal Facility Council. *Deferred Maintenance Reporting for Federal Facilities.* Federal Facilities Council Technical Report #141. (Washington: National Academy Press, 2001).

U.S. Government Accounting Office, Military *Infrastructure: Real Property Management Needs Improvement* (Washington: U.S.Government Printing Office, September 1999) NSIAD-99-100.

National Research Council, *Stewardship of Federal Facilities: A Proactive Strategy for Managing the Nation's Public Assets* (Washington: National Academy Press, 1998).

Civil Engineering Research Foundation, *Level of Investment Study of Facilities and Infrastructure Maintenance and Repair* (Washington: Civil Engineering Research Foundation, 1996).

Melvin, Eric *Plan. Predict. Prevent. How to Reinvest in Public Buildings* (Chicago: American Public Works Association, 1992).

National Research Council, *Committing to the Cost of Ownership: Maintenance and Repair of Public Buildings* (Washington: National Academy Press, 1990).

Rush, Sean and Johnson, Sandra, *The Decaying American Campus, A Ticking Time Bomb* (Washington: Association of Physical Plant Administrators of Universities and Colleges, and the National Association of College and University Business Officers, 1989).

Department of Defense, *Renewing the Built Environment; Department of Defense Report to Congress* (Washington: Department of Defense, 1989).

M&R Funding Strategies

National Research Council, *Investments in Federal Facilities: Asset Management Strategies for the 21st Century* (Washington: National Academy Press, 2004).

Adams, Matthew, *Successful Funding Strategies for Facility Renewal* (Alexandria: APPA, 1997)

Federal Facility Council. *Budgeting for Facilities Maintenance and Repair Activities.* Federal Facilities Council Technical Report #131. (Washington: National Academy Press, 1996).

Rush, Sean, *Managing the Facilities Portfolio: A Practical Approach to Institutional Facility Renewal and Deferred Maintenance* (Washington: National Association of College and University Business Officers, 1991).

Critical Issues in Facilities Management, Volume 4: Capital Renewal and Deferred Maintenance, Steven Glazner, Editor (Washington: Association of Physical Plant Administrators of Universities and Colleges, 1989).

Hatry, Harry and Steinthal, Bruce, *Guide to Selecting Maintenance Strategies For Capital Facilities* (Washington: The Urban Institute Press, 1984).

M&R Cost Surveys and Analysis

Statistics Canada, *Survey on Capital and Repair Expenditures*. (Ottawa: Author, Annual).

Building Owners and Managers Association International, *Experience Exchange Report* (Washington: BOMA, Annual).

APPA: The Association of Higher Education Facility Officers, *Comparative Costs and Staffing Report for Colleges and Universities* (Alexandria: APPA, Bi-Annual).

Urban Land Institute (ULI), Dollars & Cents of Shopping Centers (Washington: ULI, Tri-Annual)

Institute of Real Estate Management (IREM), *Income/Expense Analysis: Conventional Apartments* (Chicago: IREM, Annual).

Institute of Real Estate Management, *Income/Expense Analysis: Office Buildings* (Chicago: IREM, Annual).

Department of Defense, DoD Facilities Cost Factor Handbook Version 6.0 (Washington: Office of the Undersecretary of Defense (Installations), April 2004). Also see www.whitestoneresearch.com/fsm/.

Whitestone Research, *Whitestone Report: An Alternative View of Facility Depreciation and Recapitilization Costs*. (Santa Barbara: Whitestone Research, January 2004).

Building Management Information. *Review of Maintenance costs 2002*. (London: RICS Building Cost Information Services Ltd. 2002)

Ottoman,G., Nixon, W., & Lofgren, S.. "Budgeting for Facility Maintenance and Repair, I: Methods and Models." *Journal of Management in Engineering*, (American Society of Civil Engineers, 1999, July/August) 15(4), 71-83.

Loranger, Robert, *What We Learned from a 50-Hospital Collaborative Facilities Survey* (Chicago: American Society for Healthcare Engineering, 1997).

Cole, Chuck, *Engineering Department Physical Plant Staffing Requirements* (Chicago: American Society for Healthcare Engineering, 1997).

Whitestone Research, "Notes on M&R Cost Benchmarks" in *Midyear Update* (Santa Barbara: Whitestone Research, 1996).

Lufkin, Peter and Desai, Anand, *The Estimation of Building Maintenance and Repair Costs* (Santa Barbara: Science Applications International Corporation, 1992).

International Facility Management Association (IFMA), *Benchmarks II*: Research Report #13 (Houston: IFMA, 1995)

Neely, Edgar and Neathammer, Richard, *Maintenance Resources by Building Use for U.S. Army Installations, Volume I: Main Text; Volume II: Appendices A-H; Volume III: Appendices I-P.* Technical Report P-91/29 (Champaign, IL: U.S. Army Corps of Engineers, 1991).

Au-Yeung, Peter, "The Uses of Maintenance Cost Records in Life Cycle Cost Predictions," in *Building Maintenance Economies and Management*, Alan Speeding, Editor (London: Spon. 1987).

Presentation of Cost Analyses

American Society for Testing and Materials, *Annual Book of Standards: Section 4, Construction* (Philadelphia: ASTM, Revised annually).

Tufte, Edward, *The Visual Display of Quantitative Information* (Cheshire, Connecticut: Graphics Press, 1983).

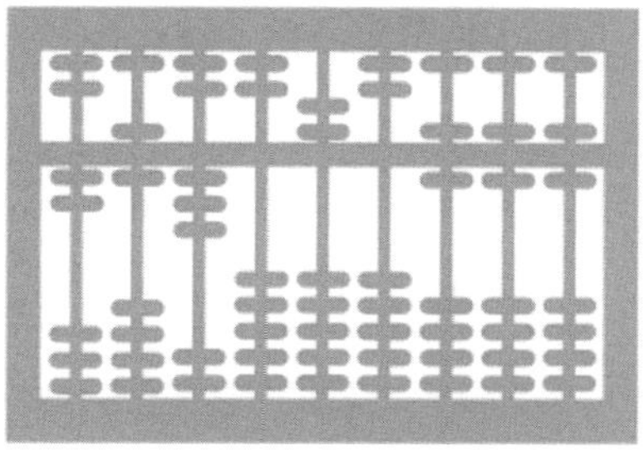

Appendix

Building Model Replacement Costs

M&R Labor Requirements by Building Type

Uniformat II

Service Lives of Buildings and Infrastructures

Table A-1
Building Model Replacement Costs, 2004

Building Type	Gross Sqft.		Replacement Cost*		Repl. Cost per Sqft.
Aircraft Hangar	32,000	$	2,563,618	$	80.11
Apartments, 1-3 story	22,500	$	2,333,481	$	103.71
Apartments, 4-7 story	60,000	$	6,326,889	$	105.45
Apartments, 24 Story	220,000	$	17,883,074	$	81.29
Auditorium	24,000	$	2,636,255	$	109.84
Bank	4,100	$	658,879	$	160.70
Bowling Center	20,000	$	1,324,651	$	66.23
Bus Terminal	12,000	$	1,147,611	$	95.63
Car Wash	800	$	172,111	$	215.14
Childcare Center	12,000	$	1,413,236	$	117.77
Church	17,000	$	2,015,066	$	118.53
College Classroom	90,000	$	9,651,343	$	107.24
College Student Union	25,000	$	2,663,038	$	106.52
Community Center	10,000	$	993,655	$	99.37
Country Club	6,000	$	824,059	$	137.34
County Jail	318,455	$	126,898,222	$	398.48
Court House, 1 Story	30,000	$	4,068,162	$	135.61
Department Store	94,000	$	6,502,987	$	69.18
Dormitory, 50 Room	25,000	$	2,157,174	$	86.29
Elementary School	47,000	$	3,426,265	$	72.90
Fast Food Restaurant	4,000	$	475,169	$	118.79
Fire Station	6,000	$	627,169	$	104.53
Funeral Home	10,000	$	893,983	$	89.40
Garage, Auto Sales	21,000	$	1,500,604	$	71.46
Garage, Parking	110,808	$	370,931	$	3.35
Garage, Service Station	1,400	$	169,882	$	121.34
General Hospital	125,000	$	15,963,874	$	127.71
Gymnasium	40,000	$	3,818,650	$	95.47
Laundromat	3,000	$	364,953	$	121.65
Light Manufacturing Plant	45,000	$	2,684,893	$	59.66
Medical Clinic	13,000	$	2,284,251	$	175.71
Motel, 18 Units	8,000	$	778,158	$	97.27
Motel, 40 Units	18,000	$	1,963,380	$	109.08
Movie Theater	10,000	$	1,097,944	$	109.79
Office Building, 2 Story	83,000	$	7,816,035	$	94.17
Office Building, 15 Story	250,000	$	30,492,533	$	121.97
Office Park	65,000	$	3,159,728	$	48.61
Post Office	13,000	$	1,123,638	$	86.43
Public Library, 3 Story	60,000	$	6,603,063	$	110.05
Religious Education	10,000	$	1,110,196	$	111.02
Research Hospital	540,200	$	150,422,976	$	278.46
Restaurant	10,000	$	1,263,065	$	126.31
Rink, Hockey	30,000	$	3,362,789	$	112.09
Self-storage Warehouse	24,000	$	1,569,209	$	65.38
Social Club	22,000	$	2,148,832	$	97.67
Store, Convenience	4,000	$	299,732	$	74.93
Store, Retail	8,000	$	623,999	$	78.00
Supermarket	96,000	$	6,853,088	$	71.39
Telephone Exchange	5,000	$	549,219	$	109.84
Town Hall, 1 Story	11,000	$	998,001	$	90.73
Warehouse	80,000	$	3,782,837	$	47.29

* Washington DC Metro Area

Table A-2
Average Annual M&R Labor Requirements by Building Type, 2004

Building Type	Gross Sqft.	Direct Hours[1] In-house	Direct Hours[1] Contract	Direct Hours[1] Total	Overhead[2]	Total	FTE[3]	Sqft / FTE
Aircraft Hangar	32,000	250	521	771	1,542	2,313	1.1	28,776
Apartments, 1-3 story	22,500	776	900	1,676	3,352	5,028	2.4	9,308
Apartments, 24 Story	220,000	1,997	7,031	9,028	18,056	27,084	13.0	16,896
Apartments, 4-7 story	60,000	1,944	2,385	4,329	8,658	12,987	6.2	9,610
Auditorium	24,000	401	433	834	1,668	2,502	1.2	19,952
Bank	4,100	122	110	232	464	696	0.3	12,253
Bowling Center	20,000	266	273	539	1,078	1,617	0.8	25,727
Bus Terminal	12,000	241	238	479	958	1,437	0.7	17,370
Car Wash	800	50	41	91	182	273	0.1	6,095
Childcare Center	12,000	191	194	385	770	1,155	0.6	21,610
Church	17,000	355	352	707	1,414	2,121	1.0	16,671
College Classroom	90,000	1,074	1,788	2,862	5,724	8,586	4.1	21,803
College Student Union	25,000	352	543	895	1,790	2,685	1.3	19,367
Community Center	10,000	214	243	457	914	1,371	0.7	15,171
Country Club	6,000	115	203	318	636	954	0.5	13,082
County Jail	318,455	2,026	7,528	9,554	19,108	28,662	13.8	23,110
Court House, 1 Story	30,000	578	653	1,231	2,462	3,693	1.8	16,897
Department Store	94,000	1,133	1,039	2,172	4,344	6,516	3.1	30,006
Dormitory, 50 Room	25,000	484	698	1,182	2,364	3,546	1.7	14,664
Elementary School	47,000	617	988	1,605	3,210	4,815	2.3	20,303
Fast Food Restaurant	4,000	81	143	224	448	672	0.3	12,381
Fire Station	6,000	158	197	355	710	1,065	0.5	11,718
Funeral Home	10,000	197	175	372	744	1,116	0.5	18,638
Garage, Auto Sales	21,000	331	284	615	1,230	1,845	0.9	23,675
Garage, Parking	110,808	144	633	777	1,554	2,331	1.1	98,876
Garage, Service Station	1,400	64	49	113	226	339	0.2	8,590
General Hospital	125,000	2,295	3,489	5,784	11,568	17,352	8.3	14,984
Gymnasium	40,000	639	666	1,305	2,610	3,915	1.9	21,252
Laundromat	3,000	106	51	157	314	471	0.2	13,248
Light Manufacturing Plant	45,000	600	568	1,168	2,336	3,504	1.7	26,712
Medical Clinic	13,000	281	396	677	1,354	2,031	1.0	13,314
Motel, 18 Units	8,000	240	347	587	1,174	1,761	0.8	9,449
Motel, 40 Units	18,000	413	627	1,040	2,080	3,120	1.5	12,000
Movie Theater	10,000	179	260	439	878	1,317	0.6	15,793
Office Building, 15 Story	250,000	1,330	3,738	5,068	10,136	15,204	7.3	34,202
Office Building, 2 Story	83,000	696	1,127	1,823	3,646	5,469	2.6	31,567
Office Park	65,000	650	869	1,519	3,038	4,557	2.2	29,669
Post Office	13,000	247	276	523	1,046	1,569	0.8	17,234
Public Library, 3 Story	60,000	465	1,982	2,447	4,894	7,341	3.5	17,000
Religious Education	10,000	251	333	584	1,168	1,752	0.8	11,872
Research Hospital	540,200	11,027	13,271	24,298	48,596	72,894	35.0	15,414
Restaurant	10,000	251	283	534	1,068	1,602	0.8	12,984
Rink, Hockey	30,000	577	338	915	1,830	2,745	1.3	22,732
Self-storage Warehouse	24,000	834	491	1,325	2,650	3,975	1.9	12,558
Social Club	22,000	289	433	722	1,444	2,166	1.0	21,127
Store, Convenience	4,000	87	89	176	352	528	0.3	15,758
Store, Retail	8,000	163	152	315	630	945	0.5	17,608
Supermarket	96,000	1,117	1,084	2,201	4,402	6,603	3.2	30,241
Telephone Exchange	5,000	114	138	252	504	756	0.4	13,757
Town Hall, 1 Story	11,000	205	249	454	908	1,362	0.7	16,799
Warehouse	80,000	977	797	1,774	3,548	5,322	2.6	31,266
Average	**54,829**	**749**	**1,171**	**1,919**	**3,839**	**5,758**	**2.8**	**19,825**

[1]Annual average requirements based on the 50-year profiles shown in Chapter 2. Preventative maintenance, minor repair, and unscheduled maintenance tasks done by the in-house staff; renewal and replacement tasks are assumed done by contract.

[2]An overhead rate of 2 (times direct hours) approximates support & supervision, non-productive time, and other indirect hours.

[3]Full time equivalents (FTE) are estimated as total (direct & overhead) hours / 2,080 hours.

Table A-3
Uniformat II Classification of Building Elements

Level 1 Major Group Elements	Level 2 Group Elements	Level 3 Individual Elements
A SUBSTRUCTURE	A10 Foundations	A1010 Standard Foundations A1020 Special Foundations A1030 Slab on Grade
	A20 Basement Construction	A2010 Basement Excavation A2020 Basement Walls
B SHELL	B10 Superstructure	B1010 Floor Construction B1020 Roof Construction
	B20 Exterior Enclosure	B2010 Exterior Walls B2020 Exterior Windows B2030 Exterior Doors
	B30 Roofing	B3010 Roof Coverings B3020 Roof Openings
C INTERIORS	C10 Interior Construction	C1010 Partitions C1020 Interior Doors C1030 Fittings
	C20 Stairs	C2010 Stair Construction C2020 Stair Finishes
	C30 Interior Finishes	C3010 Wall Finishes C3020 Floor Finishes C3030 Ceiling Finishes
D SERVICES	D10 Conveying	D1010 Elevators & Lifts D1020 Escalators & Moving Walks D1090 Other Conveying Systems
	D20 Plumbing	D2010 Plumbing Fixtures D2020 Domestic Water Distribution D2030 Sanitary Waste D2040 Rain Water Drainage D2090 Other Plumbing Systems
	D30 HVAC	D3010 Energy Supply D3020 Heat Generating Systems D3030 Cooling Generating Systems D3040 Distribution Systems D3050 Terminal & Package Units D3060 Controls & Instrumentation D3070 Systems Testing & Balancing D3090 Other HVAC Systems & Equipment
	D40 Fire Protection	D4010 Sprinklers D4020 Standpipes D4030 Fire Protection Specialties D4090 Other Fire Protection Systems
	D50 Electrical	D5010 Electrical Service & Distribution D5020 Lighting & Branch Wiring D5030 Communication & Security D5090 Other Electrical Systems
E EQUIPMENT & FURNISHINGS	E10 Equipment	E1010 Commercial Equipment E1020 Institutional Equipment E1030 Vehicular Equipment E1090 Other Equipment
	E20 Furnishings	E2010 Fixed Furnishings E2030 Movable Furnishings
F SPECIAL CONSTRUCTION & DEMOLITION	F10 Special Construction	F1010 Special Structures F1020 Integrated Construction F1030 Special Construction Systems F1040 Special Facilities F1050 Special Controls and Instrumentation
	F20 Selective Building Demolition	F2010 Building Elements Demolition F2020 Hazardous Components Abatement

Table A-3
Uniformat II Classification of Building Elements (continued)

Level 1 Major Group Elements	Level 2 Group Elements	Level 3 Individual Elements
G BUILDING SITEWORK	G10 Site Preparation	G1010 Site Clearing G1020 Site Demolition and Relocations G1030 Site Earthwork G1040 Hazardous Waste Remediation
	G20 Site Improvements	G2010 Roadways G2020 Parking Lots G2030 Pedestrian Paving G2040 Site Development G2050 Landscaping
	G30 Site Mechanical Utilities	G3010 Water Supply G3020 Sanitary Sewer G3030 Storm Sewer G3040 Heating Distribution G3050 Cooling Distribution G3060 Fuel distribution G3090 Other Site Mechanical Utilities
	G40 Site Electrical Utilities	G4010 Electrical Distribution G4020 Site Lighting G4030 Site Communications & Security G4090 Other Site Electrical Utilities
	G90 Other Site Construction	G9010 Service and Pedestrian Tunnels G9090 Other Site Systems & Equipment

Note: Shaded elements are usually not included in component inventories for M&R cost modeling.

Source: ASTM Standard E 1557-97, with permission. Contact ASTM Customer Service at (610) 832-9585.

Table A-4
Service Lives of Buildings and Infrastructure

Facility Type	Service Life (Years)
Aircraft Hangar	31
Bridges, Trestles and Overpasses	46
Broadcast and Communication	33
Bulk Storage Tank	24
Canals and Waterways	48
Church	47
Communication Towers, Antennae Station	16
Day Care Center	42
Docks, Wharves and Piers	31
Dormitory, Student	44
Dredging and Pile Driving	29
Electrical Power, Distribution Line	45
Electrical Power, Substation and Switching	45
Farm Building	37
Fire Station	28
Garage, Auto Sales	28
Garage, Maintenance and Repair	30
Gas Mains and Services	36
Gas Pipelines	36
Grain Elevator and Terminal	30
Highways, Roads and Streets	28
Hospital and Clinic	40
Hotel, Motel or Convention Center	41
Indoor Recreational Building	36
Irrigation and Land Reclamation	48
Laboratory, Research Center	36
Lagoon	24
Light Manufacturing Plant	30
Lighting, Exterior Lines	45
Museum	49
Natural Gas Processing Plant	21
Nursing Home	37
Office Building	41
Oil Pipeline	20
Oil Refinery	21
Other Communication Construction	19
Other Industrial and Commercial Construction	26
Other Institutional Construction	36
Other Marine Construction	37
Other Oil and Gas Facilities	34
Other Sewage System Construction	24
Other Transportation Construction	18
Other Waterworks Construction	33
Outdoor Recreational Facility	25
Parking, Lots and Garages	32
Passenger Terminal	33
Penitentiaries and Courthouses	41
Pollution Control	24
Power, Stand-By/Emergency	45
Public Library	35
Pumping Stations, Gas	34
Rail Track, Roadbeds and Signals	40
Railway and Engine House	30
Reservoir, Including Dam	39
Residential, Single Family Housing	80
Residential, Mobile Home	20
Restaurant, Fast Food	27
Runway, Including Lighting	20
Sanitary and Storm Sewer	34
Schools, Colleges and Universities	41
Service Station	22
Sewage Treatment and Disposal Plant	29
Shopping Center	38
Telephone and Cable Lines	28
Theater, Performing Arts	37
Treatment Plant and Pumping Station	43
Trunk and Distribution Main	33
Tunnels	60
Warehouse and Freight Terminal	36
Waste Disposal Facility	19
Wildlife Management Area	60

Sources: Statistics Canada, U.S. Bureau of Economic Analysis and Whitestone Research.